Human Factors in Design, Safety, and Management

Edited by

D. de Waard
K.A. Brookhuis
R. van Egmond
Th. Boersema

2005, Shaker Publishing

Europe Chapter of the Human Factors and Ergonomics Society

europechapter@hfes-europe.org

http://www.hfes-europe.org

© Copyright Shaker Publishing 2005

All rights reserved. No part of this publication may be reproduced, stored in a retrieval system, or transmitted, in any form or by any means, electronic, mechanical, photocopying, recording or otherwise, without the prior permission of the publishers.

Printed in The Netherlands.

Dick de Waard, Karel A. Brookhuis, René van Egmond, and Theo Boersema (Eds.).

Human Factors in Design, Safety, and Management

ISBN 90-423-0269-0

Shaker Publishing BV
St. Maartenslaan 26
6221 AX Maastricht
Tel.: +31 43 3500424
Fax: +31 43 3255090
http://www.shaker.nl

Contents

Preface 9
   Dick de Waard, Karel Brookhuis, René van Egmond, and Theo Boersema

Safety & Transport

A study of driver visual behaviour while talking with passengers, and on mobile phones 11
   Rachel. K. Smith, Toni Luke, Andrew M. Parkes, Peter C. Burns, & Terry C. Lansdown

Visual distraction of paper map and electronic system based navigation 23
   Toni Luke & Nick Reed

Interaction between car drivers: a diary study 35
   Maura Houtenbos, Marjan Hagenzieker, & Andrew Hale

The number of participants required in occlusion studies of in-vehicle information systems (IVIS) 39
   Nick Reed, Nikki Brook-Carter, & Simon Thompson

User friendly terminology for driver assistance systems 49
   Marion Wiethoff, Alan Stevens, & Nikki Brook-Carter

Designing out terrorism: human factors issues in airport baggage inspection 63
   Alastair G. Gale, Kevin Purdy, & David Wooding

Visual strain of the underground train driver 67
   Hilda Herman

HUMANIST: Networking in the European Union for a human-centred design of new driver support systems 69
   Sascha M. Sommer

Different congestion indications on dynamic route information displays 73
   Ilse M. Harms, Matthijs Dicke, & Karel A. Brookhuis

The relation between seat belt use of drivers and passengers 81
   Erik Nuyts & Lara Vesentini

An exploratory study on crew actions as a precondition to use remote experts for preventing shipping disasters 93
   Elsabé J. Willeboordse, Wilfried M. Post, & Anthony W.K. Gaillard

Does Automatic Identification System information improve efficiency and safety of Vessel Traffic Services? 97
   Erik Wiersma & Wim van 't Padje

Risk & Design

Turn on the lights: investigating the Inspire voice controlled smart home system . 105
   Heleen Boland, Jettie Hoonhout, Claudia van Schijndel, Jan Krebber, Sebastian Möller, Rosa Pegam, Martin Rajman, Mirek Melichar, Dietmar Schuchardt, Hardy Baesekow, & Paula Smeele

Is there such a thing as a mental representation for interface layouts? 121
   Francesco Di Nocera, Michela Terenzi, Giovanni Forte, & Fabio Ferlazzo

Risk assessment in industry and offices 129
   Piia Tint & Karin Reinhold
Improvement of risk management in the marble industry 135
   Abdelaziz Tairi & Ahmed Cherifi
Prevention of Repetitive Strain Injuries (RSI) at Delft University of Technology . 139
   Marijke C. Dekker & Kineke Festen-Hoff
Application of Human Factors Engineering to an Italian ferryboat 153
   Antonella Molini, Stefania Ricco, Manuela Megna, & Dario Boote
Observed risk awareness in user centred design 157
   Freija van Duijne, Heimrich Kanis, Andrew Hale, & Bill Green
Iteration in design processes 173
   Daniëlle M.L. Verstegen

Alarms & Sound
Acceptance of mobile phone text messages as a tool for warning the population .. 187
   S. (Simone) Sillem, J.W.F. (Erik) Wiersma, & B.J.M. (Ben) Ale
Ways of representing sound 201
   Kirsteen Aldrich, Judy Edworthy, & Elizabeth Hellier
The effect of phonetic features on the perceived urgency of warning words in
   different languages 205
   Mirjam van den Bos, Judy Edworthy, Elizabeth Hellier, & Addie Johnson
A contextual vision on alarms in the intensive care unit 219
   Adinda Freudenthal, Marijke Melles, Vera Pijl, Addie Bouwman,
   & Pieter Jan Stappers
The influence of perceptual stability of auditory percepts on motor behaviour 235
   René van Egmond, Ruud G.J. Meulenbroek, & Pim Franssen

Anthropometry
Working postures and physical demands on a utility vehicle assembly line 247
   Laura K. Thompson, Thorsten Franz, & Heinzpeter Rühmann
Ergonomic rucksack design for elementary school students in Indonesia to
   minimise low back pain 251
   Johanna Renny Octavia Hariandja, Bagus Arthaya, & Nana Suryani
Comparison of genetic and general algorithms of MLPs in posture prediction
   based on 3D scanned landmarks 263
   Bing Zhang, Imre Horváth, Johan F.M. Molenbroek, & Chris Snijders
The integration of anthropometry into computer aided design to manufacture and
   evaluate protective handwear 275
   Gavin Williams, Simon Hodder, George Torrens, & Tony Hodgson
Enhancing the use of anthropometric data 289
   Johan Molenbroek & Renate de Bruin

Usability & Performance
Differential usability of paper-based and computer-based work documents for
   control room operators in the chemical process industry 299
   Peter Nickel & Friedhelm Nachreiner

Analogue presentation of flight parameters on a head-up display 315
   Antoine J.C. De Reus & Harrie G.M. Bohnen
The effects of party line communication on flight task performance 327
   Helen Hodgetts, Eric Farmer, Martin Joose, Fabrice Parmentier,
   Dirk Schaefer, Piet Hoogeboom, Mick van Gool, & Dylan Jones
Yellow lessens discomfort glare: physiological mechanism(s) 339
   Frank L. Kooi, Johan W.A.M. Alferdinck, & David Post

Automation & Trust
Psychophysiological predictors of task engagement and distress 349
   Stephen H. Fairclough & Louise Venables
Consequences of shifting from one level of automation to another: main effects
   and their stability 363
   Francesco Di Nocera, Bernd Lorenz, & Raja Parasuraman
The Ergonomics of Attention Responsive Technology 377
   Alastair G. Gale, Kevin Purdy, & David Wooding
ACC effects on driving speed – a second look 381
   Nina Dragutinovic, Karel A. Brookhuis, Marjan P. Hagenzieker,
   & Vincent A.W.J. Marchau

Medical Systems
Ergonomics in surgery 387
   Martine A. van Veelen & Richard H.M. Goossens
The use of colour on the labelling of medicines 397
   Ruth Filik, Kevin Purdy, & Alastair Gale
Data display in the operating room 401
   Noemi Bitterman, Gideon Uretzky, & Daniel Gopher
Simulation and assessment of a North Sea rescue vessel 405
   Alistair Furnell, Matthew Mills, & Paul Crossland

Aviation
Task analysis, subjective workload and experienced frequencies of incidents
   in an airport control tower 417
   Clemens M. Weikert & Suzanne A. van Ham
Measuring head-down time via area-of-interest analysis: operational and
   experimental data 427
   Brian Hilburn
Aviation incident reporting in Sweden: empirically challenging the
   universality of fear 437
   Kyla Steele & Sidney Dekker
Effects of workload and time-on-task effects on eye fixation related brain
   potentials in a simulated air traffic control task 451
   Ellen Wilschut, Piet Hoogeboom, Ben Mulder, Eamonn Hanson,
   & Berry Wijers
Fz theta divided by Pz alpha as an index of task load during a PC-based air
   traffic control simulation 465
   Matty Postma, Jan Schellekens, Eamonn Hanson, & Piet Hoogeboom

Designing safety into future Air Traffic Control systems by learning from
   operational experience 471
   Deirdre Bonini & Tony Joyce

Management
How do we find safety problems before they find us? 479
   John A. Stoop
Developing a safety culture in a research and development environment:
   Air Traffic Management domain 493
   Rachael Gordon & Barry Kirwan
Moving closer to Human Factors integration in the design of rail systems:
   a UK regulatory perspective 505
   Deborah Lucas
Problems of limited context in redesign of complex situations in infrastructures ... 513
   Ellen Jagtman & Erik Wiersma

Acknowledgement to reviewers 527

Preface

Dick de Waard[1], Karel Brookhuis[1,2], René van Egmond[2], and Theo Boersema[2]
[1]University of Groningen, [2]Delft University of Technology
The Netherlands

From October 27 to 29, 2004, the Annual Meeting of the Europe Chapter was in Delft, The Netherlands. This book is published on the occasion of that meeting and contains peer-reviewed contributions based on presentations that were given.

The city of Delft is more than 750 years old and is renowned for its Delft Blue, a typical example is depicted on the cover. We thank the two hosts of the meeting, the Faculty of Technology, Policy and Management, and the Faculty of Industrial Design Engineering of Delft University of Technology, for their hospitality and support. We also thank the reviewers who helped to take care that each manuscript received the attention it deserved.

The conference was supported by the European Office of Aerospace Research and Development of the USAF, Air Force Office of Scientific Research, Air Force Research Laboratory, under Award No. FA8655-04-1-5035. We are very grateful for their contribution.

In D. de Waard, K.A. Brookhuis, R. van Egmond, and Th. Boersema (Eds.) (2005), *Human Factors in Design, Safety, and Management* (pp. 9). Maastricht, the Netherlands: Shaker Publishing.

A study of driver visual behaviour while talking with passengers, and on mobile phones

Rachel. K. Smith[1], Toni Luke[1], Andrew M. Parkes[1],
Peter C. Burns[2], & Terry C. Lansdown[3]
[1]TRL Limited
[2]Transport Canada,
[3]Heriot Watt University
UK

Abstract

This paper presents the findings of a driving simulator study that assessed the impact of talking with a front seat passenger, or over a hands-free mobile phone, on driver visual behaviour. The visual behaviour was also compared to that observed during a series of typical in-vehicle tasks, such as adjusting the climate control or entertainment systems, and to a baseline control where no additional tasks were conducted. Structured conversations, both with the passenger, and over the mobile phone, resulted in significantly fewer glances away from the road ahead, towards either dashboard displays or surrounding traffic. It also appears that although the drivers look more at the forward road scene, they do not actively process it, in the way they would, if they were not engaged in conversation at the same time. One of the driving measures included reaction times, where participants were required to respond to a specific road sign in the dynamic driving scene. Reaction time was slowest for the hands-free mobile phone condition, and showed reliable difference to talking to a front seat passenger, conducting other in-vehicle tasks, or driving with no other tasks.

Introduction

Background

The Independent Expert Group on Mobile Phones report (Stewart, 2000) for the UK concluded that drivers should be dissuaded from using phones while driving. This is because an increased risk of motor vehicle collision has been associated with mobile phone use while driving as shown by real world collision data (Redelmeier & Tibshirani, 1997). This has also been supported by experimental research showing mobile phone conversations to impair driving performance both in driving simulators and in real road trials (Fairclough, Ross, Ashby & Parkes, 1991; Parkes, 1991ab; Parkes, Fairclough & Ashby, 1993; Burns, Parkes, Burton, Smith, & Burch, 2002). The IEGMP report stated a need to compare the distraction by hands-free phone conversations with other current driver distractions.

Mobile phone use has increased dramatically over the past few years. This increase has led to a rise in the number of individuals using their mobile phone when driving. Recent measures to reduce the possible dangers of driving and using a mobile phone in the UK include new legislation to ban the use of hand-held mobile phones whilst driving. However, this legislation is based on the assumption that the distraction that is produced is from the structural demands on the driver in operating the mobile phone as opposed to the cognitive demands placed on the driver whilst conversing. However, studies have been published that suggest there is more to it than merely the structural demands (McCarley, Vais, Pringle, Kramer, Irwin, Strayer 2001, Burns, Parkes, & Lansdown 2002). Conversation alone can impair driving performance. This study looked at the conversation itself on the visual behaviour of drivers whilst engaging in conversations both with a front seat passenger and on a hands-free mobile phone.

Prior research into mobile phones and driving has established that the manual manipulation of equipment (for example answering the phone, dialling a number etc.) has a negative impact on driving performance (e.g., Brookhuis, De Vries, & De Waard, 1991). Recent research has further documented the potential negative impact of the conversation itself on visual scanning of traffic scenes when conversing on a hand-held mobile phone (McCarley et al., 2001). Studies in simulators have demonstrated that drivers' reaction times for braking and generally responding to events happening on the road were increased when driving and conversing on a mobile phone (Alm & Nilson, 1995; Fairclough et al., 1991; Parkes et al., 1993). Recent meta-analysis has shown that there are clear costs on driving performance due to using a mobile phone (Horrey & Wickens, 2004).

Strayer, Drews, Albert, and Johnston (2002) investigated the use of hand-held and hands-free mobile phones on driving performance, and found no significant difference in driving performance between hand-held and hands-free mobile phone use. They instead found that it was the conversation itself that was distracting the driver. The study indicated that subjects engaged in cell phone conversations were more likely to be involved in traffic accidents, missed more signals, and reacted slower to events in the driving environment than when they were not engaged in cell phone conversations. This study also demonstrated that it is the active engagement in the cell phone conversation that causes the dual task interference rather than simply attending to verbal material. However, the subjects used in this study were of a very narrow age bracket and do not represent the full age range of the driving population who would be driving and using a mobile phone.

Dual task studies have assessed the effects of cellular telephone conversations on performance of a simulated vehicle (Strayer & Johnton, 2001). Findings indicated that performance was not disrupted by listening to the radio, or on a continuous shadowing task using a hand-held mobile phone. However, the findings did indicate that there was an interference with driving performance in the word-generation variation of the shadowing task. This therefore suggests that the distracting element of conversations is the active participation in the conversations rather than the element of listening.

Further research conducted by McCarley et al. (2001) looked in to the effects of conversations on visual scanning of traffic scenes. The method used to measure visual scanning was a change detection task. The results showed that error rates for change detection tasks were higher during conversations than under single task control conditions. The study demonstrated that even simple conversations can disrupt attentive scanning and representation of a visual scene.

This study's aim was to investigate the visual behaviour of drivers as a way of understanding the distraction caused by conversations. Visual behaviour of drivers while talking to a front seat passenger was measured as well as when talking over a hands-free mobile phone, to benchmark the distraction caused by the conversation.

The results were expected to indicate that the conversation over a hands-free mobile phone would potentially be more distracting than conversation with a front seat passenger because passenger conversations can be expected to be more sympathetic to the driving scene. A simulator study was chosen as a safe environment for the participants to carry out the driving tasks without risk of injury. Another advantage of simulated driving environments is that the traffic conditions can be kept consistent across all drives and conditions.

Method

Participants

Thirty drivers aged between 21 and 64 (M = 40.9, SD = 12.39) participated in this study. They all had experience with using mobile phones. The sample was split evenly by gender. The sample of mobile phone users was randomly selected from the TRL volunteer database.

Procedure

The participants were invited to the laboratory for a pre-trial session. During this session they were asked to provide background information on their driving history, health and some details about their phone usage. They were given a description of the experiment and gave informed consent. They were introduced to the simulator and given a test drive to allow them to become more familiar and comfortable with the environment and the displays and controls of the simulator vehicle.

The trial was scheduled within a week of their pre-trial session. This trial started with a warm-up drive in the simulator where they were given a chance to practice the in-vehicle reference tasks. They were then asked to continue with the experiment when they felt comfortable with the simulator and tasks.

For the experiment, the participants were asked to drive as they normally would and to respond to the requests of the experimenter at various times during their drive.

The four conditions were:

- Passenger – Conversation with the experimenter as a passenger while driving.

- <u>In-vehicle</u> – Adjustment of the in-car controls while driving (no passenger present).
- <u>Hands-free</u> – Conversation with the experimenter through a hands-free mobile phone system while driving.
- <u>Driving control</u> – Driving only, no simultaneous tasks.

The order of the conditions was balanced and video recordings were taken of the participants' visual behaviour in all four conditions.

Equipment

The TRL driving simulator, with right-hand driver position, consisted of a medium size saloon car surrounded by 3 x 4 meter projection screens giving 210 degree front vision and 60 degree rear vision, enabling the normal use of vehicle mirrors. The road images were generated by advanced Silicon Graphics computers and projected onto the screens. The car body shell was mounted on hydraulic rams that supply motion to simulate the heave, pitch and roll experienced in normal braking, accelerating and cornering. The provision of car engine noise, external road noise, and the sounds of passing traffic further enhanced the realism of the driving experience.

A professionally fitted Nokia hands-free phone kit was used with a Nokia 3310 phone in this study. The phone bracket was mounted on the upper left side of the centre stack within easy reach and view from the driving position. A CCD colour video camera (PULNiX TMC-X) was mounted on the cowl above the instrument cluster in a position to capture a clear view of the drivers' eye movements. The camera images were recorded on a video (Panasonic AG 7350) mixed together with the forward view of the simulated road scene and the speedometer.

The in-vehicle radio tasks were performed on an aftermarket Radio/ CD player (Sony CDX - CA600). The original climate controls of the Rover 400 series car were used for the climate tasks.

Route and traffic scenarios

Participants drove a 17 km route that was composed of four different segments. The route started with a car following task on a motorway. Drivers were instructed to maintain their initial distance from the lead vehicle

After completing the car-following task, drivers were instructed to drive as they would normally do on a motorway. The motorway had 3 lanes and a moderate amount of traffic. The speed limit was 70 mph (113 km/h), the standard speed for motorways in the UK. A section of curved road was used to measure the driver's ability to control the vehicle on a more demanding type of rural road. Drivers were instructed to maintain a speed of 60 mph (96.6 km/h) and a central position within the left lane.

The curves were followed by a section of dual carriageway (2 lane road), which ended with traffic lights. During this section, drivers had to respond selectively to 24 warning signs at various points along the dual carriageway. They were instructed to flash their headlights whenever a particular target sign appeared. There were 4 different warning signs in this choice reaction time task: Elderly pedestrians, Pedestrian crossing, Cyclists and Roadwork. Each sign appeared 6 times.

Hands-free mobile phone and passenger conversation tasks

Questions from the Rosenbaum Verbal Cognitive Test Battery (RVCB) were administered by the experimenter. The RVCB measures judgement, flexible thinking and response times (Waugh, Glumm, Kilduff, Tauson, Smyth & Pillalamarri, 2000). The battery is composed of a 30-item remembering sentences task (e.g., repeat the sentence: "Undetected by the sleeping dog, the thief broke into Jane's apartment) and 30 verbal puzzle tasks (e.g., Answer the question: "Felix is darker than Antoine. Who is the lighter of the two?"). The test battery has five levels of difficulty with six items within each level of both tests. These questions were split across the conditions and also included short monologues on familiar topics (e.g., forty seconds describing a recent holiday). The conversation task was chosen to provide a complex conversation that would require active participation from the subject. In a survey of mobile phone users, Mcknight and Mcknight (1991) reported that 72% of mobile phone conversations are for business purposes. The content of these calls has potential to be important, complex and urgent and may increase driver distraction.

The experimenter administered the conversation material over the hands-free mobile phone during the hands-free test condition. The experimenter administered the questions at certain points during the drive. During the passenger conversation condition the experimenter administered the conversation material whilst in the front passenger seat of the simulator. This was designed to replicate a real passenger who may be sensitive to the driving scene, by adjusting the pace of the conversation according to the complexity of the driving scene.

In-vehicle tasks

Table 1. Summary of in vehicle tasks

Climate controls:	Adjust fan
	Change fan mode
	Adjust temperature
Audio system:	Turn on/off
	Adjust volume
	Change station (track)
	Find station (track)

In a separate trial run, participants were asked to perform a set of embedded tasks in the vehicle (see Table 1). Full instructions for the use of the in-vehicle systems were given to the participant. The participant was also given the opportunity to practice using all of the controls during the baseline drive. During the baseline and the in-vehicle task conditions, the instructions to operate the climate and entertainment controls were given to the participant by the experimenter via an intercom system installed in the car. Instructions were given at specified points along the route at intervals comparable to the spacing of the conversation tasks in the other conditions.

Driving control condition

In a separate trial run drivers were given the same performance tasks (car following, curve negotiation, choice reaction, and so on), but were not distracted by either conversation or in-vehicle tasks.

Visual behaviour and driving performance were the two dependent variables measured. One useful measure of distraction is to look at eye movement of the driver whilst driving and conversing. Reaction time to signs was chosen as another useful method of measuring driver distraction. Table 2 displays the measures that were taken.

Table 2. The dependent variables

	Measures
Visual behaviour	
Eye Glances	Mean duration of glances off road
	Total percentage of time looking off road
	Glance behaviour to mirrors, centre stack, phone and passenger
Driving performance	
Reaction times to signs	Time in seconds

Results

Visual behaviour

Visual behaviour of the drivers was sampled across the whole test route in each driving condition. For each of the four elements of the drive a one-kilometre section was identified and analysed in detail. All glances to the mirrors, the speedometer, the climate and entertainment controls, or to any 'other' region inside the vehicle were recorded. Glances were operationally defined as occupying the period that starts with a movement of the eye away from the road ahead, and includes the dwell time in one of the zones of interest. The glance finished at the point the eye started to move to another location. To aid data scoring and analysis, glances were categorised in 0.5 second intervals. The results from each of the four elements of the drive were aggregated to provide a single composite score for each participant and task condition.

Glances away from road ahead

Glances to mirrors
A one-way repeated measures ANOVA was calculated for the number of glances and the mean number of glances. There were no significant differences between

conditions in either the number of glances towards the mirrors, or in the average duration of those glances.

Glances to speedometer

A one-way repeated measures ANOVA was calculated for the mean glance duration (Table 3). There was an overall effect of task condition on the average duration of a glance to the speedometer ($F(3,78)=3.0$, $p<.05$).

Table 3. Mean duration of glances to the speedometer

Task	Mean (seconds)	SD
Control	0.505	0.018
Passenger	0.503	0.009
In-vehicle	0.517	0.031
Mobile phone	0.507	0.018

The mean duration of glances indicates that the longest mean glance duration was in the in-vehicle task conditions. The shortest glance duration was for in the passenger condition. A post-hoc Fisher test revealed that there was a significant difference between the duration of glances to the speedometer when participants were involved in completing in-vehicle tasks when compared to any of the other conversation or control conditions ($p < 0.05$). There was no difference between the other conditions.

There was also a significant effect of task condition on the number of glances made to the speedometer during the trials (Table 4; $F(3,78) = 24.9$, $p<.0001$).

Table 4. Number of glances to the speedometer

Task	Mean	SD
Control	12.2	7.5
Passenger	6.1	4.6
In-vehicle	11.3	5.2
Mobile phone	5.5	4.1

A post-hoc Fisher test revealed that there was no significant difference between the control and in-vehicle tasks, or between the mobile phone and the passenger tasks, but there was for all other comparisons (Table 5).

Table 5. Mean difference in the number of glances

Comparison	Mean difference	Critical difference	p value
Control and Passenger	5.9	2.9	.0001
Control and In-vehicle	1.1	2.9	.4704
Control and mobile phone	6.5	2.9	<.0001
Passenger and in-vehicle	-4.9	2.9	.0010
Passenger and mobile phone	0.5	2.9	.7153
In-vehicle and mobile phone	5.4	2.9	.0003

The conversation conducted both with the passenger and over the mobile phone resulted in significantly fewer glances to the speedometer.

Other regions included glances to the mobile phone, the passenger, the radio or climate controls, or any other target of interest inside the vehicle. There was no significant difference in either frequency or duration of glances to these other regions as a function of task condition.

Total percentage time looking off road ahead

There was a significant difference in the percentage of time spent looking away from the road ahead in the various task conditions. (Table 6; $F(3,87) = 90.9$, $p<.0001$).

Table 6. Mean percentage of time spent looking away from the road ahead

Task	Mean	SD
Control	6.5	3.1
Passenger	4.4	2.4
In-vehicle	10.2	3.0
Car phone	3.6	1.8

A post hoc Fisher test revealed an interesting set of comparisons. There was no significant difference between the conversation conditions, but both these were significantly lower than the control condition, whilst the in-vehicle condition was significantly higher (Table 7).

Table 7. Comparisons using a post hoc Fisher test

Comparison	Mean difference	Critical difference	p-value
Control and Passenger	2.0	1.4	.0045
Control and In-vehicle	-3.8	1.4	<.0001
Control and mobile phone	2.6	1.4	.0003
Passenger and in-vehicle	-5.8	1.4	<.0001
Passenger and mobile phone	0.6	1.4	.3888
In-vehicle and mobile phone	6.4	1.4	<.0001

Choice reaction times

A one-way repeated measures ANOVA was calculated for the median reaction time ratings across the four conditions (Figure 1). The mean reaction time data were skewed, so the median reaction time for the six events was used. These median data were normally distributed. There was a significant main effect by condition for median reaction time ($F(2.4, 84) = 24.4$, $p < 0.001$, MSE = 0.47). There was a significant problem of sphericity with the data, so a Huynh-Feldt correction was used. Post hoc tests were run to compare the reaction times. Reaction time was significantly slowest for the hands-free phone condition in comparison to the in-vehicle tasks ($p = 0.046$, one-tailed), talking with a passenger ($p = 0.03$, one-tailed) and the control drives ($p < 0.001$). Reaction times in the control drive were also significantly faster than during the in-vehicle task ($p < 0.001$) and passenger drives ($p < 0.001$).

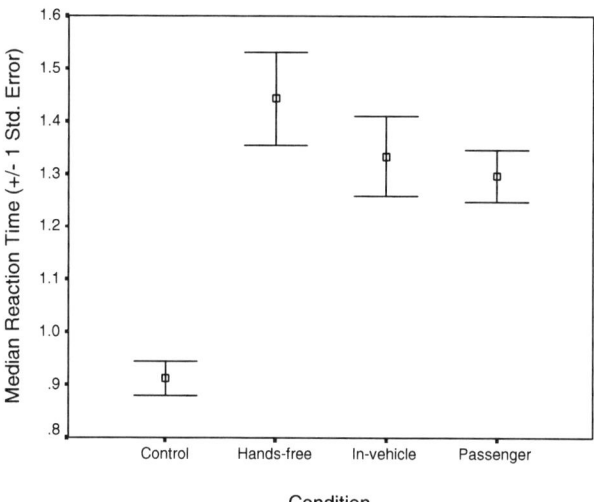

Figure 1. Median reaction times (seconds)

False Alarms

There were no significant differences in the number of false alarms across the four conditions

The greatest number of missed targets occurred in the hands-free drive (n = 31) followed by the in-vehicle task drive (n = 27) and passenger conversation drive (n = 22). Only one target was missed during the control drive. The missed target data were significantly skewed, so a nonparametric Friedman's test was used to compare conditions. The number of misses differed significantly across the four conditions ($\chi^2 = 21.6$, $p < 0.001$). Post-hoc comparisons showed significant differences between the control drive and other drives. There were no significant differences among the other conditions.

Discussion

Reaction times of selected road signs were faster in the control condition than all the other task conditions. Reaction times in the hands-free condition were slower than the other conditions and there was no difference in reaction times for the passenger and in-vehicle conditions. In general, reactions are slower while driving and performing a simultaneous task, and talking on a hands-free phone has the greatest effect on performance. The number of missed target signs was also affected by condition. The greatest number of missed targets occurred in the hands-free condition, followed by the in-vehicle and then passenger conditions.

The analysis of visual behaviour revealed an interesting and consistent picture. The presence of conversation tasks, whether presented by the front seat passenger or over the hands-free mobile phone, had an impact on visual scanning. The number and duration of glances to mirrors were low in the simulator during this experiment because of the low volume of traffic in the test scenarios. Glances to other regions of interest in the analysis were also low and there was no significant effect of task condition evident. However, glances to the speedometer region were affected differently depending on the concurrent task. Around half as many glances were made when the participant was having a conversation when compared to both in-vehicle tasks and control conditions. It appears that the act of talking and thinking in the structured conversations reduces the capability of the driver to monitor the vehicles displays as effectively as normal.

When considering total percentage of time spent looking away from the road ahead there is the expected result that engagement in tasks such as adjusting the radio or climate controls diverts the eyes from the road ahead, and the total percentage of time spent looking away from the road ahead increases. In the scenarios adopted in this study the figures increased from 6.5 to 10.2 percent. What is equally interesting is the result that the conversation tasks reduced the time off road to 4.3 percent when talking to a passenger, and to 3.6 percent when talking on the mobile phone. So here the distraction is operating in a different way. The in-vehicle tasks provoke a physical interference that diverts gaze from the road ahead. The conversation tasks appear to produce an additional mental load that interferes with normal vehicle monitoring functions.

Interpretation and Implications

The measures of visual behaviour produced results that need careful interpretation. There was the clear and intuitive result that performing in-vehicle tasks such as operation of the entertainment or climate controls drew visual attention away from the road ahead, as drivers needed to look at the device as they manipulated it. When compared to the control condition, it is clear that having a conversation demands a level of concentration (given the complexity of task in the structured conversations used here) that disrupts normal scanning of vehicle functions on the dashboard display. The results show that the drivers spent more time than usual looking at the road ahead, because they adjusted their normal monitoring behaviour. We find that the increased time spent looking forward is not reflected in a benefit to the ability to respond quickly or accurately to traffic signals. It appears that although the drivers look forward at the road scene, they do not actively search it in the way that they do when not engaged in a conversation at the same time. This result is entirely consistent with other work (Parkes & Hooijmeijer 2001) that has shown a reduction in Situation Awareness produced by both hand-held and hands-free mobile phone conversations. This result is also consistent with research (McCarley et al., 2001) indicating that conversations interrupt the normal scanning of the traffic scene, which in turn may lead to poorer driving performance.

Reaction times are more seriously affected by talking on a hands-free phone than performing other tasks. It can be argued that reactions govern most aspects of driving. Reaction has an even greater role to play in more complex road situations such as navigating junctions and responding to hazards. In these situations the performance deficit caused by hands-free conversations is likely to be further emphasised.

The conversations chosen for this study were designed to be challenging and to encourage active participation from the subject. This type of conversation could represent the type of conversation that drivers may engage in when using hands-free mobile phones for business purposes. Usual passenger conversations are unlikely to be of this nature, and passenger conversation behaviour can show consideration for the current traffic situation at the current time. For example, passengers may pause slightly before asking a question if a driver is approaching a complex driving scene. Driver conversation behaviour between passenger and hands-free conversations may vary. Drivers may find it more acceptable to pause in conversation in complex driving areas when conversing with a passenger as opposed to when they are conversing on a hands-free mobile. One possible reason for this could be that the driver is aware that passengers can see the driving scene and will be aware of the reasons for the pause in conversation.

Future research should consider the effects of different types of conversation on driver visual behaviour and driving performance whilst using hands-free mobile phones.

Conclusion

This paper compares the distraction by hands-free phone conversations with other current driver distractions. The findings from this work supports the IEGMP report (Stewart, 2000) in concluding that conversations especially mobile phone conversations disrupt the normal visual scanning that takes place during driving, leading to an impairment of driving performance.

References

Alm, H. & Nilson, L. (1995). The effects of mobile telephone task on driver behaviour in a car following situation. *Accident Analysis and Prevention*, 27, 707-715.

Brookhuis, K.A. De Vries, G., & De Waard, D. (1991). The effects of mobile telephoning on driving performance. *Accident Analysis and Prevention*, 23, 309-316.

Burns, P.C. Parkes, A.M. Burton, S. Smith, R.K., & Burch, D. (2002). How dangerous is driving with a mobile phone? Benchmarking the impairment to alcohol. TRL Report 547. Crowthorne, Berkshire, UK: Transport Research Laboratory.

Fairclough, S.H. Ross, T. Ashby, M.C., & Parkes, A.M. (1991). Effects of handsfree cellphone use on driver behaviour. *Proceedings of ISATA Conference* (pp. 403-410). Florence, Italy. Croydon, England: Automotive Automation Limited.

Horrey, W.J. & Wickens, C.D. (2004). The impact of cell phone conversations on driving: A meta-analytic approach. Technical Report AHFD-04-2/GM-04-1. Aviation Human Factors Division Institute of Aviation. University of Illinois.

Parkes, A.M. (1991a). Drivers decision making ability whilst using carphones. In T Lovesey (ed.) *Contemporary Ergonomics* (pp. 427 – 432). London: Taylor and Francis.

Parkes, A.M. (1991b). The effects of driving and handsfree telephone use on conversation structure and style. *Proceedings of Human Factors Association of Canada Conference* (pp. 141-147). Vancouver. Canada.

Parkes, A.M. Fairclough, S.H. & Ashby, M.C. (1993). Carphone use and motorway driving. In T. Lovesy (Ed.) *Contemporary Ergonomics* (pp. 403-408). London: Taylor and Francis.

Parkes, A.M. & Hooijmeijer, V. (2001). Driver Situation Awareness and carphone use. 1st Human Centred Transportation Simulation Conference. University of Iowa. Iowa City USA, Nov, 4-7.

McCarley, J.S. Vais, M., Pringle, H. Kramer, A.F. Irwin, D.E. & Strayer, D.L. (2001). Conversations Disrupt Visual Scanning of Traffic Scenes. Beckman Institute, University of Illinois at Urbana-Champaign.

McKnight, J. & McKnight, A.S. (1991). The effects of cellular phone use upon driver attention. National Public Services Institute.

Redelmeir D.A. & Tibshirani, R.J. (1997). Association between cellular-telephone calls and motor vehicle collisions. *New England Journal of Medicine*, 336, 453-458.

Stewart, W. (2000). Mobile Phones and Health, The Independent Expert Group on Mobile Phones, Sir W. Stewart (Chairman), (http://www.iegmp.org.uk).

Strayer, D.L. Drews, F.A. Albert, R.W., & Johnston, W.A. (2002). Why do cell phone conversations interfere with driving? 81st Annual Meeting of the Transportation Research Board. Washington, DC: National Academy of Science.

Strayer, D.L. & Johnston, W.A. (2001). Driven to distraction: dual-task studies of simulated driving and conversing on a cellular phone. *Psychological Science: 12*, 462-466.

Waugh, J.D. Glumm, M.M. Kilduff, P.W., Tauson, R.A., Smyth, C.C., & Pillalamarri, R.S. (2000). Cognitive workload while driving and talking on a cellular phone or to a passenger. *International Ergonomics Association Conference*, San Diego, USA.

Visual distraction of paper map and electronic system based navigation

Toni Luke & Nick Reed
TRL, Wokingham, Berkshire, UK

Abstract

This report presents the findings of a comparison of three navigation methods in terms of visual behaviour. The study employed a repeated-measures design where participants drove three routes in real traffic, under three different conditions: electronic system navigation with auditory instructions; electronic system navigation without auditory instructions and paper map based navigation. Participants were all experienced drivers and users of the navigation system. Drivers eye movements, participants subjective ratings of workload and trip time were collected as measures of distraction, workload and performance.

The aim of this work was to examine the visual behaviour of drivers using a paper-map and a navigation system in a real-world situation and results are interpreted in this context. Analysis of visual behaviour suggests that neither the map nor the navigation system is a major source of visual distraction if used responsibly by experienced users. The results did not provide convincing evidence that using a paper map is more visually demanding than using an electronic navigation system. Mean glance times to the map and system were less than a second in all conditions.

Introduction

The distracting effect of In-Vehicle Information Systems (IVIS) is a growing source of concern as more systems and functions become available. The primary concern is with visual distraction. Mapping navigation systems in particular require the driver to look at a display that may contain complex information. The time spent looking at a system is "eyes-off-road" time and this may result in potential failure to detect lethal hazards.

In-vehicle navigation systems are generally considered as an alternative, or a supplement, to paper maps. For this reason several studies have conducted a direct comparison between system navigation and paper-based map navigation. Parkes, Fairclough and Ross (1991) compared navigation using a paper map to navigation using an electronic text-based system, on the road in an area unfamiliar to the participants. The key differences between the methods were: the presentation method (i.e. diagrammatic versus text); the level of automation and the organisation of the information. The electronic system presented the instructions automatically and in chunks which were controlled by the subject where the paper map continuously presented all the information throughout the task. Measures of visual behaviour, heart rate, driving speed and subjective workload all indicated that a higher load was imposed upon the driver in the map condition.

Daimon (1992) also compared paper map navigation to electronic system navigation in both real and simulated environments. The electronic system in this case presented an electronic map. Various measures of workload were collected including, secondary task performance, heart rate variability and eye movements. Secondary task performance and heart-rate variability both indicated higher workload when using a paper map compared to the electronic system. However, measures of visual behaviour indicated more frequent glances to the electronic system in comparison to the paper map. Similarly, Burnett and Joyner (1993) reported a higher percentage of glance durations to an electronic route guidance system than to a paper map with the required route highlighted. More recently, Uang and Hwang (2003) compared paper map based navigation to electronic map navigation using a desk-top driving simulator. Uang and Hwang examined different scales of the electronic map display and the effects of congestion information. None of the measures of subjective workload revealed significant differences between paper map and system navigation. Other measures of performance in this study relate to the efficacy of the various electronic system configurations as navigational aids. Only the small scale map with congestion information produced a reduction in trip time and large scale maps actually increased trip time regardless of the provision of traffic information.

The overall pattern of results appears to indicate that paper map navigation is associated with higher workload than electronic system navigation. However, these studies have used a variety of different maps and electronic systems. The design of the system and the map has a large role to play in the drivers' experience in terms of workload and way-finding performance. The context of use is also important. The purpose of the present study was to attempt to measure drivers' visual behaviour when using a real, production status in-vehicle navigation system in real traffic, and to compare this to using a paper map under similar conditions. The trial attempted to replicate realistic use of the navigation methods by professional drivers and compared the visual behaviour when using a system with auditory instructions, the same system without auditory instructions, and a paper map.

Method

Participants

Ten male drivers participated in the study. All participants were private hire vehicle drivers and frequent users of the navigation system. All had normal or corrected vision. Participants were aged between 26 and 55 years old.

Equipment

The basic equipment for the trials included the vehicle with the navigation system installed, a map, a video camera, monitor and recorder. The trials were all conducted in Mercedes A Class vehicles. Each participant completed the trials in their own vehicle. The system was the VDO Dayton MS 6000 with a 6.5 inch colour LCD in-vehicle display screen. The display screen of this system is integrated with the vehicle audio system and slides out and folds up into position from the normal DIN slot. The system screen was therefore positioned approximately level with the bottom of the steering wheel. The input controls were mounted on a remote control which was sited on the centre stack of the dashboard next to the system. During the navigation system conditions the system was set to display a split screen, which showed a map and a direction arrow with the map on the largest scale of 125m. During route guidance the map was displayed with a heading-up orientation.

The map consisted of an A4 colour copy of a standard London A-Z map. The scale was 1:19,000. On this type of map motorways are depicted in blue, "A" roads are depicted in dark yellow, "B" roads are depicted in light yellow and more minor roads are white. This type of map presents street name and other landmarks such as schools, hospitals and bodies of water. The view of the driver's face was recorded onto VHS format tapes using a video recorder that was secured in the back of the vehicle. A monitor for checking the camera view was also secured in the back. The camera was mounted clear of the drivers view above the centre stack of the dashboard.

Design

The study employed a repeated measures design and compared the visual behaviour of drivers under three conditions. The three conditions were:

A. System navigation with audio instructions
B. System navigation without audio instructions
C. Paper Map Navigation

Each driver completed three routes. All drivers completed the routes in the same order. The order of conditions was counterbalanced to prevent learning effects. For consistency and safety, all drivers were instructed to observe the speed limit and follow the rules of the road strictly. In the system navigation conditions drivers were instructed to follow the instructions provided by the system as long as it was safe and legal to do so. In the paper map condition drivers were instructed to select their own route. Drivers were free to select any route providing they observed the rules of the road.

Routes
Three routes in South London were selected for the trial. The routes were designed to be approximately the same length and to have starting and finishing places where the drivers could safely stop. This area was chosen as it was identified as an area where the drivers did not often work and would therefore have limited familiarity with. The routes consisted primarily of A and B class urban roads with high traffic volumes.

Procedure

Participants were met at their normal place of work. Participants were initially presented with information about the trial. They were also asked to fill in a questionnaire to collect basic demographic information. Participants were then escorted to their vehicle. At this point participants were given standardised instructions relating to the trial in general.

Participants were next instructed to drive to the starting location using whatever navigation method they preferred. This was not part of the trial and no data were collected. The vehicle was stopped in a safe position at the first location. Before starting each route a calibration of the video equipment was conducted. The driver was asked to look at each of the glance targets in turn. This calibration was recorded as a reference for the later analysis. In the map condition, the map was placed on the passenger seat within reach and view of the driver. In the map condition, a video calibration was completed in exactly the same way. Drivers looked at the map target in this position.

The experimenter configured the system correctly for the condition. The driver was then informed of what the next condition would be and given details of the next destination on a card. The drivers were instructed to enter the destination into the system or to familiarise themselves with the route on the paper map. When ready, the participants began the journey to the next destination and pulled up at a safe location once they had arrived. At the destination they completed a short workload questionnaire (NASA TLX, Hart & Staveland, 1988). This procedure was repeated for each condition. When drivers had completed the three routes they returned to their place of work.

Results

Analysis procedure

Video data were collected and analysed according to the procedures set out in BS EN ISO 15007-1: 2002 and BS EN ISO 15007-2:2001. The last ten minutes of footage from each condition was analysed for the glance behaviour of the drivers. The videos of the drivers' eye movements were observed frame by frame to achieve the greatest possible accuracy. Each frame was classified according to glance target. There was occasional data loss if the driver's viewpoint was unclear. However, this occurred in less than 5% of the frames analysed. Glance targets of interest were; wing and interior rear view mirrors, vehicle instrumentation (referred to in analysis as 'Other in veh.'), the navigation 'system' (electronic or paper map) and the road ahead.

A glance was defined to begin at the moment when the eyes began to move away from their current target towards the subsequent target. Therefore, the initial transition time to a new target was included in the duration of the glance (BS EN ISO 15007-1, 2002). Glance behaviour when the vehicle was stationary was excluded from analysis. The glance frequency, and the total and mean durations on each target were analysed to compare the visual behaviour of drivers in each condition. The three conditions under test were system navigation with audio instructions (A), system navigation without audio instructions (B), and paper map navigation (C).

Glance frequency

Glance frequency was examined using the total number of glances made by a driver to each target area in the ten minute trial period. Figure 1 shows the mean number of glances made by all drivers to each target in the three conditions.

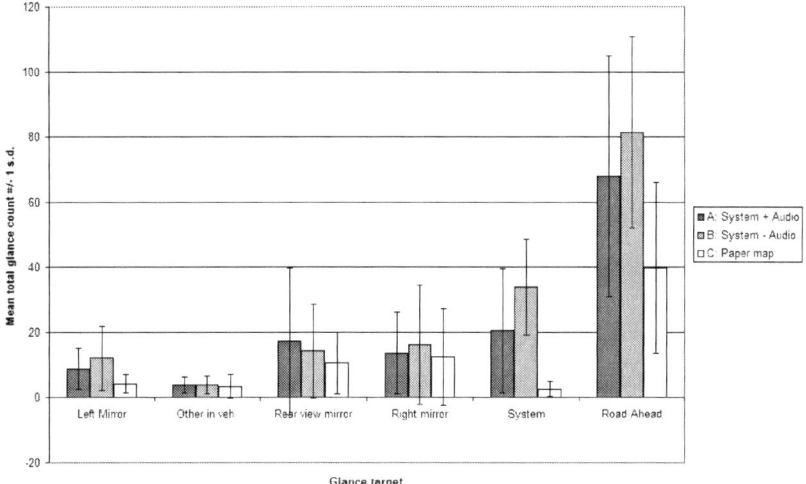

Figure 1. Mean number of glances to each target across conditions

It is clear that drivers made the greatest number of glances to the road ahead. However, it is only in conditions A and B in which glances to the system were the second most frequent glance target. The chart suggests that drivers made fewer separate glances to the various targets in condition C, suggesting greater consistency of viewpoint. A within-subjects ANOVA test found a highly significant difference in the number of glances made across conditions ($F(1, 71) = 15.8$, $p < 0.001$). Post-hoc pairwise comparisons revealed that drivers made significantly fewer glances in condition C than in conditions A or B ($p < 0.001$ in each case). A within-subjects ANOVA with pairwise comparisons was used to compare the number of glances made to each target across the three conditions. There was a significant difference for glances made to the system ($F(1, 8) = 10.4$, $p < 0.05$), with the fewer glances made in condition C significantly different to those made in conditions A or B ($p < 0.05$ in each case). There was also a significant difference between conditions in the number of glances to the road ahead ($F(1, 8) = 12.5$, $p < 0.01$). The pairwise comparisons revealed significant or near significant differences between all three conditions (A vs. B: $p = 0.053$; A vs. C: $p = 0.027$; B vs. C: $p = 0.001$).

In summary, drivers made significantly fewer glances overall when navigating using a paper map compared to using an electronic navigation system, with or without auditory instructions. Drivers made significantly fewer glances to the map than to the navigation system, with or without auditory instructions. Finally, there were significant or near significant differences in the number of glances to the road ahead in all conditions.

Total glance duration

The total glance duration is the time that the driver spent looking at a particular target, while the vehicle was in motion, over the ten minute trial period. Figure 2 shows the mean of the total glance duration for each driver to each target across the three different conditions. Figure 2a, shows the mean of the total durations for targets other than the road ahead. Due to the massive difference in scale, the mean of the total glance duration to the road ahead is shown separately in Figure2b.

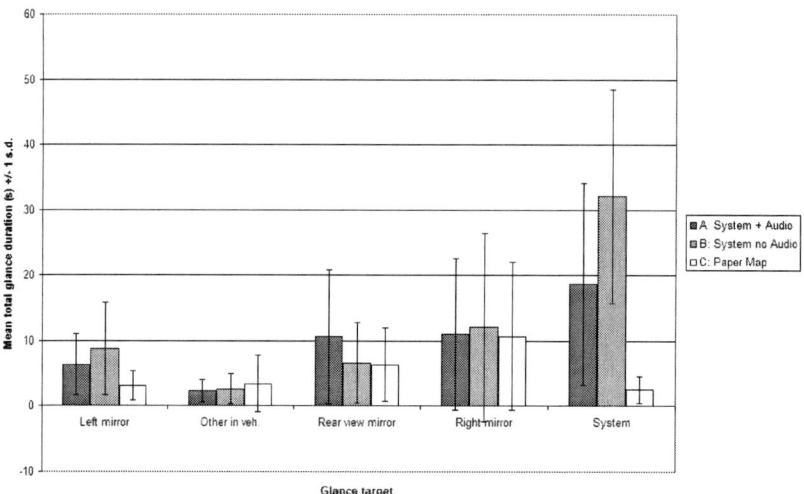

Figure 2a. Mean total glance duration for each target across conditions

The charts demonstrate that drivers spent the clear majority of their time viewing the road ahead. Figure 2b illustrates that there are some apparent differences in the total glance duration for other targets. The differences between conditions were investigated using a within-subjects ANOVA procedure with pairwise comparisons. They revealed that there was a significant difference in total glance duration to the system across conditions ($F(1, 8) = 10.8$, $p < 0.05$), and that condition C was significantly different to conditions A and B ($p < 0.05$ in each case). This reinforces the previous finding that drivers made fewer glances to the route information in condition C. Interestingly, although the greater number of glances and total glance time suggests increased visual distraction by the electronic navigation systems (conditions A and B) drivers spent no less time looking at the road ahead using these systems.

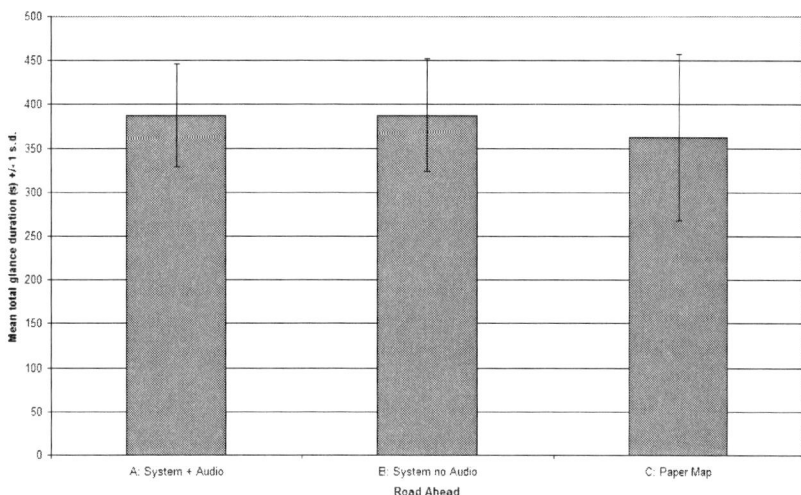

Figure 2b. Mean total glance duration for each target across conditions

In summary, the average total time the drivers spent looking at the route information was significantly different in the paper map condition than in the electronic system conditions, with or without auditory instructions. Finally, there was no difference between conditions in the total time spent looking at the road ahead.

Mean Glance duration

The mean glance duration is the mean time that a driver spent looking at each target in a given glance instance. Figure 3a shows the mean glance duration to targets other than the road ahead. Again, the difference in scale dictates that mean glance duration to the road ahead is displayed separately in Figure 3b.

The mean duration that drivers spend looking at the road ahead in a single glance is less consistent across conditions than the total length of time spent looking at the road ahead. This result is unsurprising given the differences in glance frequency and similarity in total glance duration. Glance duration to the system is the next longest target for drivers in conditions A and B but not for those in condition C. Again the within-subjects ANOVA with pairwise comparisons procedure was used to compare mean glance duration across conditions. The mean glance duration for drivers looking at the system neared significance ($F(1, 8) = 4.05$, $p = 0.079$). The pairwise comparisons revealed that this was due to differences between in condition C and conditions A and B ($p < 0.09$ in each case). Drivers tended to take glances of nearly a second to the electronic navigation system but less than 0.6 seconds to the map. There was a significant difference in the mean glance duration to the road ahead across conditions ($F(1, 8) = 17.4$, $p < 0.005$). In contrast to the previous result, drivers tended to look at the road ahead for much longer in condition C than in conditions A or B ($p < 0.01$ in each case). These two results suggest that drivers paid more attention to the road in condition C. The difference between conditions A and

B was also significant (p = 0.026). This suggests that drivers tended to glance for longer at the road ahead in condition A when audio instructions were given.

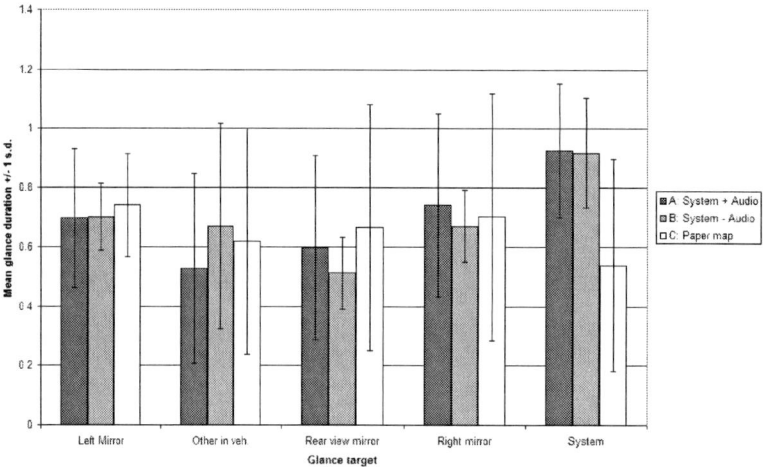

Figure 3a. Mean glance duration for each target conditions

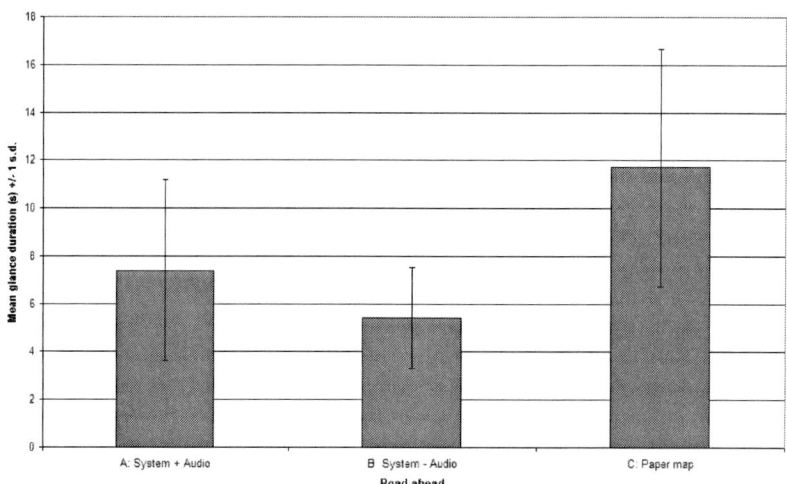

Figure 3b. Mean glance duration for each target conditions

The results suggest that the mean duration of a single glance to the system was somewhat lower in the map condition than in the system conditions although comparisons did not quite reach significance. The mean glance duration to the road ahead was longer in the map condition than in either of the electronic navigation conditions. The mean glance duration to the road ahead was also longer in the system condition with auditory instructions compared to the system condition without auditory instructions.

Glance duration frequency distribution

The glance duration frequency distribution to the route information was examined for each condition. These distributions show that although the majority of the glances were less than 2 seconds in duration there were occasional examples of long glances away from the road ahead. In condition A 1.9% of glances exceeded 2 seconds, in condition B 2.3% of glances exceeded 2 seconds and in condition C 3.8% of glances exceeded 2 seconds. There were not enough examples of long glances to analyse statistically.

Perceived Workload

Drivers' perceived workload was measured using the NASA TLX. This measure looks at perceived workload on several dimensions: mental demand, physical demand, temporal demand, performance, effort and frustration. Drivers' responses on each dimension in each condition were compared for each condition using paired t-tests.

Table 1. Mean subjective ratings for NASA TLX workload measures

Workload dimension		A: System + Audio	B: System no Audio	C: Paper map
Mental demand	Mean	2.61	3.79	2.51
	s.d.	2.35	3.16	2.65
Physical demand	Mean	1.88	2.18	2.43
	s.d.	2.36	2.36	2.31
Temporal demand	Mean	1.34	2.96	1.43
	s.d.	1.02	3.12	0.78
Performance	Mean	2.54	2.83	2.25
	s.d.	3.34	3.07	2.72
Effort	Mean	1.87	3.22	2.21
	s.d.	2.33	3.17	2.36
Frustration	Mean	2.52	3.08	1.54
	s.d.	2.48	2.70	0.79

There was a significant difference in the rating of effort between condition A and condition C (A vs. C, $t(8) = -3.29$, $p = 0.008$). The map was rated by participants as requiring more effort than the navigation system with auditory instructions. There were no other significant differences in the measures of perceived driver workload.

Discussion

The aim of this study was to compare the visual behaviour of experienced professional drivers, and frequent users of the navigation systems, when using different navigation methods to find their way in real traffic in an unfamiliar area. As such the results must be interpreted with the limitations of this type of applied research in mind, as the controls which are expected in a traditional experimental context are not possible. Analysis of the drivers' glance behaviour in each of the conditions leads to the initial conclusion that navigation using a paper map is less visually demanding than using a navigation system while moving, either with or without auditory instructions. Drivers made fewer and shorter glances at the map than at the navigation system. However, there was no difference between conditions in the total time spent looking at the road ahead. The percentage of long glances also conflicts with this conclusion as the map condition produced the greatest proportion of long glances.

In general, the results of the analysis of glance behaviour are consistent with the findings of previous research which also demonstrated that the percentage of eye fixations to a navigation system was greater than fixations to the map. A possible explanation for this pattern of results is that drivers automatically manage their glances to reduce the safety risk.

This study was conducted in real traffic conditions and as safety was the primary focus, time while the vehicles were stationary was not included in the analysis. There were a large number of occasions where the vehicle was stopped, for example, while waiting in queues of traffic or at traffic lights. It is reasonable to assume that drivers prefer to make glances to the map or the system when stationary. In addition, the overall navigation strategy when using a map may be different. Drivers might consult the map at the beginning of the journey and attempt to remember the route. Drivers then only need to make short glances for confirmation, whereas with a navigation system they can rely on it for turn by turn instructions. Neither of these assumptions can be validated on the basis of this research. In safety terms, the duration of glances at the system or the paper map during driving is the most important measure, as this represents time spent not attending to the traffic situation. The mean duration of glances to the map was shorter than glances to the system either with or without auditory instructions. It should be noted that the mean durations of glances were less than a second in all conditions. Glances of a second or less are similar to what would be expected when changing radio stations and are unlikely to constitute a safety risk (Zwahlen and DeBald, 1986). However, there were also examples of long glances in all conditions; some of the glances to the system exceeded 3 seconds in length. The greatest percentage of long glances (over 2 seconds) was observed in the map condition. The smallest percentage of long glances was observed in the electronic system with the auditory instructions on. Therefore, even though the map appears to require less visual attention in general, it seems that there is also greater incidence of long glances while driving. More research would be required to confirm this as the total number of long glances observed in this trial was too low to perform statistical analyses and the results may be biased by the fact that the map was positioned further away from the drivers' normal line of sight.

The average length of a glance to the system was longer when the auditory instructions were on than when they were off. In contrast, the percentage of glances to the system which exceeded 2 seconds in duration was slightly less when the auditory instructions were used. On the whole, there seemed to be little benefit for reduction of visual demand in having the auditory instructions. This could be a product of the participant group of this study. All participants were experienced users of the system who would use the system with the sound off for the majority of the time. In addition, in this study a split screen was used which displayed a large simple pictogram of the junction as well as the map. This diagram is easy to assimilate; differences may be observed if a more complex display were used, such as a detailed map. Drivers' perceived effort was greater in the map condition than in the system condition with auditory instructions. There were no differences on any of the other dimensions of the NASA TLX. As previous studies have found differences in workload using other methods it is possible that the NASA TLX was not sensitive to differences in driver workload in this instance. All drivers were experienced professional drivers and users of the system in question. This may be an explanation for the fact that ratings of workload on all dimensions across all conditions were typically low.

Overall, the results do not provide convincing evidence that paper map based navigation is more visually demanding than using an electronic navigation system. Measures such as the glance duration and frequency suggest that using a map is less visually demanding than using a system but the number of long glances and reduced attention to the left mirror suggest the opposite is the case. The mean glance duration in all conditions was less than one second and the total time spent looking at the road ahead was equal in each condition. It is likely that the characteristics of the display information in each condition place qualitatively different demands on the driver. The results of this study are therefore a function of the properties of the particular navigation system and map that was employed. In this study, with the display set to show the junction pictogram and a large scale map, there was little difference between the conditions with the auditory instructions and without the auditory instructions. It is likely that drivers moderated their behaviour and regarded the system or the map when they were stationary. There are also several factors that may have affected the behaviour of the drivers. For health and safety reasons the drivers had to be explicitly reminded that they were ultimately responsible for the safety of the vehicle and its occupants. The participants were also aware that they were being filmed. Although no instruction was given, participants did not tend to manually interact with the system while driving, for example, they did not tend to change the map display or use the menu functions while driving. On this particular navigation system all functions are accessible while driving and in a more natural situation they may have felt comfortable interacting with the system while in motion. Finally, all the participants were experienced users of the particular system. The system is likely to require more visual attention for a period of time for a new user.

The implication of these findings is that both paper maps and electronic navigation systems similar to those studied here are equally suitable navigation methods. While in motion the majority of glances away from the road were less than a second. However, the findings of this study should be viewed as an example of the driving behaviour of experienced drivers and users of the navigation system who are paying particular attention to safety.

References

BS EN ISO 15007-1 (2002). *Road Vehicles-Measurement of driver visual behaviour with respect to transport information and control systems, Part 1: Definitions and Parameters*. International Organisation for Standards.

BS EN ISO 15007-2 (2001). *Road Vehicles- Measurement of driver visual behaviour with respect to transport information and control systems, Part 2: Equipment and Procedures*. International Organisation for Standards.

Burnett, G.E., and & Joyner, S.M. (1993). An investigation on the man-machine interfaces to existing route guidance systems. *Proceedings of the IEEE-IEE vehicle navigation and information systems conference* (pp. 395-400). Piscataway, NJ: Institute of Electrical and Electronic Engineers.

Daimon, T. (1992). Driver's Characteristics and Performances when using In-Vehicle Navigation System, *Proceedings of 3rd International Conference on Vehicle Navigation and Information systems* (pp. 251-260). Location: Publisher.

Hart, S.G., & Staveland, L.E., (1988). Development of NASA-TLX (Task Load Index): Results of empirical and theoretical research. In P.A. Hancock and N. Meshkati (Eds.) *Human mental workload* (pp. 139-183). Amsterdam: North-Holland.

Parkes, A.M., Fairclough, S.H., & Ross, T. (1991). Multi-level evaluations of in vehicle route information systems. In T. Lovesy (Ed.) *Contemporary Ergonomics* (pp. 157-162). London: Taylor and Francis.

Uang, S., & Hwang, S., (2003). Effects on driving behaviour of congestion information and scale of in-vehicle navigation systems. *Transportation Research: Part C, 11*, 423-438.

Zwahlen, H.T., & DeBald, D.P. (1986). Safety aspects of sophisticated in-vehicle information displays and controls. In *Proceedings of the Human Factors and Ergonomics Society 30th Annual Meeting* (pp. 256 - 260). Santa Monica, CA: Human Factors Society.

Interaction between car drivers: a diary study

Maura Houtenbos[1,2], Marjan Hagenzieker[1], & Andrew Hale[2]
[1]SWOV Institute for Road Safety Research, Leidschendam
[2]Delft University of Technology
The Netherlands

Abstract

To gain a deeper understanding of expectations of (young) drivers a diary study was performed. A group of young drivers (n = 211) was asked to describe any unexpected situations that they encountered in traffic in a web-based diary. The 528 coded diaries showed that the majority of situations could be regarded as interaction situations. Also, the majority of scored behaviours took place at the perception/attention and decision making level of information processing. This is in line with a model of information processing in interaction, where expectancy is placed between perception and decision making. Future research will involve comparing these findings to in-depth studies on accident causation using the violation and error concepts.

Introduction

When participating in traffic, it is very likely that a driver will encounter other road users on this trip. In interaction with other road users, time is often a limiting factor in deciding what is the appropriate action in that particular situation. More often than not, crossing an intersection is performed successfully and safely, despite the limited time available to make appropriate decisions. Therefore, it seems plausible that drivers must have some kind of expectation of what is about to happen in the next moments in order to be able to react in time.

In a study into the information processing of drivers involved in interaction situations, a model has been formulated (Houtenbos, Hagenzieker, Wieringa, & Hale, 2004, see Figure 1). In this model, two kinds of expectancy are distinguished; 'long term' and 'short term' expectancies. Long term expectancies are derived from the driver's mental model of the situation and are thus based mainly on experience and education. An example of a long term expectation is the expectation that road users on a motorway will all drive in the same direction. Subsequently, short term expectancies are based on these long term expectancies and include information from the situation at that particular point in time. An example of such an expectancy is the expectation that another road user will be at a certain position in the next moment. A concept that might also be useful when investigating expectancy is Situation Awareness (SA, Endsley, 1995). The three levels of SA (Perception, Interpretation, and Projection) could help to gain insight into the expectations of drivers in interaction situations (Figure 1). At the third level of SA (Projection) the driver has an expectation of what is about to happen in the near future. It is important that these expectations do not conflict with expectations of other road users as this could lead to unsafe situations. The driver is able to make this projection by perceiving (SA1) and interpreting (SA2) what is happening in the environment. An obvious limitation of drivers on the first level of SA is that drivers are physically not able to really look around the corner (e.g. if there is an obstacle blocking their line of view), which provides the driver with less time to formulate an expectation about the road user around the corner. On the second level of SA, drivers can have trouble with interpreting the implications of an amber light. In some cases, the driver will come to a stop, whereas in other cases the driver might increase driving speed to try and pass the intersection before the light becomes red. Rules (formal and informal) are important for interpreting the environment, which includes the infrastructure as well as other road users.

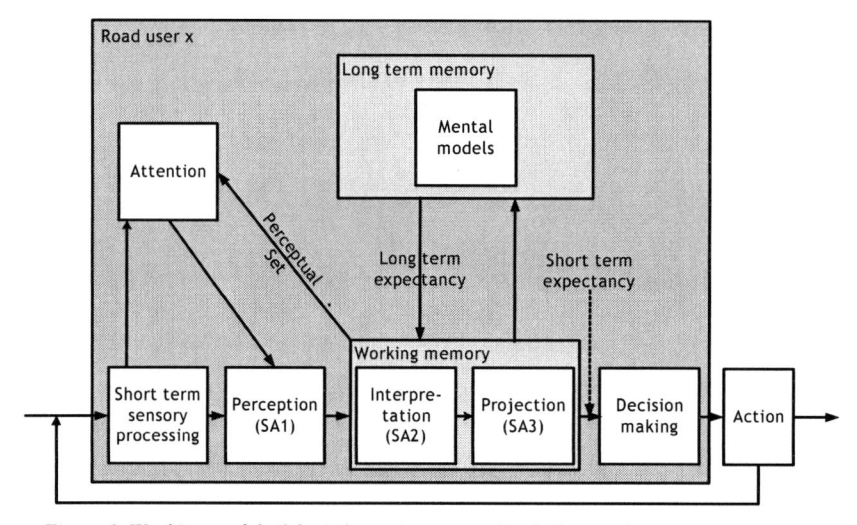

Figure 1. Working model of the information processing in interactions

As formulating expectations is usually a rather unconscious process it is difficult to find out what people expect. However, when expectations are violated, that is, something unexpected happens, we seem to gain (limited) access to these expectations. The aim of this study is to gain a deeper understanding of expectations of (young) drivers. The following questions are addressed:

- What kind of situations are reported the most?
- At what kind of locations are unexpected interactions most likely to happen?
- What kind of behaviour is reported in unexpected interactions the most?

Method

Participants were 211 young novice drivers (18-25 years of age), who held their drivers licences for about six months. As part of a larger study on young drivers (De

Craen, Vissers, Houtenbos, & Twisk, 2005), which was focused on the effects of a calibration course, participants were asked to keep a web-based diary during two periods of three weeks and were asked to describe any situation they encountered in traffic that startled them or that they found unexpected. The 528 diaries were coded by trained reviewers on various aspects of the situation reported.

Results

The majority of situations (416 of the 528 situations reported) included other road users and could be regarded as 'interaction situations' (Table 1). As Table 1 shows, this majority can be found at every (coded) location. This could imply that particularly aspects of the other road user were unexpected compared with other aspects of the situation.

Table 1. Types of road users involved and locations of reported situations

Area	None	Pedestrians (P)	Motorised road users (M)	P & M
Urban	46 (20%)	48 (21%)	137 (59%)	1 (0%)
Rural (Motorway)	18 (15%)	0 (0%)	103 (85%)	0 (0%)
Rural (Other)	39 (25%)	16 (10%)	103 (64%)	1 (1%)
Unknown	9 (21%)	0 (0%)	7 (44%)	0 (0%)
Total	112 (21%)	64 (12%)	350 (66%)	2 (0%)

The behaviours that contributed to the 'unexpectedness' were also coded. A distinction was made between the perception/attention level, the decision-making level and the action level for the reporting driver in order to determine at what level the situation became unexpected. For the behaviour of other road users involved in the situation this distinction was not made. Scored most frequently were 'driving too fast', 'not giving right of way', and 'cutting in on someone'. Table 2 shows how often these behaviours were mentioned in interaction situations.

Table 2. Frequency of behaviour mentioned in reported interaction situations

	Perception /attention	Decision making	Action	Other road user
Driving too fast	6	20	7	39
Not giving right of way	50	12	2	70
Cutting in on someone	44	17	5	111

The results also showed that 'driving too fast' and 'cutting in on someone' mostly took place on a straight stretch of road. 'Not giving right of way' is most frequently reported at an unsignalised intersection, which is not a surprising location for priority problems to occur. Table 2 shows that the majority of scored behaviours took place at the perception/attention and decision making level. This is in line with the earlier mentioned model of information processing of the road user in interaction (Houtenbos, Hagenzieker, Wieringa, & Hale, 2004, see Figure 1). In this model, expectancy is placed between perception and decision making.

Conclusions and Further research

Based on these results, no firm conclusions can be made yet. However, the notion that expectancy plays an important role in traffic interactions is supported by the (preliminary) results of this study. The finding that when asked to report unexpected situations, the majority of situations involved other road users provides us with an indication that expectancy and interaction are somehow associated. To be able to understand more precisely what aspects of the reported situation were experienced as unexpected, the diaries will be studied in greater depth. The results of this study will also be compared to in-depth studies focusing on accident causation. The distinction between errors and violations in the driving task will be taken into account in this comparison as they are said to contribute significantly to road accidents (Sabey & Taylor, 1980). Violations are characterised as deliberate, whereas errors are usually unintended. Errors are often broken down into slips, lapses and mistakes (e.g. Norman, 1983) but they are all concerned with cognitive processes whereas violations are more concerned with the social context. According to Reason, Manstead, Stradling, Parker, & Baxter (1991), female drivers tend to be more involved in accidents as a result of perceptual and judgmental errors, whereas male drivers are more frequently involved in accidents in which violations (e.g. speeding or drinking) were committed. Thus, it might be interesting to investigate if this effect can also be found in interaction situations that were experienced as unexpected. If this is the case, could it be that particular 'types' of unexpected behaviour are more (or less) easy to recognise early on and thus, to compensate for? For example, are deliberate violations made by other road users easier to compensate for than perceptual errors? This and other hypotheses that are derived from the results of this study will have to be explored in future research.

References

De Craen, S., Vissers, J.A.M.M., Houtenbos, M, & Twisk, D.A.M. (2005). *Young drivers experience: the results of a second phase training on higher order skills.* Report SWOV-R-2005. Leidschendam, The Netherlands: SWOV.

Endsley, M.R. (1995). Measurement of Situation Awareness in Dynamic-Systems. *Human Factors, 37*, 65-84.

Houtenbos, M., Hagenzieker, M., Wieringa, P., & Hale, A. (2004). Modelling Interaction Behaviour in Driving. In D. de Waard, K.A. Brookhuis, and C.M. Weikert (Eds.) *Human Factors in Design* (pp. 35-45). Maastricht, the Netherlands: Shaker Publishing.

Norman, D. (1983). *Position Paper on Human Error.*, NATO Advanced Research Workshop on Human Error., Bellagio, Italy, 1983.

Reason, J.T., Manstead, A.S.R., Stradling, S.G., Parker, D., & Baxter, J.S. (1991). *The Social and Cognitive Determinants of Aberrant Driving Behaviour* (Contractor Report 253): Transport and Road Research Laboratory.

Sabey, B.E., & Taylor, H. (1980). *The Known Risks We Run: The Highway* (Supplementary Report 567). Crowthorne, Berkshire: Transport and Road Research Laboratory.

The number of participants required in occlusion studies of in-vehicle information systems (IVIS)

Nick Reed, Nikki Brook-Carter, & Simon Thompson
Transportation Research laboratory, TRL
Berkshire, UK

Abstract

The aim of the trials conducted in this research was to address gaps in occlusion research. Twenty participants completed training trials on four In-Vehicle Information System (IVIS) tasks (1: Enter destination, 2: Place of Interest (PoI) search, 3: Change radio frequency, 4: Dial telephone) whilst wearing occlusion goggles that intermittently blocked their vision. Participants achieved a level of proficiency at which no further improvement was observed. Participants then completed three test instances of each task. Task completion times for the fastest and slowest N subjects were compared to times for all participants using independent sample t-tests. The value of N was increased sequentially until no significant difference existed between groups, thus providing an estimate of the group size required to obtain a representative sample. Results suggested that a group size of seven participants was sufficient for task 4, ten participants were required for tasks 1 and 3, and fourteen participants were required for task 2. Inferences are made about the reasons for the discrepancy in the number of subjects that are required to assess each task and recommendations are proposed for the trial protocol of studies that assess IVIS tasks using occlusion.

Introduction

In-vehicle information systems (IVIS) can provide the driver with a vast array of useful information. Navigation systems, road traffic information systems, mobile telephones and in-vehicle entertainment systems are all becoming widely available to the consumer. However these secondary tasks can also distract the driver from the primary task of safely controlling the vehicle. The level of distraction imposed by an IVIS task is dependent on the visual demand of the task, the resumability of the task, and the driver's behaviour. Although it is difficult to control driver behaviour, the visual demand and resumability associated with an IVIS task can be controlled for through the system's design.

Visual occlusion has been employed since the mid-sixties (Senders et al., 1966) with the aim of understanding and modelling driver visual behaviour. It can be used to assess both the visual demand and resumability of an IVIS task and has been found to be effective in this assessment (Burns, Lansdown, and Parkes, 2004). The

In D. de Waard, K.A. Brookhuis, R. van Egmond, and Th. Boersema (Eds.) (2005), *Human Factors in Design, Safety, and Management* (pp. 39 - 48). Maastricht, the Netherlands: Shaker Publishing.

technique is used for simulating the shift in visual attention associated with driving where a driver has to share visual resource between the driving scene and the in-vehicle information system (IVIS). This is either done using a shutter to hide and expose the IVIS from view, or goggles that also block or reveal the IVIS. The amount of time (duration and frequency) that the IVIS is visible or occluded (blocked from view) is controlled during occlusion. The objective is to investigate whether in-vehicle tasks can be carried out in short bursts of visual attention towards a display screen (typically 1 to 2 seconds).

This report details the methods and findings from a trial carried out using the occlusion method measuring the visual load imposed by IVIS. The trial was carried out under the Occlusion project UG407, funded by the Department for Transport and was designed to establish the appropriate sample for robust and repeatable results. This is an important issue; if a large sample size is required for robust, repeatable results then the method will not be a cheap, low resource method for assessing IVIS. It will therefore not appeal to manufacturers. The trial described in this report involved twenty participants who carried out specific IVIS tasks. Participants received different types of training on these tasks and were then required to perform them whilst their vision was periodically occluded by PLATO Occlusion goggles. This is a widely used technique employed by Bengler and Rosler (2001), Fichtenberg (2001), Karlsson and Fichtenberg (2001), Weir et al. (2003), amongst others. This paper details the methodology used during the trial, how the results were analysed, and the findings in relation to suitable sample sizes for assessment of IVIS tasks.

Methodology

Participants

Twenty licensed drivers from the TRL participant database were involved in this trial. Individuals considered as 'technophobes', or those who were highly experienced with route guidance systems and new technology, were excluded from the study. Participants' ages ranged from 28 to 55 years (with a mean of 41.3 years, SD 9.0). Their driving and technology experience were ascertained using pre-trial questionnaires and are shown in Table 1.

Table 1: Driving, computer, and mobile experience of the twenty participants recruited to take part in this study

	Driving	Computer	Mobile telephone
		Experience (yrs)	
Mean	21.3	10.5	6.1
Minimum	10	0	1
Maximum	35	24	13
SD	8.2	6.1	3.2

Participants reported either using a stereo whilst driving frequently or very frequently. Ten of the participants reported never using a mobile telephone whilst driving, six very infrequently, one occasionally, two frequently and one very frequently. Finally, one participant reported using a navigation system whilst driving occasionally, one very infrequently and the remaining 18 participants never.

Equipment

PLATO Occlusion goggles, as shown in Figure 1, were used to block participants' vision intermittently as they completed test sessions. The goggles fit like spectacles and there is padding on the arch for the nose and on the inside of the arms to provide some comfort.

Figure 1. PLATO Occlusion goggles

The goggles are operated by a Translucent-Technologies control box linked to a computer that controls the duration and frequency of occlusion periods. The lenses are simultaneously switched from their light scattering occluding state to their transparent state in just under 1 millisecond. In the transparent state, up to 90% of incident light is transmitted. Transition back to the occluded state takes about 5 milliseconds. When the display closes, luminance is maintained to prevent effects of light and dark adaptation. The occlusion interval is the period of time in which the goggles are closed and vision is occluded. In relation to driving, this period is assumed to represent the time when the driver's visual attention is directed to the surrounding road environment in the primary task of controlling the vehicle. It is also referred to as the 'goggle close', 'shutter close', or 'display off' interval. The vision interval is the period of time in which it is assumed that the driver has transferred visual attention from the primary task of driving to the secondary (i.e. IVIS) task. It can also be referred to as the 'goggle open', 'shutter open', or 'display on' interval. The current (pre-CD) draft of the occlusion standard submitted by the International Organisation for Standardisation (ISO N3XX, 2002) proposes a method of occlusion for assessing the visual or visual manual interfaces accessible to the driver whilst the vehicle is in motion. The draft standard states that the vision interval shall be 1.5 seconds and the occlusion interval shall be 2.0 seconds. Consequently these were the intervals used in this study.

Experimental design

The four tasks were:

Task 1: Destination entry
Participants were given a street name and town name and asked to enter them into the navigation system as the destination.

Task 2: Place of interest search
Participants were given a specific place of interest, such as a golf course or hotel in a given town, and were asked to search for that place of interest within the navigation system and input it as the destination.

Task 3: Changing radio frequency
Participants were given a specific radio frequency and station name and asked to search for and select it using the search arrow buttons (not the pre-selected station buttons).

Task 4: Dialling a telephone number
Participants were asked to enter a telephone number that was familiar to them from memory (so that they did not have to read a number written down) into the cradled mobile telephone and select the call button.

The vehicle in which participants completed the tasks was a BMW 7-Series (E38 saloon, year of manufacture 1998) and was stationary at all times. Tasks 1-3 were carried out using its factory equipped GPS-Navigation and entertainment system. Task 4 was carried out using a Nokia 7250i mobile telephone placed in a dashboard-mounted cradle, meeting current UK legislation on mobile telephone use within vehicles. When the trial tasks began the investigator remained quiet unless asked a specific question. An additional investigator sat in the rear of the vehicle to record task completion times and system response delays. For each attempt, the observers recorded the Total Task Times, any system response delays and the number of errors made.

Trial groupings

All participants completed five training sessions for each of the four IVIS tasks. For the training sessions, the twenty participants were each allocated to one of two groups. As shown in Table 2, the ten participants in Group A performed each IVIS task in the five training sessions without any form of occlusion (static training). The ten participants in Group B performed five training sessions of each of the four tasks whilst wearing occlusion goggles (occluded training). There then followed three test sessions of each task in which all participants wore occlusion goggles. Throughout training and testing, participants completed the IVIS tasks in a pseudo-random order to counterbalance and minimise any possible system learning and order effects.

Table 2. Occlusion conditions for the two groups completing training and test trials

	Training sessions	Test sessions
Group	1-5	1-3
A	No occlusion	Occlusion
B	Occlusion	Occlusion

The independent variables were the IVIS tasks and the training technique (Group A (static) vs. Group B (occluded)). The dependent variables were Total Task Time static (TTTstatic), Total Shutter Open Time (TSOT), and errors.

Before beginning the trial, participants were asked to read an information sheet containing instructions relating to the nature of the trial. If participants were happy with what was required of them and wished to continue with the trial, they were asked to fill in a consent form. Participants were then administered the driver profile questionnaire in order to obtain details relating to the driver's demographics and experience both in terms of driving and in terms of using new technology. This included the questions on how long they had used equipment such as mobile telephones and how often they would use them whilst driving. Each participant occupied the driver's seat of the vehicle throughout their test session. When the participants were carrying out the sessions wearing occlusion goggles, the goggles were connected to a laptop which was placed to the left side of the driver's seat.

Analysis

Total Task Times (static) were recorded during the five training sessions for the ten participants in Group A (static training), where Total Task Time (static) was the time taken to complete a task when not wearing the occlusion goggles. Total Task Times (occluded) were recorded during the study for the training sessions completed by the ten participants in Group B (occluded training) and for all participants during the three test trial times. Total Shutter Open Times were calculated based on the Total Task Time (occluded) and the vision interval. During the trial, system response delays were timed on each occasion a participant attempted a task. A system response delay was classed as a delay in the IVIS response to driver input. This delay cannot be controlled by the driver and was therefore removed from Static Total Tasks Times by deducting a proportion of the total system delay relative to the total vision interval:

$$\text{System response delay} \times \{\text{vision interval}/(\text{vision interval} + \text{occlusion interval})\}$$

No system delays were encountered in task 3 or 4 and so this adjustment was only made to task completion times for tasks 1 and 2. Independent samples t-tests were performed on the TSOT recorded in test trials and demonstrated that there were no significant differences across the times for Group A and Group B. Hence, training method (with/without goggles) did not appear to have an effect on test times. Consequently, data from all twenty participants (Groups A and B) were used in the analysis of the test trial task completion times.

Participants required

The mean task completion times for all twenty participants' test trials in each task were compared with the N participants who completed each task in the fastest time and also with the N participants who completed each task in the slowest time (where N is a number less than the total number of participants used in trials; in this case, twenty). The value of N was varied systematically to find the number at which the difference in test times between the fastest N participants and all twenty participants is non-significant and the difference between test times of the slowest N participants and all twenty participants was non-significant ($p > 0.05$). At this point, N gives an indication of the smallest number of participants that could be used to match the results of the larger group. A conventional power analysis could not be used to determine the number of participants that were required to complete each task because the size of the effect of interest was unclear. Independent-sample t-tests were used to compare the mean test times in each task across the fastest, slowest, and all trial participant groupings for the range of N values tested.

Figure 2 shows the task completion times for task 1, which were the longest for any of the tasks. However, the task completion times were relatively consistent across participants and so only 10 participants were required to find no significant difference between either the fastest 10 participants and the overall mean nor the slowest 10 participants and the overall mean.

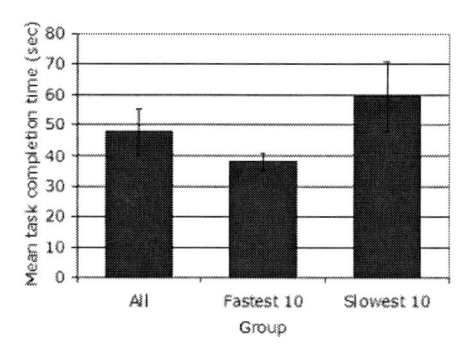

Figure 2. Comparison of the mean task completion time in task 1 (Destination entry) for all participants to completion time for the fastest 10 and the slowest 10 participants (error bars show 95% confidence intervals on the mean).

Like the destination entry task, the place of interest search required 10 participants in each group to stand comparison with the overall mean. However, task completion times for this task were significantly faster (see Figure 3). This reflects the simplicity of the task where the participant was able to scroll through options rather than enter data as was required in task 1.

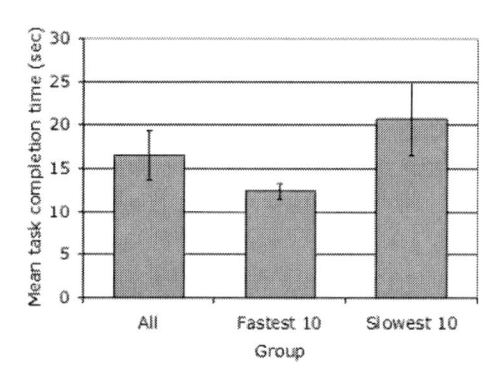

Figure 3. Comparison of the mean task completion time in task 2 (PoI search) for all participants to completion time for the fastest 10 and the slowest 10 participants (error bars show 95% confidence intervals on the mean).

The change radio frequency task required 14 participants for the fastest and slowest groups to demonstrate mean task completion times that did not differ significantly from the overall mean (see Figure 4). This was the greatest number of any of the tasks. This is likely to be a reflection of the differing task demands across task instances – sometimes the target frequency was a short distance from the start frequency, sometimes it was further away. This gave increased variability to task completion times and therefore, more participants were required to obtain a representative result. Whilst it is preferable to have consistent task demands, one must take into account the predictability of the task and the likely real world operation of the system.

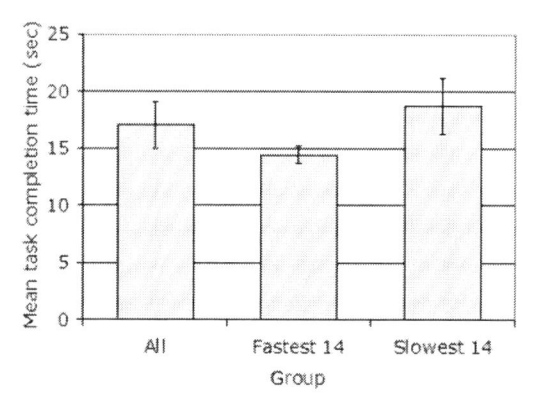

Figure 4. Comparison of the mean task completion time in task 3 (Change radio frequency) for all participants to completion time for the fastest 14 and the slowest 14 participants (error bars show 95% confidence intervals on the mean).

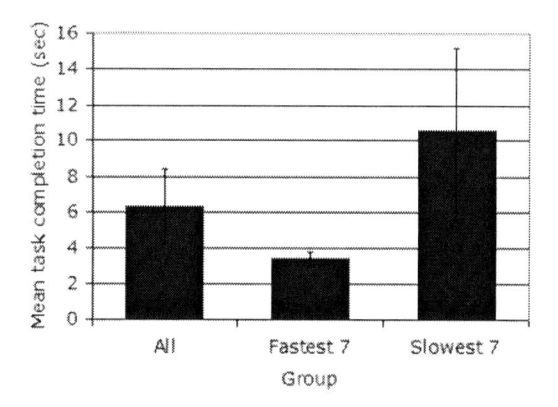

Figure 5. Comparison of the mean task completion time in task 4 (Dial telephone number) for all participants to completion time for the fastest 7 and the slowest 7 participants (error bars show 95% confidence intervals on the mean).

In task 4, participants were required to dial a familiar number on a hands-free mounted telephone. The simplicity and familiarity of the task reduced the variation in task completion times. This explains why this task required the fewest participants to obtain a representative sample. This task had clearly the fastest task completion times (see Figure 5). The fastest 7 participants could complete the task, on average, in less than 5 seconds and the overall mean task completion time was around 6 seconds. Interestingly, this was the only task where the mean task completion time for all participants was less than the '15-second rule' recommended by the JAMA guidelines (2000) for use of in-vehicle information systems.

The results of this analysis suggest that a minimum sample size of ten participants, each completing three task replications, would be required to obtain a reasonable representation of the performance of the population for the tasks involving programming of the navigation system (tasks 1 and 2). For task 3 (changing the radio frequency), a minimum of fourteen participants completing three task replications would be required. However, for task 4 (dialling a telephone), only seven participants completing three task replications would be required.

Conclusion

Based on findings from this trial, the minimum number of participants required to repeat this occlusion study would be fourteen. This is the number of participants that would be necessary in order to obtain results representative of the larger sample of twenty in all of the four tasks undertaken. Examining the tasks individually, the analysis suggests that a minimum sample size of ten participants, each completing three task replications, would be required to obtain a reasonable representation of the performance of the population for the tasks involving programming of the navigation system (tasks 1 and 2). For task 3 (changing the radio frequency), a minimum of fourteen participants completing three task replications would be required. However, for task 4 (dialling a telephone), only seven participants completing three task replications would be required.

The variation in minimum participant numbers suggests differing task demands. Dialling a memorable telephone number on a mobile telephone requires the entry of eleven digits on a familiar keypad. Task completion times are unlikely to vary greatly between participants. Hence, fewer are required to obtain a representative sample. The protocol of the trials in task 3 dictated that the requirement to change the radio frequency varied significantly between trials as participants had to cycle through different frequency ranges to reach the target station in different trials. It is likely that greater task variability is responsible for the increase in the number of participants required to obtain a suitable sample and that the number of participants required could be reduced by standardising the task across test instances. This must be balanced against the risk of trial predictability having a detrimental effect on the validity of the results. It was surprising to find that the navigation tasks seemingly require fewer participants than the radio task. With the increased complexity of the navigation tasks, one might expect greater variability in task performance and hence a requirement for greater numbers of participants. However, that the navigation tasks require fewer participants than the change radio frequency task suggests that they are made up of relatively consistent discrete tasks; e.g. enter street name, enter postcode, search for address. Completing all of these discrete tasks leads to greater variability than dialling a telephone number but less than changing the radio frequency by differing amounts.

The technique used for finding the minimum number of participants required to obtain results representative of those achieved in this experiment is sensitive to the number of participants that were selected to take part. For instance, if the original group had comprised a hundred rather than twenty participants, the recommendations for the number of participants required to achieve results representative of the original group may have been very different. However, since twenty participants is the approximate maximum number that could be employed workably as a test group, the recommendations made here represent an assessment of the number of participants required to achieve reliable results within constraints of practicality and workability.

Two ways to increase confidence that task times are representative for a given group of participants are to either increase the group size participating in the test or to increase the number of test trials completed by the group (or both). Since asking fewer participants to perform many trials is more cost effective than asking many participants to complete fewer trials, this is likely to be an appealing strategy for industry. Whilst these experiments cannot be used as a basis for statistically robust conclusions on the trade-off between sample size and test sessions (number of participants and number of measurements), they can be used as a basis for a reasonable practical compromise. Using engineering judgement, rather than statistical evidence, a practical recommendation would be 10 participants, each completing 5 training trials, followed by 5 test trials.

References

Bengler, K. & Rosler, D. (2001) *Evaluation of ITSWAP-HMI using occlusion*. Workshop on Occlusion, Torino, Italy, 12-13th November 2001.

Lansdown, T.C., Burns, P.C., & Parkes, A.M. (2004). Perspectives on occlusion and requirements for validation. *Applied Ergonomics, 35*, 225-232.

Fichtenberg, N. (2001). *Occlusion Experiment #2. (Follow on from 'Karlsson and Fichtenberg-how different occlusion intervals affect total shutter open time'*. Volvo Car Corporation. Workshop on Occlusion, Torino, Italy, 12-13th November 2001.

ISO N 3XX (Draft standard, 2002) *Road vehicles – Ergonomic aspects of transport information and control systems – Occlusion Method to Assess Visual Distraction Due to the Use of In-Vehicle Systems.*

Japan Automobile Manufacturers' Association (JAMA) (2000). *Guideline for in-vehicle Display Systems.*

Karlsson, R. & Fichtenberg, N. (2001). *How different occlusion intervals affect total shutter open time.* Volvo Car Corporation (August 2001). Workshop on Occlusion, Torino, Italy, 12-13th November 2001.

Senders, J.W., Kristofferson, A.B., Levision, W., Dietrich, C.W., & Ward, J.L. (1967). The attentional demand of automobile driving. *Highway Research Record, 195,* 13-15.

Weir, D.H., Chiang, D.P., & Brooks, A.M.(2003). *A study of the effect of varying visual occlusion and false duration conditions on the driver behaviour and performance or using secondary task human-machine interface.* ADE technical papers series 2003-01-0128. 2003 SAE world congress, Detroit, Michigan, March 3-6, 2003.

User friendly terminology for driver assistance systems

Marion Wiethoff[1], Alan Stevens[2], & Nikki Brook-Carter[2]
Delft University of Technology
Department of Work and Organizational Psychology
Delft, The Netherlands
Transport Research Laboratory, UK

Abstract

ADAS (Advanced Driver Assistance Systems; technology based systems that provide the driver with assistance in the driving task) may help to make the entire driving experience safer and more comfortable for road users. However, the number of different systems and the variations in functionalities is changing and increasing. Misunderstandings between developers' technical terms and users' expectations and assumptions are making the conversation between the providers and the end users ineffective, and there are also differences between authorities'/ administrations' (i.e. policy makers') and users' expectations and preferences. These misunderstandings may lead to safety risks of various kinds and will lead to legal obstacles for implementation and safe use. Therefore, within the ADVISORS project (GRD1-2000-10047), a first draft of a user friendly terminology has been developed which is understandable for all stakeholders, including all types of consumers. In a subsequent study, several types of stakeholders have evaluated a relevant subset of the terminology proposed via a web based questionnaire.

Introduction

Relevance

A wide variety of Advanced Driver Assistance Systems (ADAS) currently exist and many more will become available in the near future. These systems aim to help the driver by providing information, for example about routes, traffic congestion and accidents. Furthermore, some of these systems aim to support the driving task by actively interacting with elements of the driving task, and in some cases in such a way that the system cannot be overruled by the driver. Examples are tactile feedback to the driver via the accelerator pedal or automatic control of the distance to the vehicle in front. Because these systems can provide drivers with real-time information (and are becoming increasingly sophisticated and thus more useful) there is a growing concern that they may interfere with the primary driving task and as a result compromise safety.

Drivers may not be aware of the potential hazards relating to secondary tasks involved in interacting with an ADAS whilst driving. Instructions need to communicate critical information (how to use a system in the safest and most effective way) effectively and in a way that can be easily understood, learnt and remembered. One aspect of this is ensuring that the driver understands the terminology and phrasing used in all the relevant communications. Understanding the terminology requires that the terms arouse the correct annotations. The terms may be used within the user manual, but also in other communications with the driver.

Commonly, ADAS user manuals are provided to drivers to assist them in using a system and also to warn them of potential hazards and system misuse. Information can also be provided to drivers in the form of on-line help, a quick start leaflet and advertising. The information provided in all of these formats influences how the driver interacts with an ADAS and may consequently impact on driving safety. Therefore, in each case, the terminology and phrasing used to communicate information to the driver should be carefully considered. Designing effective information to aid drivers in correct, safe and appropriate system use is a complex task. If the information presented is poorly designed it can be off-putting and be misinterpreted, ignored, forgotten or discarded and thus compromise usability.

Glossary and definitions in user manuals

System user manuals are documents that assist users in their interactions with devices or systems. User manuals contain instructions and warnings to assist users. An instruction communicates procedures and critical contextual information; it tells users what to do, how to do it and provides reasons for acting. In addition, warnings can be provided which communicate to users that hazards exist, the nature and severity of these hazards and how they can be avoided through correct system use.

User manuals give written and visual instructions for the use of a wide range of products and systems. User manuals are 'how to' books for the system users and, although often only referred to when initially using the system or when coming across problems or questions, are a critical part of system use and effectiveness. ADAS user manuals therefore need to effectively communicate instructions on operation, installation, maintenance, repair of the system, efficient use and, in addition, provide safety warnings and details of user responsibilities. It is up to the driver to then act on the information within the manual. This behaviour cannot be guaranteed; however, without having access to this critical information there is even less chance of correct behaviour and the probability of an accident is increased.

There are two separate sections in the user manual to be considered: (1) the instructions on how to operate the system, and (2) the glossary or definitions of the terms used in the manual. The instructions part is an active part to be read by users whenever they have to know what to do and how. The glossary with the definitions is a more passive part to be read by users to learn the meaning of the term. In the present paper, mainly part (2) is considered.

Guidelines for choice of terminology and phrasing

The guidelines described below relate to the appropriate choice of terminology and phrasing used to communicate information to the driver on system use. These are taken from the Design Guidelines for Usability and Safety of In-Vehicle Information System (IVIS) User Manuals (Brook-Carter, 2004). Most of the work is based on earlier research, e.g. by Chapanis (1965), who demonstrates by very colourful examples how instructions can be written in such a way that they are unintelligible or even deceptive, just because the writer obviously did not consider *what action* is actually expected from the reader.

Extent of information
Lengthy explanations should be avoided. There should not be too much information and the information should not be overly complex, but kept simple and concise. Complex explanations should be broken down into short simple steps. Only useful information that is required by the user in order to operate the system efficiently and safely should be provided.

Example of a simple sentence:
> "The navigation CD provided contains a digitised street map. The street map details highways, national, regional and district roads in different colour. The scale of the map can be decided by the driver."

Example of a complex sentence:
> "The navigation CD provided contains a digitised street map, which details highways, national, regional and district roads in different colours and the scale of the map can be determined by the driver."

Example of superfluous information:
> "The GPS system is based on the reception of navigation signals from a total of 26 GPS satellites, which orbit the earth at an altitude of 21,000 km in approximately 12 hours"

The user only needs to know how the navigation signals work if it is necessary i.e. it has an effect on how to use the system.

Abbreviations and jargon
Explanations of abbreviations should be provided either every time they first appear in each section of the user manual and/or within a glossary at the back of the user manual. It is preferred to include both methods of explanation (text and glossary). It should be assumed that the user is a novice and does not have previous knowledge of the system or any technical information/knowledge associated with the system. Numerous abbreviations or jargon words that may be unfamiliar to the novice user should be avoided where possible.

Example of undefined abbreviations:
> "Once in contact with the RAM Network the letters M, T or W will be displayed…"

This statement does not define RAM.

Example of Defined abbreviations:

"The system uses a PIN number; ['Personal Identification Number']"

Language
The driver population should be considered in relation to the likely and intended use of the system as well as the native languages and other languages spoken and read. Therefore, one should consider the type of use that one can expect from the driver. Published statistics on language proficiency by country could be used as a reference; as a minimum, instructions should be provided in the majority language of the country in which the system is being sold, but official and minority languages should ideally also be considered. The language used in the manual should also be the normally spoken, everyday language of the user population.

Illiteracy and declining literacy should be considered when designing the manual. If the system is complex, unusual or hazardous then reliance on visuals or pictures is essential, particularly with respect to safety warnings. In addition to illiteracy, the users' reading ability should be considered (Chapanis, 1965). Although no specified level of reading ability is recommended in the production of user manuals, the basic underlying idea behind readability is that the longer the sentences and the more complex the vocabulary (or abbreviations), the more difficult to read will the text be. Therefore, long sentences with complex vocabulary or unfamiliar abbreviations should be avoided in order to make the text readable for the majority of the user population (Wright, 1971, Broadbent, 1977). Wright (1971) concludes that short sentences are best to be used for information that needs to be remembered. Other types of prose can be very appropriate for easy understanding, as long as it is not written in a bureaucratic style.

As members of the general public, ADAS users may be technically sophisticated or naive. The manual should be directed at the naive user, hence should be written avoiding jargon and unfamiliar words and abbreviations (Chapanis, 1965). The manual instructions should be kept simple, clear and non-technical. Furthermore, in general, sentences which are formulated in active and affirmative style are generally more easily understandable than passive, negative sentences (Broadbent, 1977), although there are exceptions (e.g. whenever the reader will have a contradictory assumption to what the writer wishes to explain to the reader).

In the ADVISORS project, these guidelines for terminology and phrasing were collected and applied in order to enhance appropriate communication between the ADAS stakeholders, e.g. drivers, experts, industry, and policy makers.

ADVISORS

The ADVISORS project[1] formulated as its overall objective to "develop a comprehensive framework to analyse, assess and predict the implications of a range of ADAS, as well as to develop implementation strategies for ADAS which are expected to have a large positive impact on one or more of these effects". Several studies have been performed in the project, and the comprehensive framework developed. One part of the project involved dissemination of the results to the broad public, and the main aim of this part of the study was to define user-friendly terminology (UFT) for some of the most important ADAS.

In the first phase of the UFT an exhaustive table of user-friendly terminology of several ADA systems and their technical sub-systems was produced, (Wiethoff, 2003b, Mankkinen et al.., 2001). The terminology has been based on a literature and standards review done by Dr. Alan Stevens from TRL, as part of the ADVISORS project. In addition to the review, the ADVISORS work (Mankkinen et al.., 2001) described the technical functions of different ADAS, especially those of system manufacturers. In the second phase, a few terms, relevant to systems that were piloted within ADVISORS, were evaluated.

First version of the User Friendly Terminology (UFT)

The UFT is to be considered as a glossary with a list of terms commonly used in the ADAS field, as in user manuals but also in articles from magazines and newspapers. Each term is accompanied by an "official", or technical, definition and a proposed more user-friendly definition. Sources for the original definitions included dictionaries (e.g. NVF 53), existing guidelines (e.g. Battelle, 2000) and international technical standards from groups such as Vehicle Roadway and Control, ISO TC204 WG14, (e.g. Adaptive Cruise Control ISO 15622; Collision Warning ISO 15623). Through involvement of members of the ADVISORS consortium in standardisation activities it was also possible to examine unpublished developing drafts in a number of ADAS functional areas.

Procedure for establishing the first version

The first UFT, containing 282 items, was established through the following process: first, a series of published sources of information were gathered (including international standards, glossaries, European Community documents and institute reports) that were related to advanced driver assistance systems. Second, the sources were examined and an alphabetical list of terms and definitions produced.

---

[1] Action for advanced Driver assistance and Vehicle control systems Implementation, Standardisation, Optimum use of the Road network and Safety; GRD1 2000 10047), is a project co-funded by the European Commission (DGTREN), in which governmental and other research institutes, a transport company, insurance companies, and industry of ten different European countries participated. ADVISORS finalised in 2003. Research and management of the ADVISORS project has been performed for SWOV Institute for Road Safety, The Netherlands

Third, human factors specialists, without an in-depth knowledge of the technical aspects of implementation of ADAS functions, then examined the definitions and re-wrote them. Typically this involved using more common and less technical terms and concentrating on the function as presented to the driver (rather than the implementation of the function). And finally, at least two human factors specialists reviewed each "proto-user-friendly" definition before a first draft table was disseminated. The first draft table contained 282 items, including 10 items which were presented more than once with different definitions. Some of the items are very generally used (e.g. "risk", or "user needs", or "subject vehicle") but most of the items are very specific to ADAS, e.g. "Regulatory sign information", or "Obstacle warning coverage zone", or "Maximum longitudinal deceleration (amax)".

For each term, the official definition is given, and the user friendly definition, as well as the source for the user friendly definition. The user friendly definitions are derived or developed on the basis of the criteria (mentioned in the Introduction).

Some example items

Term	Dynamic route selection information
Definition	Refers to any route selection function which is performed during the drive. The purpose of this ATIS function is to provide the driver with a mechanism for recovering once they have left or wandered from the intended route. When a driver makes a wrong turn and leaves the intended route, the dynamic route selection function can generate a new route which will accommodate the driver's current position.
User Friendly Definition	Information to advise the driver to change direction in the event of a deviation from the route provided by a navigation system.
Source	Battele Guidelines

Term	Enforcement
Definition	Measures or actions performed by enforcement authorities to achieve compliance with laws, rules and regulations
User Friendly Definition	Actions taken by the authority to ensure regulations are obeyed
Source	NVF 53

Term	Headway (h)
Definition	Time between which the leading surfaces of two consecutive vehicles pass the same location along the roadway. Headway is related to subject vehicle speed (v), clearance (c), and length (L) by the formula: $h = (L + c)/v$.
User Friendly Definition	Distance to the vehicle in front.
Source	SAE J2399CD 5-3

ADVISORS analysis

In the ADVISORS project, drivers participated in empirical studies: amongst others a Lateral Support System (LSS) on-the-road pilot, a laboratory simulator study on (inter urban) Adaptive Cruise Control system (ACC) and a laboratory simulator study on Adaptive Cruise Control + Stop & Go system (ACC+S&G). The participants were asked three open questions (Nilsson et al.., 2002) whose purpose was to collect these drivers' views of the functionality of the ADAS they had experienced in the pilots, as described by their own words. One assignment was: "Try to explain in your own words the functionality of the system". The responses regarding wording were used in a later stage to propose alternative definitions of the terms. Two other questions were: "What would you call the system?" and "Did the system do anything that you had not expected?" The descriptions for the piloted ADAS follow below.

Lateral Support System (LSS)

The following (summarised) descriptions of ADAS functionality were mentioned by the subjects participating in the pilot (Tango et al., 2002). The LSS is:

> a system that could complete the driver's perception / enhance the driver's attention / increase the driver's vigilance when s/he makes an unintended manoeuvre; a system that could warn the driver if s/he starts overtaking and is not able to see an approaching vehicle; a supporting system to make overtaking more safe; a support that makes distractions during driving less dangerous.

It should be noticed that according to the judgement of many drivers the main function of the LSS is not only to warn the driver in dangerous situations, but also to generally enhance his/her attention. Some participants felt controlled by the system, and some of them mentioned the risk that the driver might rely solely on the system and for example might stop looking backwards over his/her shoulder.

Conclusion: the drivers associated the system with enhancement of attention rather than with lane-keeping functionality. They considered the system generally as a safety device. However, for a definition of the system, it is not an option to mention the enhancement of attention (in the definition), but it is a favourable side-effect, not what the system does initially. It can be mentioned, of course in the description of the system. Therefore, an alternative definition could be: "System to assist the driver to avoid leaving unintentionally the lane on either the left or right."

Inter Urban Adaptive Cruise Control (ACC)

Some of the following (summarised) descriptions of the functionality were mentioned by respondents (Törnros et al.., 2002). The ACC: helps the driver to control the speed; keeps the speed, but helps the driver to reduce speed automatically if the distance to the vehicle in front gets too short; controls automatically the distance to vehicle ahead with automatic speed adjustment; lets the car take over speed control in order to drive more relaxed and focus on the road instead of on gear and pedals; is good with automatic speed adjustment when approaching another car from behind; good to be able to speed up and slow down with two buttons; driving was more calm and safe since the cruise control kept a set speed.

Conclusion: most of the definitions given, focus on the set speed and add the automatic functionality of adjustment of the speed when approaching the car in front. An alternative definition could be: "An in-vehicle system that will maintain speed, but helps the driver to reduce speed automatically if the distance to the vehicle in front gets too short. "

Urban Adaptive Cruise Control with Stop and Go (S&G) Functionality
The following (summarised) descriptions were mentioned (Brook-Carter et al., 2002) "The ACC with S&G: keeps the car a safe distance from the car in front; allows the driver to relax or makes driving more comfortable; keeps the car at a constant speed".

Some participants wondered whether the system is safe and mentioned things like "driver can become complacent", "limitations include need for extra vigilance for traffic lights etc.", or "it removes concentration from driving allowing the driver to become distracted". Other participants however described the system as being safer or efficient and allowing the driver to observe more freely what is going on around her/him.

Conclusion: there was a tendency to opt for a definition focussing on the safe distance to the vehicle in front. This is understandable, since the system differs from interurban ACC because in urban areas there may be little opportunity to keep the car at a constant speed. An alternative definition could be: "An in-vehicle system that will keep the car at a minimal distance to the vehicle in front but maintains speed and works also at lower speeds."

Validation study

Method and Analysis

For the validation study, it was decided to select those items that were related to the ADAS functions which were the focus of interest within ADVISORS: the items related to Adaptive Cruise Control (ACC), Intelligent Speed Adaptation (ISA), Driver Monitoring (DM) and Lane Departure Warning Systems and Lane Support Systems (LDWA and LSS) and some very generic terms, such as "Forward vehicle", "Auditory message length" or "Border".

Respondents
Respondents were contacted by e-mail with a request to visit a website where they could fill out the questionnaire and, if they were interested, could receive the results. All respondents were contacted because of their apparent interest and expertise in the field of ADAS and the automobile industry, so that they would be able to evaluate the quality of the definitions in terms of validity and have enough contacts with other people outside the field (e.g. policy makers, customers, lay men) to be able to judge the definitions on user friendliness. In total 17 respondents completed the

questionnaire and rated each user-friendly definition and sometimes added comments or proposed different wordings.

The group of respondents was composed as follows:

Affiliation: Automotive industry (2), Car sales (1), Public Authority (4), Research Organisation (9), European Commission (1). In total 17. Position: Manager/Sales (2), Policy maker (2), Researcher (10), Designer/Engineer (2), Administrator (1).

Rating procedure
For the rating the "Surveyor Programme" was used (Object Planet). This is a web based programme, and for each item, the official definition and the user friendly definition was given (Figure 1). For each item the respondents rated the degree of appropriateness of the user friendly definition in a 5-point scale (Extremely poor, Poor, Neutral, Rather appropriate, Very appropriate).

Furthermore, respondents were invited to add comments or rephrase the wordings.

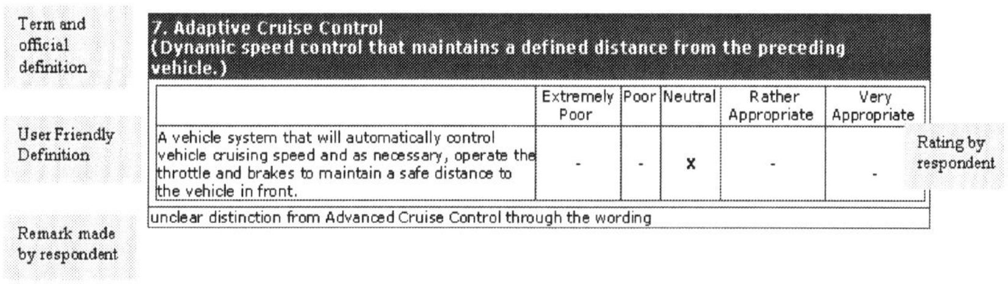

Figure 1. Example of an item and single response

Sorting procedure
The user friendly definitions were sorted on appropriateness. The sorting procedure was designed in such a way that the higher quality definitions are at the top of the list, and the definitions which aroused "poor" ratings and "very poor" ratings are towards the bottom. The rationale is that a bad rating of a definition is also useful information.

The procedure was as follows: (1) Sort on percentage of Rather appropriate" or "Very appropriate", (2) then *inversely* on percentage of responses of either "very poor" or "poor", and (3) then on percentage of responses "very appropriate". In this way, the sorting procedure also accounted for negative responses. The sorting procedure was designed to support the alterations to be made on the basis of a textual analysis of the user friendly definition and the comments.

Textual analysis of the comments
The lower the apprpriateness (low in rating), the more the need for a rephrasing was acknowledged, and the more the rephrasing was allowed to differ from the original proposed user friendly definition. The comments and the rephrasing, proposed by the respondents were analysed and a choice was made on the best rephrasing.

The following rules were applied (in accordance with a.o. Hartley, 1994):

- Simple sentences instead of one long, complex sentence.
- Abbreviations should be explained
- No jargon or expert's language or complex words
- The higher the rating of the definition, the smaller are the alterations that need to be made to improve the definition.
- Whenever items relate to each other (e.g. ACC-off state and ACC-stand-by state), rephrasing can occur, but always in such a way that the definitions will be phrased consistently, and the rephrasing will be most close to the higher rated definition.
- In general, a short definition is preferred above a long definition.
- A definition concerning a system always starts with "an in-vehicle system ..." or "an externally controlled system ...." or "a system .....". or "a device"
- On the basis of the comments rephrasing of the definitions took place.

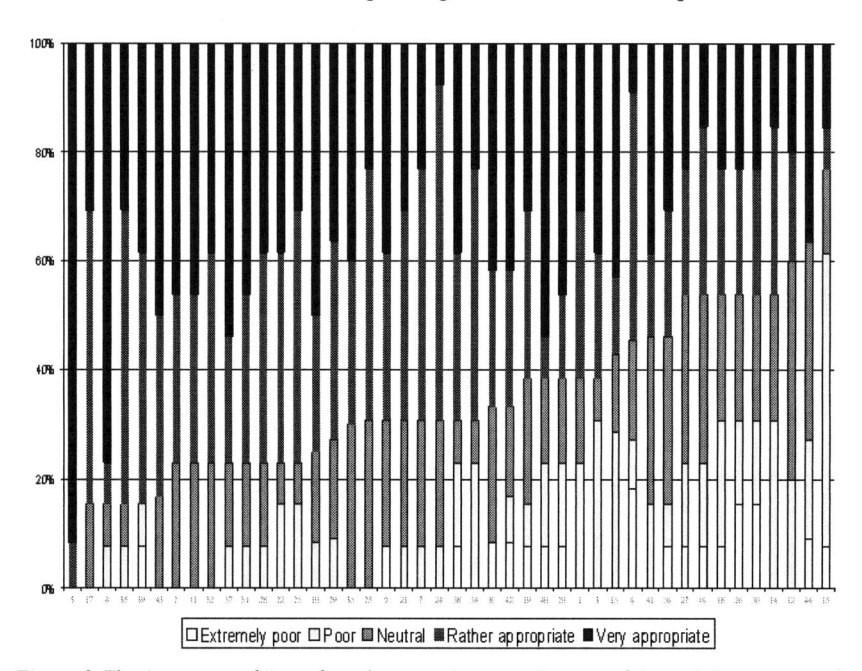

Figure 2 The items, sorted in order of appropriateness. Best rated items left, worst rated items right. On the Y-axis: percentage of ratings

Results

The sorting procedure based on the quality of user friendly definitions resulted in the following overview: (please refer to Figure 2). On the X-axis the items are listed. On the Y-axis, the cumulative ratings are given, in percentage of the classified quality. For instance, for the item presented at the left, all 100% of the ratings were "very appropriate". For the item presented at the very right part of the table, the frequency was as follows: Extremely poor rated by 8 %, Poor: 54%, Neutral: 15%, Rather appropriate: 8%, Very appropriate: 15%.

An example of a high rating (in the highest 25% percentile):

Term and official definition	User friendly definition
Maximum selectable time gap (Tmax) (Largest available value of time gap,T, that can be selected by the driver, in systems that permit driver selection of time gap.)	Longest selectable time gap to the vehicle in front
Remark Selectable … awkward word	*Proposed user friendly rephrasing* Longest time gap to the vehicle in front, which can be set by the driver."

(In the rephrasing, the fact that the driver sets the time gap is made clear without using the word 'selectable')

An example of a moderately high rating (in the second 25% percentile):

Term and official definition	User Friendly Definition
Lane Keeping Assistance Systems (with vision systems) (The system always assists the driver to keep the vehicle almost in the center of the lane using on-board vision systems or dedicated lane markings such as magnetic nails or magnetic tapes.)	System to assist the driver in avoiding leaving his lane on either the left or right.
Remark The definition misses the active intervention after the driver crossing the border of the lane. Is direction important??	*Proposed user friendly rephrasing* System to assist the driver in avoiding leaving his lane unintentionally on either the left or right

(The word "unintentionally" is added, and the definition is indicating the active intervention.)

An example of a moderately low rating (in the third 25% percentile):

Term and official definition	User friendly definition
Maximum longitudinal deceleration (amax) (Maximum permissible sustained level of longitudinal deceleration that can be achieved by an ACC system, independent of initial velocity.)	Maximum automatic braking rate of the ACC system.
Remark (Prefer "deceleration" above "rate")	*Proposed user friendly rephrasing* Maximum deceleration under automatic braking of the ACC system."

(The word 'deceleration' is used)

An example of a low rating (lowest 25% percentile):

Term and Official definition	User Friendly Definition
Lane departure (Any part of the body or the wheel which crosses the lane marking.)	Moving from the traffic lane to another part of the road.
Remark Add: Lane boundary, unintentionally, and add the direction	*Proposed User Friendly rephrasing* The vehicle is apparently unintentionally crossing the lane boundary on either the left or right.

(All proposed words are used).

For the final version of the UFT, a set of proposed alternative definitions (including the proposed alternatives presented above), was formulated. These alternative definitions will be presented to a number of respondents in order to be rated and prioritised. Only high ratings will be accepted for the final version of the UFT.

Discussion and conclusions

Apparently, there is a clear need for User-friendly terminology and definitions. The sheer number of items (272!), the type of definitions, the number of abbreviations, definitions containing jargon, being complex and with very technical terms indicate that there is a problem. Consumers, sales officers, managers and policy makers need to be able to understand each other. Terminology should always be appropriate. Sometimes, technical terminology and jargon is an efficient means of communication between informed groups, e.g. researchers, technical employees and experts-policy makers. But in all other areas, a UFT is effective and helps to avoid misunderstandings. Existing terminology (even from well-respected sources) is often unnecessarily unfriendly. For instance, in nearly all cases one can reconstruct complex sentences into a few simpler ones and explain abbreviations. Most terminology can be improved towards a UFT.

In general, systems developers are not experienced in understanding the needs of users; they have a different focus. Human factors personnel are generally better placed to understand the needs of users, although even UFT developed by human factors professionals can be improved. In general, there should be more awareness of developing a UFT, in particular for manuals, help functions and product information. Human factors professionals will invariably say "ask the users". Usually, they are right, because the users have their own aspects they find important.

The ADVISORS study revealed that the definitions given by drivers for the Interurban ACC and the ACC combined with S&G indicated a different type of experience. The drivers experienced *constant speed* to be essential for Interurban ACC, and *minimal distance to the vehicle in front* for the S&G system. However, the example of the definitions given for Lane Support System also showed that drivers should not be the only group to be considered. The drivers' definitions focussed on "attention enhancement system" which is definitely not the intention of the system, and would only cloud the term further. It is, therefore, important to include *all* user stakeholders in determining a terminology, also technical specialists. From the questionnaire study, it was clear that this is an efficient way of gaining outside knowledge. Even with small numbers of respondents, which is a limitation, the questionnaire proved highly useful, through high quality responses. Of course, this is only possible if the questionnaire is well designed and well targeted.

It can be concluded that the process used here was effective and a significant step has been made in improving the UFT. In the next phase, the alternative definitions will be rated and prioritised. Finally, it should be noted that terminology is an evolving and living endeavour, and that the UFT should be periodically evaluated and updated.

References

Battele Guidelines, NHTSA (2000) Development of safety principles for in vehicle information and communication systems

Broadbent, D.E. (1977) Language and Ergonomics. *Applied Ergonomics, 8*, 15-18

Brook-Carter, N. (2004). *In-Vehicle Information System (IVIS) User Manual Usability and Safety Checklist.* (Report TRL PPR009). Crowthorne, UK: The Transport Research Laboratory.

Brook-Carter, N., Parkes, A. M., Burns, P., & Kersloot, T. (2002) Evaluation pilots of an Urban Adaptive Cruise Control (ACC) system. In: *Proceedings 9th World Congress on intelligent transport systems*, ITS America, ERTICO, Chicago, October 2002.

Brook-Carter, N., Parkes, A. M., Burns, P. & Kersloot, T. (2002) An experimental assessment of an urban adaptive cruise control (ACC) system. In: *Proceedings of International Congress on IT Solutions for Safety and Security in Intelligent Transport,* ERTICO, European Commission, ITS France, Lyon, September 2002.

Chapanis, A. (1965). Words, Words, Words. *Human Factors*, 7, 1-17

Conrad, R. (1962). The design of information. *Occupational Psychology*, 36, 159-162

Hartley, J. (1994). *Designing instructional text* (3rd edition). London: Kogan Page.

ISO 15622:2002[E] Road vehicles – Adaptive Cruise Control (ACC) Systems – Performance requirements and test procedures. Published October 2002.

ISO 15623:2002[E] Road vehicles – Forward Vehicle Collision Warning (FVCWS) Systems – Performance requirements and test procedures. Published October 2002.

Mankkinen, E., V. Anttila, V., Penttinen, M., Marchau, V., & A. Stevens: (2001) Actor interests, acceptance, responsibilities and users' awareness enhancement. (Public Deliverable D2.2v2.0 ADVISORS project; GRD1-2000-10047).

Nilsson, L, J. Törnros, J. Parkes, A., Brook-Carter, N., Dangelmaier, M., Brookhuis, K., Roskam, A-J., De Waard, D., Bauer, A., Gelau, C., Tango, F., Damiani, S., Jaspers, I., Ernst, A., Yannis, G., Antoniou, C., & Wiethoff , M. (2002) *An integrated methodology and evaluation checklist and ADA design. Pilot Evaluation Results.* (Public Deliverable D4/5.2 ADVISORS project; GRD1-2000-10047).

NVF 53, Road transport informatics terminology dictionary

Object Planet Surveyor Programme, http://www.surveyor.tbm.tudelft.nl

Tango, F., Damiani, S., Wiethoff, M., Bauer, A., & Gelau, C. (2002) Evaluation of the Lateral Support System. A Pilot study within the scope of the ADVISORS Project. In: *proceedings of 9th World Congress on intelligent transport systems*, ITS America, ERTICO, Chicago, October 2002.

TC204 WG14 N143-15 Standardization status and future procedure of driving support systems

TC204 WG14 PWI 180100, SAE J2399CD 5-3 ACC human operating characteristics and user interface

Törnros, J., Nilsson, L., Östlund, J., & Kircher, A. (2002). Effects of ACC on Driver Behaviour, Workload and Acceptance in Relation to Minimum Time Headway. In: *Proceedings 9th World Congress on intelligent transport systems*, ITS America, ERTICO, Chicago, October 2002.

Wiethoff, M (2003a). *ADVISORS Final Publishable Report*. (ADVISORS project; GRD1-2000-10047). .

Wiethoff, M (2003b). *ADVISORS User Friendly Terminology, special Annex to ADVISORS Final Publishable report*. (ADVISORS project; GRD1-2000-10047). .

Wiethoff, M., Stevens, A., & Brook-Carter, N. (In prep.) A validated User Friendly Terminology for Driver Assistance Systems.

Wright, P. (1971). Writing to be understood: why use sentences. *Applied Ergonomics, 2*, 207-209

Designing out terrorism: human factors issues in airport baggage inspection

Alastair G. Gale, Kevin Purdy, & David Wooding
Applied Vision Research Centre, ESRI
University of Loughborough, Loughborough, UK

Abstract

All air passenger baggage is screened at airports by means of 2-D X-ray imaging which results in a computer display of each luggage item that is then visually searched by an operator (screener) for the presence of potential threat items (e.g. knifes, guns, improvised explosive devices [IED]). Despite improvements in screener training and available technology (e.g. image enhancement functions, threat image projection, 3-D X-ray imaging) the performance of screeners is variable which leads to the potential for terrorist threat to aircraft and passengers. A new training scheme to improve performance in baggage screening is under development (EPAULETS: Enhanced Perceptual Anti-terrorism Universal Luggage Examination Training System) and some of the initial human factors issues that underlie variable screener performance are considered.

Imaging interpretation

The detection and recognition of potential threat items within cabin baggage involves a human operator inspecting and interpreting a two-dimensional (static or dynamic) X-ray image of hand luggage or clothing item. Consequently this process is similar to the task of medical image interpretation, where a radiologist inspects an image (generally two-dimensional but can be three-dimensional) of human anatomy, which may be static (e.g. chest radiograph) or dynamic (e.g. ultrasound scans). Therefore the medical image inspection task is very similar to aircraft baggage inspection and data and theoretical approaches found applicable in the former, which has a larger research base, should be able to be applied to the latter. More strictly, baggage inspection relates to a typical radiological screening situation where many normal images are inspected which leads to the detection of relatively low numbers of abnormalities. For instance, in breast screening in the UK, circa 1.5 million women are screened each year, which leads to the detection of some 10,000 cancers.

Ergonomic differences between the two imaging interpretation situations mainly concern the viewing conditions themselves. In medical imaging the image is viewed in a quiet and darkened room with no extraneous distractors, whereas in airports the image is viewed in normal room lighting with many potential distractors present. There are specific reasons for this arrangement which are not considered here.

In both imaging domains, even when individuals are working to the best of their abilities, errors can still occur primarily due to the very large range of appearances of potential targets. Errors can be false positives or false negatives. False positive detections in baggage inspection are dealt with by subsequent manual examination of the bag and can cause delay to the passenger as well as adding to the workload of security personnel. False negative decisions are very problematic as a potential threat item is then allowed to be carried on to the aeroplane. Consequently it is these errors which primarily need to be addressed.

Modelling image inspection

A theoretical model has been developed to adequately describe and account for these errors in medical imaging (c.f. Gale, 1997). The model can also be applied to baggage inspection (Gale et al., 2000). This model emphasises the various stages of the visual inspection process. These are that the individual first approaches any new image with a potential hypothesis (schema) about its nature (i.e. normal, abnormal etc.).

An initial glance at the image yields a rapid global processing which leads on to serial detailed visual inspection of the image. This serial process entails the dynamic process of bringing particular image areas on to the fovea (or specifically an area of visual attention somewhat larger than the fovea – termed the Useful Field Of View, UFOV) for detailed inspection. The eye moves over the image in a series of very rapid saccadic eye movements coupled with intervening fixations. Vision is largely inhibited during an eye movement and so the perception of the image can be conceived of as a series of glances of varying fixation times interleaved with ballistic saccadic eye movements. Precisely where the eye fixates and then moves to fixate next is considered to be as a result both of the cognitive plan for looking coupled with the information gleaned from each glance.

The search process can be monitored by suitably recording the screener's eye movements. This allows false negative errors to be characterized as due to the process of search, detection or interpretation. Search errors are where the observer has clearly failed to look directly at, or near to, the potential threat area. Consequently the threat was not encompassed within focal vision. Detection errors are where the observer has actually looked at, or near to, the potential threat but has failed to detect the abnormal features. Typically research has demonstrated that detection errors are accompanied by eye fixations at, or near to, the abnormality of less than one second. Similar in nature to detection errors are interpretation errors where the observer has clearly looked at or near to the abnormality but for more than one second, yet has still failed to interpret the information there appropriately. It is taken that the observer has actually had time to detect the appropriate visual information but has then failed to interpret this information adequately (i.e. an error of cognition).

In general most errors in imaging interpretation are found to be due to detection or interpretative processes rather than search per se, although visual search is a necessary part of the overall inspection process. Therefore, recording visual search of these images appropriately is key to determining whether errors are due to visual or cognitive factors which can be elicited on the basis of fixation time measures. The EPAULETS project utilises eye movement recording in experimental designs employing MRMC (multiple reader multiple condition) ROC analyses to quantitatively assess error performance.

Discussion

Previous work (Gale et al., 2000) studied performance in an IED identification task and demonstrated that more interpretation errors were found with baggage screening than have typically been found in medical imaging interpretation. This was surprising and may well indicate that examining baggage items for potential IEDs is a very difficult cognitive task. However, an alternative explanation may be that the particular IEDs used in this study were very difficult or of low visual conspicuity.

Building on this research the EPAULETS project will develop an image database of air passenger baggage items, some of which contain potential threat items. By presenting these to screeners and monitoring their visual search behaviour then their skills, together with the precise nature of false negative errors will be detailed. Information from numerous empirical studies will then be used to inform advanced training strategies with the overall aim of minimising the possibility for baggage screeners to make errors.

Acknowledgements

This research is supported by the EPSRC 'Technologies for Crime Prevention and Detection' programme. The research is performed in collaboration with Tobii Technology and QinetiQ.

References

Gale A.G. (1997): Human response to visual stimuli. In W. Hendee & P. Wells (Eds.) *Perception of Visual Information* - second edition, New York: Springer Verlag.

Gale A.G., Mugglestone M., Purdy K.J., & McClumpha A. (2000). Is airport baggage screening just another medical image? In E.A. Krupinski (Ed.) *Medical Imaging 2000: Image perception and performance* (pp. 184-192) Bellingham: SPIE.

Visual strain of the underground train driver

Hilda Herman
Institute of Public Health
Bucharest, Romania

Abstract

The present study focuses on the visual strain of the underground railway train driver. The train lines and the lighting conditions, the driver's activity and complaints were studied. Results showed great differences in illumination and luminance between the tunnel and the stations along the tunnel. The lighting values change very frequently and quickly from dark to light and from light to dark. The visual acuity, retinal adaptation are permanently strained. Technical, organisational, personal protection and medical preventive interventions are necessary.

Method

In large cities the underground plays an important role in public transport. The underground runs only or predominantly underground and as a result, the lighting conditions under which the engine driver has to work and secure safety of passengers may be unfavourable. In the present study, the train lines, and the driver's tasks were analysed, and the lighting conditions in the driver's cabin, the tunnel and at stations (including light sources, illumination, and brightness) were assessed. The frequency at which lighting changes was timed, and the driver's subjective reports were collected. Tunnel illumination was measured in the cabin through the windshield, hence this was the value perceived by the driver.

Results

Most of the underground line and stations are underground, with exception of the end stations and turning places, which are at the surface level. The driver's task is to operate speed and check the instruments on the instrument panel, check signals and respond to these if required. At stations the driver has to monitor passengers getting in and out of the underground train and operate the doors. As the largest part of each journey is traveled underground, lighting is artificial. Sources are the incandescent lamps placed on the cabin ceiling, and in the tunnel tubular fluorescent lamps with a protective screen on the tunnel walls at the right and left; and the exterior incandescent lamps of the train itself. In stations tubular fluorescent lamps with protective screen are placed on the platform ceiling and on the walls.

The illumination levels were: in the cabin, 5-10 lx, in the tunnel and 10-15 lx in stations; in the tunnel, 0.25-5 lx at the exterior lamps of the cabin, 130 lx in the

In D. de Waard, K.A. Brookhuis, R. van Egmond, and Th. Boersema (Eds.) (2005), *Human Factors in Design, Safety, and Management* (pp. 67 - 68). Maastricht, the Netherlands: Shaker Publishing.

zones lighted by the tunnel lamps; 15-200 lx in stations. In the exterior zones, the illumination is very high by daylight. There are great differences in illumination and luminance between the tunnel and the stations, and also along the tunnel which may cause glaring, a potential risk of getting an accident. The lighting values change very frequently and quickly from dark to light and from light to dark, at very short intervals, between 60-150 seconds, the time to cover the distance between stations. In the tunnel large differences in illumination occur as a result of meeting trains and if the lamps placed on the walls are not well covered, these represent bright light sources. A similar condition is by day the end of the line above the ground surface, especially when the sun shines and at snow in winter. The arrival and the departure of the train in and from the stations expose the driver to great, sudden and frequent variations of illumination and luminance from low to high and from high to low values (Figure 1).

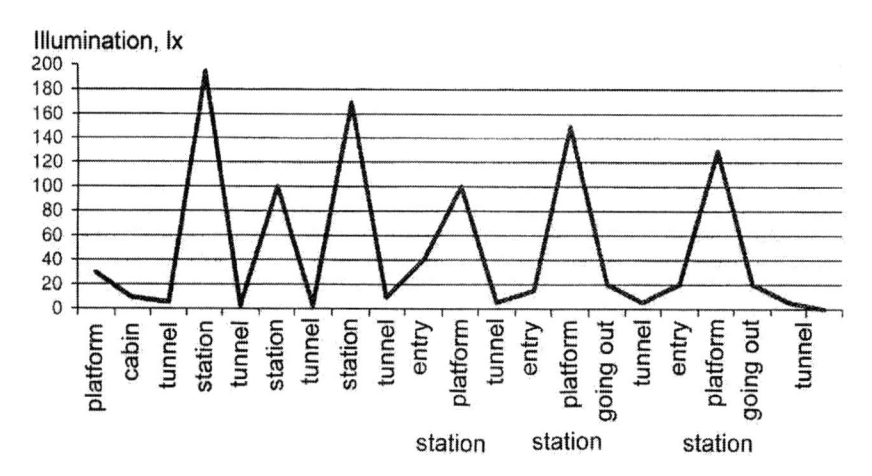

Figure 1. Change in the illumination during an underground journey

The lamps of the tunnel perform light ray cones on the walls in the dark tunnel. In those lighting conditions the luminance of the lamps has a higher physiologic value against the darkness of the underground. Therefore the charge of the visual acuity and retina adaptation and also the visual attention load are high and permanent. The remarks of the train drivers indicate ocular and visual fatigue.

Conclusions

The underground environment with its lighting characteristics puts high demands on the driver's visual strain. Technical, organisational, personal protection and medical preventions are necessary, such as adequate screening of the tunnel and station lamps, of the exterior drive cabin lamps, and adapting the illumination in stations both for the travellers but also for the train driver. Driver might protect their eyes with spectacles with adequate lenses; ophthalmologic examination at appointments and at regular intervals is required, respecting the medical contra-indications.

HUMANIST: Networking in the European Union for a human-centred design of new driver support systems

Sascha M. Sommer
Institute for Occupational Physiology at the University of Dortmund, IfADo
Germany

Abstract

Since March 2004, the European Commission is co-funding the Network of Excellence HUMANIST. The project is about human-centred design of new road transport telematics. The main objective of the network is to structure and organise this research field in order to integrate the activities of its members.

The HUMANIST approach

Information and communication technologies provide the means for road telematics and advanced driver assisistance systems. Applications of pre-trip and on-trip as well as nomadic assistance aim at the support of the driver's safety and mobility. However, every new secondary information source may distract from the primary driving task and increases driver workload, thereby possibly counteracting the potential benefits new support systems may have. In order to reduce the negative side-effects of driver support systems their design has to be centred on human factors. To strengthen and focus the competencies existing in Europe for transport systems and human factors research, the European Commission's Directorate General for the Information Society is co-funding the Network of Excellence HUMANIST (Human-centred design for information society technologies; applied to road transport). The activities concentrate on the development of ergonomic design criteria, the improvement of existing methodologies to assess the effects of new driver support systems, and the specification of the needs of different driver groups. The connection and co-operation between the partners is enhanced by a researcher exchange programme, access to the research infrastructure of other network members and the development of new media to exchange knowledge and data. Training programmes both for new researchers and professionals in the field of transportation human factors are currently in preparation and will be offered in later project stages.

The workplan for the Network of Excellence (NoE) is divided into two main areas of activities: integrating activities and activities to spread the excellence of the network. The integrating activities manage and consolidate the NoE. The activities to spread the expert knowledge and experience of the network include co-operations with other relevant initiatives like standardisation bodies and policy makers. The workplan contains in total 15 different, but related task forces dedicated to the following topics: Identification of driver needs in relation to intelligent transport systems (ITS), evaluation of potential benefits of ITS, cognitive models of the driver-vehicle-environment system, impact analysis of ITS on driving behaviour, development of innovative methodologies to evaluate safety and mobility, education and driver training for ITS use, use of ITS to train drivers, establishment of an international researcher exchange programme, research infrastructure sharing, electronic means for knowledge sharing, transfer of knowledge to stakeholders and the industry, training programmes, diffusion of knowledge as well as network management and assessment.

Expected benefits of the network

Research in the area of road transport telematics is quite scattered. HUMANIST has a strong interdisciplinary background, focusing on the integration of the different competencies available in the network. The main scientific aim is to structure and organise the existing knowledge to harmonise future research activities. During this process, significant knowledge and research gaps will be identified based on the current state-of-the-art. The main scientific goal is to study the impact of new driver support technologies on driver behaviour and road safety in a prospective way, i. e. the effects of new systems should be known well before a system is introduced to the market. Therefore it is crucial to assess both the benefits and the risks for specific driver groups that result as a consequence of the implementation of a particular driver support tool into the human-vehicle-environment system. E.g., it has to be taken into account that, on the one hand, older drivers with an age-related decline of information processing capacities could benefit strongly from additional driver support information. Due to the reduction in processing capacities this driver group is, on the other hand, probably the most susceptible to negative side-effects of additional information like overload or distraction (Sommer et al., 2004).

The development of driver training programmes for the actual application of different types of support systems is likewise associated with the different needs of each driver group. E.g., younger drivers tend to have too much confidence into the skills they learned during driver training programmes and into the reliability of new technological systems. An efficient training programme to teach younger drivers the use of new support systems would have to take into account such particular characteristics of the driver group in question (Twisk, 2004).

In addition to the immediate scientific progress, further technological and societal benefits are expected from the network. They include the contribution to standardisation bodies, to policy developments and decisions, the improvement of new driver support systems by promoting design practices based on human-centred approaches (e.g. inclusion and universal access principles) and the development of new assessment procedures for usability and safety evaluation of in-vehicle systems. The expected societal benefits are, mainly, an enhancement of road safety, but also e.g. the improvement of the mobility of drivers with special needs by taking into account the heterogenity and diversity of the driving population as a whole.

The financial support by the European Commission includes 11 grants for Ph.D. students. A central objective of the training programme for these young researchers is to teach both the technological and the behavioral dimension of road transport telematics in a complimentary way. The participation in the courses, seminars and lectures of the new curriculum will be also open to external researchers and transportation professionals. The events will be regularly announced on the network's homepage (http://www.noehumanist.org/).

References

Sommer, S. M., Falkmer, T., Bekiaris, E. & Panou, M. (2004) Toward a Client-Centred Approach to Fitness-to-Drive Assessment of Elderly Drivers. Scandinavian Journal of Occupational Therapy, 62-69.

Twisk, D. (2004) Inventory of ITS Functionalities According to Driving Task Models. HUMANIST Deliverable F.1. Leidschendam, Netherlands: SWOV.

Different congestion indications on dynamic route information displays

Ilse M. Harms[1], Matthijs Dicke[1], & Karel A. Brookhuis[1,2]
[1]*Department of Psychology, University of Groningen, Groningen*
[2]*Delft University of Technology, Delft*
The Netherlands

Abstract

A new type of dynamic route information display is the Full Colour Information Panel (FCIP). The advantage of an FCIP is that it can display a route map, travel time and congestion at the same time. There are different ways to depict maps with travel times and congestion, for instance, with colour, a congestion sign or a combination of both. An experiment was carried out to study various designs. No difference in reaction time between the different congestion representations was found when the participants were required to choose the quickest route. However, there were differences in reaction time if extra signs were included on the FCIP. Contrary to what was expected, reaction time was not affected by FCIPs with a congestion depicted on a route with less travel time than the other routes. However, if shown successively these FCIPs did affect reaction time for FCIPs on which the congestion was consistent with travel time.

Introduction

The present experiment is a follow-up of the EU TravelGuide project (Roskam et al., 2002), involving the design and layout of a specific dynamic route information display, the Full Colour Information Panel (FCIP), positioned over the motorway at the entrance of the city of The Hague. FCIPs are dynamic and able to display any graphical information. This specific FCIP can display a route map and the travel time of each of the three routes. The route-choice that motorists will make depends upon the information they receive. People may not only choose the route with the least travel time, they could also consider the (potential) presence of congestion (Bovy, 1989). In order to make such a choice, these elements should be depicted on the FCIP. There are different ways to depict a congestion on a FCIP, e.g. by means of a colour, with a congestion sign or a combination of both.

Congestion sign and colour use

The obvious way to depict a congestion is using a standard congestion sign, a queue of cars in a red triangle. The congestion sign is widely used, so motorists are familiar with the sign and its meaning. A second option to indicate congestion is using colour, which has obvious advantages. It enables an indication of the size of a congestion and where it is exactly located. Another advantage is that colour is predominantly processed, in the first stage of information processing (Johnson & Proctor, 2004). The use of colour can help to direct attention and provide additional information (Derefeldt et al., 2004; De Waard et al., 2002; Roskam et al., 2002). Colour coding can inform the driver about the type of congestion: is the traffic moving slowly or is it completely standing still? Colour-coded routes serve as eye-catchers which immediately direct attention to the road network. Colour coding can also reduce the complexity of the FCIP (De Waard et al., 2002). Objects with colour coding can be located faster within fewer fixations. Reaction time reduces especially when colour coding is used in displays that are cluttered or complex and contain a lot of information (Christ, 1975; Hughes & Creed, 1994). However, prudence is called for; previous research has shown that only few colours should be used. Usage of colour must be functional and should be kept at a minimum (Fokkema, Hazevoet & Bijlsma, 2002; Roskam et al., 2002).

In the present experiment a congestion is depicted as a coloured rectangle at the location of the congestion. Research by Fokkema, Hazevoet and Bijlsma (2002) showed that a rectangle that covers the whole width of the road is preferred. The majority of their participants interpreted a red rectangle as a congestion. Their research also showed a preference for making a distinction between the type of congestion: moving slowly or being stagnant. In the present experiment colour coding is also used in this way as well. Yellow represents slow moving traffic and red represents stagnant traffic. It is expected that colour, contrary to a congestion sign, will not interfere with other traffic signs on a FCIP. Finally, colour and a congestion sign can be combined. On the one hand, the familiarity of the congestion sign could be enhanced by the advantages of colour. On the other hand, using extra elements might increase the information processing time, i.e. reaction time could increase (Sternberg, 1966). If the extra element is a redundant element, reaction time may not increase that much (Miller, 1982). The depiction of a congestion is likely to be a redundant element, if the congestion is consistent with travel time. In order to determine the increase in reaction time as a consequence of an extra element, redundant or not redundant, extra traffic signs are added to the FCIP. These traffic signs are not supposed to influence route-choice. When extra traffic signs are added, the effects of congestion representations in more complex displays can also be examined. It is expected that extra traffic signs diminish the redundancy effect of a congestion sign and make it less conspicuous. This would result in longer reaction times.

The basic design of the FCIP originates from research on reducing uncertainty in route-choice, based on road maps (Dicke et al., 2004) and the TravelGuide project (Roskam et al., 2002). The aim of the present experiment is to examine the different ways of depicting a congestion. Does one way of depicting result in shorter reaction times than the other? Will depicting a congestion induce a redundancy effect? What are the effects on reaction time of adding extra elements to FCIPs indicating congestions and will displaying congestions inconsistent with travel time have negative influences on reaction time?

Method

Thirty volunteers participated in this experiment. The majority of them are students and graduates with different fields of study. The mean age is 23 years, with the youngest participant being 16 and the oldest being 30. None of them were colour-blind. The FCIPs used in this experiment were divided into two groups. Group one, which will be referred to as the consistent group, consists of FCIPs without congestion indications and FCIPs on which the congestion indications are consistent with travel time (see Figure 1). Group two, the inconsistent group, consists of FCIPs on which the congestion indications are inconsistent with travel time (see Figure 2). The groups were divided over two blocks. Block 1 only included the consistent group. Block 2 included both groups. The order in which these blocks were shown has been counterbalanced between subjects.

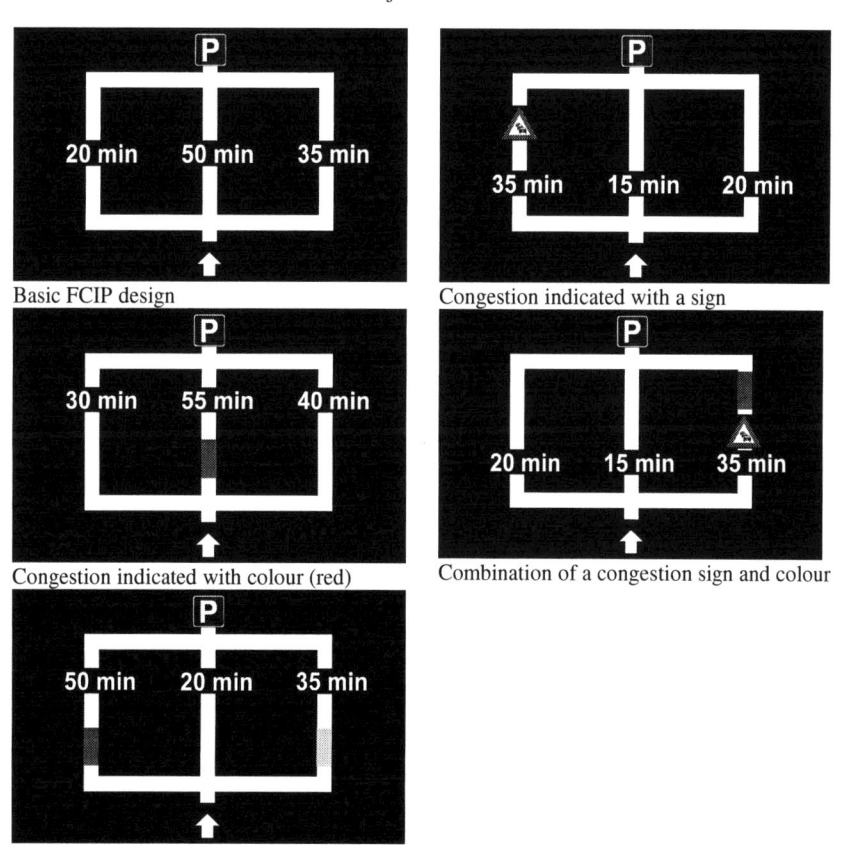

Basic FCIP design

Congestion indicated with a sign

Congestion indicated with colour (red)

Combination of a congestion sign and colour

Congestion indicated with red, extra information through colour coding (yellow)

Figure 1. Five common designs of consistent FCIPs in the experiment. See also http://extras.hfes-europe.org *for these FCIPs in colour*

FCIPs from the consistent group are FCIPs without congestion and FCIPs depicting a congestion with a congestion sign, with one colour, with two colours and with a combination of colour and a congestion sign (see Figure 1). FCIPs from the inconsistent group are FCIPs depicting a congestion with a congestion sign and with one colour (see Figure 2). Also, extra elements were added to all the different types of FCIPs. There were four levels of adding extra elements, i.e. no traffic sign added, one traffic sign added, two traffic signs added and three traffic signs added. The extra elements were traffic signs for different purposes (e.g. no parking signs) that were not supposed to influence route-choice. To discourage participants to look at the travel times only, their position varied between trials. Seven different travel times were used to prevent participants from reacting on a certain amount of minutes instead of looking at the FCIP as a whole. For example, 20 minutes could indicate the quickest route, but it could also be the second quickest route. The FCIPs were shown on a computer screen with a resolution of 1024 x 768. The instruction was to select the quickest route. Block 1 consisted of 270 frames and block 2 consisted of 373 frames. The experiment was self-paced within a maximum of four seconds, which is the same amount of time a motorist has to perceive the FCIP when he or she is driving underneath with a speed of 100 km/h (De Waard et al., 2002). Participants had to look at a fixation dot for one second, after which they had a maximum of four seconds to respond to the stimulus. The stimulus was succeeded by a blank screen during one second. Participants were instructed to use the keyboard keys V, B and N with their preferred hand, corresponding respectively with the left route, the centre route and the right route.

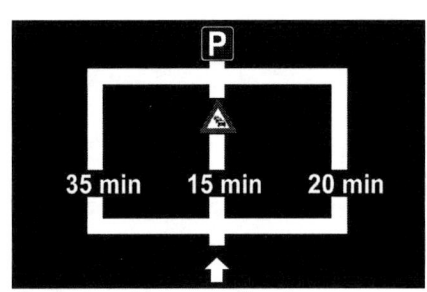

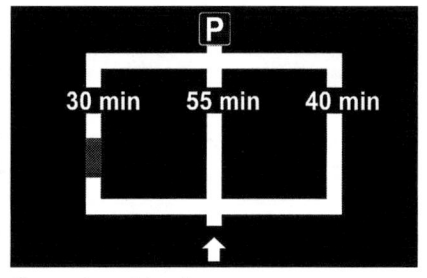

Congestion with a congestion sign   *Congestion with colour*

Figure 2. Two common designs of inconsistent FCIPs in the experiment. See also http://extras.hfes-europe.org *for these FCIPs in colour*

Results

Raw averages of all reaction times (mean RT = 890 ms, sd = 272) changed slightly after removal of extremes (below 200 ms and more than 3x sd above average, Stevens, 2002), leading to an adapted average (mean RT = 876 ms, sd = 241). Despite initial training, reaction time still dropped significantly for the second part of the experiment, from 928 ms to 824 ms (t(1, 29) = 8.4, p<0.001). Since the blocks are counterbalanced this will not affect the outcomes. It also appeared that the data of block 1 could not be compared with the data of block 2 (see below). This had to be taken into account with the statistical analysis. Reaction times in block 1 were significantly lower than those in block 2.

As the FCIPs in block 1 are more common to be found in real life, the outcomes of data from block 1 will be discussed more extensively. There was no difference in reaction time between no congestion and any of the different ways of depicting a congestion if no extra traffic signs are added. Adding extra signs showed that only when three extra traffic signs were added to the basic FCIP, reaction time was significantly faster compared to the basic design ($F(1,530) = 4.9$, $p<0.027$). For the FCIPs with a congestion displayed there were no significant differences in reaction times between displaying the FCIP without an added sign and displaying it with any amount of extra signs.

Neither a congestion sign, nor colour or its combination interfered with other traffic signs. There was no significant difference in reaction time between FCIPs without a congestion and FCIPs with a congestion depicted in colour, with a congestion sign or a combination of both. Displaying extra traffic signs does not interfere with displaying a congestion. However, if colour coding was used the reaction time dropped significantly if three signs were added, compared to FCIPs with the same amount of added signs but without a congestion depicted ($F(1,528) = 4.5$, $p< 0.034$).

In order to find redundancy effects, FCIPs without a congestion plus one extra sign were compared with FCIPs with a congestion. With the exception of the FCIP with two colours plus two added signs ($F(1,527) =3.9$, $p<0.047$), reaction time for displaying a congestion was not significantly faster. So there is no redundancy effect.

There was a significant difference between FCIPs displayed in block 1 and the same FCIPs displayed in block 2 in general ($t(1,62) = -2.7$, $p<0.009$). When the FCIPs with two colours displayed in block 1 (mean RT = 861) were compared to the same FCIPs displayed in block 2 (mean RT = 881), the difference was significant ($t(1,53) = -2.9$, $p< 0.005$). If the FCIPs with a congestion sign and colour in block 1 (mean RT = 855) were compared to block 2 (mean RT = 884), the difference was significant ($t(1,44) = -3.7$, $p<0.001$). The FCIP on which a congestion was depicted with one colour was the only one that did not differ significantly. The two kinds of inconsistent FCIPs do not significantly differ from their consistent counterparts either.

Discussion and conclusion

Participants were asked to select the quickest route instead of their preferred route, potentially leading to a higher reaction time on the FCIP if it is displayed amongst inconsistent FCIPs. It is possible that participants prefer to avoid congestions instead of taking the quickest route which may have affected their reaction time. In a subsequent experiment participants will be experienced car drivers. They will not only be asked to select the fastest route, but also to select their preferred route. One of the remaining research questions is to be addressed as well; whether route-choice is determined by travel time or the presence of a congestion.

The reason that colour coding causes faster reaction times in complex displays may also be that two congestions are depicted on the FCIP. On the one hand, this might make it easier to detect the correct answer. On the other hand, this would also mean that the correct answer is easier to detect in less complex FCIPs. However, this effect has not been observed. Moreover, displaying a congestion means displaying an extra element which would increase reaction time (Sternberg, 1966), which has not been observed either. Further research should provide a better understanding of this phenomenon.

Displaying congestion does not result in longer reaction times. Reaction times are considerably faster if colour coding is used in addition to three added signs and there is also a redundancy gain concerning colour coding in complex displays. There is no reason not to include information concerning congestions on FCIPs. Since there are no disadvantages of displaying information about congestion it easily justifies the advantage of extra traffic information. There is no meaningful difference between the different ways to depict a congestion. It should be kept in mind, however, that reaction time decreases when colour coding is used in displays that are cluttered or complex, containing a lot of information (Christ, 1975; Hughes & Creed, 1994). When colour coding is used, reaction time drops significantly when three extra signs were added to the FCIP.

Adding extra signs raised reaction time only for FCIPs without a congestion and only when at least three extra signs were added, contrary to what was expected (cf. Sternberg, 1966). It turned out that neither a congestion sign, nor colour nor the combination of both interfered negatively with any added signs. There were also no redundancy effects for these three groups of FCIPs. The only effect of redundancy shows up in FCIPs with colour coding in complex displays. There are no redundancy effects of displaying a congestion consistent with travel time. As the range of reaction times was narrow, this may be due to little or no effect of the number of signs added. Nevertheless, displaying congestions that are inconsistent with travel time should be avoided as the reaction time on the consistent FCIPs suffers considerably from displaying inconsistent FCIPs in the same condition.

References

Bovy, P.H.L. (1989). Routekeuze van reizigers. In C.W.F. van Knippenberg, J.A. Rothengatter, and J.A. Michon, *Handboek sociale verkeerskunde* (pp.153 - 170). Assen, the Netherlands: Van Gorcum.

Christ, R.E. (1975). Review of analysis of color coding research for visual displays. *Human Factors, 17,* 542-570.

Derefeldt, G., Swartling, T., Berggrund, U., & Bodrogi, P. (2004). Cognitive color. *Color Research & Application, 29*, 7-19.

De Waard, D., Roskam, A.J., Uneken, E., & Brookhuis, K.A. (2002) TRAVLler and traffic information systems: GUIDElines for the enhancement of integrated information provision services, GRD1-1999-10041, Task report 5.3.3. *Travelguide,* University of Groningen.

Dicke, M., Bouwers, F.-L., Koenders, E., Kok, R., Van der Veer, R., & Brookhuis, K.A. (2004). Human factors in the design of travel and traffic information. In D. de Waard, K.A. Brookhuis, and C.M. Weikert (Eds.), *Human factors in design* (pp. 95-107). Maastricht, The Netherlands: Shaker.

Fokkema, J., Hazevoet, A., & Bijlsma, M. (2002). *Pilot GRIPS: Experiment met Grafische Route Informatie Panelen, Rapportnummer TT02-091.* Traffic Test bv, Veenendaal.

Hughes, P.K., & Creed, D.J. (1994). Eye movement behaviour viewing colour-coded and monochrome avionic displays. *Ergonomics, 37*, 1871-1884.

Johnson, A., & Proctor, R.W. (2004). *Attention: Theory and practice.* Thousand Oaks: Sage Publications.

Miller, J. O. (1982). Divided attention: Evidence for coactivation with redundant signals. *Cognitive Psychology, 14*, 247-279.

Roskam, A., Uneken, E., De Waard, D., Brookhuis, K., Breker, S., & Rothermel, S. (2002). Evaluation of the comprehensibility of various designs of a Full Colour Information Panel. In D. de Waard, K.A. Brookhuis, J. Moraal, and A. Toffetti (Eds.), *Human factors in transportation, communication, health, and the workplace* (pp. 231-244). Maastricht, The Netherlands: Shaker.

Sternberg, S. (1966). High speed scanning in human memory. *Science, 153,* 652-654.

Stevens, J.P. (2002). *Applied multivariate statistics for the social sciences.* Mahwah, New Jersey: Lawrence Erlbaum Associates.

The relation between seat belt use of drivers and passengers

Erik Nuyts & Lara Vesentini
Policy Research Centre of Traffic Safety
Universitaire Campus, Diepenbeek, Belgium

Abstract

The present study gives input to the theme Safety and Transport, by pointing out a target group for campaigns to reduce the severity of accidents. Seat belt use and sex of both the driver and the passenger were observed. A logistic regression model was used to analyse the relation between seat belt use and the position in the car (driver or passenger), sex of the driver and passenger, time of day and location of driving.

Results show that men used their seat belt less often than women. Drivers with passengers did not wear the seat belt more often or less often than drivers without passengers. However, drivers and passengers often behaved the same. They both wore or did not wear a seat belt. Seat belt use of a male driver depended also on the sex of the passenger. In control condition (outside the centre of the city, and outside of peak hours of a working day), the probability that a male driver without passenger wore a seat belt was 48%. Accompanied by a female passenger (keeping the other variables constant), the probability increased to 59%. Accompanied by a male passenger, the seat belt use of the male driver decreased to 34%. Female drivers seat wearing behaviour was not influenced by the presence nor sex of passenger. Male passengers use the seat belt less often if they are sitting next to a male driver. At present, it is not clear which human factor causes the differences found. It is possible that wearing a seat belt may be due to direct social influence. In that case, the differences are intra-individual. Alternatively, the differences found could be inter-individual. E.g., when two men sitting together in a car they are more often members of social groups that use a seat belt less than average, while male drivers accompanied by female passengers are more often members of social groups that use a seat belt more than average.

In both cases, men are a target group for campaigns for seat belt use. Especially if two men are sitting together in a car, since in that case both driver and passenger have extremely low seat belt use.

Introduction

In Belgium the use of seat belts has been legally compulsory for drivers and front seat passengers since 1975. Yet, based on counts, in 2001 seat belt use in Belgium was only 63% on motorways, 58% on rural roads, and only 47% on urban roads (Belgian Institute for Traffic Safety -BIVV, 2001). Attempts have been made to increase this percentage by seat belt campaigns. Since it is proven that campaigns are more effective when aimed at a specific target group (Green & Kreuter, 1991; Bartholomew, Parcel, Kok, & Gottlieb, 2001) it is worthwhile to look for differences in seat belt use under different conditions, and to focus on target groups that may emerge from these differences.

Several studies find differences between drivers and passengers (e.g., Van Bekkum et al., 2000; Koornstra et al. 2002; In't Veld, 2003), differences between working days and weekend days (e.g., BIVV, 1999, 2000, 2001; Eby et al., 2001; in't Veld, 2003; Kim & Kim 2003) and between different types of roads (e.g. BIVV, 1999, 2000, 2001; Van Bekkum et al., 2000; Koornstra et al., 2002). Seat belt use also differs between sexes, women more frequently wear the seat belt than men (e.g., Reinfurt et al., 1996; Eby et al., 2001; Koornstra et al., 2002; BIVV, 2003; Cedersund, 2003; Kim & Kim, 2003; Transports Canada, 2003). It is also found that driver and passenger often both use the seat belt, or both do not use the seat belt (e.g. Van Bekkum et al., 2000; Kim & Kim 2003; in't Veld, 2003).

The present study investigates whether drivers wear their seat belt more when accompanied by a front seat passenger. Koornstra et al. (2002) mentioned that "Drivers alone in a car use the seat belt less than if he/she has a passenger...". This would imply that drivers alone are a target group for campaigns. This also suggests that there is a social influence stimulating seat belt use when people are not alone

Material and Methods

Data collection

The traffic police collected data on seat belt use in Antwerp, Belgium in April 2003. In order to avoid a direct observation effect with as result people putting on suddenly their seat belts, policemen were not in uniform. A total of 8039 drivers and 2621 passengers were observed when stopping or slowing down at traffic lights and at a roundabout. Data were collected during peak-hours of working days (7-9 am and 4-6 pm), off-peak hours of working days (10-11 am and 2-3 pm) and during the weekend (8 am - 6 pm). During every observation period 500 cars were observed.

Observation included: seat belt use of the driver, sex of the driver, presence or absence of front seat passenger, and if a passenger was present, seat belt use and sex of the passenger. Back seat passengers seat belt use was not registered. An attempt was made to classify a broad age category of the occupants. Since too many were difficult to classify (e.g., what is the boundary between 'adult' and 'old' ?), this attempt was not continued. As a pilot study make and type of car were scored. After scoring 126 cars no relevant tendencies were found yet and recording the make of cars was not easy for some observers. As this increased the risk on incorrect classification car make was no longer scored.

Observations took place at three streets in the centre of Antwerp, and at three access roads towards the centre of the city. Observations were (relatively) equally spread over the six roads and over the three different time periods. The roads had been selected as representative for general traffic flow. To avoid selection bias of e.g. commuters, data were also collected during off-peak hours. None of the roads had a high percentage of lorries, and seat belt use of lorry drivers was not recorded in this study.

Logistic regression

Since the dependent variable 'seat belt use' is 'yes' or 'no', stepwise logistic regression is used for the main analysis. In a logistic regression, a logistic transformation is performed on the dependent variable, resulting in:

$$\ln\left(\frac{P}{1-P}\right) = b_0 + b_1 X_1 + b_2 X_2 + ..b_n X_n \tag{1}$$

P is the probability that the person under consideration is wearing the seat belt, X_1, X_2... X_n are independent variables such as 'being passenger' or 'sex', and b_0, b_1, …b_n are constants calculated by the statistical package.

Equation (1) is equivalent with

$$\textit{The probability of wearing a seat belt} = \frac{1}{1 + e^{-(b_0 + b_1 X_1 + b_2 X_2 + ..b_n X_n)}} \tag{2}$$

which is sometimes easier to interpret.

In the final regression, only the significant (p<0.05) independent variables are retained. E.g., "female" and "passenger" are in the regression, but the interaction variable 'female x passenger' is not. This is not because the variable was not evaluated, but because it had no significantly added value above the variables already in the regression.

Results

Seat belt use of drivers with and without passengers

Seat belt use of drivers without passenger was 58,7 %; seat belt use of drivers with passengers was 59,2 % (Table 1, Z-test for proportions, N(0,1) =0.38, NS). From these results one cannot conclude that there is some social pressure from passengers on drivers to wear a seat belt. More detailed results will be discussed when performing a logistic regression (see below).

Yet, there is a clear connection between driver seat belt use and passenger seat belt use. Drivers and passengers showed in most situations the same behaviour. They both wore the seat belt (48.6 %) or they both did not wear the seat belt (33.0 %) (Table 2). Hence, in 82% of the situations, their behaviour was the same. When seat belt use would be independent, the expected percentages should be 33% and 18%, together 51%. The difference with the 82% found is significant (Z-test for proportions, N(0,1) =31.75, p<0.001).

Table 1. Presence of passengers and the relation with driver seat belt use

Frequency Row percentage Column percentage	Passenger absent	Passenger present	Total
Driver wearing seat belt	3182 67.2 % 58.7 %	1551 32.8 % 59.2 %	4733 58.9 %
Driver not wearing seat belt	2236 67.6 % 41.3 %	1070 32.4 % 40.8 %	3306 41.1 %
Total	5418 67.4 %	2621 32.6 %	8039 100 %

Table 2. Passenger seat belt use compared with driver seat belt use

Frequency Percentage Row percentage Column percentage	Passenger wearing seat belt	Passenger not wearing seat belt	Total
Driver wearing seat belt	1273 48.5 % *82.1 %* 86.2 %	278 10.6 % 17.9 % 24.3 %	1551 59.2 %
Driver not wearing seat belt	204 7.8 % 19.1 % 13.8 %	866 33.0 % *80.9 %* 75.7 %	1070 40.8 %
Total	1477 56.4 %	1144 43.6 %	2621 100 %

Combined effect of passenger, sex, period and location on seat belt use

In Table 3 a logistic regression is given of seat belt use depending on different situations (period, with or without passenger,…), and different combinations (male/female, driver/passenger,..). The reference situation is a male driver, sitting alone in his car, during an off peak hour of a working day, on an access road towards the centre.

Table 3 shows that female drivers used the seat belt more often than in the reference situation (male person) since the parameter estimate of female is positive (0.52). In the weekend drivers wore less often the seat belt (parameter estimate = -0.14) but more often during peak hours (0.35). If, during a working day, the driver was a man accompanied by a male passenger, the probability that he wore the seat belt was much lower (-0.58). But if the passenger of a male driver was a woman, seat belt use of the male driver increased (+0.44). And, if during the weekend two men drove together, seat belt use was higher than expected based on the correction of weekend (-0.14) en the correction for two men (-0.58). Above both corrections a third one was significant for the interaction variable (+0.66), increasing seat belt use to a level not to far below that of the reference situation (-0.14 -0.58 +0.66 =-0.07).

Table 3. Logistic regression of seat belt use, depending on sex, period, location, driver or passenger, sex of the passenger (if a passenger is present) and interaction variables of those (e.g. passenger and weekend).
A variable in italics is an interaction variable: e.g. pass_weekend is an extra correction if the person under consideration is a passenger and the trip was made during the weekend.
Analysis of Maximum Likelihood Estimates N = 10660; -2 Log L (Intercept only) = 14485.982; - 2 Log L(intercept en covariates)= 14107.331

Parameter	df	Estimate	Std. Err	Wald χ^2	Pr > χ^2
Intercept	1	-0.08	0.05	2.1832	0.1395
Female	1	0.52	0.05	99.6056	<.0001
Weekend	1	-0.14	0.06	5.4829	0.0192
Peak hour	1	0.35	0.06	37.2136	<.0001
male driver with male passenger	1	-0.58	0.11	27.1278	<.0001
male driver with female passenger	1	0.44	0.07	46.0959	<.0001
male driver with female passenger in weekend	*1*	*0.66*	*0.17*	*14.2288*	*0.0002*
passenger	1	-0.16	0.10	2.6331	0.1047
pass_weekend	*1*	*0.26*	*0.12*	*5.1416*	*0.0234*
pass_peak hour	*1*	*-0.32*	*0.13*	*6.4733*	*0.0110*
male passenger with male driver	1	-0.33	0.11	9.7417	0.0018
centre of the city	1	0.29	0.04	50.4960	<.0001

Seat belt use by female drivers was not significantly correlated with the presence or absence of passengers as such, nor with their sex (if applicable). It seems as if passengers wore less often the seat belt than drivers (-0.16). Yet, this difference was not significant (p = 0.1047). Since some interaction variables with passenger were significant, the variable 'passenger' is retained in the model. Male passengers wore significantly less often a seat belt if the driver was a man too (-0.16 – 0.33 = - 0.49).

For passengers, the effect of the period was the opposite to the effect on drivers, or it is negligible. During the weekend passengers wore the seat belt more than during off peak hours of a working day (-0.16 -0.14 +0.26= -0.04 is greater than the coefficient of passengers without other correction: -0.16). On the other hand, passengers wore the seat belt only slightly more during the peak hour than during off peak hours (-0.16+0.35 -0.32= -0.13 is quite comparable with –0.15). There was no interaction variable including 'passenger' and 'female'. Since the effect of passenger as such is not significant, the present study provides no hard arguments to state that female passengers wore less often the seat belt than female drivers.

People wore more often seat belts in the centre roads than in the access roads of Antwerp (+ 0.29). The discussion above is presented in a table in such a way that it becomes more clear which change in the reference situation has the biggest impact on seat belt use[*].

Situations with the most extreme changes from the reference situation were found at the top and bottom rows of Table 4. The most extreme results were correlated with the sex of the occupants of the car. The most positive result was found for female drivers. The second best result was found for male drivers accompanied by a female passenger. The most negative result was found for men, both driver and passenger, when two men drove together. But if two men drove together during the weekend, the difference with the reference situation was much smaller.

Table 4. Situations ordered according the magnitude of the impact of seat belt use, compared to the reference situation (male driver, no passenger, off peak hour on a working day, access road to the centre).

Variable	coefficient of	change in coefficient from reference situation
female driver	*driver*	*0.52*
male driver with female passenger	*driver*	*0.44*
peak hour	driver	0.35
centre of the city	driver	0.29
Intercept	**driver**	**0.00**
2 persons during the weekend	passenger	-0.04
2 men during the weekend	driver	-0.06
2 persons during peak hour	passenger	-0.13
weekend	driver	-0.14
2 men	*passenger*	*-0.49*
2 men	*driver*	*-0.58*

Discussion

In the present study seat belt behaviour of drivers and passengers in Antwerp, Belgium, were analysed. Seat belt use was also related with several variables such as the presence of a front seat passenger, driver versus passenger, sex of driver and passenger, time of day and location. Here the potential effect of sex and of passengers on drivers' seat belt use is discussed. Traffic variables with less social dimensions are discussed in the original report (Nuyts & Vesentini, 2004).

Three main results are.

(1) Driver and passenger often behave in the same way: both wear seat belts, or both do not. This finding is in line with literature. In different countries it was found that if the driver uses the seat belt, then passengers often also do: 92-96% in the Netherlands (Van Bekkum et al., 2000; In 't Veld, 2003), 89% in Hawaii (Kim & Kim, 2003), and as was found in this study, in Belgium 82% (Table 2). Moreover, if

---

[*] We prefer not to start with Table 4, in order to provide the reader the basic information. In this way the reader has the possibility to draw also his own conclusions.

the driver does not use a seat belt, passengers most often do not either: 66-75% in the Netherlands (Van Bekkum et al., 2000; In 't Veld, 2003), 91% in Hawaii (Kim & Kim, 2003), in Belgium: 81% (this study, Table 2).

(2) Men less often wear the seat belt than women. Again, this is in agreement with national and international literature (e.g., Reinfurt et al., 1996; Eby et al., 2001; Koornstra et al., 2002; BIVV, 2003; Cedersund, 2003; Kim & Kim, 2003; Transports Canada, 2003;).

(3) Seat belt use of drivers with passengers does not differ significantly from seat belt use of drivers without passengers in Antwerp. But if sex is taken into account, significant relations appear. Male drivers with female passengers use the seat belt more than male drivers without passengers. During working days, male drivers with a male passenger use the seat belt less often than male drivers without passengers. In the weekends, the difference in seat belt use is smaller. For female drivers, no significant differences in seat belt use are found between women alone in a car, with a male passenger or with a female passenger. Male passengers use seat belts less if the driver is a male too. This result differs from the he SUNflower report, which states: *"Drivers alone in a car use the seat belt less than if she/he has a passenger..."* (Koornstra et al., 2002). Unfortunately, the figure in the SUNflower report to which is referred in the text only provides data on 'drivers', 'front seat passengers' and 'back seat passengers' (Koornstra et al., 2002). No data are given about the difference between 'drivers with passengers' and 'drivers without passengers'. This makes comparison with the data from Antwerp difficult.

Due to the methodology of observation, recording seat belt use in passing cars, the data in this study do not prove causal relationships, only correlations. Therefore, the behaviour observed could be caused by two mechanisms. It is possible that seat belt use is triggered by direct social influence, or that the results found are mainly due to internalised attitudes. In the first case, drivers and their passengers often behave in the same way, implicit social pressure to wear (or not to wear) a seat belt is found when people adapt their seat belt use to behaviour of the other occupant. More explicit social influence could be that one of both asks the other to wear the seat belt. Then the same male drivers could wear the seat belt when driving alone, but not when accompanied by another man (e.g. because it is not 'cool' to wear a seat belt). In Belgium, at the time of obtaining the data, seat belt use enforcement was not a priority for the traffic police. There was hardly police enforcement, and fines were relatively low. Speeding or alcohol abuse was perceived as far more important issues, both by the police as by the civilians. Because of this, more room is left for social pressure in relation with seat belt use. When one is encouraged to speed up, the driver can use the excuse of getting a fine. For seat belt use, this excuse is much less valid. With no 'accepted' excuse, it is more difficult to resist social pressure. Alternatively, seat belt use could be mainly an internalised attitude. Roughly spoken, the population is split up in seat-belt-users and non-seat-belt-users. Then the results found are caused by the fact that people with the same attitude drive more often together than people with different attitudes. A typical example are families. Families travelling together have similar backgrounds with respect to attitudes and knowledge, and are likely to make similar decisions about wearing seat belts. For instance, it is reported that seat belt use is higher among high school students whose parents wear their seat belt more (Shin et al., 1999; Eby et al., 2001). On a different level one can expect that people of the same socio-economic status use their seat belt in a more similar way. From travel surveys we know that persons with higher social status more often own the car they drive. People with a lower social status have less often a car of their own, and thus they have to ride more often together. From this, we can assume that people alone in a car more often have a higher social status than people who are together in a car. Braver (2003) has pointed out that persons of a low socio-economic status wear their seat belt less often. If the assumption is true that on average, people in a car with a passenger have a lower status, it could explain why men who drive together often do not wear their seat belt. But research is needed to prove this assumption. Probably both social pressure and the socio economic status have an influence on the results found. Those two elements can also influence each other. Perhaps persons of a low socio-economic group are more vulnerable for social pressure or maybe educated spouses have more impact on their husband.

It was found that men use the seat belt less often than women. But seat belt use of young people (4 to15 years) did not differ between boys and girls (Eby et al., (2001); Diamantopoulou et al., 1996 – cited in Eby et al. (2001); Eby, Vivoda, Fordyce, 1999 – also cited in Eby et al., 2001). This suggests that parents/adults do not make a difference between the sex of their passenger child when putting on or not putting on his/her the seat belt. Hence, the difference in seat belt use appears when the children grow up. Seemingly, at the age that young persons start making their own decisions the percentage of boys that decides to wear the seat belt is much lower than the percentages of girls deciding to use the seat belt. To be sure about this, one should perform a longitudinal research to see how seat belt use evolves when children of both sexes grow up from 12 year till 22 years.

For adults (from 15 years old) direct social influence is reported by BIVV (2003): seven of ten Belgian drivers who use the seat belt say that they ask their passenger to use it too, if the passenger does not do it spontaneously. For female drivers and drivers older than 55, this number is even higher (BIVV, 2003). The report does not state anything on passenger asking drivers to wear the seat belt. It is expected that passengers will ask this less easily to the driver, since within a car, the 'status' of the driver is often higher than that of the passengers (often the car belongs to the driver, the driver most often decides the speed of the car, which road is to be taken, etc.). If such a direct social pressure from passenger to driver exists, it is so limited that it does not show up in the results of the present study, since no change of seat belt use was found between drivers with our without passengers. However, note that such a result was stated in the SUNflower report (Koornstra et al., 2002).

Cunill et al. (2003) report that the opinion on seat belt use of friends and family is significantly related with and a good predictor for seat belt use of the person under consideration. This suggests that the general social environment is important. However, since driver and passenger are often friends or family, this does not exclude direct social impact.

For campaigns to improve seat belt use, it is not important what exactly triggers the results found. They clearly point out that 'men' are a target group, and especially 'two men together in a car'. In Belgium, it seems generally accepted that the seat belt can protect someone when an accident happens. But a large part of the population thinks that wearing a seat belt or not is a personal decision. They argue that –if an accident happens- the only one who suffers from not wearing the seat belt is the person who does not wear the seat belt. This argument is not completely right. In Belgium, a large part of the hospital costs are paid by taxes and not by the patient. Hence, the cost of a person becoming needlessly injured is also paid by the total population. Apart from this economical argument, it is clear that also relatives and friends suffer when someone gets hurt or killed, especially if this injury or dead could be avoided. Therefore, it is worthwhile to do campaigns in which people are suggested to help other people to wear the seat belt. The Belgium Institute for Traffic Safety (BIVV) has already conducted several campaigns to increase seat belt use. The campaign of the year 2000 "The seat belt: a natural reflex" uses the family as the group in which social influences refer to (Figure 1). It suggests that within a good family, people stimulate each other to wear the seat belt. The results of the present study point out that two men are a target group for seat belt campaigns. For this target group, the 'sweet family' probably has no large appeal. Campaigns could be set up around an idea of male companionship, with slogans like "sworn friends wear both the seat belt" or "If you climb a mountain with your friend, and he forgets to put on his climbing sling, you will tell him. Friends should also take care for each other's 'climbing slings' in traffic." Campaigns like this one could decrease the threshold of involving in the seat belt use of other people.

Figure 1. Poster of a campaign for seat belt use by Belgian Institute for Traffic Safety: "The seat belt. A Natural reflex."

Future research in this context is twofold. Firstly, the study should be expanded with a rural and a motorway survey. It is known that seat belt use is lower in urban areas compared to rural areas and motorways. Since all the data of this paper are collected in a city, it is possible that the results will not only differ quantitatively, but also qualitatively in rural areas or on motorways. Secondly, a research should be set up to evaluate the effect of a seat belt campaign for a limited target group. Pointing out a target group is only the first step to improve traffic safety. The next step is to do a campaign. In Belgium, evaluation of seat belt campaigns mainly involves testing if the campaign was noticed. Up till now, hardly any research is done to measure the actual seat belt increase after a campaign.

Acknowledgements

Work on this subject has been supported by grant given by the Flemish Government to the Flemish Research Centre for Traffic Safety. We are also grateful to the Traffic Police of Antwerp, especially Hubert Ruypers, Frank Vangeel and Karen Penneman for their co-operation with this study. Thanks also to an anonymous referee whose suggestions improved the discussion.

References

Bartholomew, L.K., Parcel, G.S., Kok, G. & Gottlieb, N.H. (2001). *Intervention mapping: Designing theory and evidence-based health promotion programs.* London: Mayfield Publishing Company.

BIVV (1999). *Heeft iedereen klik gedaan? Evaluatie.* No report number. Brussel: Belgian Institute for Traffic Safety.

BIVV (2000). *Gordeltelling, februari 2000.* No report number. Brussel: Belgian Institute for Traffic Safety.

BIVV (2001). *Tot ziens? Klik ze vast, altijd! Evaluatie.* No report number. Brussel: Belgian Institute for Traffic Safety.

BIVV (2003). *Even met de wagen? Gordel dragen! Evaulatie 2de campagneluik.* No report number. Brussel: Belgian Institute for Traffic Safety.

Braver, E. (2003). Race, Hispanic origin and socioeconmic status in relation to motor vehicle occupant death rates and risk factors among adults. *Accident Analysis and Prevention 35 (3).* 295–309

Cedersund, H.A. (2003). *Car seat belt usage in Sweden 2002.* Meddelande 945, VTI. Linköping: National Road and Transport Research Institute.

Cunill, M., Gras, M.E., Planes, M., Oliveras, C. & Sullman, M.J.M. (2003). An investigation of factors reducing seat belt usage amongst Spanish drivers and passengers on urban roads. *Accident Analysis and Prevention, 36,* 439-445.

Eby, D.W., Kostyniuk, L.P. & Vivoda, J.M. (2001). Restraint use patterns for old child passengers in Michigan. *Accident Analysis and Prevention, 33,* 235-242.

Green, L.W. & Kreuter, M.W. (1991). *Health Promotion Planning. An educational and environmental approach.* London: Mayfield Publishing Company.

In 't Veld, R., (2003). *Gebruik van beveiligingsmiddelen in auto's.* Heerlen, The Netherlands: Ministerie van Verkeer en Waterstaat, Directoraat-Generaal Rijkswaterstaat, Adviesdienst Verkeer en Vervoer.

Kim, S. & Kim, K. (2003). Personal, temporal and spatial characteristics of seriously injured crash-involved seat belt non-users in Hawaii. *Accident Analysis and Prevention, 35,* 121-130.

Koornstra, M., Lynam, D., Nilsson, G., Noordzij, P., Pettersson, H., Wegman, F. & Wouters, P. (2002). *SUNflower: A comparative study of the development of road safety in Sweden, the United Kingdom, and the Netherlands.* Leidschendam: SWOV.

Nuyts, E. & Vesentini, L. (2004). *De relatie tussen de gordeldracht van autobestuurders en passagiers.* Report RA-2004-33. Diepenbeek, Belgium: Policy Research Centre of Traffic Safety,

Reinfurt, D., Williams, A., Wells, J. & Rodgman, E. (1996). Characteristics of drivers not using seat belts in a high belt use state. *Journal of Safety Research, 27,* 209-215.

Shin, D., Hong, L. & Waldron, I. (1999). Possible causes of socioeconomic and ethnic differences in seat belt use among high school students. *Accident Analysis and Prevention, 31,* 485-496.

Transports Canada (2003). *Results of transport Canada's september 2002 survey of seat belt use in rural areas of the country.* Road Safety Fact Sheet, RS-2003-02E, TP2436E. Ottawa (Ontario): Transports Canada.

Van Bekkum, P.H.G., Wagemakers, J. & Hiddinga, S.K. (2000). *Gebruik van beveiligingsmiddelen in 2000. Onderzoek naar het gebruik van autogordels, hoofdsteunen en kinderzitjes in personenauto's en bestelauto's.* Report V&I - 99019040_3, AVV. Heerlen: Ministerie van Verkeer en Waterstaat, Directoraat-Generaal Rijkswaterstaat, Adviesdienst Verkeer en Vervoer.

An exploratory study on crew actions as a precondition to use remote experts for preventing shipping disasters

Elsabé J. Willeboordse, Wilfried M. Post, & Anthony W.K. Gaillard
TNO Human Factors
Training and Instruction
The Netherlands

Abstract

On behalf of the Dutch Maritime Knowledge Centre, TNO studied the possibilities of using remote experts in addressing serious non-routine problems onboard. The idea is that this might contribute to the captain maintaining a better control of the situation on board and to the prevention of disasters at sea. The objective of this study was to do a first exploration about any possible obstacles to involving remote experts from the crew point of view. In two cases the actions were investigated that were taken prior to a disaster by the captain and his crew and that in retrospect appeared to have been unsafe. Results showed that more than 24 hours elapsed between the moment of first discovery of irregularities until the point of no return. In theory, this would leave sufficient time to involve remote expertise. However, results also showed that while the crews noticed irregularities, they did not fully understand underlying problems or implications of these irregularities. Obstacles in involving external guidance might be that ships' captains cannot always recognise the significance of structural failures onboard, and that they may be hesitant to disclose problems to anyone beyond the crew.

Introduction

The objective of this investigation, which is part of a wider study, is to perform an exploration regarding the feasibility of involving remote experts or external specialist knowledge in addressing serious problems onboard. The aim of this concept is that it contributes to a better control of the situation. In shipping industry, safety specialists are normally not part of the crew onboard. The Bureau Enquêtes Accidents / mer (BEAmer) recently recommended the use of remote experts by stating that: 'The classification societies set up round-the-clock- 'safety watches' able to answer questions from a captain or ship owner about what action to take, when a vessel has suffered damage affecting its structure and/or stability' (BEAmer, 2003, p. 147). To do a first tentative research on the question whether this concept would be feasible, we investigated possible obstacles to involving remote experts from the crew point of view.

In D. de Waard, K.A. Brookhuis, R. van Egmond, and Th. Boersema (Eds.) (2005), *Human Factors in Design, Safety, and Management* (pp. 93 - 96). Maastricht, the Netherlands: Shaker Publishing.

Background

This study is part of a larger study in which TNO researches human factors involved in catastrophes at sea. In recent years, catastrophes at sea have not happened very frequently. Still, when they happen the consequences are ruinous, with (sometimes heavy) loss of life or extensive environmental damage. For example, the sinking of the ERIKA in 1999 resulted in the pollution of hundred kilometres of French coastline. A better knowledge of the human factor in disasters can be useful to researchers and officers responsible for safe transport. It also can shed light on the question which measures should be taken to prevent catastrophes in future (Wagenaar, 1990).

Research questions and method

Two cases were studied using accident reports by official accident investigation officers. To contribute to assessing the feasibility of remote expert involvement, we investigated which actions and decisions were taken by the crew prior to the catastrophe. The research questions were: How did the process that led up to the catastrophe develop? Did the crew perform actions or made decisions that contributed to the major accident (i.e. unsafe acts), and if so, what actions did they perform? Against a timeline the unsafe acts were presented. The process was reconstructed from the moment when the crew first discovered deviations from the normal condition related to the disaster, until when the catastrophe was inevitable. The method was based on a methodology for investigating and analysing the human factors involved in accidents and incidents: TRIPOD (Wagenaar, Groeneweg & Hudson, 1994). The starting point of TRIPOD is the understanding that human behaviour plays a crucial role in accidents and is influenced by its work-environment (Wagenaar & Van der Schrier, 1997). The Tripod method consists of a large number of diagnostic questions related to humans and tools, which together constitute the work-environment.

Results

Once deviations were detected, it took approximately 24 hours before a major disaster became inevitable. The data of both cases as presented in the official investigation reports suggest that unsafe acts of the crew played a major role in the occurrence of the disaster. One of the factors involved may have been that the crew experienced the situation to be complicated. For instance, in the case of the ERIKA, when the vessel started listing, the listing was a difficult problem to solve; many complex hypotheses had to be verified and falsified (BEAmer, 2003). Although the captain even knew about the procedures prescribing that a mayday signal should be given, he continued trying to solve the problem onboard, instead of sending the mayday signal. BEAmer stated that 'In spite of the captain being well aware that the Erika is an old ship and that several port and ballast tanks are corroded heavily, he is unable to draw any conclusions from this concerning the structural integrity of this vessel' (BEAmer/CPEM, 2003, p. 119). It has been established that while the crews noticed irregularities, they did not fully understand underlying problems or implications of these irregularities. On top of that, they did not foresee the events triggered by the deviations they observed. The captains made errors in judgement and – in disregard of known safety procedures - waited too long before informing authorities. After the two captains recognised the seriousness of the situation, they remained overly reluctant to alter the ships' original direction and speed. Moreover, the captains' communication with rescuers and authorities was imprecise. In short, the main unsafe acts were:

- Not asking for assistance or a decision to secure crew while facing a huge structural failure.
- No precise or clear communication towards important people, leaving them confused. Subsequently omitting an obliged (according to regulations) act crucial to safety.
- No effort to communicate information when this is most necessary to prevent a catastrophe, even when being asked to do so. Means to do this were available.
- No decision to evacuate crew[*], no taking responsibility or lead in the problem when this is most necessary. Seemingly attempt to make coastguard accountable for assistance.

Discussion

The results suggest that captains cannot always distinguish fundamental problems from regular problems. Most likely, captains are used to problems on a daily basis and naturally handle problems as part of daily routine. However, it should be noted that the conclusions presented here are based on data regarding actual behaviour as recorded in the investigation reports, which do not take into account the intentions and perceptions of the captain. Additional contextual information would result in a more complete picture. The results of this study highlight the need for research into the context, motives and intentions for crew behaviour prior to major disasters. Unfortunately, the accident reports mainly focus on technical factors and do not include detailed information about the intentions and reasoning of the crew. Consequently it was not always possible to determine unambiguously whether actions performed by the crew played a significant factor in shipping catastrophes. To better understand the actions performed by the crew, these should be seen in the context of the situation: the information available to the captain, the perception of the captain about the actual situation, his interpretation, his goals, and consequently, the measures taken to achieve his goals (Hendy, 2003). It is recommended that accident investigators make this reconstruction. To obtain information it is crucial that accident investigators talk to the people who were involved. Even though this involves a very difficult task, it is necessary to gain their trust and avoid discussion who is to blame.

Conclusion

The main obstacle for involving remote experts in addressing problems onboard may be that the problem onboard, which potentially leads to a disaster, is not recognised as such by the captain. Therefore the captain cannot be expected to timely initiate the

---

[*] In the case of the BRAER this fault was corrected by coastguard

process of accessing remote knowledge. In addition, captains may be hesitant to disclose problems to anyone beyond the crew.

Only when we know a person's intentions and perception of his environment we can learn about how to prevent accidents in the future (Taylor, 1987). At present, TNO continues to investigate current situational risk factors crucial to behaviour at sea, interviewing crewmembers and sending out a questionnaire to shipping crews. This can help to better understand actions and decisions of the captain and his crew at sea.

References

Hendy, K.C. (2003). *A tool for Human Factors Accident Investigation, Classification, and Risk management.* (Technical Report TR2002/057). Toronto, Canada: Defence R&D.

Taylor, D.H. (1987). The Hermeneutics of Accidents and Safety. In J. Rasmussen, K. Duncan, and J. Leplat (Eds.), *New Technology and Human Error*. Chicester, UK: John Wiley & Sons.

Wagenaar, W.A. (1990). Risk evaluation and the causes of accidents. In K. Borcherding, O.L. Larichev, and D.M. Messick (Eds.), *Contemporary Issues in Decision Making* (pp. 245-260). Amsterdam: North-Holland.

Wagenaar, W.A. & Van der Schrier, J. (1997). Accident analysis the goal, and how to get there. *Safety Science, 26*, 25-33.

Wagenaar, W.A., Groeneweg, J., & Hudson, P.T.W. (1994). Promoting safety in the oil industry. *Ergonomics, 37*, 1999-2013.

Does Automatic Identification System information improve efficiency and safety of Vessel Traffic Services?

Erik Wiersma[1] & Wim van 't Padje[2]
[1]Safety Science Group
Delft University of Technology, Delft
[2]Maritime Simulation Rotterdam B.V., Rotterdam
The Netherlands

Abstract

A universal GPS-based Automatic Identification System (AIS) on board ships is being implemented rapidly. The system is expected improve safety of navigation by providing ships with accurate information about their position and the position of other ships. The system is also expected to improve efficiency by providing to ships and to competent authorities information about a ship and its cargo. A study of consequences of AIS for Vessel Traffic Services (VTS) was carried out using a test-bed that consists of a mobile VTS simulator and other demonstrations of the implications of AIS for VTS systems. VTS operators in a number of European ports participated in the study. Operators ran test scenarios and discussed opportunities and threats posed by AIS in a port environment. Several issues were addressed in the study including Resource Management, Incident and Calamity Abatement, presentation of information, and data reliability. These issues were related to tasks and responsibilities of operators in the different ports. The results of the study form the basis of recommendations for the implementation of AIS in a future VTS system were AIS information is combined or merged with currently available information from radar and other sources.

Introduction

Novel technologies are being introduced in the maritime domain to support in ship navigation aboard ships. Implementation of a GPS-based Automatic Identification System (AIS) is required under SOLAS (SOLAS V, regulation 19.2, as amended 12/13/02) for all new ships built after 1 July 2002, and is mandatory being fitted on new and existing ships in the next years. The required AIS system must be capable of providing information about the ship to other ships and to coastal authorities automatically. AIS is an autonomous and continuous broadcast system, operating in the VHF maritime mobile band. It is capable of exchanging information such as vessel identification, position, course, speed, etc. between ships, between ship and shore through information broadcasts. It is expected that the system can provide many benefits, including increased situational awareness, improved navigational

In D. de Waard, K.A. Brookhuis, R. van Egmond, and Th. Boersema (Eds.) (2005), *Human Factors in Design, Safety, and Management* (pp. 97 - 104). Maastricht, the Netherlands: Shaker Publishing.

safety and automatic reporting in areas of mandatory and voluntary reporting schemes. (IALA, 2002).

AIS is also going to have impact on shore-based services such as Vessel Traffic Services (VTS). According to the IMO Guidelines (Res. A857(20)), a VTS is defined as "A service implemented by a competent authority, designed to improve the safety and efficiency of vessel traffic and to protect the environment. The service should have the capability to interact with the traffic and to respond to traffic situations developing in the VTS area." VTS operators are responsible for a safe and efficient handling of vessel traffic. They perform this function in harbours, rivers, and approach areas all around the world. They monitor traffic, provide information on request and co-ordinate movement of ships in (emerging) conflict situations (Wiersma, Butter, & Padje, 2000).

A study is being performed to determine the potential benefits and constraints in application of AIS information for Vessel Traffic Services. A test-bed has been created to demonstrate the potential effects of AIS on the VTS work. This demonstrator is used in a round trip to a number of European ports. The objective of this trip is to discuss the effect of AIS on the VTS work with VTS operators. On the basis of these discussions recommendations will be made on presentation of derived data in order to optimise common interpretation between VTS and ships.

Method

Test-bed

A test-bed has been developed to illustrate the effects of AIS information on VTS work and to study the potential benefits and constraints in application of AIS information for VTS. The test-bed consists of an interactive portable VTS simulator, on which traffic scenarios are run. The simulator can be controlled in two ways: through predefined scenarios or through on-line control. Reactions of the VTS operator are assessed through a number of assessment methods, including performance measurements and interviews. This makes both the operator and the facilitator part of the test-bed environment (Figure 1).

Issues

Traffic scenarios and demonstrators have been developed to illustrate issues relating to AIS in a port environment. Some of the issues are described below:
- *Changes in communication.* With AIS, ships receive information from other ships, including information where other ships are going. As a result, they do no longer need the intervention of a VTS operator to get this information. Therefore it is expected that oral communication will decrease. A scenario was developed to demonstrate how communication may change as a result of the introduction of AIS.
- *Different levels of information in one display.* At present AIS information is at some VTS stations displayed on a different system than the radar information, at other VTS stations the AIS information is combined with radar information in

one display. A scenario was developed that presents different types of information in one display to study if and how operators differentiate between information from different sources.
- *Potential advantages of AIS information.* Situations exist where AIS may present information that radars can not detect, for instance because an area does not have radar coverage or the radar image is blocked. A demonstration of such situations was used to discuss the potential of AIS in such situations.
- *Reliability of AIS information.* At present the use of the AIS information is limited because the information provided is not reliable. At this moment of introduction of AIS, we see that many ships do not have the AIS system well implemented, and that on board many ships there is a lack of discipline to fill in the AIS information correctly. This may be partly due to the fact that AIS is new and that new users need time to acquire the discipline for regularly updating the AIS system. A number of situations were developed to illustrate reliability issues (Figure 2).

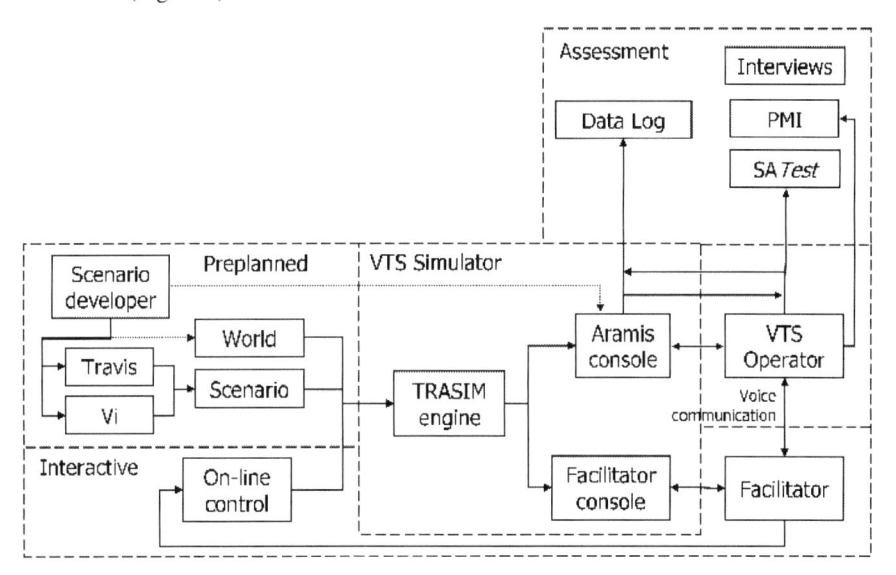

Figure 1. Schematic overview of the EMBARC Test-Bed

Ports

A wide variety of ports spread throughout Europe participated in this project. The ports were chosen to represent all geographical parts of the European Union; North, East, South and Western Europe, to include both large and small ports, and ports with different geographical conditions. This paper presents the results of visits to the ports of Cork in Ireland, the ports of Rotterdam and Vlissingen in the Netherlands, to the port of Genoa in Italy, and to the port of Helsinki in Finland.

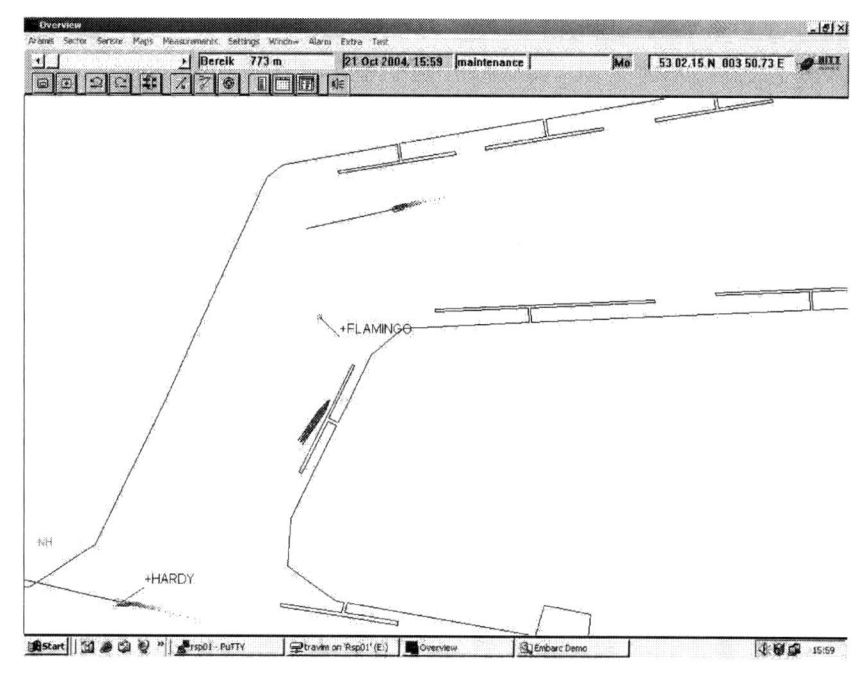

Figure 2. Example of unreliable AIS data: A mistake in the setting of the GPS Geographical Reference Datum. The radar image of the ship and the AIS image of the ship are projected 120 m apart

Experts

The different ports and conditions under which these ports are operated guarantee a wide variety of experts participating in the study. Some Vessel Traffic Services are operated by VTS operators alone, while in one port VTS was operated by a combination of VTS operators and pilots and in one port by the coast guard. Sessions were held all of the above and with their supervisors. Furthermore interviews were conducted with technical staff and management.

Results

In all discussions it was very clear that Vessel Traffic Services all over Europe are aware of the introduction of AIS, however the potential impact thereof was seen to vary. During the course of the study the number of ships equipped with AIS has increased enormously and internet pages with actual AIS information have been introduced. VTSes all around Europe are preparing for AIS. The approach taken in different ports is very different. Some ports have already integrated AIS in their systems. Some ports are considering what to do with AIS information in a VTS environment. This study contributes to a balanced approached towards integration of AIS information in current and new VTS systems.

The tests and discussions show that the differences in ports and traffic that the VTS operators are used to, are large. Nevertheless, the results are fairly consistent:

- Operators from ports all over Europe are well capable of operating the test-bed and handling the traffic presented in the scenarios.
- Whilst the use of ARAMIS based software was easier for some operators than others, the type of interface was familiar to all.
- Discussions on the issues under investigation in this study were not influenced by the choice of interface. The operators are well able to separate their skills with the equipment from the real issues at hand.
- AIS can provide useful information for VTS: Operational advantages are seen in the identification of ships in an early stage and in the presentation of the positions of ships that fall outside of radar coverage.

Reliability of AIS information

Tests and discussions make clear that potential advantages of AIS information is generally recognised. AIS offers opportunities that exceed the possibilities of radar data alone. The main advantages are in approach areas and in areas with low radar coverage. However, to be a useful addition to the information currently available, AIS data have to be reliable. Currently the reliability of AIS data is a widespread concern. The unreliability of AIS data can be a significant limitation for introduction of AIS information in VTS systems. The issues presented need to be solved if AIS is ever going to be a source of information that can be trusted by VTS operators. Unreliable AIS information has been addressed in several of the presented scenarios used at the tests. The scenarios invoked good discussions about current practises with data reliability and possible solutions.

The study shows that in the VTS, information of different sources, such as radar or AIS, is used indiscriminately when it is presented in a VTS display. Most VTS operators do not differentiate between radar and AIS information. In our study operators were asked after the scenario run if they had been aware of the differences in information presented. Most operators were not; all information was treated equally. For the data reliability issue this must have consequences. One possible solution discussed at the test was a technical solution where unreliable or incorrect data is omitted automatically from the VTS display. This does not change the situation on board, but only in the VTS environment.

This study suggests that such a solution will introduce new problems. The issue of data reliability should not be solved automatically; without the intervention of the operator and without notification of the operator of the changes or omissions. Communication between ships and VTS will be severely hampered if by some technical solution information is filtered out of the VTS system that is still present in the data presentation on board ships. The AIS information as presented on board ships should be accessible in the VTS display. It is undesirable for the operator not to be aware of relevant information. In that case an operator will not be able to build situation awareness of the fact that there may be discrepancies in information presentation between ship and shore. There may be additional tools available in the

VTS system to support recognition of errors in AIS data, but filtering out of this information altogether is not a good option. Therefore this option has been rejected in the discussions over and over again.

The issue of dealing with reliability of AIS data in general needs to be addressed, and in the case of VTS, preferably in a designated area at the border of the VTS area. VTS operators in different ports do not agree whether the validation of AIS data is a VTS task. The port of Helsinki has experience with this task. For several years operators in the port of Helsinki have contacted ships displaying incorrect AIS information, and experience has shown that some ships do correct their AIS data, whilst others ignore the request. After repeated instructions to ships (on several trips) VTS operators leave it at that and just make up a report.

All operators agree that the best solution for the problem of data unreliability is to solve the problem on board ships: if ships transmit the correct information in the first place many problems may be solved.

Presentation of information

A further objective of this study was to determine recommendations on presentation of derived data in order to optimise common interpretation. Common interpretation has been taken in this study to be the interpretation of traffic between VTS and ships. This issue is especially important in situations where AIS data are unreliable as has been demonstrated in the scenarios presented in this study.

The operators use information from different sources, often without discrimination; as soon as AIS information becomes available, operators accept this information as valid. AIS simplifies identification of ships, making it easier to address ships. AIS makes it possible to present in a display the positions of ships that fall out of radar coverage. This information is considered useful. Other information presented by AIS labels, such as port of destination is not automatically used. It may require a period of familiarisation before this information will be used. Changes in communication that are a result of the introduction of AIS on board ships were not yet evident to operators. Only in the evaluation of the scenarios did operators recognize that such changes may take place. Apparently the changes are too implicit in the scenarios for operators to notice and the differences between ports are such that a "standard" level of communication across European ports is hard to define.

In some of the ports visited, implementation of AIS data in VTS systems has been chosen where only the best information available is presented in the VTS display; a technical solution that integrates radar and AIS and presents only the results of this integration to the VTS operator is not a good solution. VTS operators are unaware of how the data that are presented are derived. Based on the discussions carried out in this study the authors advise against such an approach since it takes the operator out-of-the-loop. It is preferable to present the incorrect data in the VTS display, along with other types of data (e.g. radar image), so the operator is able to inform the affected vessel traffic in the area. Images may be enhanced to indicate the discrepancies between radar and AIS data, but the raw data should be accessible where deemed necessary in the display. For optimal common interpretation it is necessary to minimise the discrepancy between ship and VTS of information displayed. Communication is best facilitated if both parties share the same information.

Conclusions

The introduction of AIS is an issue that Vessel Traffic Services all over Europe are very much aware of. The approach taken in different ports is very different: some ports have already implemented AIS in their VTS systems, whilst others are still studying the consequences of AIS information for their situation. This study contributes to a balanced approached towards integration of AIS information in current and new VTS systems. AIS can provide useful information for VTS: Operational advantages are seen in the identification of ships in an early stage and in the presentation of the positions of ships that fall outside of radar coverage. However, reliability of AIS data is a widespread concern.

The unreliability of AIS data can be a significant limitation for introduction of AIS information in VTS systems. Reliability issues need to be resolved before AIS can be a trustworthy source of information for VTS operators. The best solution for the problem of data unreliability of AIS is to solve the problem on board ships; if ships transmit the correct information in the first place the problems may be solved at the source. The issue of dealing with reliability of AIS data in a VTS needs to be addressed explicitly as a separate task, preferably in a designated area at the border of the VTS area. VTS operators in different ports do not agree whether the validation of AIS data is a VTS task.

When presented in a VTS display, information of different sources, such as radar or AIS, is used indiscriminately. Most VTS operators do not consciously differentiate between information from radar and information from AIS. The information presented to the operators should not be of substantially different quality from the data displayed on board. This would lead to miscommunication between ship and shore. It is better to present the incorrect data in the VTS display, along with other types of data (e.g. radar image). Images may be enhanced to indicate the discrepancies between radar and AIS data, but the raw data should be available in the display. For optimal common interpretation, presentation of information in VTS displays should share the same information. This facilitates communication best. When managing traffic, operators do not question the reliability of AIS information. The information is integrated in the situation awareness of the operator without distinction with information from other sources, such as radar. Technological solutions in the VTS equipment (e.g. omitting unreliable or incorrect data from the VTS display automatically) can solve the problems of unreliable AIS data in the VTS display. However, from a Human Factors perspective this is not desirable. In the VTS systems of some ports only the best information available is presented in the VTS display. An algorithm integrates radar and AIS and presents the results of this integration to the VTS operator. This is not a good solution.

Recommendations

- Where appropriate, ensure that both radar and AIS data are accessible within the VTS display, with a clear distinction between the two.
- Avoid taking the VTS operator out-of-the-loop by solving the problem of reliability of (AIS) data with technical means without informing the operator
- Designate an area at the border of the VTS where reliability of AIS data is addressed.
- To ensure adequate awareness training programs during the transition by VTS authorities towards AIS.

Acknowledgements

This paper is based on a study carried out as part of a European research project EMBARC funded by the European Commission DGTREN Transport and Energy (Contract No: GRD1-2000-25500). Many parties have participated in this project to get a broad, European perspective on the potential benefits and limitations of AIS for Vessel Traffic Services around Europe. Our special thanks go to the Port Authorities of Cork, Genoa, Helsinki, Rotterdam and Vlissingen for their contribution to the discussions.

References

IALA. (2002). *Recommendation on the Provision of Shore Based Automatic Identification Systems (AIS)*. IALA Recommendation A-123. Saint Germain en Laye.

Wiersma, E., R. Butter, & Padje, W. v. 't. (2000). *A Human Factors Approach to Assessing VTS Operator Performance*. Paper presented at the VTS 2000 Symposium, Singapore.

Turn on the lights: investigating the Inspire voice controlled smart home system

Heleen Boland[1], Jettie Hoonhout[1], Claudia van Schijndel[1], Jan Krebber[2], Sebastian Möller[2], Rosa Pegam[2], Martin Rajman[3], Mirek Melichar[3], Dietmar Schuchardt[4], Hardy Baesekow[4], & Paula Smeele[5]

[1] Philips Electronics Nederland B.V., Eindhoven, The Netherlands
[2] Institute of Communication Acoustics (IKA), Ruhr-University Bochum, Germany
[3] LIA, Ecole Polytechnique Federale de Lausanne (EPFL), Switzerland
[4] ABS Gesellschaft für Automatisierung, Bildverarbeitung und Software mbH, Jena, Germany
[5] TNO Human Factors, Soesterberg, The Netherlands

Abstract

The Inspire project aims to realise an interactive spoken dialogue system for wireless command and control of home appliances. In terms of the user interaction approach adopted, this could be done in basically three ways: via an embodied agent mediating between the user and the devices, via an invisible agent (ghost), which mediates between the user and the devices, or by directly interacting with individual devices. In the test described in this study, human factors aspects of this system, and of the three different interaction metaphors, were investigated. Questions that were addressed included: appreciation of the different metaphors, suitability of voice control for the different devices that were included in the test set-up, and other aspects of the system that are of interest regarding the actual interaction of the users with the systems. Data were collected through observations regarding the participants' opinions on various aspects of the Inspire system. Furthermore, data for various dialogue parameters were collected using a specially developed annotation tool. The main conclusions are that the interaction with the individual devices was preferred most compared to the other interaction approaches; that there was a difference in preference of interaction approaches between older and younger users; and that in developing these types of interaction systems it would probably be more beneficial to focus on the size of the vocabulary and the flexibility rather than 'human-human' like interaction in terms of grammar and social standards.

Introduction

The technological complexity of electronic devices in and around the house, or rather the way to operate these devices, is increasing fast. Nowadays more and more users of such devices, certainly user groups such as elderly and technically non-inclined people, are faced with difficulties even when performing daily tasks. Intelligent user interfaces that can function as 'home assistants' might facilitate these highly complex operations (Wahlster, Reithinger, & Blocher, 2001). The aim of the INSPIRE (INfotainment management with SPeech Interaction via REmote microphones and telephone interfaces; IST-2001-32746) project is to develop such an assistant using speech technologies. The Inspire system aims to provide an interactive, natural speech dialogue-based support that facilitates the wireless command and control of several common devices in the house, e.g. the television, the answering machine and lamps. It can be directed both from within the house, through a microphone array or wireless microphones, and from remote locations such as the office or the car, via the telephone network. Currently the system supports interactions in Greek and German, but its modularity ensures that it can easily be extended for other languages. The system consists of several different components that have been developed and evaluated separately (e.g. acoustic pre-processing and adaptive noise cancellation, speech and speaker recognition, speech output and dialogue management; see Ganchev 2003ab, and Bui & Rajman 2004, for details on these separate tests).

Speech control has been adopted in a variety of application areas. For example, in industry, speech control is used in inspection tasks, feeding inspection information directly into the computer system, thus speeding up the process (Baber & Noyes, 2001). Speech control is also used in office applications such as word processing and operating software applications – often recommended to users with severe symptoms of repetitive strain injury. Speech technology is also more and more used to handle relatively simple telephone-based inquiries, for example for rail travel scheduling information.

A number of usability tests and marketing studies have indicated that users generally appear to be positive about voice control for consumer electronics (e.g. Hongli Ma, 1998; Vogten, Kaufholz, Bekker, & De Ridder, 1998). Reasons for liking voice control that are often mentioned by participants in such studies are:

- no need to search for the control device;
- one can control devices while doing other things (hands busy with other tasks, walking around);
- expected to be easier to use than for example a remote control with all its buttons;
- admiration of new technology.

For example, in a usability test of a voice-controlled consumer device (Vogten et al., 1998) 73% of the participants were positive about voice control before the test, and 90% afterwards. However, what is important to take into account here is that in this study (as in most other usability tests) the participants used the device for an hour at the most. It is impossible to say what participants would have said (and done) if they could have used the device for a much longer period of time. Furthermore, a Wizard of Oz procedure was adopted in this test, i.e., the users were not working with an actual speech recognition system, but with a simulation. This will have resulted in a much better system performance than would have been possible with an automatic speech recognition system.

Generally, what one often sees is that when users can "talk" to a system, their expectations about the capabilities of such a system tend to be unrealistically high (Dryer, 1999). Furthermore, it is usually assumed by users that using speech control is not going to require any training, since talking is such a natural, long practiced, skill.

Although users thus seem to be almost in awe of voice control, and it certainly does have a number of benefits and advantages, it is good to realize that speech interfaces are not always the most suitable approach for every application. This can be ascribed to at least two factors. First, speech recognition systems are far from flawless, and approaches to correct recognition errors are difficult, and usually result in diminishing user satisfaction (Krahmer et al., 2001). Human speech tends to be imprecise and variable, making it difficult to match it with the requirements of the currently available technology. One possible approach to control recognition performance, is to greatly reduce the vocabulary that can be used in interactions with the system – however, this has a large impact on the "natural" feel of the dialogue with the system, and it will require some training effort on the part of the user to learn the endorsed vocabulary. Secondly, for some tasks (e.g., ones that involve presenting lists), speech is simply a less suitable modality. Also certain environments might be less suitable for speech interfaces, e.g., noisy environments, or public areas. One can imagine that users might feel embarrassed if they could be seen talking to a machine in public areas.

The study described in this paper was set up in order to investigate suitability of voice control to operate devices in the home, and in particular to learn about requirements regarding the dialogue style that would optimally fit with operating smart home applications. More concretely, an integrated prototype of the Inspire voice controlled smart home system was evaluated, focusing on different aspects of the user-system interaction. With respect to the basic user interaction style that can be adopted for a spoken speech dialogue system approach three different interaction metaphors were considered for the Inspire system. The three metaphors are listed below, together with the major advantages and drawbacks that were predicted prior to this study.

- The "intelligent devices" metaphor allows the user to interact with the devices directly. Advantages are that the user can direct the attention to the device that is being operated and that the user can receive direct feedback from the device. This requires, however, that the user is in the vicinity of the device.
- In the "talking head" metaphor an assistant that is shown on a display, mediates between the user and the devices. It is, again, clear to the user where to address commands, but the user might not be able to see the direct feedback from the devices.
- The "ghost" metaphor allows the user to operate the devices with help of an invisible assistant. The assistant can be addressed from anywhere in the house, but the user does not have a fixed point to address and the user might not be able to see direct feedback from the devices.

The metaphors were implemented by assigning the system's output to different loudspeakers for each metaphor (see subsection 'design'). At the time of the study the performance of the automatic speech recogniser was not yet evaluated. To minimise the effects of a possibly poor recognition rate on the flow of the interaction, the speech recognition was taken care of using a Wizard of Oz design.

The study described in the following section was designed to evaluate the usability and acceptability of these three different interaction metaphors to potential end-users. Other aspects that were addressed in this study were the acceptability of the system in terms of the devices it can address, the dialogue style adopted in the interaction, and to what extent it meets the expectations of the users regarding voice control.

In order to provide a realistic setting for the Inspire system during the test, the system was integrated in the living room of Philips' HomeLab. The HomeLab provides a realistic home environment and is ideal for studies such as the one described here, because it combines laboratory facilities, such as easy integration of system prototypes, with facilities for direct observation and recording of video and audio data, in a setting that is designed to make participants feel at ease – at home if you like.

Method

Participants

In this study 28 native German speakers participated, 12 women and 16 men. One participant was physically disabled and moved around in a wheelchair. The age of the participants ranged from 21 to 65 years old, 7 of them being 50 years or older. The participants had different professions, ranging from housewives, to students, to engineers. Their education ranged from high school to academic. Most participants had experience with artificial dialogue systems like Tele-banking and other telephone services. Only 3 participants did not have any such experience at all.

Inspire set-up

The Inspire system was set up in such a way that the participants could engage in a realistic scenario of 'coming home'. They were asked to imagine, for example, finding the house dark and too warm upon entering. Through natural dialogue interaction, they could then ask the Inspire system to turn on the fan and the lamps, and to adjust the brightness of the lamps. The following is an example of a dialogue with Inspire (originally the dialogue was in German):

> System: welcome to the Inspire Smart Home System. What can I do for you?
> User: please turn on all lamps.
> System *(turns on all the lamps)*: what else can I do for you?
> User: Inspire, it is warm here
> System *(turns on the fan)*: What else can I do for you?

The prototype of the Inspire system as used in this study consisted of the following main components: a speech input module, a dialogue module, a device interface, and a speech output module. To eliminate the effects of a possible poor recognition rate, the automatic speech recogniser was replaced by a wizard who was instructed to literally transcribe all user utterances. The speech understanding, i.e. the mapping of the transcribed words to the words in the system's vocabulary, was still done automatically. The output component consisted of the interface between the speech component and several devices in HomeLab and was run on a PC.

Figure 1. a) the dining room with the 'talking head' screen and the table lamp, b) the researcher acting as the wizard in the observation room, and c) the living room with the two standing lamps. See also http://extras.hfes-europ.org

Inspire could be used in the living room, dining room and kitchen of the HomeLab (all located on the ground floor). The participants were encouraged to move around the house, by including as a first task in each scenario, the instruction to fetch either a drink or something to eat, or to read a postcard that could be found in the kitchen. The devices that could be operated were placed in different corners of the living room and dining room in such a way that there was not one location from which all the devices were visible. The devices that could be operated were:

- Three lamps: a table lamp that was placed on the dining room table, a white standing lamp and a yellow standing lamp that were placed on either side of the couch in the living room
- A tall-standing fan that was placed between the dining room and the living room

- An answering machine that was placed in the far corner of the dining room
- Automated blinds covering the windows in the dining room
- A television (including recording functions and an Electronic Program Guide) in the living room

There was at least one loudspeaker near every device and they were used in such a way that, in the intelligent devices metaphor, each device was assigned a different loudspeaker. All the devices were automatically controlled by the device interface except for the blinds that were controlled manually by the wizard from a remote location. The functions of the answering machine as well as the contents of the electronic program guide (EPG) were simulated.

Apart from a home environment the HomeLab also contains a fully equipped observation room, which is located on the ground floor, directly behind the living area. During this study the observation room was used by the wizard to keep track of the interaction between the participant and the Inspire system through several monitors presenting the scenes captured by cameras and microphones located in the living room and dining room area. Figure 1 contains pictures of the rooms in HomeLab and of the observation room.

Design

A single factor within-subject design was used in this study with the type of interaction metaphor as independent variable. Three levels were defined corresponding with the three interaction metaphors:

Condition A: the "intelligent devices" metaphor. In this metaphor the users were required to interact directly with the different devices, as appropriately, and the output of the system was delivered through the loudspeaker that was situated nearest to the device that was being addressed.

Condition B: The "talking head" metaphor. In this metaphor the users interacted with a visible assistant (see Figure 2). This assistant was displayed on a computer screen, which was situated in such a way that it was visible from as many locations in the living room as possible. To avoid synchronisation problems between the mouth and the voice, the assistant was not designed as an animated human face but as a moving puppet-sock against a dark background. The puppet only moved (mimicking facial expressions) when the system was giving output. The output of the assistant was delivered through a designated loudspeaker, which was positioned next to the screen.

Condition C: the 'ghost' metaphor. In this metaphor the system was presented as an invisible assistant. However, the ghost was otherwise addressed in much the same way as the visible assistant. The output of the invisible assistant was given through loudspeakers mounted in the ceiling of the room.

The participants took part in the study individually and they used the system in three separate trials, with the three different conditions (metaphors) balanced over the trials. At the start of each trial they were handed a set of tasks, described on cards.

The tasks that they were asked to perform included operating all devices at least once: turning lamps on or off, and dimming them (in total three different lamps could be controlled), putting the blinds up or down, turning the fan on and off, manipulating messages on the answering machine, consulting the EPG on the television and recording a film, or setting a reminder for the start of a film.

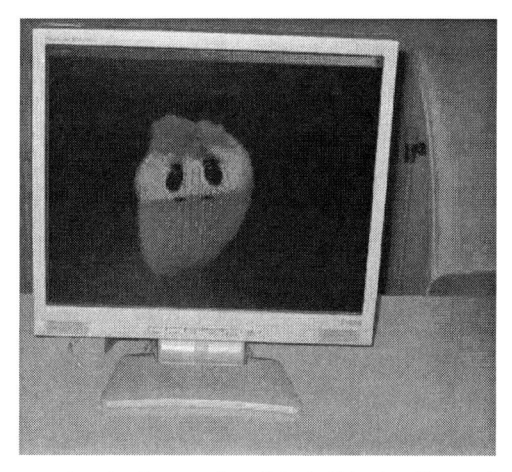

Figure 2. The puppet-sock as implementation of the 'talking head' metaphor

The tasks were presented in the form of a scenario, which aimed to provide an appropriate context for performing the tasks, e.g. "it is warm in the room so you decide to use the fan". The scenarios were written in such a way that the wording did not give away clues as how to exactly address the system. Each scenario began with a different starting position of the lamps and the blinds, to simulate different home situations. To avoid an effect of repetition, three different scenarios were used with a different order of tasks. Both the order of presentation of the conditions as well as the order of presentation of the three scenarios was fully counterbalanced across participants.

Evaluation measures

To determine the user judgements of the three metaphors a paper-and-pencil version of the Subjective Assessment of Speech System Interfaces (SASSI) questionnaire was used as developed by Hone and Graham (2000). The SASSI questionnaire distinguishes six dimensions: system response accuracy, likeability, cognitive demand, annoyance, habitability, and speed. For this test the items of the SASSI questionnaire were translated from English into German. The items were formulated as statements, which were accompanied by a 7-point Likert scale (with 1 meaning 'strongly disagree' and 7 meaning 'strongly agree'). Furthermore, a ranking for the three interaction metaphors was obtained through a paired comparison test. For this test, the participants were asked to judge, for each of the three possible pairwise combinations of metaphors, which metaphor they preferred to use.

The expectations of the participants prior to the test regarding the concept of voice control were determined on the basis of a 7-item questionnaire (developed for this study) and a semi-structured interview, addressing different aspects of voice control such as advantages and disadvantages compared to more traditional modes of operation, and judgements about its ease of use. The items of the questionnaire were again presented at the end of the experiment, to be able to determine to what extent the Inspire system met the expectations of the participants.

After the trials a semi-structured closing interview was conducted with questions on the acceptability of voice control as an interaction concept, the acceptability of the interaction metaphors, the experienced ease of learning how to use the system and the sufficiency and adequacy of the system feedback messages. In this interview the suitability of voice control for the specific set of devices and the suitability of the speech commands for performing the specific tasks were also addressed.

The interactions were further analysed on the basis of 65 dialogue parameters describing the behaviour of the user and of the system during the dialogue (Möller, 2002, 2003). Fourteen parameters were automatically logged during the experiments. The remaining parameters were determined based on the transcription and annotation of the dialogues by a human expert. Part of the parameters were related to dialogue and communication (e.g. the dialogue duration, the number of system and user turns in each dialogue, and the number of questions per dialogue from both user and system). Part of them were related to meta-communication (e.g. the number of times the system gave an error message to signal inconsistencies in the dialogue, and the capacity of the system to recover from user utterances for which the speech recognition or understanding process partly failed) and co-operativity (e.g. the appropriateness of the system utterances). Part of them were related to task success (e.g. the number of tasks that were fulfilled per dialogue) and, finally, speech input (e.g. the number of correctly recognised words and the number of utterances that the system could not 'hear' or understand).

Procedure

The sessions took approximately 2 hours per participant. The wizard function was carried out by a native German speaker who was fully experienced with using the specific wizard interface for the Inspire tests. Another experimenter took care of instructing and debriefing the participants. Participants could address their questions to this experimenter, and this experimenter could provide help and explanation to the participant in case of technical problems with the system. Furthermore, two experimenters made observations, using the monitors in the observation room, and kept a general overview of each session. Finally there were two experimenters who kept a logbook of technical events and issues, and helped to debug the system in case of technical problems.

At the start, the participants were given a brief introduction about the Inspire system, the HomeLab and the procedure of the experiment. After the introduction, the participants were asked about their expectations of voice control. Next, they were given a tour of the living- and dinning room and the kitchen, and they were shown the locations of the devices they were to operate. During the tour it was emphasized that the participants should feel at home as much as possible and for this purpose it was made explicitly clear that they could use the kitchen to their own liking, e.g. take a drink whenever they felt like it. The participants used the system once for each metaphor. Each time, before using the system, they were given a situation sketch that explained the metaphor. They were also given a set of task cards that described the scenario.

Before using the system for the first time, the following additional instructions were given: a) the user has to wait for the system to welcome him/her before being able to use it, b) it is possible to stop the system giving output by saying "stop Inspire" and c) if the tasks are all performed the participant can come back to the entrance hall, outside the dining room. When the blinds and lights were set according to the specifications for that particular scenario the participant was asked to enter the dining room alone and to use the system to complete the tasks. When the participant had returned to the hall the experimenter presented the SASSI questionnaire in the reception room.

The second and third trials were conducted in the same way. Each trial lasted approximately 15-20 minutes. After the third trial, the participants were presented with the paired comparison test, with a questionnaire addressing again the same items as in the expectations questionnaire and a semi-structured closing interview.

Results

Expectations

The expectations of the participants regarding voice control were analysed and compared to their opinions of voice control after using the Inspire system. Both before and after using Inspire, the participants rated the following aspects of voice control: how useful it is, how difficult and fun to use, how it is preferred compared to more traditional modes of control, how ubiquitous it will be in the future, how easy it is to learn and how useful it is as universal control (see Figure 3). Any difference between the ratings of before and after would indicate that the Inspire system does not match these expectations. However, paired samples t-tests showed that there were no significant within-subject differences.

Metaphors

The usability and acceptability of the three interaction metaphors were compared in terms of subjective user judgements. In addition, the dialogue parameters provided feedback on actual behaviour towards the system. The effects of the type of metaphor on the subjective judgements and the interaction parameters were analysed using repeated measures. For the analysis of the subjective judgements the data of all 28 participants were used. However, for the analysis of the interaction parameters only 26 dialogue sets were used, because for two participants the data set was not complete (for each of the two participants, one dialogue was missing).

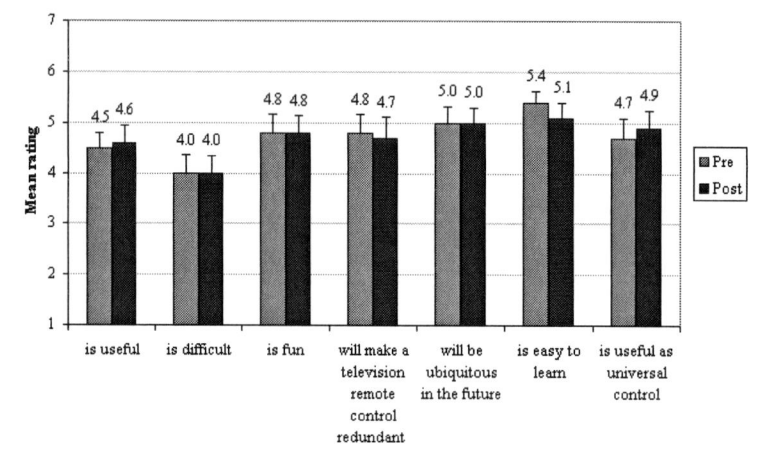

Figure 3. mean ratings and standard errors (N=28) on a 7-point Likert scale (with 1 meaning 'strongly disagree' and 7 meaning 'strongly agree') for seven aspects related to the concept of voice control, rated before (pre) and after (post) using Inspire

First, a ranking for the three metaphors was obtained with the paired comparison test. In this test the participants were asked for each pairwise combination of metaphors, which metaphor of the pair they preferred. Using Thurstones Law of Comparative Judgement (Meerling, 1988) the results can be presented on a one-dimensional scale. Overall, the participants preferred the intelligent devices metaphor (z-value of proportion = 0.34) and liked the talking head metaphor least (z-value of proportion = -0.34). The ghost metaphor was ranked relatively in the middle (z-value of proportion = 0.00).

Table 1. means ratings and standard deviations on a 7-point Likert scale (with 1 meaning 'strongly disagree' and 7 meaning 'strongly agree') for the general impression of the system overall and the three interaction metaphors

Rating	N*	Mean	Std. Dev.
Overall	27	3.52	1.50
Intelligent Devices	26	3.73	1.46
Talking Head	26	4.00	1.52
Ghost	25	4.24	1.33

*Due to missing values, N is not equal for all conditions

Next, the metaphors were compared in terms of the overall impression made on the participants. This comparison showed that there was no main effect of type of metaphor (see Table 1).

Third, the metaphors were compared in terms of the SASSI dimensions. For that, the assumption of the SASSI scale that the items within each dimension are correlated was checked. This assumption was satisfied for all but the dimension 'habitability' (Cronbach's alpha for habitability was > 0.50, whereas Cronbach's alpha for the other dimensions was > 0.70). Therefore the three metaphors were compared in terms of the five dimensions 'system response accuracy', 'likeability', 'cognitive demand', 'annoyance' and 'speed'. The items of the dimension 'habitability' were analysed separately. All the results are presented in Figure 4.

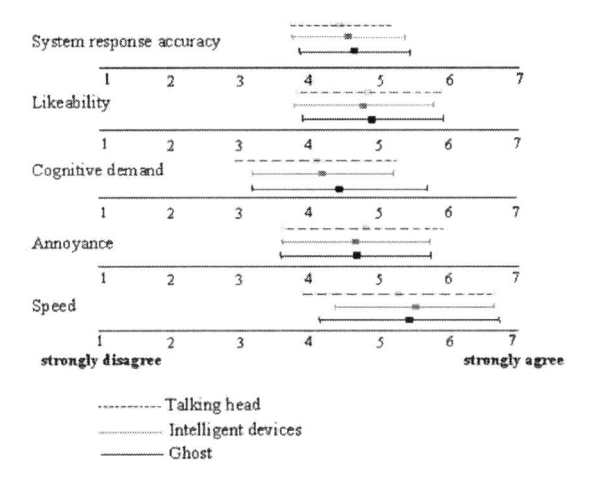

Figure 4. mean ratings and standard deviations of the three metaphors on a 7-point Likert scale (with 1 meaning 'strongly disagree' and 7 meaning 'strongly agree') for five dimensions of the SASSI scale

In terms of the dimensions there are, again, no significant differences between the metaphors. However, for one of the items of the dimension habitability there was a significant main effect of metaphor ($F(2,52)=5.99$, $p<0.005$); the participants thought it was easiest to lose track of the interaction with the system when using the intelligent devices metaphor (on a 7-point Likert scale with 1 meaning 'strongly disagree' and 7 meaning 'strongly agree', mean rating = 3.57), and least easy when using the ghost metaphor (mean rating = 2.96).

Finally, the interaction metaphors were compared in terms of the interaction parameters. The analysis of the results of the 26 participants of whom the dialogues were annotated, showed that there were again no significant differences.

Whereas the questionnaires and the interaction measures showed no differences between the three interaction metaphors, the results of the closing interviews gave an insight into their (possible) advantages and disadvantages and the suitability of the way they were implemented. First of all, sixteen participants thought the differences between the current implementations of the metaphors were not very clear. Most of these participants mainly failed to see a difference between the intelligent devices and the ghost metaphor, because the intelligent devices' *default* system output, which was active whenever no specific device was selected, used the same output source as in the ghost metaphor, i.e. the ceiling loudspeakers. On the one hand, eight participants said to prefer the concept of the intelligent devices, because it would require less interaction compared to an assistant functioning as an intermediary between the user and the devices. On the other hand, nine participants said to prefer the concept of the ghost, because it would allow remote operation. Addressing something invisible was, however, regarded as unnatural and uncomfortable by two participants. With regard to the talking head metaphor, three participants said to have higher expectations of its intelligence compared to the other metaphors, but they did not experience it as being more intelligent. Nine participants had the feeling of being forced to look at the screen on which the personal assistant was shown or even felt the need to walk up to the screen. They thought it was very unpractical to have one fixed screen, because they felt it took away some freedom of mobility. Only three participants preferred to be able to address a fixed point. Eleven participants disliked the design of the talking head itself.

Learning effect

In general, the participants expected that voice control is easy to learn (on a 7-point Likert scale with 1 meaning 'strongly agree' and 7 meaning 'strongly disagree' the mean rating for expecting it to be easy to learn was 5.4). After using Inspire their opinion had not changed significantly. Also, in the closing interview many participants (17) said that they found it easy to learn how to use the system. However, many participants (16) also said that they felt forced to learn a specific vocabulary. Although the participants could use whole sentences, the system would only react to keywords. The set of keywords was limited and the participants had to find the limitations through trial and error, e.g. many participants tried to use 'dim' to reduce the brightness of the lamps; however, this command was not included in the system's vocabulary.

The data of the interaction measures agreed with what the participants had said in the interviews. They showed a significant learning effect. The duration of the dialogues ($F_{(2,25)}=26.38$, $p<0.001$) as well as the number of turns per dialogue ($F_{(2,25)}=17.60$, $p<0.001$) were higher for the first session compared to the second and third session and the differences were significant. The number of words that could not be understood, i.e. the number of times the system did not find one or more words of a user utterance in its vocabulary, decreased significantly over the three sessions ($F_{(2,25)}=11.29$, $p<0.001$). Also, the number of times the system posed a question for clarification ($F_{(2.25)}=14.23$ $p<0.001$) and the number of times it provided help when the dialogue required it ($F_{(2,25)}=3.89$, $p<0.027$) decreased significantly. Finally the percentage of turns needed to make corrections in the dialogue also decreased over the sessions for both the system ($F_{(2,25)}=14.89$, $p<0.001$) and the user ($F_{(2,25)}=8.66$, $p<0.001$), and the percentage of appropriate system utterances increased ($F_{(2,25)}=5.17$, $p<0.001$). These changes indicate a learning effect for the participants. As they learned how to formulate the commands appropriately, the system had less trouble interpreting them.

For most of the parameters, the biggest learning effect occurred between the first and second session. The exceptions are the parameters denoting 'percentage of user correction turns' and 'percentage of appropriate utterances', for which the differences between the first and the second session and between the second and the third session were similar.

Suitability of voice control for the devices

In the interview, the participants were asked for each device whether they thought it was easy to operate it by voice and whether the interaction style was suitable. Quite a few participants were satisfied with the way the blinds (9) and the fan (14) could be operated, except that the noise of the blinds sometimes interfered with the interaction and that it took some time to stop the blinds going up or down. Regarding the operation of the lamps, many participants (11) thought it was inefficient that the lamps could only be operated one by one or all together (which were the only possibilities in this set-up). Instead they would prefer to operate them in groups, e.g. all the lamps in the living room together. The operation of the television, and mainly that of the EPG, were considered by quite a few participants (9) to be laborious and time-consuming due to the inflexible dialogue structure (this reflects a problem with the interaction style of this element of the system rather than a problem with the concept of an EPG itself). Some participants remarked that they found it easier to use voice control to operate the VCR compared to traditional modes of control of a VCR, because it required less technical knowledge. Finally, many participants appreciated the possibility of remote operation of the answering machine through speech.

Acceptability of the Inspire system

The acceptability of the current version of the Inspire system as a means of operating electronic household devices was mainly discussed in interviews with the participants. The interview data also provided useful feedback on the design of the Inspire system. Most participants (17) said that they thought it would be easy to learn how to use the Inspire system to do the tasks. Even though most participants (16) agreed that they felt forced to learn a specific vocabulary, all but one of them found this acceptable, provided that that would make the interaction efficient. They did not think that the dialogues resembled a human-human interaction, but they did not find this unacceptable. However, according to many participants, the system did not give the user sufficient control of the dialogue. Most importantly it did not allow the user to barge in, i.e. to interrupt system feedback when it is clear which command to use. Furthermore, by asking the user what it should do next, the system implicitly forced the user to react, instead of allowing the user to control the interaction. Secondly, all the participants were very dissatisfied with the slow speed of the system in performing actions and giving feedback. For voice control to be acceptable in these settings and with these devices, it has to work considerably faster than it does now. Due to these imperfections, the interaction style was found to be unacceptable for complex tasks. Moreover, in a situation with more than one user, voice control would be acceptable only if it proves to be robust: it should not be disrupted by speech that is not directed to the system, and it should not interrupt conversations between people. Finally, the fact that quite a few participants (9) explicitly indicated remote controls (or buttons) to be a necessary fallback solution indicated that voice control as implemented by the current system is not considered to be reliable enough as an alternative to more traditional modes of operation of household devices.

As far as the general concept of voice control was concerned, there were two aspects that some participants regarded as possibly unacceptable drawbacks. Firstly there was the security issue: quite a few participants (9) were concerned that the system would facilitate misuse of the devices by unauthorised persons, e.g. intruders but also children. Secondly, there was the health issue: two participants, of whom one was the physically disabled participant, expressed the concern that users of voice control might become too lazy, no longer needing to get out of their chair for these common tasks.

Discussion

The results of this study showed little difference between the three interaction approaches, intelligent devices, talking head and ghost, in terms of subjective judgements and interaction measures. For several reasons, however, it seems premature to draw the conclusion that the interfaces would be equally suitable and preferred by the users.

First of all, the smoothness of interaction influenced many aspects on which the metaphors were rated, like friendliness, response accuracy and annoyance. In this study, the smoothness of the interactions between the users and the Inspire system was severely impeded by the slow speed and the lack of flexibility of the system. So, without this smooth interaction, the participants were most likely not able to appreciate, or even acknowledge, the aspects on which the metaphors were rated and compared.

And secondly, the participants seemed to have specific expectations of the metaphors. In the paired comparison test, some participants explicitly stated that they preferred the intelligent devices based *on their expectations* of them, rather than *on their experiences* with them. In fact, most participants were able to come up with specific advantages and disadvantages of all three metaphors, but they did not seem to have experienced these when using the Inspire system.

Based on the results of this study it is difficult to conclude whether voice control is more suitable for simple tasks, like turning the lamps on and off, or for more complex tasks, like operating the VCR. The participants were satisfied with the simple tasks that required the least interaction. However, they expected voice control to be most useful and suitable for technically complex tasks, especially in meeting the needs of technically non-inclined people. Although it was not always explicitly mentioned by many participants, they expected a system using voice control to have a significantly higher level of intelligence than switches or buttons, e.g. for allowing the operation of groups of lamps or predefining settings. It is probably a general issue, that people have a relatively high expectation of the intelligence of voice controlled systems (Dryer 1999). The expectations could change with long term usage of the system when users become more acquainted with voice control. Unfortunately, for this study it was not possible to test long term usage of Inspire.

In addition to the analyses reported above, the data of the study were also analysed taking into account the age of the participants. The results of those who were 50 years of age or older were compared to the results of those who were younger. These analyses were not reported here, because the number of 'elderly' participants was rather small (7), and given the large variance in the data, also difficult to interpret. Nevertheless, the results seem to indicate that there might indeed be certain differences between the age groups. For example, the different age groups seem to differ in terms of preferences for the type of interaction metaphor; this is certainly worth further investigating.

Conclusion

This study proved to be useful in several ways. First of all, the results will be (and already partly have been) used to improve the next version of the Inspire system. Secondly, the study has shown that it is probably more worthwhile for developers of these types of speech based interaction systems to focus on enlarging the vocabulary and flexibility of the dialogue structure, rather than to strive for 'human-human'-like interactions in terms of using natural language sentences. The participants preferred what some called a master-slave relationship: they wanted to give short commands and expected no system initiative. They highly valued flexibility in terms of a large set of synonyms and the possibility to combine commands. Thirdly, for certain devices and tasks there might be more efficient ways of control than voice control, e.g. using an automatic detection system to switch on lights. Finally, several tendencies have been found among all participants and among older and younger participants separately with respect to how they (prefer to) operate household devices using voice control. Future research could investigate the robustness of the results found in this study over larger groups of people, specifically larger target groups as older and physically challenged users.

Acknowledgements

This study was onducted in the context of an ITS project funded by the European Union
Inspire: http://www.inspire-project.org

References

Baber, C., & Noyes, J.M. (2001). Speech Control. In K. Baumann and B. Thomas (Eds.), *User interface design for electronic appliances*. London: Taylor & Francis.
Bui, T.H. & Rajman M. (2004, confidential). Rapid dialogue prototyping methodology and Wizard-of-Oz interface generator manual. *"Report Inspire WP4: deliverable 4"*. Lausanne, Switzerland: EPFL.
Dryer, D. C. (1999). Getting Personal with Computers: How to Design Personalities for Agents. *Applied Artificial Intelligence, 13*, 273-295
Ganchev, T. (2003b, confidential). Signal Processing Components. *"Report Inspire WP2: deliverable 2.2"*. Patras, Greece: WCL.
Ganchev, T. (2003a, confidential). Speaker recognition component. *"Report Inspire WP2: deliverable 2.3"*. Patras, Greece: WCL.

Hone, K.S. & Graham, R. (2000). Towards a tool for the subjective assessment of speech system interfaces (SASSI). *Natural Language Engineering, 6*, 287-305

Hongli Ma. (1998). *Automatic Speech Recognition Consumer Research Project*, Philips Consumer Electronics Internal Report, no. FF1037 (confidential).

Krahmer, E., Swerts, M., Theune, M., & Weegels, M. (2001). Error detection in spoken human-machine interaction. *International Journal of Speech Technology, 4*, 19-30.

Meerling (1988). *Methoden en technieken van psychologisch onderzoek. Deel 2. Data-analyse en psychometrie* Meppel/Amsterdam: Boom.

Möller, S. (2002). A New Taxonomy for the Quality of Telephone Services Based on Spoken Dialogue Systems. In Proc. *Third SIGdial Workshop on Discourse and Dialogue* (pp. 142-153), USA-Philadelphia PA.

Möller, S. (2003). *Quality of Telephone-Based Spoken Dialogue Systems*, Habilitations Thesis. Bochum, Germany: Institut für Kommunikationsakustik, Ruhruniversität.

Vogten, L., Kaufholz, P., Bekker, M., and De Ridder, H. (1998, confidential). *Voice control of an audio set. A user study and some user-interface design implications*. Eindhoven, Philips Research, Internal report Nat.Lab. 7042.

Wahlster, W., Reithinger, N., Blocher, A. (2001). SmartKom: Multimodal Communication with a Life-Like Character. In *Proceedings of the 7th European Conference on Speech Communication and Technology*, Vol. 3 (pp. 1547-1550). Aalborg, Denmark: Eurospeech.

Is there such a thing as a mental representation for interface layouts?

Francesco Di Nocera, Michela Terenzi, Giovanni Forte, & Fabio Ferlazzo
Cognitive Ergonomics Laboratory, Department of Psychology
University of Rome "La Sapienza", Italy

Abstract

The Cognitive GeoConcept was introduced as a method for finding geometrical associations between meaningful objects (links or functions) in web pages and it is supposed to elicit users' spatial schemata or representations underling the way people look for information within a web page. The procedure itself is based on the analysis of users' click responses to verbal labels indicating web objects on a large number of trials. In this study, ten words indicating links often found in the navigational menu of Italian academic web sites were used as stimuli, and eye movements were collected together with clicks. Results showed that fixations were better suited to differentiate experts' patterns from novices'. Moreover, differential groupings were found depending on expertise. Particularly, experts organise the stimuli according to a prototypical interface deployment, whereas novices seem to organise stimuli according to personal criteria that get lost in the overall pattern. Overall, results confirmed the usefulness of the procedure as a technique for eventually supporting information architects' decisions.

Introduction

Assessing whether individuals might expect particular objects at specific locations in a web page is a necessary step for effective design. Indeed, interfaces that are designed consistently with the type of organisation the user expects will be likely more accessible, easy to browse, and satisfactory. Although this concern over users' expectations is common among designers, there is a lack of sound theory and methods, leaving this assessment to rather casual approaches.

Research attempts in this direction are sparse, and cannot be considered conclusive. Bernard (2001), for example, asked a large number of subjects to arrange pictures of web objects (internal and external links, advertisement banners, and the like) on a depiction of a browser window, finding regularity in the arrangement of most of them, and no differences in the deployment between Internet experts and novices. Contrarily, Di Nocera, Capponi, and Ferlazzo (2004) found differences due to expertise when analysing users' click responses to verbal labels indicating web objects on a large number of trials. Particularly, expert individuals responded to the verbal labels by clicking in a clearly interpretable spatially ordered fashion, whereas novices showed a more variable and less meaningful pattern. Taken together, these results suggest that expectations about the location of web objects exist and may be founded upon high-level schemata or representations, mainly based on navigation experience. However, these studies could not disentangle the activity of low-level, universally shared schemata and higher level, late effects due to memory or context processing Indeed, both picture placement and clicks may be viewed as just the final outcome of complex processing mechanisms, and might be not well suited to address them with sufficient precision.

Experiment

The Cognitive GeoConcept (CG) procedure was introduced as a method for finding geometrical associations between meaningful objects (links or functions) in web pages. The CG is supposed to elicit users' spatial schemata or representations underling the way people look for information within a web page, and allowing the optimisation of users' behaviours in navigation tasks. Such a prototypical organisation of the information would contain rules and specifications for the location of 1) objects within the page layout, and 2) contents within the site structure. The present study is aimed at generalising those results to more similar objects than those previously used, and at investigating whether eye-movements provide the same information as clicks. Eye-movement data have been frequently used in human factors research (e.g. Kramer & McCarley, 2003), also for investigating mechanisms underlying visual search in complex tasks (e.g. Grahame, Laberge, & Scialfa, 2004; Patrick & James, 2004; Renshaw et al., 2004; Underwood, Jebbett, & Roberts, 2004). Especially, eye-movements to salient areas of the screen can be viewed as a more direct index of the pattern of visual exploration than click or picture placement, as they reflect moment-to-moment the cognitive processes engaged in various tasks (see Rayner, 1998, for a review). Moreover, the present study employs a reduced number of labels respect the ones used by Di Nocera et al. (2004). Indeed, the largely heterogeneous stimuli used in that study might have affected the overall meaningfulness of the patterns, particularly in novice users' performance.

Method

Participants
Eighteen students (13 females) volunteered in this experiment. Their mean age was 22.6 years. Eight subjects reported to use Internet every day, and were classified as experts. Ten subjects classified as novices reported to navigate rarely. All users reported being right-handed, with normal or corrected to normal vision, and were naïve as to the purpose of the experiment.

Stimuli
Ten words indicating links often found in the "Students" menu of Italian academic web sites (*Basic degree, Classes, Dissertation, Erasmus, Exams, Masters, Old degree* -Italy recently switched to a different academic system-, *Schedule, Advanced degree, Stages*) were used as stimuli.

Procedure

Participants sat in front of a 17" computer monitor mounting the Tobii 1750 eye-tracking system, and followed a calibration procedure for the eye-tracking. After this procedure, they received instructions about the task and the meaning of the stimuli (that is what the labels pointed to): they had to respond as quickly as possible to the stimuli by clicking on the area of the (blank) screen where they would expect to find the corresponding link in a web site. This task can be conceptualised as a "blind" navigation in a prototypical web site. Stimuli were presented centrally, white on black, for 250 ms and inter-stimulus interval was 2000 ms. On any trial the mouse pointer returned to the centre of the screen. Fifty repetitions of each stimulus were randomly administered to the subjects. Overall, 500 trials were administered and they were divided in two homogeneous blocks.

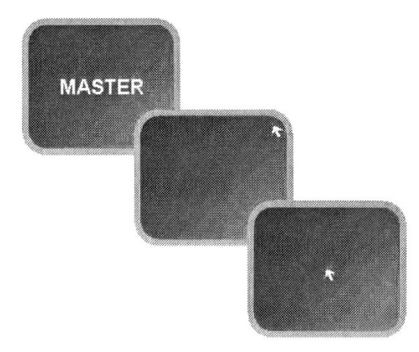

Figure 1. The Cognitive GeoConcept procedure: 1) stimulus administration, 2) subject's click, 3) automatic return of the pointer to the centre of screen.

Analyses

Complete Spatial Randomness (CSR) hypothesis was tested separately for experts'/novices' clicks and eye-movements distributions using the Nearest Neighbour Index (NNI: see Appendix). Median coordinates of both clicks and eye-movements were used as points. The same data were successively analysed using Cluster Analysis (Ward's method: Ward, 1963). Input distance matrices (for experts and novices) were created using average point-to-point Euclidean distances. Positioning responses were further examined using quadrat counts. A 1024 x 768 - 4 x 4 grid (see Table 1) was used to divide the area in 16 quadrats. Click coordinates and the coordinates of the last fixation preceding the clicks were used as measures. Angular transformations ($2 \cdot \sqrt{\arcsin(x)}$ where x is the proportion) of proportion of clicks and (last) fixations within the quadrats were analysed. Angular transformed data were used as proportions cannot be analysed through ANOVA (e.g. Hinkelmann & Kempthorne, 1994). Eye-movements from and towards the centre of the screen were removed from the dataset because central fixations were only related to label reading and were not informative to our aims (links are rarely found in the centre of a web page). Thus, in order to compare clicks and eye-movements, ANOVA designs Expertise (Experts vs. Novices) by Row (1st vs. 4th) by Cell (1st vs. 2nd vs. 3rd vs. 4th) by Measure (clicks vs. fixations), and Expertise by Column (1st vs. 4th) by Cell

(1st vs. 2nd vs. 3rd vs. 4th) by Measure (clicks vs. fixations) were run for each stimulus. Given the high number of tests (twenty: two for each stimulus), the alpha level of each individual test was adjusted using the Bonferroni correction procedure (p<.003).

Table 1. The ideal 4 x 4 grid used to divide the screen area in 16 quadrats from q1 to q16.

q1	q2	q3	q4
q5	q6	q7	q8
q9	q10	q11	q12
q13	q14	q15	q16

Results

Spatial test
Novices' clicks distribution was found to be regularly dispersed (NNI=1.67; p<.0001), whereas CSR test failed to show grouping for experts' clicks distributions (NNI=.92; p>.05). CSR test for fixations distribution showed regularity (NNI=1.15; p<.05) for novices and grouping (NNI=.52; p<.0001) for experts.

In order to assess whether these results were affected by subjects' experience with the stimuli, the NNI was also computed for the two blocks of trials respectively including repetition of the stimuli 1 to 25 and 26 to 50. Figure 6 shows that no critical differences can be observed between the two blocks.

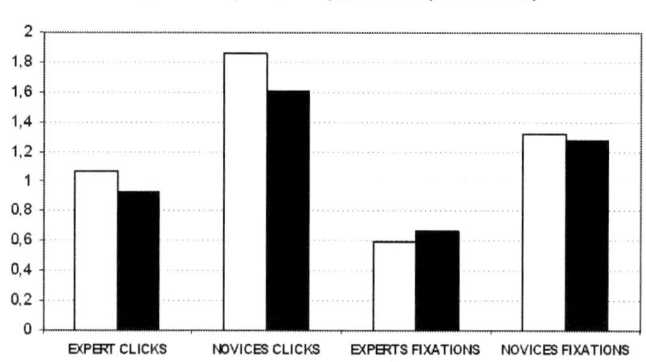

Figure 2. Nearest Neighbour Index values for early (1-25) and late (26-50) trials.

Cluster analysis
Cluster Analysis on clicks showed partially different patterns for the two groups. Experts showed three clusters: "Graduates' Activities" (Masters, Stages and Erasmus), "Degrees" (Basic, Advanced, Old), "Info" (Dissertation, Exams, Classes, Schedule). Novices also showed three clusters, one of which -Info- is similar to the

experts' with the exception that it does not include the label dissertation, whereas the other two are not clearly interpretable. Cluster Analysis on fixations showed the same results. Experts showed the very same clusters, whereas Novices showed two (or four according to the linkage distance considered) not clearly interpretable clusters.

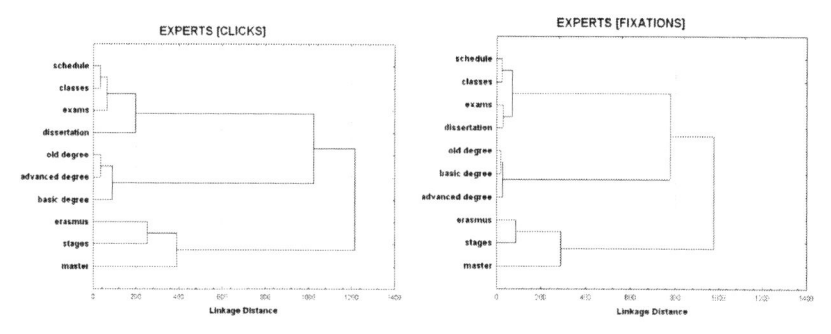

Figure 3. Dendrogram showing Cluster Analysis results for experts' clicks.

Figure 4. Dendrogram showing Cluster Analysis results for experts' fixations.

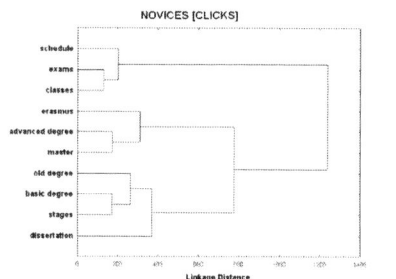

Figure 5. Dendrogram showing Cluster Analysis results for novices' clicks.

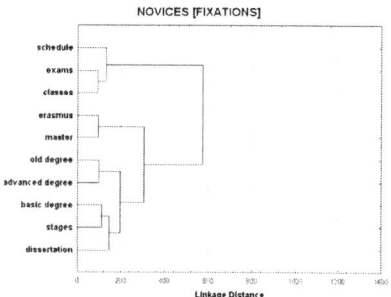

Figure 6. Dendrogram showing Cluster Analysis results for novices' fixations.

Quadrat counts
All the analyses showed a main effect of Measure (p<.001). Clicks were significantly more frequent than fixations. However, this is a bias due to the removal of clicks and fixations from the centre. This led to a specific removal of eye-movements towards the centre of the screen that are more likely to happen rather than clicks (i.e., because subjects wait for the next target appearing in that position).

The Expertise x Row x Cell x Measure Analysis of Variance showed a main effect of the Cell factor for "Exams" ($F_{1,16}$=6.93; p<0.001). Clicks and fixations were more frequent in uppermost-leftmost (q1) and in the lowermost-leftmost (q13) quadrats. A main effect of the Row factor for "Schedule" ($F_{1,16}$=71.45; p<0.001) was found. Clicks and fixations were more frequent in upper row rather than in the lower row. A statistically significant interaction between Cell and Row was found for "Classes" ($F_{3,48}$=11.04; p<0.001). Duncan testing showed that clicks and fixations were equally distributed on the lower row, but were more frequent in the upper row, particularly in the uppermost-leftmost (q1) quadrat. A statistically significant interaction between Cell and Measure was found for "Stages" ($F_{3,48}$=7.34; p<0.001). Duncan testing showed that fixations were equally distributed on all the cells, but clicks were more frequent in corners (q1, q4, q13, q16).

The Expertise x Column x Cell x Measure Analysis of Variance showed a main effect of the Cell factor for "Advanced degree" ($F_{3,48}$=19.35; p<.001), "Old degree" ($F_{3,48}$=9.28; p<.001), "Dissertation" ($F_{3,48}$=14.24; p<.001), "Exams" ($F_{3,48}$=12.56; p<.001), "Stages" ($F_{3,48}$=18.69; p<.001), and "Schedule" ($F_{3,48}$=19.44; p<.001). The stimuli "Advanced degree", "Old degree", "Dissertation", and "Stages" elicited more clicks and fixations in the central part of both columns, whereas "Exams" elicited a similar pattern limited to the upper cells of the middle portion of both columns (q5 and q8). Additionally, "Schedule" elicited clicks and fixations more often in the upper half of the two columns (q1, q4, q5, q8). A main effect of Columns was found for "Exams" ($F_{1,16}$=22.31; p<.001). Clicks and fixations were more frequent in the left rather than in the right column. A significant Column by Cell interaction was found for "Classes" ($F_{3,48}$ =7,85; p<.001). Duncan testing showed that clicks and fixations were uniformly distributed on the cells belonging to the rightmost column, but they were more frequent in the upper-half of the left column (q1 and q5). Finally, a statistically significant interaction Cell x Measure x Expertise was found for "Basic degree" ($F=_{3,48}$=11,13; p<.001). Duncan testing showed that clicks were significantly higher than fixations for both experts an novices (see above for a description of this effect). However, novices fixate and click more often on the central part of both columns (q5, q8, q9, q12), whereas experts show a similar pattern including a higher frequency of clicks in q9 and q12.

Maps depicting the frequency of clicks and fixations for each stimulus are available at http://extras.hfes-europe.org.

Discussion and Conclusions

This study was aimed at providing support to the hypothesis that both spatially- and semantically-based schemata, as well as schemata based on navigation experience, underlie the cognitive organisation for web pages layout. The most important innovation in this study compared to the previous one (Di Nocera et al., 2004) was the use of eye-movements as an additional source of information. CSR tests indicated that fixations were better suited to differentiate experts' patterns from novices'. Semantic grouping for experts was indeed found. Moreover, computing the NNI on two blocks of trials (from 1 to 25 and from 26 to 50) showed that the results obtained on the whole dataset do not vary along time, suggesting that these effects do not depend on the experience with the procedure (that is, subjects do not "build up" their representation on the basis of stimuli repetition). Rather, such effects would be due to schemata that exist prior to engage the task.

Cluster Analyses results also provided support to the idea of differential groupings in experts and novices. Particularly, experts tried to organise the stimuli according to a typical interface deployment (i.e., the three types of degree were grouped), whereas novices seemed to organise stimuli according to personal criteria that got lost in the overall pattern. ANOVAs results failed to show differential spatial organisations except for a limited number of stimuli. A preference towards the middle part of the lateral portion of the screen was found, and the upper row and the leftmost column where the most important attractors for stimuli such as "Exams", "Schedule", "Classes". In general, only small differences between experts' and novices' spatial deployments were found. However, one should consider that in this case comparisons were made using an arbitrary number of cells distributed on the overall screen area, which is wider than the space actually used by subjects (some of the areas defined by the cells received few -if any- clicks and fixations). With that in mind, it is possible to conclude that quadrat counts is not a suitable way to analyse this type of data. This is a useful lesson learned for successive studies using the present procedure.

Concluding, these preliminary results confirm the usefulness of the procedure as a technique for eventually supporting information architects' decisions. Indeed, the CG may be used not only for gathering information about the preferred location of links, but also as a spatial equivalent of the card sorting technique (limited to the screen space). Euclidean distances are indeed direct indicators of the relationship between words, concepts, and the like. Moreover, the use of statistical tools (such as Cluster Analysis) is strongly facilitated by a matrix of real distances. Eventually, guidelines may be derived from the CG procedure. However, at this time it is not possible to gather clear indications for design. More research is needed, and the procedure itself should be tested with a wider sample, and with different stimuli. Nevertheless, these preliminary results indicate that the Cognitive GeoConcept is a candidate method for investigating the role of spatial schemata in web navigation.

References

Bernard, M.L. (2001). Developing schemas for the location of common web objects. *Proceedings of the Human Factors and Ergonomics Society 45th Annual Meeting*, *1*, 1161-1165.

Clark, P.J., & Evans F.C. (1954). Distance to nearest neighbor as a measure of spatial relationships in populations. *Ecology*, *35*, 445-453.

Di Nocera, F., Capponi, C., & Ferlazzo, F. (2004). Finding geometrical associations between meaningful objects in the web: a geostatistical approach. *PsychNology*. 2(1), 84-98.

Grahame, M., Laberge, J., & Scialfa, C. T. (2004). Age differences in search of web pages: the effects of link size, link number, and clutter. *Human Factors*, *46*, 385-398.

Hinkelmann, K. & Kempthorne, O. (1994). *Design and Analysis of Experiments, Volume 1*. NewYork: Wiley & Sons.

Kramer, A. F., & McCarley, J. S. (2003). Oculomotor behaviour as a reflection of attention and memory processes: Neural mechanisms and applications to human factors. *Theoretical Issues in Ergonomics Science, 4*, 21-55.

Patrick, J., & James, N. (2004). Process tracing of complex cognitive work tasks. *Journal of Occupational & Organizational Psychology, 77*, 259-280.

Rayner, K. (1998). Eye movements in reading and information processing: 20 years of research. *Psychological Bulletin*, 124, 372-422.

Renshaw, J.A., Finlay, J.E., Tyfa, D., & Ward, R.D. (2004). Understanding visual influence in graph design through temporal and spatial eye movement characteristics. *Interacting with Computers, 16*, 557-578.

Underwood, G., Jebbett, L., & Roberts, K. (2004). Inspecting pictures for information to verify a sentence: Eye movements in general encoding and in focused search. *Quarterly Journal of Experimental Psychology: Human Experimental Psychology, 57A*, 165-182.

Ward, J. H. (1963). Hierarchical grouping to optimize an objective function. *Journal of the American Statistical Association, 58*, 236.

Appendix

The Nearest Neighbour Index (Clark & Evans, 1954) is one of the most widely used distance statistics. Computing it is very easy, and many other distance statistics are founded on it. As a first step, the nearest neighbour distance or d(NN) should be computed as follows:

$$d(NN) = \sum_{i=1}^{N} \left[\min \frac{(d_{ij})}{N} \right]$$

where min(dij) is the distance between each point and the point nearest to it, and N is the number of points in the distribution.

This index is nothing more than the average of the minimum distances. The second step is to compute the mean random distance or d(ran), that is the d(NN) one would expect if the distribution were random.

$$d(ran) = 0.5\sqrt{\frac{A}{N}}$$

where A is the area of the region (the measurement unit of the index is related to the one used here), and N is the number of points.

The final step is the actual computation of the Nearest Neighbour Index as follows:

$$NNI = \frac{d(NN)}{d(ran)}$$

This ratio is equal to 1 when the distribution is random. Values lower than 1 suggest grouping, whereas values higher than 1 suggest regularity (i.e. the point pattern is dispersed in a non-random way

Risk assessment in industry and offices

Piia Tint & Karin Reinhold
Tallinn Technical University
Tallinn, Estonia

Abstract

The working conditions in textile and wood processing industries and in offices have been investigated using a simple risk assessment method (Tint & Kiivet, 2003). The main complaints in textile industry are high temperature in the workroom, bad ventilation, intense work and the dependence of workers' work results from the others. The main risk factors in wood processing industry are tools and equipment, heavy physical load, noise, wood dust and odours of chemicals originating from polishes. The working conditions in offices differ greatly in winter and in summer time. In summer the main complaints are high air temperature and sharp sunlight from windows, noise from the streets; but in winter the dry air and odours from the new type of floorings are the most disturbing problems for workers.

Introduction

Risk assessment in the work environment has been the topic for the Estonian researchers in work safety and health from 1996, when the EU document "Guidance of risk assessment at work" became accessible. The Estonian Occupational Health and Safety Act (on the basis of EU Dir. 89/391), which demands risk assessment at every workplace, was adopted in Estonia in 1999. In this context the main problem for managers has been finding a suitable risk assessment method. Labour inspectors are not satisfied with the majority of risk assessments carried out by employers, but they cannot improve the situation, as they have no better proposals. Considering the situation in Estonian labour market, the Labour Inspectorate of Estonia as the main institution dealing with practical risk assessment at workplaces, thinks that there have to be two different types of risk assessment methods, one for industrial activities and the other for offices. It seems that the latter might be easier but in this field different new hazards have arisen, like electromagnetic fields from mobile phones or odours of chemical materials used nowadays in offices, schools or hotels as flooring materials or by cleaning firms. So there can be problems in working conditions as in industrial rooms as in offices. The main distressing problems in industrial rooms are noise, dust, and lack of air, draught.

Method

The used in the investigation simple/flexible risk assessment method (Tint & Kiivet, 2003) is based on a two-step model (Figure 1) that could be enlarged into a six-step

In D. de Waard, K.A. Brookhuis, R. van Egmond, and Th. Boersema (Eds.) (2005), *Human Factors in Design, Safety, and Management* (pp. 129 - 133). Maastricht, the Netherlands: Shaker Publishing.

model (Figure 2). The two-step model has one boundary (redline), which is a stable, widely accepted figure such as a norm or standard. The no/yes principle is used or corresponds to the norms/does not correspond to the norms or justified/unjustified risk. The model suits to small enterprises and to these that have not a complicated combination of hazards or have rather inexperienced personnel (also in work safety). The two-step model could be enlarged to the right side and also to the left. So three, four and five-step models form. The enterprise can choose the model that considers best the scale of hazards. The six-step model is shown in Figure 2.

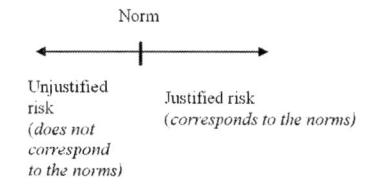

Figure 1. Two-step model for risk assessment in the work environment

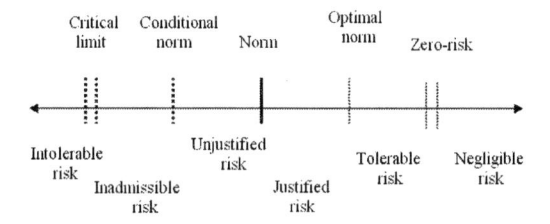

Figure 2. Six-step model for the risk assessment in the work environment

Risk assessment in textile industry

Clothing industry plays an essential role in Estonian manufacturing providing 4.1% of the total output of Estonian industry leaving behind only from food and wood products, textiles, chemicals, metal and non-metal products manufacturing. In October 2001, there were 73 companies registered in clothing industry in Estonia (70.9% of produced garments were exported). This means that many foreign companies (from Finland, Sweden, Great Britain, Italy, Germany etc) use Estonian clothing companies for subcontracting work because of the cheaper labour force. The investigated textile company employs 300 people. Its main production is work clothes (jackets, trousers, smocks, overalls, winter clothes and specific work clothes). Of the production, 85% is for subcontracting work to Finland.

Repetitive motions and awkward, uncomfortable working positions often characterize work in the clothing industry. As a result, garment workers are among the highest risk occupational groups for ergonomically related disabilities. Besides that, there are many other hazards for the health of garment workers – noise, poor lighting, dust etc. The psychological side can't be underestimated either, because sewers often face a very monotonous work and a constant time pressure (their wage is calculated by the price of certain operation minute). The hazardous chemicals used in textile industry are: formaldehyde, dust of textiles etc. The contamination of hazardous gases and particles in the air of work environment was investigated. The most common microclimate problem in the controlled companies was insufficient ventilation. The main disturbing factors were dust, lack of air, draught. It is especially problematic in summer when the doors and windows are opened to get more air which causes general sicknesses (bronchitis) as well as sicknesses caused by compulsory work posture (overload sicknesses). In most of the checked companies, there was no risk analysis carried out which means there was no certain plan framed to improve work conditions. The average noise level in the sewing department was between 67, 0…80, 0 dB (A). The lighting in the sewing department was sufficient and good. There were general lights as well as tasks lights. Most work places were well lighted with no shadows created. The graphical risk assessment is shown in the Figure 3.

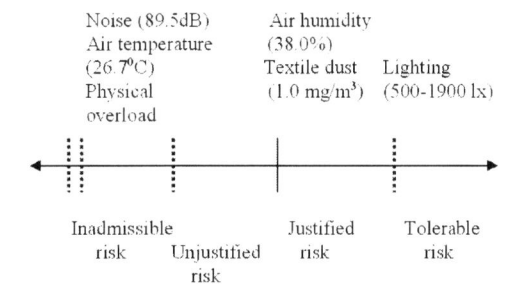

Noise (89.5dB)   Air humidity
Air temperature (38.0%)
(26.7°C)   Textile dust   Lighting
Physical   (1.0 mg/m^3) (500-1900 lx)
overload

```
  :::      :        :         :
←――――+――――――+―――――――+―――――――→
  :::      :        |         :
```

Inadmissible   Justified   Tolerable
risk   Unjustified risk   risk
          risk

Figure 3. Assessment of working conditions using simple risk assessment method in textile industry

Risk assessment in wood-processing industry

The work environment in a large wood-processing firm (1000 workers) – in a medium-sized town in Estonia was analyzed. The list of hazards was compiled before the investigation by the work environment specialist of the firm.

The main risk factors in that kind of industry are tools and equipment, also heavy physical load (moving the wheelbarrow), noise, wood dust and in some places odours of chemicals (mostly formaldehyde) originating from polishes.

The measurements of the hazards were carried out in the department were polishing and varnishing take place. A simple/flexible risk assessment method, worked out by the researchers of Tallinn Technical University was used for determination of safety level at the enterprise (Figure 4).

The microclimate in the wood-processing department was rather good (considering that there is room for improvement by raising moisture content of the air). The safety of machines has to be taken into consideration when buying new equipment. Experience shows that even the machines with CE-mark can be sources for traumas. Noise was above the limits (85 dB) in every workplace, but breaks were taken and earmuffs were used. So the total amount of noise during an 8-hour workday is not over the permissible level (dose: 85dB x 8h). The phenol-formaldehyde varnish is a source for allergic reactions in workers. From the view-point of possibility of accidents /traumas originating from machines it was declared that one protective metallic covering component had been removed and afterwards substituted by a cardboard for protection against cutting wounds of fingers. The last type of accidents predominates in the Estonian work traumas spectra nowadays (~500 cutting traumas of fingers a year, including amputations).

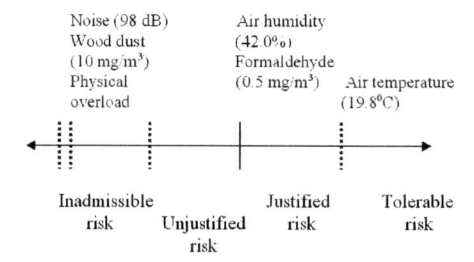

```
              Noise (98 dB)         Air humidity
              Wood dust             (42.0%)
              (10 mg/m³)            Formaldehyde
              Physical              (0.5 mg/m³)    Air temperature
              overload                             (19.8°C)

        ←─────┼┼┼─────────┼─────────┼─────────┼─────→

              Inadmissible          Justified    Tolerable
              risk       Unjustified risk         risk
                         risk
```

Figure 4. Assessment of working conditions in wood processing industry

Risk assessment in offices

One part of the investigated offices belongs to the car-selling companies (160 workers), the other part to the educational institutions (kindergartens, schools, also Tallinn Technical University included, 300 workers). The working conditions differ very much in winter and in summer in offices, where workers have to sit by their computers all the workday. Bad knowledge of ergonomics with displays is one of the most distressing problems in new, just renovated office-rooms. The other problems in renovated office-rooms are: not-suitably regulated ventilation and lighting, bad microclimate particularly in summer, noise. These problems have not been considered during the planning of renovating work. In non-renovated office-rooms the problems are: low temperature of the air, ventilation ducts not cleaned and so ventilation not working, lighting non-sufficient, bad ergonomics.

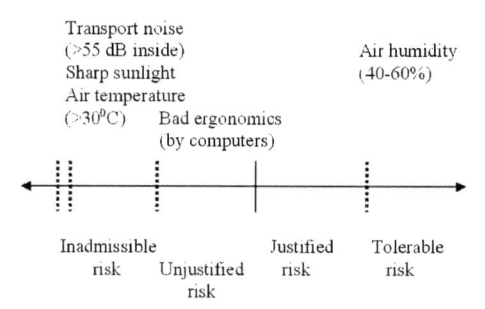

```
              Transport noise
              (>55 dB inside)
              Sharp sunlight        Air humidity
              Air temperature       (40-60%)
              (>30°C)    Bad ergonomics
                         (by computers)

        ←─────┼┼┼─────────┼─────────┼─────────┼─────→

              Inadmissible          Justified    Tolerable
              risk       Unjustified risk         risk
                         risk
```

Figure 5. Assessment of working conditions in office-rooms in summertime

The improvement of working conditions has to be considered when planning the renovating work; otherwise the problems from the side of workers will arise more strongly than they were before the repair-work. With the rearrangement of workplaces the problems could be solved; also the age of worker has to be considered. The ageing personnel have totally different problems than the young ones. The five-stage simple/flexible risk assessment model was used for the assessment of the working conditions in offices in summer (Figure 5) and in winter (Figure 6).

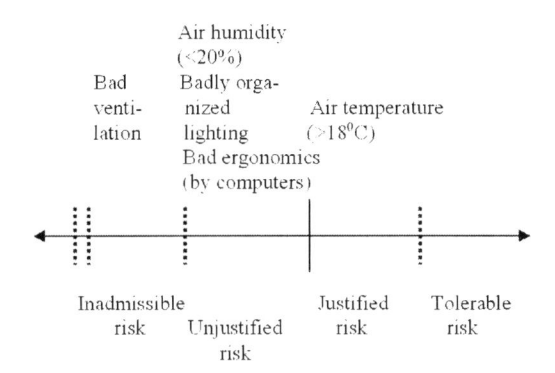

Figure 6. Assessment of working conditions in office-rooms in winter season

Conclusions

The simple/flexible risk assessment method can effectively be used as in industrial as in office-rooms. The graphical solutions give clear picture about the level of hazardous factors in the work environment. Comparison with the existing exposure limits can be carried out. As the problems at firms are complex, not everything is solved during the risk analysis carried out by the accredited laboratories or occupational health services in Estonia. The process of risk assessment is continuous; all the hazardous factors do not appear during the short time the risk analysts spend at firms. The co-operation with the local safety and health personnel (work environment specialists, ergonomists, and occupational health doctors) is much needed.

References

Tint P., & Kiivet, G. (2003). A Simple and Flexible Risk Assessment Method in the Work Environment. *International Journal of Occupational Safety and Ergonomics, 9(2)*, 237-248.

Improvement of risk management in the marble industry

Abdelaziz Tairi & Ahmed Cherifi
University of Boumerdes, Algeria

Abstract

Ergonomic aspects of the marble enterprise are presented with a focus on improving performance. The general context of the national economy and the new judicial environment of enterprises will have serious repercussions on missions of the enterprise. Activities throughout different technological processes applied in the quarry blocks, in the quarry stone and in the marble's transformation factory have been analysed. Thanks to this approach, the different factors influencing safety and productivity were observed. This permitted to establish the required measures for a better organisation and a new prevention policy.

Introduction

In a number of visited divisions the work conditions were observed with respect to quarry blocks, quarry stone, marble transformation factory and curio manufacturing (Grandjean, 1983). A first focus was on physiology, the protection of the physical integrity of persons at work. An evaluation of the average concentration of marble dust concerning some posts, made by a team of pneumo-phtisiologists in a study about the silicosis in Algeria (SAPPMT, 1988), showed that the admissible norms are widely exceeded in quarry stone and in the factory. The second focus is on psycho-sociology; the quality of life of man at work. The third focus, often undervalued, is of economic, with aim not only to increase the enterprises' adaptation capacities to face the fluctuating market, but also to decrease social costs linked to absenteeism and to work accidents. This also improves work quality and efficiency.

The subsystem "production" includes two major phases in the cycle (CME, 1990): Extraction - Transformation. This cycle begins with quarrying until curio production involving the following structural units: quarry blocks, quarry stone, and marble's transformation factory with curio production. Quarry blocks are a monolithic blocks with dimensions 150x150 by 90 meters of depth delimited in four intervention zones or fronts.

In the quarry stone, blocks are obtained after drilling and explosive shooting. They are cut up and transported to crushing stations. Stations treat the blocks stemmed from the cutting up in three products: granules (of several dimensions), dusts and the powder of marble (Lugdunum, 1999). In the factory, the cycle of marble transformation is realised in three stages: the sawing of unpolished blocks into semi-finished slabs, cutting up of slabs into finished products, and polishing and lustring of the finished products.

Methodology

The ergonomic intervention consisted of analysing the whole technological system through elements such as work posts, work conditions and motivation factors, social element and culture of the enterprise, management level, qualification and training (Tairi, 2003), and work accidents' statistics (Tairi, 1986).

Results

It has been raised in this study that a great number of the personnel working in crushing stations, in which marble blocks are transformed into gravel, dusts and powder, often suffer from throat irritations and respiratory difficulties. The free silica presence (SiO_2) in the dust of marble exposes operators to the risk of silicosis. Furthermore, the absence of automatic sacking, of dust collectors as well as of a work physician worsens the health situation of operators. The operators' declarations in the interviews confirmed the excess of the dust level as they suffered from respiratory irritations. In some workshops where humid processes are used, the dust danger is minor but on the other hand, the stagnation of water provokes falls of persons and frequent short circuits that delay the production. The high noise level is easily observable without measures. The sources of this high noise are compressors, perforators, explosions, tools-marble shocks, etc. Discomfort due to the noise was mentioned by some operators during interviews, but there auditory deficits were not reported, as there was no work physician on site. Furthermore, in some workshops the vapour of cooling water and the cloud of dust lead to the visibility weakness. Results presented in Tables 1, 2 and 3 are averages over five years. Table 1 shows that most of the accidents happen in the transformation factory.

Table 1. Repartition of frequency rate by divisions

Divisions	Quarry blocks	Quarry stone	Factory
Frequency Rate	82.4	67.7	175.7

As shown in Table 2, the most concerned parts of the body are hands and feet. This is explained by handling and by frequent displacements of blocks and fissured slabs, often done without individual protection.

Table 2. Repartition of accidents by places of lesion

Place of lesion	head	eyes	arms	hands	legs	feet	back	Total
Quarry blocks	2	1	2	6	1	11	4	27
Quarry stone	1	1	0	3	0	3	1	9
Factory	2	2	4	29	1	15	5	58
Total	5	4	6	38	2	29	10	94

The most frequent accidents are provoked by falls of objects and by falls of persons as shown in Table 3.

Table 3. Repartition of accidents by causes

Causes	% of accidents
Objects' fall	37.8
Persons' fall	21.6
Handling	10.4
Shocks	12.8
Others	17.4

Conclusions

To improve the management of the risk, it is necessary to establish a more rational and more optimum of marble blocks cutting by taking into account the positioning of fissures. Besides, marble blocks have to be normalised and classified according to their quality and their dimensions into several categories. On the other hand, a better maintenance of equipments and a better management of spare parts should be ensured. In addition to this, an automatic sacking of marble powder, dust collectors, and a central of automatic distribution of sand and water should be installed. Concerning the operators, they should be more sensitised on safety and provided with adequate means of individual protection. Moreover, a training and improvement program should be established to reach an indispensable up level as well as to adapt them to the technological evolution and to the new methods of management.

To conclude, if taken into account, the previous measures will eliminate accidents provoked by falls of blocks and slab pieces and will allow an increase in volume of the production of the finished products. Besides, it will protect the health of operators and improve productivity.

References

CME (1990). East Marble Complex (1990). Activity report. East region: Documentation Service.
Grandjean, E. (1983). *Précis d'ergonomie.* (pp. 395-399). Paris: Editions d'Organisation.
Lugdunum, B. (1999). Concassage en douceur. *Travail et Sécurité, 585,* 36-37.
SAPPMT (1988). Sociétés Algériennes de Pneumophtisiologie et de Médecine du Travail. *La silicose en Algérie* (Report 1988). Alger: Office des Publications Universitaires.
Tairi, A. (1986). Dispersional analysis of accidents' causes. *Safety and Environment Protection, 1,* 12-16.
Tairi, A. (2003). The integration of the ergonomic practice in the enterprise by the training students. *Proceedings of the Seventh International Symposium on Human Factors in Organizational Design and Management,* Aachen. (pp. 637-641). Santa Monica: IEA Press.

Prevention of Repetitive Strain Injuries (RSI) at Delft University of Technology

Marijke C. Dekker[1] & Kineke Festen-Hoff[2]
[1]Faculty of Industrial Design Engineering
[2]Faculty of Policy, Technology and Management
Delft University of Technology, Delft
The Netherlands

Abstract

RSI (especially non-specific RSI) is often classified as a syndrome because of the uncertainty of the cause and the variety of symptoms. Therefore, combined with a lack of injury evidence, RSI is a medically disputed phenomenon, not-withstanding the very serious human and financial consequences. Regrettably, there is still a great lack of knowledge about effective RSI prevention. A multi-disciplinary RSI prevention group was set up at the Faculty of Industrial Design Engineering (IDE) at Delft University of Technology. This working group, supported by the Educational Director of that faculty, organised various prevention activities for students and –to a smaller extent– for universal employees on a yearly basis. The prevention programme will be presented. In addition, surveys amongst IDE students were held to determine whether the group of students with RSI was increasing or diminishing and to establish the nature of the complaints. Although no direct relation between complaints and various prevention activities could be found in the presented surveys between 1999 and 2002, some tendencies will be reported.

Introduction

The RSI phenomenon

Repetitive Strain Injury (RSI) is a medical syndrome affecting the neck, upper back, shoulders, arm, wrist or hand, or a combination of these areas. The symptoms – tingling, numbness, stiffness, pain, loss of strength, and loss of motor function– are preceded by activities that involve repeated movements of arms or hands, and require keeping some body parts in a static position. There is indication that precision demands and mental pressure contribute to the occurrence of complaints (Visser, 2004). Most RSI cases are non-specific, this means no diagnosis can be made and there is no proof of any tissue damage. The scientific discussion about the origination of RSI leads to very diverse insights and hypotheses. There are multiple possible mechanisms, but none of the hypotheses forms a complete explanation and is sufficiently supported by empirical data (Visser, 2004). Furthermore, most RSI prevention is hardly supported by scientific knowledge.

In D. de Waard, K.A. Brookhuis, R. van Egmond, and Th. Boersema (Eds.) (2005), *Human Factors in Design, Safety, and Management* (pp. 139 - 152). Maastricht, the Netherlands: Shaker Publishing.

With the increased use of the computer in the last decades, RSI is mostly associated with visual display unit (VDU) work. But also in industry, various professional groups such as hairdressers and poultry workers are dealing with repetitive movements. For all these workers, RSI is a health problem leading to –sometimes chronic– complaints, participation problems at work and home activities, and in a few cases even disability. Research based on self-reported data shows a wide range of prevalence rates for RSI (20-40%) through occupational sectors ("Health Council of the Netherlands: RSI", 2000). In 2001, the percentage of new disability pension entries caused by RSI through all occupational sectors in the Netherlands is 6% – which is 6000 people (Bongers, 2003). The industry has relatively the highest number of disability pension entries. Still, the absolute amount of people entering disability pension is highest in the administrative sector.

Another group at risk is the student population. The intensity of VDU work is high and the work is mentally demanding. Furthermore, workplaces at the university vary from course to course, and student's workplaces at home are intensively used. Therefore, their situation is different from professionals. Universities are obliged to observe most regulations of the Health and Safety at Work and Environment Act with regard to their students, although the legal position of students is slightly different.

The Delft University of Technology and specifically the Faculty of Industrial Design Engineering (IDE) of this Dutch university is an interesting research area in this perspective. The working hours per week –40.9 hours– are the highest compared to other disciplines[*], computer use is high, and the work is mentally demanding because of the nature of designing.

RSI at Delft University of Technology

In 1998 the first students with RSI complaints due to VDU work approached the student advisors of their Delft University of Technology faculties to discuss the impact of these problems on their study progress. In the following academic years (1999-2003) a total of 274 students sought medical advice in relation to RSI complaints at the Students' Health Service (SGZ). IDE had the highest percentage of students consulting the SGZ (Figure 1) with 1.2 % (in comparison to 0.7 % of the Architecture students). In absolute numbers most new RSI cases came from the Faculty of Architecture with an average of 23 per year (in comparison to 18 cases at IDE). As can be noticed from Figure 1, the amount of new RSI students from the faculties IDE and Architecture consulting the SGZ because of RSI complaints seems to diminish gradually.

During these four years, about 43 % (118 students) reported such serious complaints, that after the physical check-up following a RSI protocol, the SGZ issued a medical

---

[*] Choice, centre higher education information for consumer and expert, Leiden, the Netherlands, 2004

statement. This paper indicated the reduced study load –for instance 100% reduction for some months or 50% for a period of one and a half year. The total time accumulated by the medical statements of these 118 students of the various Delft University of Technology faculties was 694 months (215 months by IDE and 302 by Architecture students). With this statement, the students had a stronger position in discussing the possibilities for a –temporarily– adjusted study programme with their student advisor, and a financial compensation for their RSI-related study delay. It should be noted that this study delay was in general longer than the time students were incapable of studying as indicated by the medical statements, because a hindrance of a week during the exam period could result in a study delay of a couple of months. The financial compensation for the accrued study delay, ranging from person to person from several months through to more than a year, was paid by the University or –in severe cases– by the Central Bureau for Dutch Study Grants.

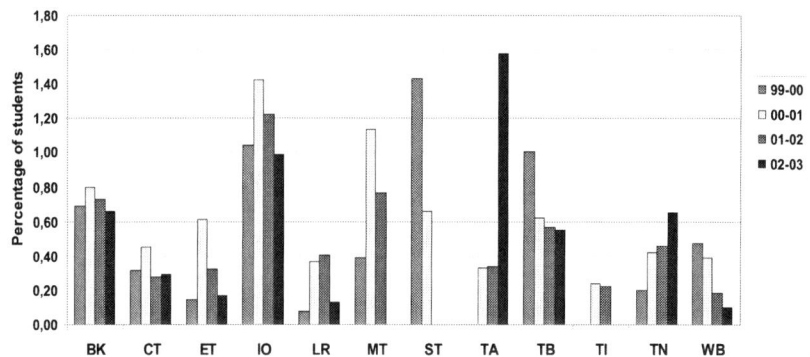

Figure 1. Percentage newly reported students with RSI per discipline at Delft University of Technology, BK = Architecture, CT = Civil Engineering, ET = Electrical Engineering, IO = Industrial Design Engineering, LR = Aerospace Engineering, MT = Marine Technology, ST = Sustainable Molecular Science & Technology, TA = Applied Earth Science, TB = Technology, Policy and Management, TI = Computer Science, TN = Applied Physics, WB = Applied Mathematics, WT = Mechanical Engineering

Of course not all students with RSI complaints contacted the Students' Health Service –an unknown amount contacted their family doctor or other medical or alternative sources– so the total numbers and percentages of students with complaints are possibly much higher than indicated. But the more serious cases were probably all registered by the SGZ, because only they are authorised to issue the medical statements. A further 11 students dropped out of their studies due to RSI-related problems through these years (of which 3 IDE and 7 Architecture students). They stopped studying, started another study or restarted their original study after some time.

Economic consequences

Apart from the human consequences, there was also financial loss due to students' incapability to study in a regular pace and the cases of study termination. The government and Delft University of Technology together compensated with extra study grants for about 270.000 Euro. In addition, the university missed bonuses from the government –estimated at 80.000 Euro– because of the few students who dropped out of their studies and never got a Delft University of Technology university degree. The students incurred financial losses due to their RSI problems as well. The financial consequences of their extra study costs and delayed career, resulted in an average individual loss of 20.000 Euro.

RSI prevention at the Faculty of Industrial Design Engineering 2000-2004

In response to the disturbing figures mentioned earlier, a working group on RSI prevention for IDE students was established by the Educational Director at the Faculty of Industrial Design Engineering in October 2000. The working group is made up of members of the Students' Health Service (SGZ), members of the Health and Safety at Work and Environment group of the Delft University of Technology, a student advisor and a student member of the Educational Management of IDE, as well a member of the section Applied Ergonomics and Design of the same faculty. The main goal of this working group is to reduce RSI amongst students by preventing them as much as possible from getting (serious) complaints. This will be actualised through informing students and –to a smaller extent– university employees about possible risk factors, training them to recognise RSI-related problems in an early stage and to react quickly towards beginning symptoms. Additionally the working group intents to reduce the fear and uncertainty– and therewith the chance of complaints getting worse– of students who experience more serious symptoms, by giving consistent information and a structure for what to do.

The creation of this prevention programme started in 2000. In science, there was still great uncertainty as to the risk factors and the effectiveness of prevention measures. Therefore, the programme was strongly based on the knowledge of the working group members such as the medical and practical experience of the SGZ director and the more theoretical and legal RSI knowledge of other working group members.

Many activities in the period 2000-2004 focused on making the risk of RSI public by means of posters, brochures, an informative website on RSI www.io.tudelft.nl/rsi, and lectures for students and teachers. Others concerned the incorporation of practical exercises in relation to RSI at several points in the IDE curriculum. In the first study year, following an introduction to RSI, new students get instructions on how to customise their workplace, practical exercises on working postures, and they get exercises for relaxation at the workplace. Third year students learn skills for time management and relaxation and are informed about their general health. Sport and relaxation workshops are organised for IDE students and employees of all years to create awareness for RSI and to provide RSI prevention tools.

Apart from disseminating RSI information and providing students tools for prevention, the working group concentrates on the reduction of risk factors within the IDE study, such as improving university workplaces and computer devices. Bulk discount programmes for office chairs and tables are organised for the home workplace as well. Lecturers are informed on 'How to deal with students with RSI complaints', because they are involved in the assessment of student work and the timing of delivery and exams. In the near future, the working group will examine the study programme to recognise peaks in the study load and in computer use. Suggestions will be made to the Educational Management for an improved distribution of these courses in future curricula. Furthermore, the working group set up an ongoing assessment by means of a biennial survey to monitor RSI amongst IDE students.

State-of-the-art knowledge on RSI risk factors and prevention

In these past 5 years that the IDE prevention programme was running, two inventories were made on behalf of the Dutch Ministry of Social Affairs and Labour as well as the Ministry of Health, Welfare and Sport about the state-of-the-art knowledge on RSI risk factors and effective prevention measures. In 2000, the Dutch Health Council ("Health Council of the Netherlands: RSI", 2000) drew the following conclusion on risk factors from a literature study on RSI:

> *"Risk factors associated with RSI include excessive use of force, working in awkward positions, working continuously in the same position (static strain) and repeated movements. Psychosocial occupational factors do not themselves lead to RSI problems, but can exacerbate physical factors. Insufficient opportunities for recovery, psychological strain (extreme pressure of work, high levels of stress, high working tempo, mentally demanding work) and inadequate social support are probably significant."*

Moreover, it went on to state the following about prevention:

> *"At present, scarcely any data are available on the effectiveness of particular preventive policies –even those that are in (wide spread) use"*

and*:*

> *"Integrated preventive strategies that address all risk factors are likely to be most effective."*

About prevention of complaints becoming ever more serious it stated:

> *"It is very important that people consulting their GP or company doctor for early symptoms of RSI receive consistent advice. In early stages, particular emphasis should be placed on information and reassurance."*

Recently the two abovementioned Dutch Ministries commissioned a second study on neck and upper limb disorders (RSI) including a determination of which preventive measures for RSI have already proved effective and the state-of-the-art concerning RSI risk factors (Blatter et al., 2004). It concluded about risk factors:

> *"High frequent movements of the arm, certainly with force exertion, increase the risk on RSI drastically. The risk in VDU work seems to be increased in case of prolonged working."*

With regard to stress and workload, it stated that several recent studies produced reasonable evidence for stress and workload being risk factors for RSI. Limited evidence was found in personal risk factors, such as personality, perfectionism, sport practising and movement– a.o. it was found that physical activity in the form of sports, lead to a reduced risk and being strongly committed to the work, lead to an improved risk of RSI complaints. The study concluded about the effectivity of preventive measures that high quality research is still lacking. Only some tentative conclusions can be drawn that training of ergonomic skills[*] seems to diminish the incidence of RSI complaints. In this second study, only limited evidence was found on the effect of office tables and chairs with arm support to reduce complaints in arm, neck and shoulders. Although stress-related interventions such as relaxation training and time management courses could be effective, evidence is lacking. Additionally measures to diversify tasks and to provide time to recover might also contribute to prevention.

Evaluation of the IDE RSI prevention programme

Comparing the IDE RSI prevention programme with the recommendations of both previous studies, the conclusion can be drawn that they have a lot in common. For instance, the working group on RSI prevention also valued the importance of consistent information and advice, resulting in easily accessible and unambiguous information sources. And the risk awareness of the combination of intense and prolonged VDU work together with psychological strain resulting from work pressure and mentally demanding tasks in the curriculum, was a central theme in the prevention programme. The emphasis was not only on the awareness but also on the creation of tools for students to reduce or avoid extreme workload and stress. Examples are the relaxation exercises and time management courses. Even suggestions on how to diversify tasks during work and to provide time for recovery after hard working are part of the prevention programme. The sport and relaxation promotion is based on the same conviction that change of body load is necessary, especially when working long hours with computers. Training of ergonomic skills and assistance on workplace optimisation at the university and at home are included already from the start of the programme.

Methods

Longitudinal survey

A longitudinal survey was set up by the working group to observe the trend, of the number of IDE students suffering from RSI and the seriousness of their complaints.

---

[*] Ergonomic skills implied, amongst others, training to reduce neck rotations and extreme postures of the wrists and to improve the workplace and devices.

Is the problem increasing, stabilising or reducing? In addition, an attempt will be made to gain more insight in the risk factors and to what extent the preventive measures taken by the working group had effect. These outcomes will help to improve the future prevention programme.

Surveys to monitor RSI amongst students at the Faculty of Industrial Design Engineering

Until today, two surveys were performed. From 2000 on, the students, as part of the first and third year RSI information sessions and instructions, manually filled in a short questionnaire on a voluntary basis. This resulted in the studies 'Students and RSI 2000' (De Bruin & Dekker, 2001), abbreviated as 'RSI2000', and 'Students and RSI 2002' (De Bruin & Dekker, 2004), abbreviated as 'RSI2002'. A third study 'RSI amongst students', abbreviated as 'RSI1999', (De Bruin & Molenbroek, 2001; De Bruin, 2001) was already completed in 1999 and was initiated by the section Applied Ergonomics and Design of IDE. It was the first research project on RSI amongst IDE students. The students voluntarily filled in a digital questionnaire. The attention of all IDE students was attracted to this project by means of an announcement in the faculty newspaper and an electronic message that appeared on screen whenever a student logged into the faculty computers. The questionnaire of RSI1999 was more elaborate than the later two studies. Furthermore the questionnaire of RSI2002 was a bit more evolved than the RSI2000 one. Consequently, not all aspects of the three surveys can be compared.

Questionnaire RSI2002

The questionnaire of the most recent study RSI2002 is discussed here. It started with an introduction, outlining the goal and emphasising the anonymous nature of it. A general part included questions about age, gender, length and weight, mainly to check the representativeness of the respondent sample in relation to the IDE student population. The part in which the prevalence and seriousness of RSI was monitored, included questions relating to the occurrence of RSI-related complaints after VDU work, their location in the body, their frequency and duration. The questions about the duration and frequency of the complaints were included to estimate the severity of the complaints. It was considered that the complaints were more serious when lasting longer and occurring more frequently, and thus the seriousness was determined by the multiplication of the duration of the complaints and their reported frequency. An alternative estimation on the severity of the complaints was based on a checklist of daily activities such as tooth brushing, hand writing, carrying a bag etc (Levine et al., 1993). The respondents had to indicate to what extent the complaints were limiting them in these daily activities. Another part of the questionnaire focused on possible risk factors and included, amongst others, a question about the amount of VDU hours per day (at home, at the university, or somewhere else). By means of an open question, the respondents were able to report the possible causes of their complaints, for instance courses in the curriculum or specific study facilities. A last part of the questionnaire focused on the effect of prevention activities. The first question was about prevention activities the students had come into contact with in the past. Nine options, such as RSI lectures, RSI website of IDE, RSI posters or the sessions on working posture were given. In answer to the second question, the perceived personal effect of these earlier activities was reported.

Results

One hundred fifty five students participated in RSI2002 (non response of 36%) and 290 students in RSI2000 (similar non response). The average age of the students in RSI2002 was 19.6 year and 20.5 year in RSI2000. In RSI2002, 52.9% was male and 47.1% female, in RSI2000 54.5% was male and 45.5% female. In the first study RSI1999, 181 students participated (non response of 88%). The average age was 22.3[*] years; 64.6% was male and 35.4% female. The general sample information of the three studies, matched well with each other and with earlier measurements of Dutch males and females of age 20-30, which also included the IDE student population (Dirken & Steenbekkers, 1998). Therefore, the samples of these three studies are considered representative for the IDE student population. On some aspects the comparison could only be made between the last two surveys, because these were not investigated in RSI1999. Some results refer specifically to the RSI2002 study, in particular the comparisons between years of study. Although the RSI2002 questionnaire was distributed as part of the first and third year RSI information sessions and instructions, the sample of this study also included 2nd, 4th, 5th, 6th and even some 7th year students, who had to redo the course or did not study on schedule (distribution: 1st 42%; 2nd 16%; 3rd 21%; 4th 13%; 5th, 6th and 7th; 9%). All following tests were conducted with the Likelihood Ratio test.

Table 1 shows that comparable percentages of students with RSI complaints were found in RSI2002 and in RSI2000. In RSI1999, the reported amount of complaints was much higher. A significant effect was found for the complaints over the years.

Table 1. Comparison of the percentages of students experiencing complaints

	RSI2002	RSI2000	RSI1999
Complaints = yes	59%	61%	82%
$\chi^2 (2) = 29.32, p < 0.001$			

The RSI2002 study shows a relatively large increase of complaints from the 2nd to the 3rd year (Figure 2). In general complaints occurred more often in higher years of study $\chi^2 (4) = 22.34, p < 0.001$. The results relating to the body location of the complaints are quite similar through the three different studies. Most complaints occur in the neck, shoulders and wrist. Figure 3 illustrates the locations of the complaints in RSI2002 (in percentages of the total respondent group).

---

[*] The higher mean age of RSI 1999 compared to the later studies is probably caused by the relative larger number of higher year students in RSI1999

Prevalence of RSI complaints

RSI complaints after VDU work

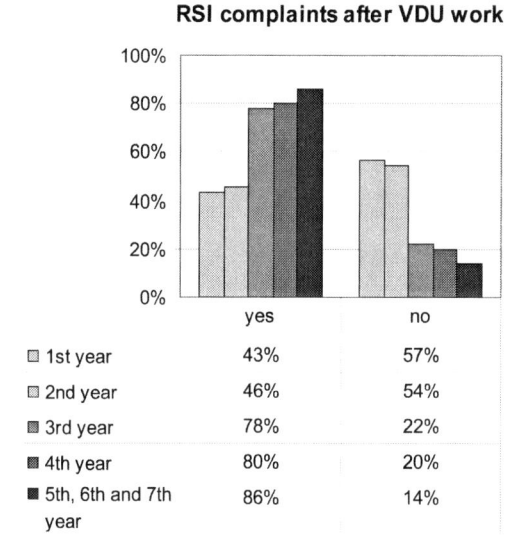

	yes	no
1st year	43%	57%
2nd year	46%	54%
3rd year	78%	22%
4th year	80%	20%
5th, 6th and 7th year	86%	14%

Figure 2. RSI-related complaints after VDU work per year of study (RSI2002)

Seriousness of RSI complaints

The frequency of the complaints (Table 2) show similar patterns over the three studies. Most of the students with RSI symptoms experienced complaints once a month or once a week, only few students experienced complaints every day. No significant effect was found for the frequency over the years. The duration of the complaints (Table 3), show similar patterns as well over RSI2000 and RSI2002. For most of the respondents, the complaints last for less than 6 hours. Only a small group suffers from complaints that last longer than 24 hours. No significant effect was found for the duration over the years.

Table 2. Comparison of the frequency of the complaints (within the group of respondents with complaints)

	RSI2002	RSI2000		RSI1999
Once a year	16%	13%	Sometimes	61%
Once a month	40%	45%		
Once a week	36%	35%	Often	28%
Every day	8%	7%	Very often	11%
$\chi^2 (4) = 3.14, p = 0.534$				

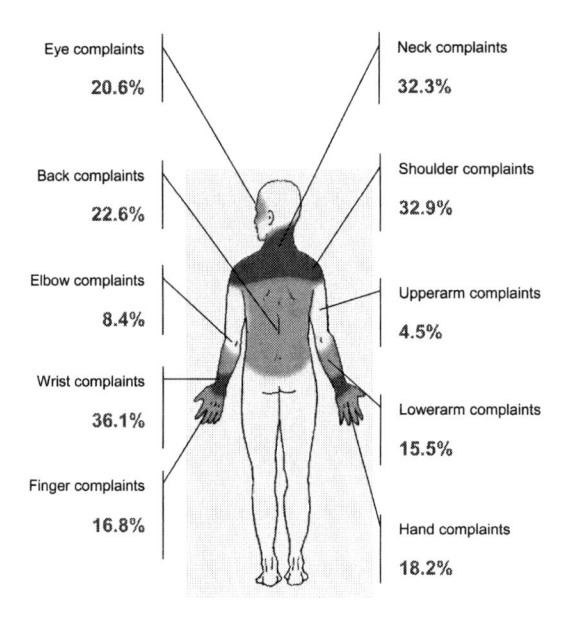

Figure 3. Locations of complaints (RSI2002)

- Eye complaints: 20.6%
- Neck complaints: 32.3%
- Back complaints: 22.6%
- Shoulder complaints: 32.9%
- Elbow complaints: 8.4%
- Upperarm complaints: 4.5%
- Wrist complaints: 36.1%
- Lowerarm complaints: 15.5%
- Finger complaints: 16.8%
- Hand complaints: 18.2%

Table 3. Comparison of the duration of the complaints

	RSI2002		RSI2000
Less than 1 hour	58%	Less than 1 hour	53%
1-6 hours	27%	Half a day	29%
6-12 hours	4%	A day	12%
12-24 hours	4%	Twenty-four hours	4%
A couple of days	4%	More than twenty-four hours	2%
Continuous	1%		

$\chi^2 (4) = 7.13, p = 0.129$

The results of RSI2000 and RSI2002 related to the seriousness of the complaints, expressed in frequency x duration, are comparable (Table 4). A large amount of the complaints are not very severe (<100 hours per year). Less than a quarter of the complaints are more serious and last for 100-800 hours. A few complaints are very serious (>800 hours per year). No significant effect was found for the seriousness over the years.

Table 4. Comparison of the seriousness of the complaints

Number of hours of complaints per year	RSI2002	RSI2000
1-50 hours	55%	51%
50-100 hours	23%	20%
100-800 hours	14%	25%
800 hours and more	9%	4%

The seriousness of the complaints per year of study are compared in RSI2002. No significant effect was found for the seriousness over the years of study $\chi^2 (8) = 8.55$, $p = 0.382$. The average score in the theoretical scaling of the severity of complaints expressed in limitation of daily activities was 1.14 (scale 1 to 5) in RSI2002. The highest score was 2. Nineteen percent has some difficulty with hand writing and 2% quite a bit of difficulty; 15 % has some difficulty carrying a bag and 5% quite a bit; 12% has some difficulty holding a book and 4% quite a bit.

Possible risk factors

As can be seen in Table 5 the number of hours VDU work per day was higher in RSI2002 than in RSI2000 (more students worked 4-6 and more than 6 hours per day). A significant effect was found for the number of VDU hours per day over the years.

Table 5. Comparison of the number of VDU working hours per day

	RSI2002	RSI2000
More than 6 hours	9%	4%
4-6 hours	24%	10%
2-4 hours	26%	28%
1-2 hours	26%	39%
0-1 hours	16%	20%
$\chi^2 (4) = 20.74, p < 0.001$		

The RSI2002 study shows that most students work both at home and at the university. The average amount of time spent on VDU work was 3.3 hours per day (1.8 hours at the home workplace and 1.3 hours at the university). Higher year students worked, on average, more hours behind the computer than lower year students $\chi^2 (4) = 11.60, p < 0.05$ (Figure 4).

VDU working hours per day

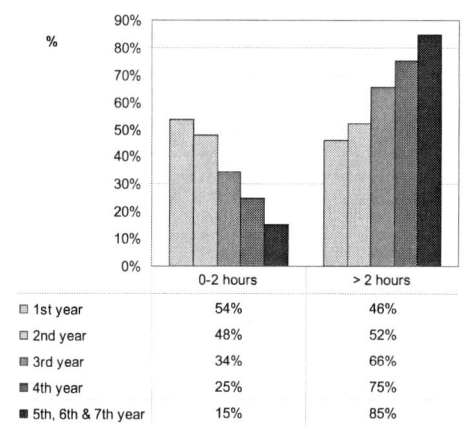

	0-2 hours	> 2 hours
1st year	54%	46%
2nd year	48%	52%
3rd year	34%	66%
4th year	25%	75%
5th, 6th & 7th year	15%	85%

Figure 4. VDU working hours per day per year of study (RSI2002)

In RSI2002 respondents most often reported the workplace as the most likely cause of their complaints. Other causes frequently reported were: the study burden, long working hours, and deadlines.

Perceived effect of prevention activities

On average, students had contact with three prevention activities in the past in RSI2002. The RSI lectures in the regular ergonomics courses, the first year sessions on working posture and relaxation exercises, the RSI brochure of the Delft University of Technology and the RSI website of IDE were the best consulted options. The answers to the open question about the perceived personal effect of these earlier activities indicated that a large part of the respondents changed their attitude (for instance, became more careful) and their intention (for instance, planned to reduce working hours) in relation to RSI prevention. 9% of the respondents even changed their behaviour (for instance, stopped working when experiencing complaints or made workplace improvements) in order to avoid RSI.

Discussion

In the time period 1999-2003, the Faculty of Industrial Design Engineering (IDE) had relatively the highest number of students experiencing RSI complaints when compared to the other disciplines of Delft University of Technology. This is hardly a surprise, because the educational programme of IDE contains many tasks that relate to risk factors for RSI as reported in literature. These risk factors are the frequent repetitive movements, required precision and static position of body parts, caused by the specific visual display unit (VDU) work, such as elaborate word processing, and the use of various graphic and 3D CAD programmes. The large number of working hours per week is an indication for a high study load at IDE, which is another RSI risk factor, especially in combination with intensive VDU work. Moreover, designing, creating new, non-existing products or environments, is –besides being attractive– also mentally demanding. It can become a 24 hours activity and the end of the job may be hard to determine.

The amount of VDU work per day has increased between 2000 and 2002. However, the results from the three surveys described in this chapter show a tendency that the occurrence of RSI complaints amongst students diminished or at least stabilised. The prevalence of RSI complaints at IDE is comparable (around 60%) in RSI2002 and in RSI2000. In 1999 the number of reported complaints was much higher (83%). This may indicate that a drastic reduction of RSI complaints between 1999 and 2000 occurred. However, it is more likely that the relative high number of higher year students in RSI1999 –who, on average experience more often complaints– contributes to this difference. Furthermore, the different ways of questionnaire distribution (see Methods paragraph) may be of influence.

The frequency and duration of the complaints, do not differ to a large extent over the years. Most of the students with RSI symptoms, experienced complaints once a month or once a week and only a few students experienced complaints every day. For most correspondents the complaints last for less than 6 hours. Only a small group suffers from complaints lasting more than a day. The seriousness, which was only calculated in RSI2000 and RSI2002, is very similar in these years as well. A large amount of the complaints are not very severe. Less than a quarter of the complaints are more serious and a few are very serious. In all three studies, complaints occur most frequently in the neck, shoulders and wrist.

From the RSI2002 survey can be concluded that students in higher years of study suffer most of RSI complaints. The number of RSI complaints in the years above the 3rd year is higher than in the 1st and 2nd years. However, no differences were found in the seriousness of the complaints between the study years. In general, students experiencing complaints have mainly 'somewhat difficulties' with daily activities.

The larger number of VDU working hours per day in higher years of study seems to be a feasible explanation for the increase of complaints in higher years, but has to be investigated in future surveys. It is very well possible that not only the amount of VDU hours but also underlying aspects, such as study load –a self-reported cause for complaints by the students– or being strongly committed to the work –responsibility towards other students or commissioners–, play a significant roll in higher years.

Since 2000, an IDE working group on RSI prevention organises various prevention activities on a yearly basis. Does this prevention programme have any effect on the IDE students? It is hard to say to what extent the IDE prevention programme contributed to the reduction or stabilisation of the amount of RSI complaints. The fact that respondents reported in the last survey that they changed their behaviour or at least their attitude and intention looks very positive.

Acknowledgement

We would like to thank Renate de Bruin for the usage of her data RSI1999 and for her strong contribution to the surveys RSI2000 and RSI2002. And we would like to express our gratitude to Wim van Donselaar, director of the SGZ, for the overview of numbers students visiting the Students' Health Service (SGZ) because of RSI complaints.

References

Blatter, B.M., Bongers, P.M., Van Dieën, J.H., Van Kempen, P.M., De Kraker, H., Miedema, H., et al., (2004). *RSI-maatregelen: preventie, behandeling en reïntegratie.* Programmeringsstudie in opdracht van de ministeries van Sociale Zaken en Werkgelegenheid en van Volksgezondheid, Welzijn en Sport. [RSI-measures: prevention, therapeutic treatment and rehabilitation. Study commissioned by the Dutch Ministry of Social Affairs and Labour and the Dutch Ministry of Health, Welfare and Sport]. Doetinchem, The Netherlands. Reed Business Information B.V.
Bongers, P.M. (2003). *Maak werk van RSI* [Put work into RSI]. Inaugural lecture, Vrije Universiteit Amsterdam, The Netherlands.
De Bruin, R. (2001). *RSI bij Studenten* [RSI amongst students] (Research Report No. 1). Delft, The Netherlands: Delft University of Technology.

De Bruin, R., & Dekker, M.C. (2001). *Studenten en RSI 2000* [Students and RSI 2000] (Research Report No. 2). Delft, The Netherlands: Delft University of Technology.

De Bruin, R., & Dekker, M.C. (2004). *Studenten en RSI 2002* [Students and RSI 2002] (Research Report No. 3). Delft, The Netherlands: Delft University of Technology.

De Bruin, R., & Molenbroek, J.F.M. (2001). *RSI bij studenten: een casestudie naar de omvang, ernst en oorzaken van RSI-gerelateerde klachten bij studenten aan de faculteit Industrieel Ontwerpen van de TU Delft.* [RSI among students; a case study on the magnitude, severity and possible causes of RSI-related problems at the Faculty of Industrial Design Engineering of the Delft University of Technology.]. Tijdschrift voor Ergonomie, 26(4), 17-28.

Dirken, J.M., & Steenbekkers, L.P.A. (1998). Project data and design applicability. In L.P.A. Steenbekkers and C.E.M. van Beijsterveldt (Eds.), *Design-relevant characteristics of ageing users; backgrounds and guidelines for product innovation* (pp.257-433). Delft: Delft University Press.

Health Council of the Netherlands. (2000). *Health Council of the Netherlands: RSI* (Publication No. 2000/22). The Hague: Health Council of the Netherlands.

Levine D.W., Simmons B.P., Koris M.J., Daltroy L.H., Hohl G.G., Fossel A.H., & Katz J.Ns. (1993). A self-administrated Questionnaire for the assessment of severity of symptoms and functional status in carpal tunnel syndrome. *The Journal of Bone and Joint Surgery. 75-A(11)*, 1585-1592

Visser, B. (2004). *Upper extremity load in low-intensity tasks*. Unpublished doctoral dissertation, Vrije Universiteit Amsterdam, The Netherlands.

Application of Human Factors Engineering to an Italian ferryboat

Antonella Molini[1], Stefania Ricco[1], Manuela Megna[2], & Dario Boote[3]
[1]*CETENA, The Italian Ship Research Centre, Genova*
[2]*Faculty of Architecture Genoa University*
[3]*Faculty of Engineering Genoa University*
Italy

Abstract

In Human Engineering human performance principles, models, measurements, and techniques are applied to systems design. Goal is to optimise system performance by taking human physical and cognitive capabilities and limitations into consideration during design. The application of this methodology as case test to an Italian ferry and the development of a software tool for the implementation of this methodology are described in this paper.

Introduction

Human Engineering (HE) deals with the application of human performance principles, models, measurements, and techniques to the design of systems. The goal of HE is to optimise system performance by taking physical and cognitive human capabilities and limitations into consideration during ship design.

Methodology

The integration of Human Engineering in the ship design Feasibility Study leads to the need of adopting a pragmatic approach, generally based long-term on-board experience of Italian ferryboat owners and designers. In Figure 1 the application of the Human Engineering Process (Malone, Bost, Molini, & Ricco, 2003) to the design of a ferryboat is shown.

Results

The main results of the HE process application to the Italian ferryboat were:

1. the optimised Scheme of Complement for this ship to guarantee ship's operability in each ship mission has been identified;
2. the skills and the necessary crew training level have been identified;
3. some possible modifications in the ship automation in order to reduce the crew workload have been underlined.

The MOST software tool has been the prototype application for the development of more complex software architectures in order to support the application of the HE methodology to other types of ships: merchant and military.

Human Factors Engineering   **Ferry Boat Application**

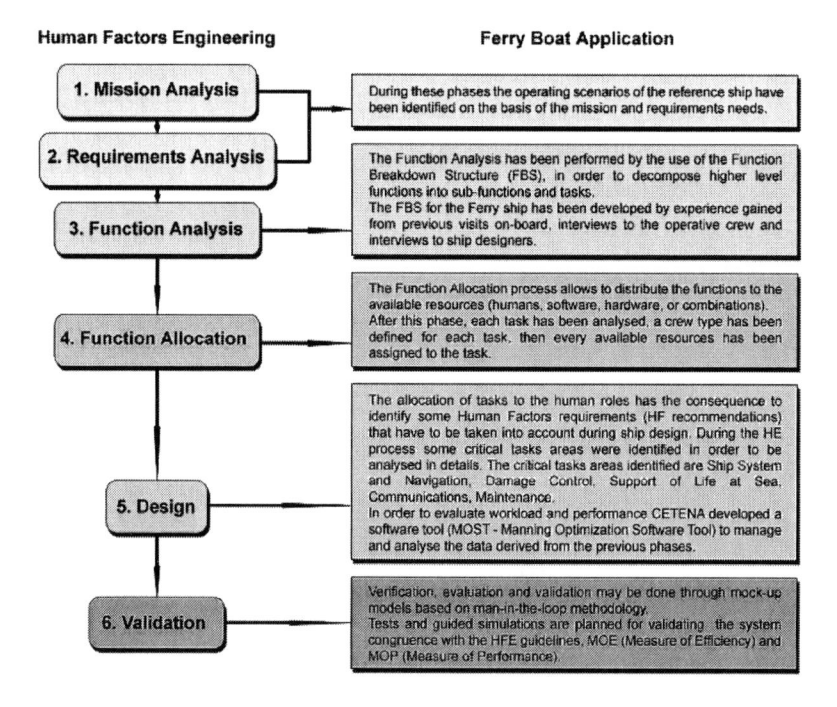

1. Mission Analysis — During these phases the operating scenarios of the reference ship have been identified on the basis of the mission and requirements needs.

2. Requirements Analysis

3. Function Analysis — The Function Analysis has been performed by the use of the Function Breakdown Structure (FBS), in order to decompose higher level functions into sub-functions and tasks. The FBS for the Ferry ship has been developed by experience gained from previous visits on-board, interviews to the operative crew and interviews to ship designers.

4. Function Allocation — The Function Allocation process allows to distribute the functions to the available resources (humans, software, hardware, or combinations). After this phase, each task has been analysed, a crew type has been defined for each task, then every available resources has been assigned to the task.

5. Design — The allocation of tasks to the human roles has the consequence to identify some Human Factors requirements (HF recommendations) that have to be taken into account during ship design. During the HE process some critical tasks areas were identified in order to be analysed in details. The critical tasks areas identified are Ship System and Navigation, Damage Control, Support of Life at Sea, Communications, Maintenance.
In order to evaluate workload and performance CETENA developed a software tool (MOST - Manning Optimization Software Tool) to manage and analyse the data derived from the previous phases.

6. Validation — Verification, evaluation and validation may be done through mock-up models based on man-in-the-loop methodology. Tests and guided simulations are planned for validating the system congruence with the HFE guidelines, MOE (Measure of Efficiency) and MOP (Measure of Performance).

Figure 1. The application of the Human Engineering to the design of a ferryboat

Conclusions and future studies

The application of the HE methodology to design in early phases is an important for ship designers to be sure that their ideas are aligned with the customer requirements and expectations. Unfortunately, not all the designers understand the importance of the role of HF studies, refusing the HF experts 'intrusions' in their activities.

The application of the HE methodology to the passenger ship analysed gave the possibility to underline the overload of some human roles on board and the risks of human errors due to bad human-machine interfaces (Boote, Bonvicini, Castelli, Megna, & Molini, 2003).

The development of the MOST tool will give CETENA a good tool to analyse very quickly different human role composition on board and to verify the impact of the application of new technical or organizational solutions.

The large experience acquired during the previous visits on board, has provided a good background to improve the adherence of theoretical studies to reality (Carta, Cataldi, Molini, Ricco, & Bisso, 2004).

References

Malone T.B., Bost. R., Molini A., & Ricco S., (2003), *Human-centered automation in total ship engineering*, Palermo, NAV2003-International Conference on Ship and Shipping Research.

Boote D., Bonvicini A., Castelli P., Megna M., & Molini A. (2003), *Evaluation of the efficiency and ergonomics of human/ship interaction*, Palermo, NAV2003-International Conference on Ship and Shipping Research.

Carta A., Cataldi A., Molini A., Ricco S., & Bisso C. (2004), *Human-System Integration issues for a low-cost, optimal-manning, high performances Frigate*, Amsterdan, INEC2004 Conference.

Observed risk awareness in user centred design

Freija van Duijne[1], Heimrich Kanis[1], Andrew Hale[2], & Bill Green[1,3]
[1]Faculty of Industrial Design Engineering, Delft University of Technology
[2]Faculty of Technology, Policy and Management, Delft University of Technology
The Netherlands
[3]Division of Health, Design and Science, University of Canberra, Australia

Abstract

Observational research on product usage aims to support design practitioners in designing safe and usable products. It is assumed that designers benefit from having insight into user-product interaction on the basis of the analysis of observational data. Knowledge of the way in which users tend to understand product characteristics and manipulate products may help them to generate successful design solutions for consumer products. This paper describes a design study in which findings from observational research were presented to novice designers. The outcomes reveal these designers' use of information and their strategies in designing a safe and usable gas lamp. Such evidence may support research and education that aims to develop (novice) designers' awareness of the relevance of information about user activities for designing safe and usable products.

Introduction

This paper addresses a design study to explore the effects of transmitting research findings on user activities to design practice: do designers pay attention to the way in which equipment is handled and do they use that information in solving the problem of how to improve safety? Very little work has been done on the communication of observational data of user activities to design practice. Findings from previous research indicate that designers consider summary results less interesting than in-depth information, such as gathered by thinking-aloud and interview transcripts (Kanis, 2002). Information that links user activities to product characteristics appeared to be the most useful in defining focal points for system properties that should be eliminated or changed. Research by Rooden (2001) indicates that designers may become focused by audio/visual descriptions, which may diminish their productivity in terms of predicting problems in user-product interaction. In Rooden's study designers were asked to consider possible usability problems. Half of them viewed a user trial with a coffee maker, the other half only got the coffee maker. It turned out that the video did not contribute to a wider insight into possible usability problems. From these data, it is uncertain whether designers benefit from the inspection of detailed descriptions of safety and usability problems.

The present study addresses the use of information about user activities by novice designers in designing a safe and usable product. The viability of this information may depend on the information format. Hence, the information on the same events is presented in three formats. These are established by reference to common practice in reporting findings from observational research, which can be presented as frequency descriptions, or narrative descriptions, or audio/visual descriptions. The study investigates whether the interpretation of the observed behaviour differs in the different formats and whether the informativeness of the material has an effect on designer activities.

Information provided

In order to yield information about risks in user-product interaction for the design project, a field study was conducted that addressed user activities and emergent risk in the replacement of pierceable gas cartridges in gas lamps used for camping purposes (Figure 1, Van Duijne & Kanis, 2002). The findings described the actions carried out by 23 users of gas equipment to replace empty cartridges of three gas lamps and indicated their awareness of risks in using a gas lamp. It was reported that some users misunderstood the sequence of actions to replace the cartridge. This may cause an unsafe situation, which was not recognised by most of the observed users who applied the unsafe sequence. The role of users' characteristics such as their previous experience with gas equipment and some anthropometric measurements was outlined by the findings. The same holds for characteristics of the observed camping units.

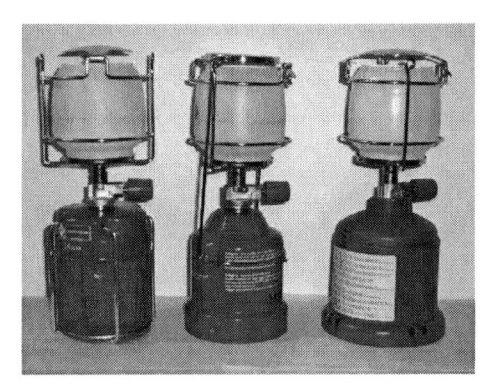

Figure 1. Three types of gas lamps that feature different mechanisms by which a gas container is fixed in the casing Figure is also available at http://extras.hfes-europe.org in colour.

The three formats for the presentation were the following:

The basic information condition
The manuals presented to designers were based on basic tabular descriptions and a storyboard. The information was 'basic' in the sense that it mentioned user characteristics, user activities and background information about the observed camping unit. Descriptions were written in an economical style that nevertheless did

justice to the observed variety within the data. The results of the anthropometric measurements (hand length, width and grip force) were displayed in graphs and tables, which show individual scores as well as central tendencies. The (absence of) relations between anthropometric measurements and user activities were presented by graphical displays. The tabular descriptions report frequency information of the observed and distinguished activities. The storyboard displays step-by-step pictures of both the intended and the unsafe way of replacing the cartridge of the gas lamps as observed in the study.

The narrative information condition
In this condition all the basic information was presented, with the same content, but was interwoven with narrative descriptions which consisted of detailed descriptions of observed user-product interaction. Again, no illustrations were provided except for the storyboard. Narratives described participants' activities and self-reports, which were generated while they were active with their replacement task. Initial explorative actions of participants were described and the observed variety in user activities was displayed. Whereas the basic information only described the sequences of participants' actions while replacing the cartridge, the narratives provided evidence of participants' understanding of the product, their concerns pertaining to the use of gas lamps, and the arguments that support the precautions that they reported taking. Participants' awareness of risks and their knowledge of the procedures for replacing a cartridge were described. Possible inferences were suggested based on the evidence presented.

The audio-visual information condition
In the third condition, audio/video descriptions were added to the basic descriptions. Using the evidence of individual cases, a selection of the audio/visual material highlighted the findings that were presented by the information that served as basic input. This selection aimed to provide a comprehensive overview of the observed problems in product use but without the use of narrative text.

Hence, all three groups had the basic data including the storyboard, whilst two had alternative presentations of richer data in either textual or visual form (see Van Duijne et al., forthcoming).

Method

Participants

Twelve novice designers participated, all of them students from the School of Industrial Design Engineering at Delft University of Technology who had reached at least the beginning of the Master's program (the final two years of a five years study). Two had recently graduated; they were allocated to different groups. No further balancing of participants could be achieved, also because the participants, who were recruited by poster advertisements, trickled in over a period of time.

Participants were asked about their experience with gas equipment that uses a pierceable cartridge, such as gas lamps, gas stoves, and soldering equipment. All designers appeared to have some sort of experience with one type of camping equipment such as lamps or cookers or with a blow torch. The nature of their experience differed, but at least one designer of each experimental condition had experience with a gas device that uses a pierceable cartridge.

Assignment and experimental condition

The designers were asked to design a gas lamp for pierceable cartridges that improves the safety for users while replacing the cartridge. The design brief mentioned that the proposed design solutions to improve safety should avoid possible negative consequences for the usability of the product. The redesigned gas lamp should be presented as a presentation drawing. The sketches and comments should be presented by a design logbook.

The designers were told to use the findings from the field study on the replacement of the cartridge in existing gas lamps in their design process. The findings were provided to them in a form defined by the three experimental conditions. Four participants were assigned to each of these conditions. All twelve participating designers had the opportunity to see and try the three products that were addressed in the field study. The designers were specifically instructed to read or watch the information presented to them. To what extent the participants complied to this instruction was addressed in the interviews at the end of the study. It was indicated that the assignment could be finished within one working day, in other words that it was only the preliminary design stage. Participants were instructed to work individually on this project, which was checked and confirmed in the interviews.

Data

The designers were asked to document which information was used at which point in the generation of ideas for improvements on safety and usability. They were asked to provide sketches and comments that explained the envisaged user-product interaction and how this redesigned product anticipated unsafe usage as observed in the field study. Shortly after finishing the assignment, the designers were interviewed. They were asked to clarify the design process as recorded in their logbooks, and what information about user activities influenced their design activities. They were also asked what other information they used in the design process. The interviews were audio and video recorded. Each interview took between forty minutes and one hour.

Results

Differences between the experimental conditions

There were little differences in designer activities found between the experimental conditions. The three presentation formats yielded similar design proposals to solve the observed problems. The basic information that includes the storyboards appeared to be sufficient to assess the problems in order to generate solutions. The designers considered the storyboards a very useful tool to understand the observed problems.

All designers, including those from the basic condition, felt well-informed of the users' perspectives on safety and usability. However, the designers in the narrative and audio/visual condition also considered the in-depth descriptions. They referred to the anecdotes they had read or to what they had seen in the audio/video compilation about user-product interaction. When proposing design solutions, they placed more emphasis on the users' perception of certain product characteristics, users' attempts of trial-and-error due to a lack of insight into the functioning of the product, and users' perception of risk.

Redesign proposals

Eleven designers (4, 4, 3)* proposed a casing that covers the cartridge, as was the case for two of the lamps in the field study (Table 1). They developed different mechanisms by which it is closed and opened. Four designers (1, 2, 1) proposed a casing that is closed by a screw thread, analogous to one of the existing lamp models (Figure 2). Some designers also mentioned the metaphor of a jam jar to describe their design solution.

Five designers (2, 2, 1) chose a concept that uses the closure that is derived from the mechanism of a preserving jar: a hinge is attached to one side and a lever can be pulled over a hook on the other side. Two of these designers (1, 1, 0) worked out this type of closure in their design concept (Figure 3). One designer (0, 0, 1) proposed a system that uses a button and a hook that is inside the casing. He knew this principle from his microwave oven.

Three designers (1, 0, 2) designed a casing that is fixed by screwing the lamp unit into the casing or the cartridge. They considered this a safe and easy mechanism to pierce the cartridge by an even movement (Figure 4). Without referring to the findings of the field study they explained that it is a gradual movement which reduces the force needed to pierce the cartridge.

Table 1. Redesign proposals

	basic	Narratives	audio/video	total
casing to cover cartridge	4	4	3	11
screw thread	1	2	1	4
hinge, click closure	2	2	1	5
screwing lamp unit	1	0	2	3

Information usage to improve usability

In order to solve the problem of the mistaken sequence of actions in replacing the cartridge of the gas lamps used in the field study, all designers indicated that they tried to come up with a design that looks simple and is understandable to users (Table 2). Four designers (0, 2, 0) explicitly mentioned that they were impressed

---

* The number of participants in each of the experimental conditions is placed in this order ('basic', 'narratives', 'audio/video'). For example, a topic indicated by (3, 0, 1) is mentioned by three designers in the 'basic' condition, none in the 'narratives' condition and one in the 'audio/video' condition.

with the observed activities of trial-and-error of many product users. They considered that users may not understand their product and that they start trying out actions by guessing.

Table 2. Design decisions to improve usability.

	basic	narratives	audio/video	total
making the lamp simple and understandable	4	4	4	12
minimise number of actions	3	2	3	8
visual cues	2	3	1	6
limit manual force needed	4	4	4	12

Eight designers (3, 2, 3) tried to minimise the number of actions needed to replace the cartridge. They reasoned, "less can go wrong if users need to perform less actions, which contributes to the safety of the activity". These designers described the integration of the lamp unit and the cover as a means to reduce use actions and to make the product look simple. Six designers (2, 3, 1) tried to provide visual cues to help users understand how to replace the cartridge. Four of these designers (1, 2, 1) tried to make the functioning of the product visually self-explanatory in order to help users to understand the product. Three designers (1, 1, 1) added symbols or words intended to help users (Figure 5). Two designers rejected the use of symbols, because they said it is not their style. In their view, the simplicity of the product should make usage understandable. Two other designers acknowledged in retrospect that the lack of obviousness of the functioning was a shortcoming in their proposal: it might confuse users, which was a reason to reject a particular concept.

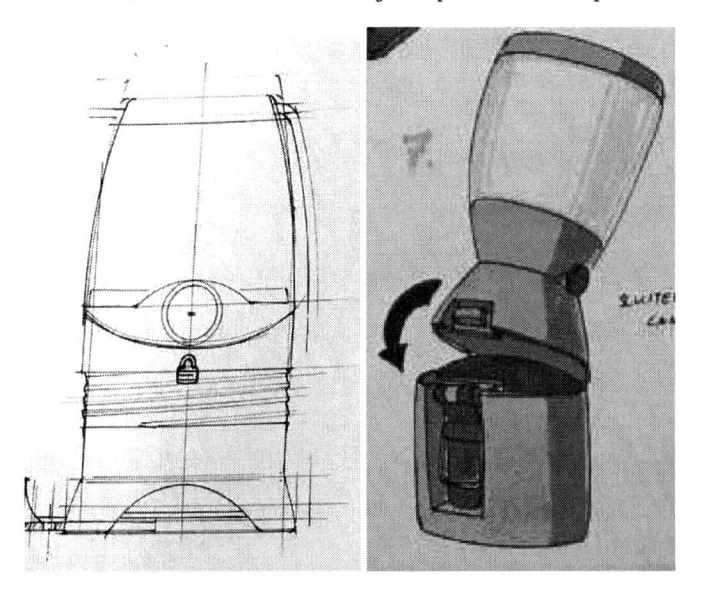

Figure 2. A screw thread separates the two parts of the casing of this gas lamp.

Figure 3. A casing that needs to be 'clicked' by a so-called click-finger.

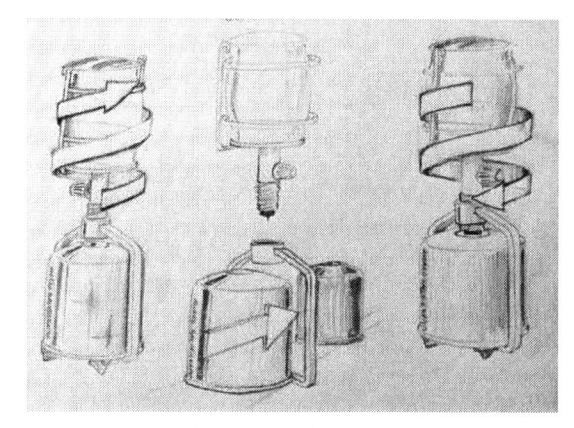

Figure 4. The cartridge gets pierced when assembling the lamp unit. Figures are also available at http://extras.hfes-europe.org in colour.

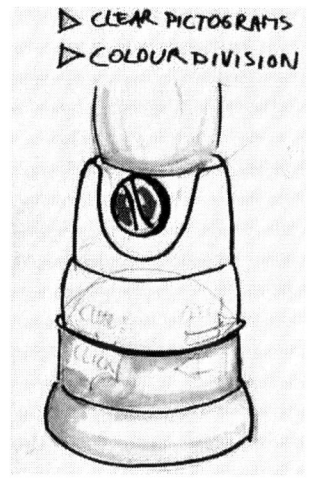

Figure 5. Arrows point to the direction in which to exert force and words ('open', 'close') indicated the mode of operation of the product. Colour division can be seen at http://extras.hfes-europe.org

All designers indicated that they tried to limit the manual force that is needed to perform the activity without making it too easy to open the casing. As mentioned by eight designers (1, 4, 3), the findings from the field research pointed to problems of some users who considered the screw thread of one type of lamp to be heavy. The designers who proposed the 'click'-closure indicated that the action associated with this closure is simple and does not require much force. In this design adjustment, the pin to pierce the cartridge is part of the lid and pierces the cartridge using the force transmitted by the movement of the lever. However, using this system that relies on a single hinge the pin would pierce the cartridge at an angle, which would cause leaks. The designers either ignored this problem or they felt unable to solve it within the limited time frame of the assignment. The designers who proposed a screw thread admitted that it would take some force to open the casing, but a closure mechanism that can be opened lightly would be unsafe: if users unintentionally open the casing, the gas would escape. Two designers (0, 1, 1) attempted to improve comfort in opening a screw thread. One of them adjusted the shape of the casing to fit the human hand and proposed ridges to provide grip, while the other designed a knob to get a good grip in order to (un)screw the casing.

Information usage to improve safety

Eight designers (1, 4, 3) took into account that users considered replacing the cartridge a risky operation, referring for instance to participants' awareness of the need to protect their children (Table 3). Two designers (1, 0, 1) thought that some people are not aware of the risk. One of them referred to his own perspective and said that "some people just act and don't think about it [the possible consequences]". He speculated that risk perception explains more about users' attitudes than about the product. The other designer referred to the audio/video tape, which showed a user who carried a package of tobacco. He considered this to be a signal that users underestimate the risk of replacing the cartridge. One designer (1, 0, 0) said that she was not interested in knowing users' perception of risk, because there would always be people who are scared of something and that she just wanted to make a safe product. One designer (1, 0, 0) did not remember what he had read about risk perception in the material provided.

Table 3. Using information to generate design solutions to improve safety

	basic	narratives	audio/ video	total
replacing a cartridge is a risky operation	1	4	3	8
some users are not aware of risk	1	0	1	2
cover danger eliciting product characteristics	4	4	3	6
gas cartridge acts as a trigger for risk	2	2	0	4
metal and a robust shape	3	2	0	5
simple looking product to make users feel at ease	1	1	0	2
raising risk awareness by using colours	1	2	1	4
insight into risk would trigger risk awareness	0	0	1	1

In developing design solutions to deal with the observed risk and perceptions of risk, eleven designers (4, 4, 3) chose to cover danger-eliciting product characteristics. They considered it important that users trust the product to be safe. Four of these designers (2, 2, 0) said that users should not be led to panic, because the gas cartridge itself acts as a trigger for risk. They said that users would be alert when using this product because it contains gas. They felt supported by the findings of the field study, which reported that some users considered gas release a risk. The designers described the following product characteristics that elicit feelings of safety: a casing that covers the 'risky' cartridge (0, 2, 1), thick materials or the use of metal and a stable robust shape for the product (3, 2, 0). Two designers (0, 2, 0) suggested that familiarity with the concept, i.e. resemblance with a jam jar, elicits trust. This idea did not stem from the findings of the field study but was assumed. Two designers (1, 1, 0) suggested that a simple looking product would make users feel at ease, which was also not described by the field study.

The designers' preference for making a design that looks safe may be disturbing if it means that they are willing to hide the dangers. The evidence from this study suggests that designers consider that 'a safe look' is as important as a safe product. However, in this study all designers were convinced that their design was safe to use. Therefore, they did not consider that they were trying to hide the dangers, but they were trying to emphasise the safety of their product.

Although they preferred a safe-looking design, four designers (1, 2, 1) wanted to raise users' awareness of risk by the use of colours: a red or orange closure and control knob would warn users not to open the product while the cartridge is still full, or not to leave the control knob open. Four (1, 1, 2) designed a closure that protects against these scenarios.

Only one designer (0, 0, 1) tried to express and communicate unambiguously the hazards that are associated with usage. He believed that providing insight into the risks would raise users' awareness of risk and evoke safe behaviour. He did not design a cover around the cartridge. Instead he proposed clamps (see Figure 4). He said that the use of less material enhances the visibility of the mechanism of piercing the cartridge, which should make users aware of the process and of the risk involved. The findings from the field study showed that users can misunderstand how the three types of existing gas lamps perform this piercing, but the observational evidence did not indicate that making the hazard visible makes users aware of risks.

Usefulness of the presented information

Three designers (2, 0, 1) indicated that they studied the material provided by this study only marginally, although they were confident that they had picked up all they needed to know. When asked if the observed usage had inspired them, they indicated that the information is in the back of your head, but it is difficult to say if it was the information or something else that influenced their proposals. When asked if they used for instance information about users' experience they denied that they used it so, at least consciously. Two of these designers (2, 0, 0) said that they did not want to know things such as anthropometric data, because these might put constrains on designs solutions and thus inhibit their creativity. The other nine designers appeared to have studied the material more conscientiously and said that they had used it to generate ideas.

The designers referred to the information provided when they described how they tried to improve the safety and usability of the product. Some designers indicated that they had experienced safety and usability problems themselves when they tried out the products and had found this experience to be confirmed by the information provided. Recognising it to be the core of the redesign task, all designers considered it to be essential to solve the problem of mistaking the sequence of activities. Furthermore, all designers wanted to limit the manual force needed to operate the lamp. For this, they used the anthropometrics without referring to the precise measurements. All designers wanted to design a stable lamp. Users' concern of an instable lamp that is used on an uneven underground is mentioned in the information provided, but some participants referred also to their own camping experience or they reasoned about the importance of stability.

Reference to other sources of information

The designers learned about the safety and usability problems not only from the information provided, they also tried the products themselves and experienced the problems that were described. Some designers referred to the problems they heard from friends or they took into account their memories on the usage of their own gas product.

The designers also explored their own ideas on user activities. They had assumptions about user activities in relation to the newly designed product characteristics such as the 'click closure' or a new type of clamps, and about the effects of certain featural and functional product characteristics, such as the use of colours to trigger risk awareness and robust materials to trigger the perception of safety.

All designers said that they would also have liked to have other information, which was not described in any of the experimental conditions, to help them design a new product (Table 4). Three designers (1, 1, 1) said that they would have liked to know about what happened during transport, two (0, 1, 1) wanted to know the company's demands regarding the production process for the lamp, and five (1, 2, 2) would have liked the opportunity to study alternative concepts of camping lights available on the market. Another five designers (2, 2, 1) felt the need to study in-depth accident data to inform them of the actual hazards and indicated that the observed usage did not inform them sufficiently on the type of activities that could lead to accidents.

Table 4. Other information that designers considered to be useful in the design process

	basic	narratives	audio/ video	total
information about what happened during transport	1	1	1	3
the company's demand during production	0	1	1	2
alternative concepts of camping lights on the market	1	2	2	5
in-depth accident data	2	2	1	5

Designers' evaluation of the presentation formats

All designers said that they read the information in order to learn the mismatches in user-product interaction. They indicated that they read it one time all through and then put it aside, once they knew the problems that they wanted to solve. Three designers (2, 1, 0) said that they considered reading difficult because the texts were in English, even though it is currently the official language in the educational program of the university (Table 5). Two (1, 1, 0) indicated that they had problems reading such a large amount of text.

As described, all designers considered the storyboard a very useful tool to give them an idea of the observed problems. Nine designers (3, 3, 3) said that they would be helped if the textual information were illustrated with pictures of the observed user activities. They considered such illustrations an additional advantage over the storyboards, because those illustrations would demonstrate the precise event that was indicated by the texts. Three of these designers (2, 1, 0) had drawn the described problems with the products in the margins of the texts in the manual.

Nine designers (3, 3, 3) appeared to have problems with statistical figures and tables, such as a simple 2 by 3 cross table. Four said that they considered it difficult to read tables and diagrams. Two said that they had only briefly looked at them. In order to understand the information in the tables these designers read the text and analysed their own experience when they tried the products. Three designers said they were searching for significant differences between the groups of users, such as experienced and inexperienced users, displayed in the tables of figures. These designers considered the tables and figures to be uninteresting if they did not show significant differences between groups. In fact, the described significance of the data was that the observed activities showed no differences between owners of a product who had experience of a similar system and those with a different system. Only three designers understood and used the information from the tables and figures. For instance, they took into account that many users of gas lamps replace the cartridge infrequently and that they may forget the procedure in the meantime, which supported their decision to design a simple looking design. This information can be derived from the tables, but also from the narratives and the audio/video compilation.

Table 5. Evaluation of the presentation formats

	basic	narratives	audio/video	total
reading English is difficult	2	1	0	3
too much text	1	1	0	2
illustrations preferred	3	3	3	9
tables and statistical figures difficult to understand	3	3	3	9
boring to read about user-product interaction	0	1	0	1
first part of the manual uninteresting	1	1	1	3
priority list of problems is missing	2	1	1	4
too small number of participants	1	1	1	3

In general, the designers evaluated the textual part of the information, that was provided to all participants, positively, although one (0, 1, 0) considered it "boring" to read about perception of product characteristics. The other three designers who read the narratives said that they appreciated the liveliness of this type of descriptions. However, none of the designers in the other conditions mentioned that they missed such user accounts. Three designers (1, 1, 1) commented on the way in which the manual was composed. They said that the first part of the manual, which explained how the study was carried out, was not interesting because it did not focus on safety or usability problems. Four designers (2, 1, 1) said that they would have liked to see a priority list of observed problems. Now, they felt that they had to decide themselves which problems they considered to be most urgent. Three designers (1, 1, 1) commented on the number of users that were observed. Despite their own experience with usability studies which always have small numbers of participants, they had doubts over the informativeness of studies that use a small number of participants.

Different ways of interpreting the information provided

Many designers misunderstood the role of experience on safety and usability problems, which was described by a table and some explanatory text. While the table and the text indicated that some experienced users forget how to replace the cartridge, six designers (3, 2, 1) said they considered that experience would coincide with proper knowledge of the procedure (Table 6). One designer even compared the activity of replacing a cartridge with bicycling and stated that it would be something that one would never forget. The designers' interpretation of the findings from the field study varied for a number of topics. The field study reported that 13 of the 23 users spontaneously mentioned the need to close the gas control knob. Because the study was done with empty cartridges, some users may not have expressed the need to close the control knob. Three designers (2, 0, 1) considered it a usability problem that some users did not mention to perform this action and aimed to solve this, while two others (1, 1, 0) were surprised that so many users closed the gas control. They said that they would forget this action themselves and expected others to do the same. Four designers (1, 2, 1) said that people immediately close the control (or light the gas) when they hear and smell the gas that is released. They referred to their own use of a kitchen gas stove.

Table 6. Differences in designers' interpretation of the information on:

	basic	nar-ratives	audio/ video	total
users' experience and knowledge of replacing cartridge	3	2	1	6
users forget to close the control knob	2	0	1	3
surprisingly high number of users close control knob	1	1	0	2
gas triggers users to close the control knob	1	2	1	4
participants must have pleased the researcher	2	0	1	3

The designers also disagreed about the findings of the field study about users' perception of the risk of smoking and other sources of ignition. The findings reported that all users said they were aware of the risk of replacing the cartridge in the proximity of sources of ignition. Three designers (2, 0, 1) said that these results might be biased because of users' wanting to please the researcher. They expected that users actually underestimate the risk and considered it likely that they would smoke or sit near a barbecue when replacing the cartridge. However, the designers indicated that, according to them, protecting against the risk of open fire was beyond the ability of a designer and that users have their own responsibility to be careful with sources of ignition.

Discussion

Presenting information about user activities

There is no clear evidence that differences in the amount and format of the information presented resulted in differences in how many ideas were generated, how many and which problems were tackled or what type of design solutions were generated to resolve them. The designers who had more detailed information did not try to solve more or different problems. The differences on these dimensions, which we see in the data, need to be sought elsewhere, in the personal preferences of the different designers for how they approach their work, in the priorities they give to different aspects of the problem, and in the weight they give to different sources and types of information and expectations in solving them.

When we turn to the way in which the different contents and formats were used, we see more differences. The designers considered the storyboards a very useful tool to understand the observed problems. In their education they were trained to use storyboards, which may explain their appreciation of this tool. The storyboards were included in the basic information package, and hence all designers had access to it. This basic package appeared to contain sufficient information for them to learn the problems in order to generate solutions. All the designers, including those from the basic condition, felt well-informed about the users' perspectives on safety and usability and agreed on the main problems to be solved in their designs. Furthermore, the designers, particularly in the basic and narrative conditions, recommended more use of illustrations in the text or with the tables, because of their clarifying properties. It is, with hindsight, a pity that there was no control group used without storyboards or the basic information, to see if this lack would have produced significantly different and poorer understanding of problems and proposed design solutions. In anticipation of this check, we can recommend that storyboards are a very valuable, if not essential, means of communicating user activities.

However, when we look at the differences between the designers in the basic condition and those in the narrative and audio/visual condition, we do see that the latter groups made considerable use of the in-depth descriptions. These designers referred often to the anecdotes they had read or what they had seen in the audio/video compilation about user-product interaction. When proposing design solutions, they explicitly took into account the users' perception of certain product characteristics and risks, as identified in those sources. The designers who had access to them valued the in-depth descriptions of the narratives, because of the liveliness of this form of description. For similar reasons, the designers who had access to the audio/video compilation appreciated watching it in addition to reading the basic information. Some designers who did not see the audio/video compilation said they wished to see it, but designers in the basic and audio/video group said they did not miss more detailed information such as narrative descriptions. The designers in the narrative condition seemed to know more about, and paid more attention to, the fact that users considered the cartridge replacement risky, and those in the audio/video group seemed most informed about the physical difficulties of some types of closure mechanism. However as described above, this did not seem to result in different solutions to the problems. We do see a slight tendency for designers from the basic condition to use their own experience to fill in questions raised by the data they had and sometimes to misinterpret it because they lacked the extra data from the narratives or audio/video to get the picture clear.

The use of information in user centred design

The tentative conclusion would seem to be that the richer data do not lead the designers to better understand the problems, or to come up with more comprehensive solutions. The data provide all of the designers with ways of justifying the choices they make, of illustrating them and the reasons for them to themselves and others, and possibly of giving more importance to some aspects and problems than others.

It has commonly been said that, early in the design process, designers do not wish to know information about usability and safety, because it constrains their creative thinking. They need to have 'mental freedom' in order to generate new ideas for a particular product. Rooden (2001) described the possibility of becoming 'mentally blocked' because of detailed information about user activities. He showed that watching an audio/video film of user activities inhibited designers' imagination of possible usability problems; the problems they suggested turned out to be plausible though (Rooden, 2001, p. 207). The inhibition effect does not seem to have happened here. If the designers' creative processes had been suffering from too much information about user activities, the designers in the 'narratives'- and the 'audio/video' condition should have produced fewer ideas and detected fewer problems compared to the designers in the 'basic' condition. The findings do not confirm this.

The novice designers used the information provided on the observed safety and usability problems in order to identify the problems that they should solve in this assignment. In general, it appeared that the designers did not read the findings from the field study very closely. Some browsed the information only superficially, studied the storyboard, and extracted a set of problems to be solved, but they did not fully understand for instance the frequency tables of observed activities despite their simplicity. They preferred to rely on their own intuitive understanding of the problems and on stories from their friends. Although this may be influenced to some extent by the presentation of the information, this finding is worrying in what it says about the skills the designers had. Earlier research indicated that designers who could not get grip on the problem to be solved appeared to ignore parts of the information provided (Christiaans, 1992). Recently, De Jong et al. (2004) reported similar observations on design students' limited abilities in applying information about product usage. They found that the students often focused on providing quick solutions for observed problems, but that they failed to look closely at the usage of their product. Their redesign proposals were often reformulations of usecues at 'button-level' instead of creative reconceptualisations based on a more comprehensive view on user-product interaction.

In sum, the present study demonstrates that novice designers may not benefit from observational evidence about user activities as previously expected. Further studies are needed to address the relevance of observational research to support design practice. For instance, while this study only looked at a preliminary stage of design, future research may study the usage of observational evidence in every stage a design project. In addition, research and design education should be concerned with developing designers' awareness of the relevance of information about user activities to design practice, and with developing their understanding of how to address and apply that information in the creation phase. Designers may be more inclined to apply the information if they do the observational research themselves, as an investigative designer, instead of receiving this information from a field researcher.

References

Christiaans, H.H.C.M. (1992) *Creativity in design. The role of domain knowledge in designing*, Dissertation, Delft University of Technology: Delft

De Jong, A.M., Boes, S.U., & Kanis, H. (2004) Usage research in the Delft design project Ontwerpen 4 *International Engineering and Product Design Education Conference*

Kanis, H. (2002) Can design supportive research be scientific? *Ergonomics, 45*, 1037-1041

Rooden, M.J. (2001) *Design models for anticipating future usage* Dissertation. Delft University of Technology, Faculty of Industrial Design Engineering, Delft, the Netherlands

Van Duijne, F.H., Hale, A.R., Kanis, H., & Green, W.S. (forthcoming) Design for safety: involving users' perspectives. In: A.R. Hale, B. Kirwan, and U. Kjellen (Eds.) *New Technology and Human Work*. Elsevier: Amsterdam

Van Duijne, F.H. & Kanis, H. (2002) *Replacing pierceable cartridges in gas lamps* Delft: Delft University of Technology, Faculty of Industrial Design Engineering

Iteration in design processes

Daniëlle M.L. Verstegen
TNO Human Factors, Department of Training and Instruction
Soesterberg, The Netherlands

Abstract

Ideally, specifications for all products should be based on a systematic analysis of what prospective users need to do, and how they can do it most effectively and efficiently. In practice, the design process is disturbed by many 'pragmatic' factors, such as conflicting constraints, interference from management, personnel changes in design teams and technological progress leading to new possibilities. Therefore, design is an iterative process. In this paper the design task of training simulator specification is used to illustrate the iterative nature of design processes. The results of two empirical studies show no clear relation between the amount of iteration and the quality of the resulting designs. Frequent iteration can be part of both an effective and an ineffective design style. In order to develop support for managing iteration during the design process it is necessary to understand in which circumstances iteration needs to occur, and how these circumstances can be predicted or recognised. For this purpose, a list of triggers for iteration is described. Finally, a number of measures to help designers to manage an iterative design process is proposed.

Introduction

Ideally, specifications for all products should be based on a systematic analysis of what prospective users need to do, and how they can do it most effectively and efficiently. During the specification process analysis, design and production can be performed in a sequential, cyclic or overlapping way. In practice, this process is disturbed by many 'pragmatic' factors, such as conflicting constraints, interference from management, personnel changes in design teams and technological progress leading to new possibilities. Reacting to these disturbances can make the process chaotic, but not reacting to them will certainly lead to solutions that do not fulfil the requirements or will not be accepted in the organisation. Therefore, design is an iterative process.

The designers of advanced products often tend to give much attention to technical aspects. This is understandable since advanced technology has to be carefully designed and constructed. The key question, however, is not what is technologically possible or most advanced, but what would be the optimal choice for future users. Thus, the challenge for designers is to design the best solution from a user's point of view, while at the same time taking other factors into account and reacting promptly to changing conditions. This has led to the formulation of the general problem as:

How can designers be supported during the iterative design process in order to design effective products?

To answer this question research has been done in the area of designing functional specifications for training simulators[*]. Training simulators are often used to train operators and maintenance personnel for technically advanced systems, such as cars, aeroplanes, power plants, computers, and radar systems. The training concerns critical and complex tasks that are not easy to learn. A large amount of training is necessary to reach adequate performance. The specification of training simulators is a complex design task. The resulting simulator should be an effective and efficient training system, but it should also be feasible within the given constraints (financial, technical, logistic etc.). This requires knowledge of the tasks to be learned and of the system to be simulated, as well as knowledge of learning processes and training methods. The latter is often brought in by domain experts who have become instructors and course designers. These probably have experience with simulator training (both as trainees and instructors), but not with the specification of training simulators. They also do not have an extensive background in (instructional) design theory (see Verstegen, 2003). Thus, the people responsible for the design of functional specifications for training simulators do often not have enough resources, expertise and experience to execute a systematic and thorough analysis of training goals and alternative training solutions.

In this paper the domain of training simulator specification will be used to illustrate why design processes are inherently iterative.. Subsequently the results of two studies regarding the relation between the amount of iteration and the quality of resulting designs will be summarised. A list of likely triggers for iteration is provided and measures to support an iterative design process are proposed.

Design is inherently iterative

Characteristics of the design task

Design tasks can be seen as a problem-solving task. Greeno et al. (1990) claim that design tasks differ in two ways from other problem solving tasks, such as calculation, proof and explanation problems. First, the problem solution space is open: it is impossible to predict which solutions a designer might come up with since design is an inherently creative activity. Second, the final state is a matter of judgement: the designer decides when the task is completed. Thus, design problems have the characteristics of ill-structured problems: the real-world conditions are ill-structured

---

[*] Functional specifications describe what the future simulator should be able to do. In other words: they specify a simulator that can fulfil the requirements of the different types of users. In the development process functional specifications are a step before the technical specifications: they specify on a behavioural level what the simulator should be capable of, not how this can or should be technically implemented.

and involve many complex variables, and there is no perfect solution. Designers can only hope to find a satisfactory solution that meets (most of) the demands. There is also no predefined solution process, no algorithm or established set of procedures. Designers have to employ heuristics, problem solving, creativity and decision-making (Nelson et al., 1988). Design tasks cannot be solved with general knowledge only, but require a substantial amount of professional, domain-specific knowledge as well (e.g. de Jong, 1986; Mettes & Pilot, 1980) On top of that, problem solvers need strategic knowledge that indicates how the problem solving process should proceed.

Descriptive research

Goel and Pirolli (1989, 1992) defined generic features of design tasks from the viewpoint of (ill-structured) problem solving. Analysing think-aloud protocols they found that these features indeed distinguish design tasks (in the fields of instructional design, architecture, and mechanical engineering) from non-design problem solving tasks. Some of these features are:

Problem structuring:	Extensive problem structuring is necessary before problem solving can begin.
Three distinct phases:	Preliminary design, refinement and detailed design; generally executed sequentially, but it is not uncommon to return to earlier phase.
Modularity:	Designers decompose the problem. Modules are, however, 'leaky', i.e. they are heavily interconnected.
Incremental design:	Interim design ideas are nurtured and incrementally developed. They are rarely discarded and replaced by new ideas.
Control structure:	Full control is impossible. Designers use a limited commitment mode control strategy with nested evaluation loops.
Reverse direction:	Designers occasionally stop, reverse the direction of the transformation function and work bottom-up.

Furthermore, Goel and Pirolli describe the tension between keeping options open and making the commitments that are necessary to complete the design task within a finite amount of time, and the complications that arise because there are no objective stopping rules: there are no right or wrong answers and direct feedback during the design process is lacking. Since it is usually not possible to directly manipulate the real world situation to try out solutions, designers construct and manipulate models of possible solutions at various levels of detail using artificial symbol systems.

Similar studies have been done for other kinds of design tasks. In the domain of architecture, for example, Hamel (1990) claims that the design task of an architect consists of five aspects: gathering information, decomposition of the design problem, solving the sub-problems, integrating sub-problems solutions and styling the solution to complete a design that meets aesthetic and professional criteria. Hamel proposes a nested model where the same categories of activities are executed repeatedly, although the concrete activities can be different in different phases of the design process. Hamel tested his model with think-aloud protocols of 15 experienced architects and found that it fitted their design process, much to the surprise of the subjects who often thought that their way of working was quite chaotic and that trying to model their design process was doomed to fail. Probably, Hamel says, these experienced architects have developed 'good habits' forced by the characteristics of architectural design tasks and their own limited memory capacity, and developed task schemata that are activated with each new design task. Experienced designers do not have to consciously manage the design process and, thus, can have more attention for solving the design problem.

Braha and Maimon (1997) discuss their field, engineering design, from a wider theoretical perspective. They claim that the design process can be viewed as a stepwise, iterative, evolutionary transformation process: as the design process develops, the designer modifies either the tentative design or the requirements based on new evidence (information), so as to remove the discrepancy between them. Braha and Maimon compare the design process to scientific research: when an anomaly is discovered in a scientific theory (or in a design), the outcome is often an adjustment of the theory (incremental redesign). The limited information-processing capacity of the designers is at least partly to blame for this. It often forces them to make decisions similar to those previously made. Innovative design, on the other hand, corresponds to a transition to a new paradigm in scientific research.

Blessing (1994) conducted a long-term case study regarding the design process in the field of mechanical engineering and compared her results with the results of eight overview studies and 66 case studies in the descriptive literature. She concludes that designers generally seem to follow a product-oriented approach focusing mainly on the refinement of an initial product idea. Designers do also not follow the advice to generate and consider alternative solutions in parallel. This is worrying, according to Blessing, since an explicit and frequent check to identify problems and failures is hardly observed either. In general, designers tend to stick to their solution and do not execute evaluation in a proper way. She derives a list of success factors, which include using a systematic approach, executing all stages and activities, and a thorough problem analysis. Blessing proposes a model of the design process as a combination of stages, activities and strategies. In general three main stages are distinguished: the problem definition stage, the conceptual design stage and the detail design stage. Activities are subdivisions of the design process related to the designer's problem solving process, for example collecting information, generating solutions, evaluating and selecting. Strategies are defined as the sequence in which design activities are planned or executed. Stages are, in principle, executed sequentially and only once (see Figure 1-a). The main flow through the activities is cyclic (Figure 1-b). Some design models explicitly combine activities and stages, resulting in a cyclic flow as illustrated in Figure 1-c: the sequence of activities is repeated in every stage. Taking into account that the part of the solution space that is considered will get gradually smaller, a concentric model can be defined as depicted in Figure 1-d.

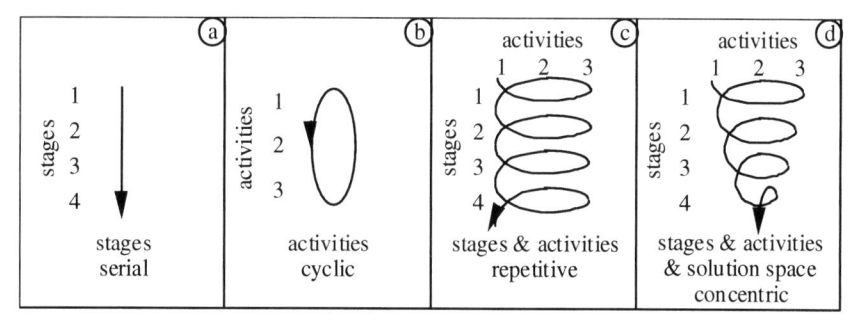

Figure 1. Main process flows (in mechanical engineering design, from Blessing, 1994, p. 41; used with permission of the author).

Conclusion

In a large-scale design process different parties are involved and they have to get to an agreement about the problem definition: what the problem is, what the demands are on the solution, which resources are available, etc. Subsequently, there are at least two ways to handle complexity: decompose the problem into sub-problems or design solutions on an abstract level first. The sub-problems are not unrelated and can, therefore, never be solved in isolation. Moreover, it is impossible to know beforehand which ideas will lead to a good solution and which will not, because:

- solutions have to be partially developed before they can be checked against the problem definition, and they may have to be adapted or discarded when they do not optimally fulfil the demands,
- good solutions for sub-problems may not be feasible when they are combined with solutions to other sub-problems, and
- practical factors, like changing resources or interruptions by other stakeholders, may actually change the problem definition during the design process.

This means that the design process can be only partially planned beforehand. Sometimes, designers will have to try out actions, assess the consequences of those actions and then decide on further actions (Rowland, 1993). Specifications are gradually improved and refined.

Thus, design tasks are ill-structured and inherently iterative problems. This is also true for the specification of training simulators: the domains are complex, every project is different, it is a long and time-consuming process, there are different kinds of -often contradictory- demands that can change over time, etc. (see Verstegen, 2003). The outcome is not only determined by what would be best given the tasks to be learned and the target trainees. More practical complications, such as conflicting interests, limited resources, personnel changes and timing problems also play a role. Moreover, designers will have to design preliminary specifications with incomplete and insecure information, e.g., when it is not completely clear which tasks will have to be learned because the operational system is not yet fully developed or because the entry level of trainees is unknown. Designers will have to review and alter their decisions frequently when new information becomes available. Experienced designers are able to adequately manage an iterative design process: generate several alternative solutions, and do not commit themselves until later on (e.g. Rowland, 1991, 1992). Novice designers, on the other hand, are inclined to spend little time in the analysis phase and to commit themselves to one solution early on. They seem to lack the necessary knowledge about possible solutions and solution procedures, and/or the strategic knowledge required to apply them to the specification at hand (Perez et al., 1995)

Iteration and quality of designs

Two empirical research studies have been done. In both studies representative users (i.e. novice designers) designed training simulator specifications using an existing design method and tool. The design problem that they solved was a real case taken from another project (van den Bosch et al., 1999). In the first study ten subjects worked on a design for 3 months, half time, at home (at their own pace). In the second study eight subjects worked on the same task for one week under supervision. In both studies, the subjects were provided with background material regarding the design problem as well as the design method that was used. They could also ask questions to different kinds of experts by e-mail. Subjects were told beforehand that the design of simulator specifications is often not a linear process, and the design tool allowed them to go back to earlier design activities at all times. Some events during the studies mimicked events that were assumed to evoke iteration in practice, e.g. feedback on the design from a peer or from an expert, or new information about the domain (tasks to be learned and/or system to be simulated). Log files show on which design activities the subjects worked and in which order. From this information, the number of iterations has been derived; i.e. how often subjects went back to design activities that they had already worked on before.

In both studies all subjects iterate. In the first study, there were large differences in the amount of time that subjects spent on the design task: roughly between 12.5 and 121.5 hours (average 62 hours and 14 minutes; standard deviation: 31 Hours and 22 minutes). However, there was no relation between time on task and the number of iterations. In the second study, time on task was controlled, and it was less than the average time on task in the first study. However, the average number of iterations is more or less the same in both studies (average 17.6 in the first study and 18.6 in the second study). In both studies some subjects made clear that they did not like to iterate, using phrases such as 'having to do the same thing again' or 'now I can throw away some of the work that I have already done'. Other subjects, however, have no problems with iteration and, in fact, the possibility to iterate is also mentioned as a strong point of the design method, especially in the second study.

In both studies, the number of iterations varies considerably between subjects, and there is no relationship with the quality of the resulting training programme designs.

Apparently, different design styles -with more or less iteration- can lead to good results. Some subjects work out a good concept and then work through the method step-by-step, going back to previous steps only when it is necessary. Others come to a good design by going back and forth between steps more often. Maybe they leave more decisions open, go to the next step(s) to elaborate the design partly, get new ideas, and then go back to work on the initial concept again, etc. Inspection of the subjects' notes shows that the number of iterations does not correspond to the amount of rethinking or reviewing done during the design period. The results of both studies show that frequent iteration can be part of an ineffective design style as well, perhaps indicating a lack of overview. Weak designers will probably make more errors and will have less insight into the possible consequences of early decisions. Good designers will have to iterate less to repair errors, but they may use iteration for another purpose: to check whether they have designed the most optimal solution or to consider possible alternatives. This would be in line with the findings of descriptive research which show that experienced designers spend more time and effort on a problem analysis, generate several alternative solutions, and do not commit themselves until later on (e.g. Rowland, 1991, 1992; Perez et al., 1995).

In conclusion, iteration is inherent to the design process and it can -but does not always- improve the result. Iteration can be used in different ways, depending on environment- and subject-related factors. There are different reasons to decide to iterate or not. As long as we do not have a good understanding of what those reasons can be, it will remain difficult to support this aspect of the design process.

Triggers for iteration

In a general sense iteration means: executing the same activities again (van Wagenberg, 1992). Van Wagenberg differentiates between four types of iteration:

1. Iteration as repetitive activity: repeating the same design activity.
2. Iteration to correct errors: reviewing the concept design and correcting errors.
3. Iteration to improve the design: fundamental revisions to get to a better design.
4. Mutual iteration: this form of iteration happens when designers decide to execute two separable activities in parallel; it can also be seen as a form of decomposition.

Reviewing these different types of iteration, the first conclusion is that some types of iteration are desirable: iteration to improve the design is obviously a good sign (provided that the required time and resources are available). Repetitive iteration can make the design process easier to manage or more efficient: designers can decide to decompose the problem into sub-problems or to work at different levels of detail. Similarly, an organisation can decide to execute different design processes in parallel, e.g., the specification of different instructional products for the same course. For these kinds of iteration, the difficulty lies in the decision about the amount and timing of iteration: which design activities are important to improve the quality of the design? To what level of detail should activities be executed at different moments in the design process? On the other hand, iteration to correct errors should be avoided as much as possible. Not all errors can be avoided, though, especially when not all information is available from the start and when constraints and demands can change during the design process. Thus, designers need to be able to recognise and correct errors as early as possible.

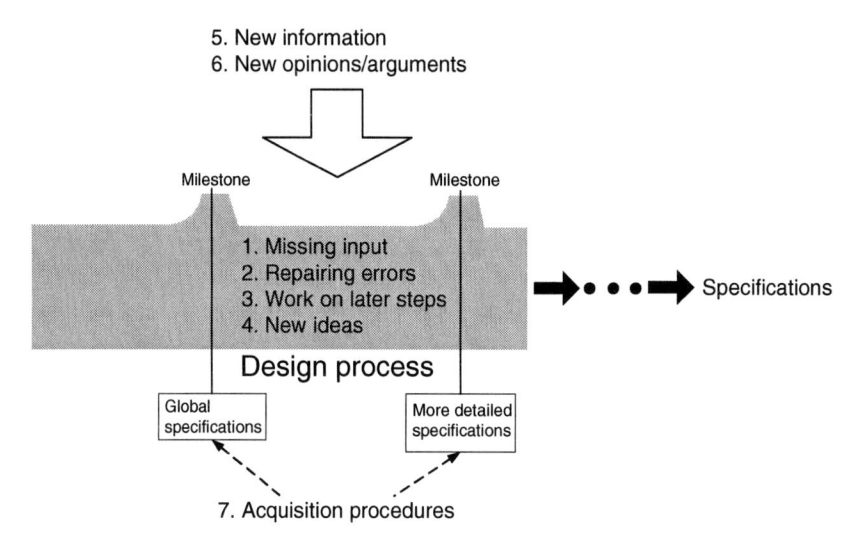

Figure 2. Seven triggers for iteration.

In order to develop support for managing iteration during the design process it is necessary to understand in which circumstances iteration needs to occur, and how these circumstances can be predicted or recognised. The triggers described below are events that are likely to occur during the design of training simulator specifications and, most probably, in design processes in general. The first four triggers are caused by, or evolve from interaction with the design process itself. They are: the discovery of missing input, the need to repair errors, new insights based on work later on in the design process, and new ideas of the designer(s). The other triggers originate from outside. New information that is relevant to the design process becomes available, or other people bring in new opinions or arguments that the designer(s) were previously unaware of. A third trigger from outside can be the procedures that are enforced in many large organisations. These procedures influence the design process through feedback on obligatory intermediate products, such as milestone documents containing preliminary versions of the design, as shown in Figure 2.

Trigger 1: Discovery of missing input
One trigger for iteration can be that the designers discover that they are missing necessary input that should have been available from an earlier design activity. This trigger is easy to recognise. It is possible to describe the input required for each design activity and to provide guidelines that direct designers back if vital input is missing.

Trigger 2: The need to repair errors
Another cause for iteration can be that designers discover that they have made an error earlier on. We saw examples of this in both evaluation studies. Dealing with errors can be supported in similar ways, e.g., by describing demands on output, or by providing guidelines that encourage designers to ask advice from peers or experts. Another possibility is to include explicit decision points with checklists at important points.

Trigger 3: New insights based on work later on in the design process
Sometimes further elaboration gives more insight into the consequences of earlier decisions and this can lead to revisions. It is not always possible to foresee if and when this will occur. However, some activities in the design process are logically related, e.g., because one provides input for the other. Experienced designers are another source of information: they will be able to predict when this trigger is likely to occur. Their input can be used to develop support, e.g. in descriptions of design activities, guidelines and warnings.

Trigger 4: New ideas of the designer(s)
This trigger is related to creativity in the design process. Designers can get new ideas at any time, triggered by seemingly unrelated events. Generating and maintaining several ideas makes the design process more difficult to manage and will probably cause more iteration. Presumably, however, it will lead to a better solution. Research studies show that experienced designers always generate a number of alternative ideas, and do not commit to one particular solution immediately (e.g. Rowland, 1991, 1992; Perez et al., 1995). Experienced designers may have a better overview of the design process, and thus more memory capacity for generating new ideas. They can probably plan their design process better and, in consequence, they might be better able to recognise other triggers for iteration and use those moments to try out new ideas as well, or they might even plan moments to contemplate and generate new ideas.

Trigger 5: New information
New information can trigger iteration, such as new information about the design problem, or the available resources. The problem is that designers might not know what information is available and that their colleagues will not always know which information is relevant for them. Guidelines can indicate what kind of information can be relevant for specific design activities, can help designers to decide what they should do with it, and encourage them to actively look for additional information.

Trigger 6: New opinions/arguments from other people
Other parties not directly involved in the design team can bring forward arguments that the designers were previously unaware of. Even when the designers do not agree with these arguments, they will have to spend some time to defend their own decisions. Guidelines or other help forms can draw the designers' attention to moments when it may be opportune to consult experts, stakeholders or more experienced peers. Another possibility would be to explicitly plan brainstorm sessions or reviews at specific moments in the design process.

Trigger 7: Acquisition procedures
The design and acquisition of new products is usually regulated by procedures that enforce an iterative design process, for example by requesting milestone documents. To obtain resources for the design and acquisition, designers usually have to present and defend their choice of this type of instructional product early on. This means that they will have to take preliminary decisions based on quick, global analyses and incomplete information. Because it is defined beforehand, this kind of iteration is easy to support with guidelines, warnings and explicit descriptions of the design cycles.

Measures to support iterative design

Given the fact that iteration is inevitable, the next question is: how can designers be helped to manage an iterative design process? In a way, designers face a double task. On the one hand, they need to work systematically in order to ensure that all important issues are addressed, and on the other hand, they should react adequately to triggers for iteration in order to ensure that their solution is optimal given the latest insights and information. They will have to choose the right level of detail, keep an overview of the evolving design product as well as the design process, decide when iteration is necessary and when it might actually be counterproductive, etc. Based on the literature and the empirical research described above, a number of measures that can help designers to deal with iteration during the instructional design process are proposed (see Table 1).

Table 1. Measures to support iteration

Category	Measure
Structure the design process	1. Use a systematic instructional development method
	2. Define design cycles
Plan triggers for iteration	3. Plan decision points regarding the design process
	4. Plan review moments
	5. Generating and compare alternative solutions explicitly
Specific advice regarding iteration	6. Strategic advice
	7. Help to recognise triggers for iteration
Measures to make iteration easier	8. Keep intermediate results easily available
	9. Make consequences of changes visible
	10. Store notes and keep them available
	11. Keep information on constraints separate and available
	12. Make assumptions explicit
	13. Have experts available for advice

Measures to support the management of iteration include explicitly structuring the design process by using systematic methods and defining design cycles:

1) Use a systematic instructional development method
When the design of training simulator specifications is clearly structured, it will be easier for designers to keep an overview of what they have done and what they still need to do, and to make sure that no important aspects are forgotten.

2) Define design cycles
The definition of design cycles is a way to make iterations explicit. Designing on a more global level first enables users to get a rough idea of the kind of product they are designing and will help them to get an overview of the design process before elaborating the design in more detail. Design cycles also force the designers to come back to their own work and review what they have done.

Other measures concern triggers for iteration that can be explicitly planned:

3) Plan decision points regarding the design process
Structuring the design process can be supported by planning explicit decision points to reflect on, and if necessary, adapt the design process. Decision points can be planned at fixed moments in the design method, prescribed by the organisation or defined for a specific project by the designers themselves.

4) Plan review moments
Another measure to support the design process is the definition of review moments where other people are invited to inspect the design and give comments or suggestions. Review moments are certainly necessary when not all of the relevant kinds of expertise are represented in the design team. Involving a wider range of stakeholders might complicate the design process, but will probably also enhance the acceptance of the chosen solution.

5) Generate and compare alternative solutions explicitly
It seems probable that the design will improve when designers generate and compare alternative ideas at important moments in the design process (as experts do). Comparing the advantages and disadvantages of alternative solutions can be supported with dedicated decision-making support systems.

Specific advice can be given about how to recognise and deal with unplanned triggers for iteration:

6) Strategic advice
Strategic advice concerns the way a certain design activity should be executed. It tries to give an answer to questions like: What should I do now? Which mistakes should I avoid? Which information do I need? Which constraints should I take into account? How do I decide whether I have enough input or information to go on? When should I stop with this and go on to the next design activity? Strategic advice can be made concrete and directly applicable by linking it to steps or activities in the design method, to decision points or to review moments.

7) Help to recognise triggers for iteration
This form of advice concerns the relationships between design activities and events that should alert users to go back and review decisions or elaborate their work further. It can be related to specific design activities or given as general guidelines stating where and how different kinds of information should influence the design. It can also explain when iteration is not desirable or necessary.

The Measures 8 to 13, described below, are forms of knowledge management that make it easier for designers to iterate. They can be seen as prerequisites for iteration.

8) Keep intermediate results available
To make it possible to reuse information that was collected earlier and to elaborate on previous work, all intermediate results should be stored and be made available to users at all times.

9) Make consequences of changes visible
Where possible designers should be helped to see the consequences of their changes, e.g. by highlighting parts of the design that are connected.

10) Store notes and keep them available
Notes are important to document the reasons that underlie decisions, the alternatives that were considered and why they were discarded. This information is important when a decision is reviewed. Notes are even more important when designers work in a team, or when other people will have to take over at some moment during the design process.

11) Keep information on constraints separate and available
The resources that are available for designing specifications and for the production of products and their use influence the design process, but they are not linked to one activity.

12) Make assumptions explicit
Often not all the required information is available, information is insecure, or designers can not foresee whether their ideas will be feasible within the given constraints. They will have to make assumptions and take preliminary decisions based on whatever information is available. This is not necessarily a problem: if an assumption turns out to be incorrect, earlier decisions can be revised. However, this is only possible when the designers have kept track of the assumptions they have made and regularly check whether these are still valid.

13) Have experts available for information and advice
When experts are not part of the design team they should be available for questions, for example, by phone or by e-mail. It might be helpful to explicitly identify experts in different areas at the beginning of the design process and to ask them when they are available.

The different measures described above should help the users to take process-oriented decisions and support and encourage an iterative design process. The challenge is to offer concrete advice at the right moment and to clearly structure the design process, while at the same time allowing and supporting users to organise their own work flow in very different ways, depending on the characteristics of the project and the domain and their own preferences. Some of the measures are quite strong, e.g., prescribed design cycles or automatic warnings. They should not be seen as algorithms or cookbook recipes since it is clear that these are not available for ill-structured instructional design problems. The risk is that designers, especially novice designers, will interpret them as prescriptions instead of the heuristics that they are. Therefore, more research is needed to investigate how designers use this kind of help, and whether it helps them to deal with iteration adequately.

References

Blessing, L.T.M. (1994). *A process-based approach to computer-supported engineering design.* PhD thesis. Enschede, the Netherlands: University of Twente.

Braha, D., & Maimon, O. (1997). The design process: Properties, paradigms, and structure. *IEEE Transactions on Systems, Man, and Cybernetics–Part A: Systems and Humans, 27*, 146-166.

De Jong, A. J. M. (1986). *Kennis en het oplossen van vakinhoudelijke problemen: Een voorbeeld uit een natuurkundig domein* [Knowledge based problem solving: an example from a physics domain]. PhD thesis. Eindhoven, The Netherlands: Technical University of Eindhoven.

Goel, V. & Pirolli, P. (1989). Motivating the notion of generic design within information-processing theory: The design problem space. *AI Magazine, 10*, 19-36.

Goel, V. & Pirolli, P. (1992). The structure of design problem spaces. *Cognitive Science, 16*, 395-429.

Greeno, J.G., Korpi, M.K., Jackson III, D.N., & Michalchik, V.S. (1990). *Processes and knowledge in designing instruction* (Report No. N00014-88-K-0152). Stanford, CA, USA: Stanford University.

Hamel, R. (1990). *Over het denken van de architect: Een cognitief psychologische beschrijving van het ontwerpproces bij architecten* [On designing by architects: A cognitive psychological description of the architectural design process]. Doctoral dissertation. Amsterdam: University of Amsterdam.

Mettes, C.T.C.W., & Pilot, A. (1980). *Over het leren oplossen van natuurwetenschappelijke problemen: Een methode voor ontwikkeling en evaluatie van onderwijs, toegepast op een kursus Thermodynamika* [On teaching and learning problem solving in science: A method for the development and evaluation of instruction, applied to a course in thermodynamics]. PhD thesis. Enschede, the Netherlands: University of Twente.

Nelson, W.A., Magliaro, S., & Sherman, T.M. (1988). The intellectual content of instructional design. *Journal of Instructional Development, 11*, 29-35.

Perez, R.S. & Neiderman, E.C. (1993). Modelling the expert training developer. In R.J. Seidel and P.R. Chatelier (Eds.), *Advanced technologies applied to training design* (pp.261-280). New York: Plenum Press.

Rowland, G. (1991). *Problem solving in instructional design.* PhD thesis. Indiana: Indiana University.

Rowland, G. (1992). What do instructional designers actually do? An initial investigation of expert practice. *Performance Improvement Quarterly, 5,* 65-86.

Rowland, G. (1993). Designing and instructional design. *Educational Technology Research and Development, 41,* 79-91.

Van den Bosch, K., Barnard, Y.F. and Helsdingen, A.S. (1999). *Taak- en trainingsanalyse beeldanalist SPERWER* [Task and training analysis image interpreter SPERWER] (Report No. TM-99-A018). Soesterberg, the Netherlands: TNO Human Factors.

Van Wagenberg, M.J.G.M. (1992). *Gericht CAD-ondersteund ontwerpen en organiseren* (Goal-oriented CAD-supported designing and organising). PhD thesis. Delft, The Netherlands: Delft University of Technology.

Verstegen, D.M.L. (2003). *Iteration in instructional design: An empirical study on the specification of training simulators.* PhD thesis. Utrecht, the Netherlands: Utrecht University.

Acceptance of mobile phone text messages as a tool for warning the population

S. (Simone) Sillem, J.W.F. (Erik) Wiersma, & B.J.M. (Ben) Ale
Delft University of Technology, Safety Science Group
Delft, the Netherlands

Abstract

In an emergency situation the community is alarmed with sirens. In The Netherlands many complaints are expressed about the audibility of this siren. This mainly concerns the audibility indoors and in big cities with a lot of background noise. Local authorities of the city of Vlaardingen, regional fire services and a text message provider have studied the use of mobile phone text messages (in The Netherlands this is called SMS, Short Message Service) as an addition to the siren. The purpose of this SMS service was to improve information and instructions to people in case of an emergency. It was determined whether this service could reach more people than the current siren and whether the community, auditory impaired, medical and teaching personnel and shopkeepers, see the service as a useful addition to the siren. The SMS service is an effective and efficient addition to the current siren. This study has shown that the SMS service is technically feasible. The range of SMS is large; the number of people that are reached can be substantially increased with use of SMS. Besides this it is possible to send differentiated messages to different target groups. SMS messages can be heard indoors as well as outdoors and can substantially enlarge the range of warning.

Introduction

Warning a community in an emergency situation does not always function properly. Recent incidents in the Netherlands, such as an incident in 2003 in Vlaardingen, have shown this imperfection (Jansen, 2003; Temme, 2003). Vlaardingen is a city in The Netherlands of 74.000 inhabitants near the Rotterdam harbour. In a period of a few months, the inhabitants of Vlaardingen had to be warned twice; once because of a fire in a fishing vessel, the other time because of a large chemical spill. Evaluation of the incidents shows that people do not know what to do if the siren sounds. There are also problems in communicating what is happening and what should be done during an emergency situation (Temme, 2003). Furthermore, a lot of people have complained about the audibility of the siren (Vos, 2003). This mainly concerns the audibility of the siren indoors and in big cities with a lot of background noise.

In the Netherlands the siren is tested every first Monday of the month at noon. In all other conditions if the siren sounds people have to do three things. 1. Go indoors or

---

In D. de Waard, K.A. Brookhuis, R. van Egmond, and Th. Boersema (Eds.) (2005), *Human Factors in Design, Safety, and Management* (pp. 187 - 200). Maastricht, the Netherlands: Shaker Publishing.

stay indoors; 2. Close all doors and windows; and 3. Listen to the local radio station or watch the local television station. These tests are carried out to familiarize the population with the sound of the siren.

There has not been much research on the audibility of the current siren. Jansen (2003) has found that the sound of the siren is 5 dB under the manufacturer's specifications. The sirens are not designed to be heard indoors. When you are already indoors, you do not need to go indoors anymore. Not being able to hear the siren indoors is a reason for many complaints (Vos, 2003). Research by De Hond (2003) concludes that on average 35 % of the interviewed people did not hear the siren on three different test occasions in 2003. This leads to recommendations made by Temme et al. (2003) to investigate the possibilities for means of communication additional to the siren.

Using text messages in addition to the siren

In recent years, the use of mobile phones for text communication has grown explosively. Two billion text messages were sent in The Netherlands in 2002. This is on average 125 messages for each inhabitant. In 1999, this number was only 500 million, which is 4 times less (Consumentenbond, 2003). Research by Vilella et al. (2004) shows that mobile phone text messages (known in The Netherlands as SMS, Short Message Service) can improve the transmission of information. Their results show that SMS can be used to increase compliance with vaccination schedules and very probably with other preventive or therapeutic measures.

The City of Vlaardingen, the Regional Emergency Services Rotterdam-Rijnmond and a company specialized in warning through mobile telecommunications wanted to study the opportunities offered by SMS as a service to inform the public in an emergency situation. Four advantages of SMS are considered in relation to conventional alarms. First, SMS on a mobile phone can be regarded as a personal alarm. Each person is alarmed individually on his own mobile telephone. It is expected that SMS messages will be perceived much better than the sirens. Second, SMS offers the opportunity to send more information than the siren. The siren is only an warning sound without further information. Instructions can be added to the SMS message. The third advantage is that SMS offers the opportunity to send differentiated messages to different target groups. It is possible for example to differentiate between professionals that have specific roles in an emergency situation and the general public. Finally, mobile phones can be equipped with a vibrating alert. This may inform people with auditory impairments about an emergency situation through the use of SMS.

SMS providers have the possibility to send large groups of selected users a specific message simultaneously. There are two main distinct methods for this. The first possibility is to send an SMS message to every mobile phone in the vicinity of a certain sending antenna (this is called Cell Broadcast). In the second method, an SMS message is sent to a predefined group of people that have enrolled for the service. For technical reasons, this study uses the latter method. An advantage of this second method is that people can apply for certain codes (in this case postal codes regions) for which they want to be informed. People can for example apply for the postal code regions of their house, their job and the school of their children. A disadvantage of this method is that people will not be alarmed if something happens when they are in an area outside their predefined postal code regions.

Target groups

Three target groups have been identified in this study, the main population, the auditory impaired and the professionals. The auditory impaired are an important group, as they are often not able to hear the siren. There are about 100.000 totally deaf people in The Netherlands, which is about 0.6 % of the population. The number of auditory impaired is much larger. They can use a mobile phone with a vibrating alert for SMS messaging. The Dutch government has promised the auditory impaired that a means of alarm would be operating in 2003. Until now they are not being alarmed. Another target group in this study are people that have a role in emergency handling different from the general public based on their profession. In this study a group of teachers, physicians and shopkeepers have participated in their role as professionals. They are a special group and different actions are expected from them when the siren sounds. For example physicians need to be informed about what kind of substance is in the air to be able to prepare treatment, and shopkeepers and teachers need to keep people inside when the siren sounds, so they need different messages. It is important to know what these special groups want and expect from this service and whether they feel they can be better reached by SMS.

Purpose of this study

The purpose of the SMS service in addition to the current siren is to improve warning and instructing people in an emergency situation. This study has been carried out in the first half of 2004. The main objectives of this study were to evaluate how many people can be reached by the SMS service compared to the siren alone; whether the SMS service is financially and technically feasible; and whether citizens, auditory impaired and professionals see the SMS service as a useful addition to the current addition.

Research question

Adressed in this paper is the question: 'Do citizens, professionals and auditory impaired accept the SMS service and do they see it as a useful addition to the current siren?'

Methods

Three thousand randomly selected citizens of the city of Vlaardingen above the age of 16 were addressed to apply for the SMS test in a letter from the mayor. They could apply by letter or on the internet. Six hundred and five people applied and participated in the study. The largest group consisted of the citizens of the city of Vlaardingen (529 participants). The second group consisted of the auditory impaired (37 participants). The last group was the professional group (39 participants).

To answer the research question, data had to be gathered on the audibility of the current siren, the range of the SMS service and the acceptance by the participants of the siren as well as the SMS service. In Figure 1 the order of all the used measurements is shown.

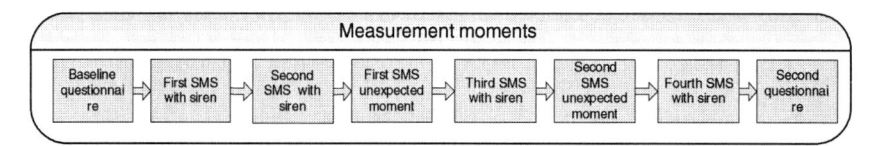

Figure 1. Timetable of moments of measurement

Six SMS messages were sent to the participants to determine the range of the SMS service. As the SMS service is meant to be an addition to the siren, four messages were sent at the same time as the monthly siren test at 12 o'clock (noon) on the first Monday of the month in March, April, May and June of 2004. The messages connected to the siren were sent a few minutes after the onset of the siren, to make sure that people could have heard the siren before receiving the SMS message. Besides these four messages connected to the siren test, two SMS messages were sent at different times to determine the range of the SMS service at unexpected moments. In every SMS message the participants were asked to respond as quickly as possible by sending a reply message. In the messages at unexpected moments they were asked to simply respond 'YES' as soon as they had read the message, in the message together with the siren they were asked to respond 'YES' or 'NO', depending on whether they had heard the siren before reading the SMS message.

Besides the SMS messages, all participants received two similar questionnaires. One before the start of the study, to determine the expectations about the SMS service, and one after the study to determine the (changes in) opinions. In total 1210 questionnaires were sent to the participants (605 prior to the start and 605 afterwards). From the first questionnaire 398 were returned and the last questionnaire had 388 returns. This is a response of 65%, which is presumed to be quite high. 288 Participants (72%) filled out both questionnaires. Questions were asked about how contented people were with the siren, how often they had heard the siren in the past 3 months, whether they expected that SMS would reach them better and by what means they would prefer to be alarmed.

The two questionnaires were kept quite short to guarantee a high response rate. To obtain some more detailed information an interview and three meetings with focus groups were carried out. The interview was carried out with two deaf people to determine their opinions and special needs concerning the SMS service and to see how they deal with the current methods of warning. Three focus group sessions were carried out with a total of 27 participants, two sessions with citizens and one with professionals, to see how they felt about the SMS service and to gain more detailed information than could be obtained by the short questionnaires. In the focus groups discussion between the participants about their opinions on the SMS service was encouraged.

Statistical analysis

The questions about the audibility of the siren and the question if SMS is seen as a usefull addition were analysed with a sign test. The sign test is chosen because this test compares the difference in averages of two individually matched groups with an ordinal variable (Nijdam & Van Buuren 1993). The questions if people can be better reached with SMS as an addition and how often people heard the siren in the last 3 months were analysed with a Wilcoxon signed-rank test. This test is used to analyse the difference in averages of individually matched groups with an interval level. For the significance level, $\alpha = 5\%$ is taken. Because the number of auditory impaired and professionals is very small, there will be no statistical analysis of their data.

Results

The current siren

In Figure 2, the number of times that participants heard the siren the *last 3 months previous to this study* can be seen in the left columns. A large number of people (80%) have not heard the siren every time in that period. 52% of the respondents have just heard the siren once or not at all. On average 49% did not hear the siren on every occasion. Figure 2 also shows the number of times the participants have heard the siren during the *last three months of the SMS test* (the right columns). It is clear that during the study the participants heard the siren more often than before the study. Only 27% of the participants that have sent a reply SMS have heard the siren once or not at all. Thirty five percent heard the siren all three times. On average 33% did not hear the siren on each occasion, compared to 49% before the study. This difference is significant ($Z=-7.66$, $N=259$, $p < .001$).

It seems that people are not very familiar with what to do when the siren sounds, 61% of the participants did not mention all three things that have to be done. An advantage of an SMS message in this situation is instructing people what to do during an emergency situation. Then people may know what is expected from them.

Figure 3 shows how contented participants were with the current siren, previous to and at the end of the study. Most participants that filled out the questionnaires were somewhat contented (31%) or somewhat discontented (29%) about the audibility of the current siren before the start of the study. A reasonable part is very discontented (18%). After the study 54% was somewhat contented or very contented and 30% is somewhat or very discontented. After the test people are more contented with the audibility of the siren ($Z=-4.88$, $N=271$, $p < .001$).

The auditory impaired were asked in the 'before' questionnaire whether and how they are now alarmed in case of an emergency. Fifty-six percent said that they are not alarmed at all and only one person said that he made arrangements about being warned by other people when the siren sounds. The rest is probably able to hear the siren, because their hearing impairment is not very severe.

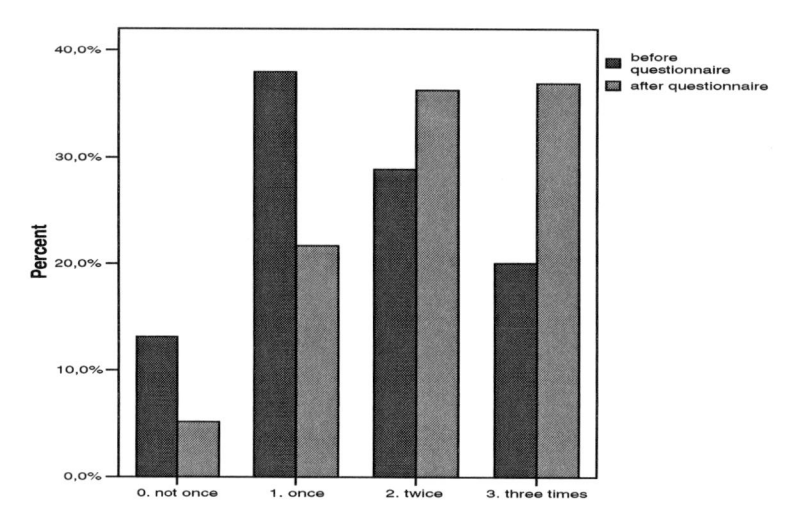

Figure 2. Question: 'How often did you hear the monthly siren test during the past three months?'

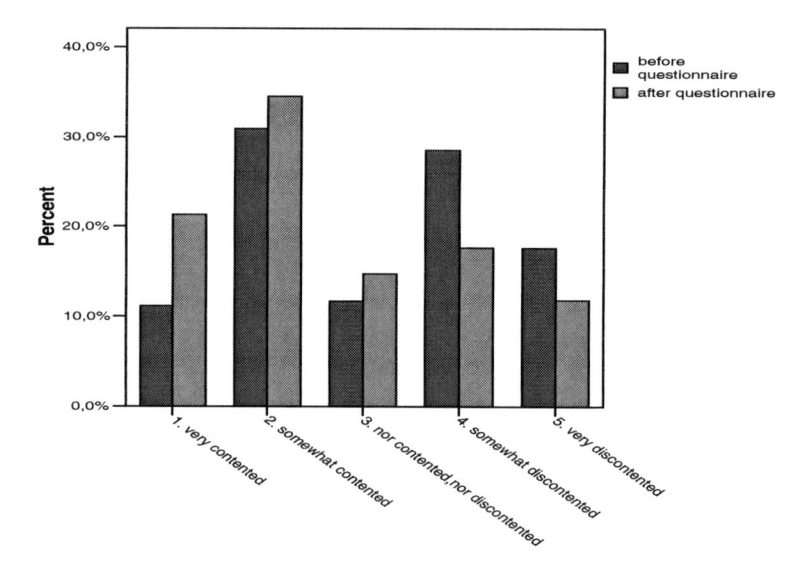

Figure 3. Question: 'About the audibility of the siren I am ....?'

The reading of and reacting to the SMS messages

It is interesting to look at the number of people that can be reached within seven minutes, because this is approximately the normal siren cycle in case of an emergency. Seven minutes is also an important time because of the need to alarm as many people as possible as quickly as possible, to be able to save lives. Forty two percent responded within seven minutes after sending the SMS messages, measured over all sent messages. Within half an hour 52% of the participants responded, eventually this number goes up to a response of 74%. On average 78% of the participants respond to the messages connected to the siren and 70% responded to the messages at unexpected moments. Figure 4 displays the cumulative distribution of the reply SMS messages. The fastest response was 200 reply messages in one minute. The late and slow sending of the first message due to some technical problems can be seen clearly by the low slope of that line.

cumulative reactions

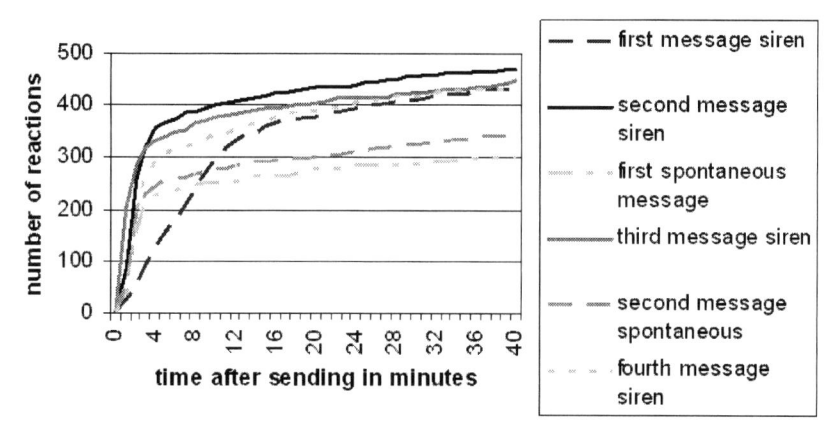

Figure 4. The cumulative number of reactions per minute. Time in minutes after sending the SMS message

Responses in the reply messages

Participants were asked in the SMS messages if they had heard the siren before they received the SMS message. There was a separate message for the auditory impaired, asking if they had been warned about the siren sounding before they read the SMS message, as they could not have heard the siren.

The number of people replying that they did not hear the siren before receiving the SMS message is a very important group. This is the group that was reached by SMS only. These people would not have been alarmed without the SMS service. On average 28% of the participants that sent a reply SMS said that they did not hear the siren before receiving the SMS message. The number of people that had not heard the siren has also been looked at per postal code region. The difference between the best and the worst postal code region is almost a factor 2 (20% compared to 37% that did not hear the siren), so the audibility of the siren is very much dependent on location.

The expectations of and opinions about the SMS service

In Figure 5 the participants' expectations about the usefulness of SMS as an addition to the siren are displayed.

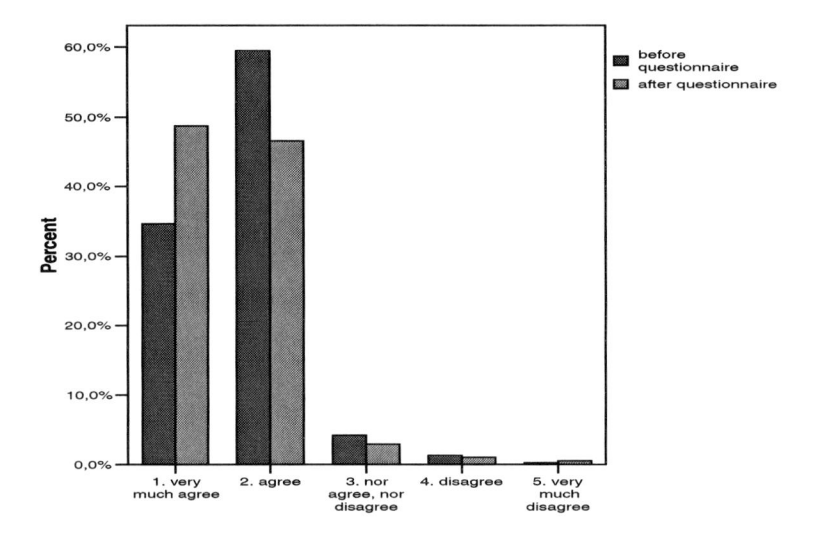

Figure 5. Proposition: 'SMS is a useful addition to the current siren'

Most people agreed very much (35.5%) or agreed (58.8%) prior to the start of the study that SMS is a good addition to the current siren. Afterwards, a lot more people very much agreed (49%) and less people disagreed (Z=–2.44, N=276, p < .01). The auditory impaired also said that SMS is a good addition, 80% very much agreed with this proposition after the study. None of the auditory impaired were neutral or disagreed. Figure 6. shows that before the test most people (91%) think that they will be better alarmed with SMS than with the siren alone. Afterwards, 84 % think that they will be better reached with SMS than with the siren alone. The difference between the two questionnaires is significant (Z=–2.96, N= 261, p < .01).

After the study, all of the auditory impaired say that they are better reached with SMS than with the siren alone. Before the study there was one person who thought he would be reached the same and one who thought he would be reached worse than with the siren alone. This high number is not unexpected, as at this moment the auditory impaired do not have any means of being alarmed when the siren sounds.

Means of warning

The participants have been asked how they would prefer to be alarmed. Table 1. shows their reactions. On the left the most preferred means of warning can be seen. The Table shows that before this study most people want to be alarmed by SMS (91%) and by the current siren (81%), although a lot of people added that the sound of the siren should be louder. At work and in leisure time most people also prefer

SMS and the siren, while during the night the siren and normal telephone are preferred. People feel they are not reached by SMS very well at night.

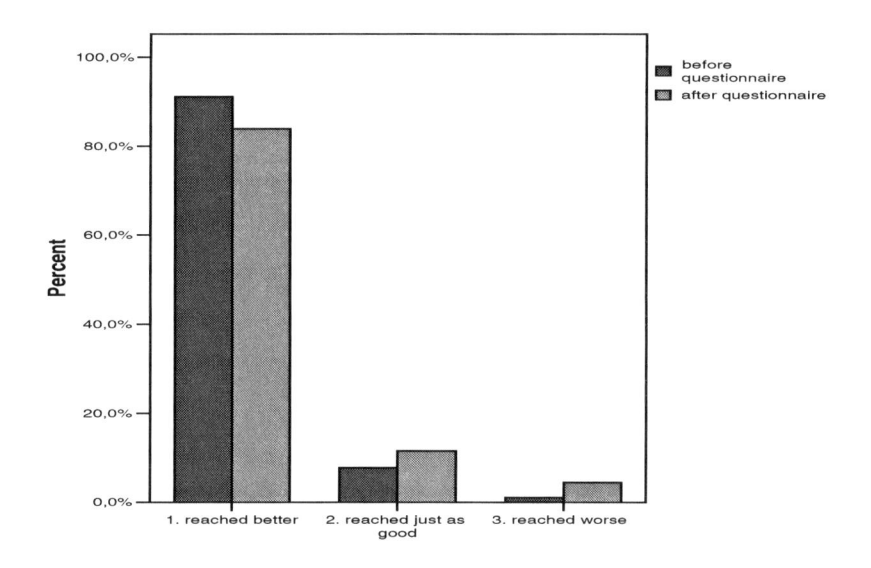

Figure 6. Question: 'With SMS as an addition to the current siren I am ...'

Table 1. Question: 'How would you prefer to be alarmed?'

baseline		SMS	Siren		Radio	TV		Normal phone	E-mail		different	
		91%		81%	63%	40%		23%	10%		10%	
afterwards		SMS	Siren		Radio	Normal phone	Car with sound		E-mail		TV	Different
	work	78%		70%	16%	13%	13%		6%		3%	2%
		SMS	Siren		Radio	Car with sound	TV		Normal phone	E-mail		Different
	leasure	81%		74%	12%	12%	9%		9%	2%		1%
		Siren	Normal phone	SMS	Car with sound	TV		Different	E-mail		Radio	
	at night	79%	43%	39%	31%	2%		2%	1%		1%	

As an extra option some people mention a car that drives around the area telling people what to do. Other means of warning mentioned are for example a louder siren or a national radio alarm that overrules all the other radio traffic.

The auditory impaired that participated in the focus groups consider SMS a good way of warning them, especially during the day and at work. At night a combination with neighbours that come to the door (which activates the flashlights that many auditory impaired have in their homes) is a good solution. According to the auditory impaired, the general public should be made aware that there are people that cannot hear the siren and that they should warn them in case of an emergency. The auditory impaired already have a vibrating alarm clock and would not like to have another apparatus under their pillow. Perhaps they could use their mobile phone's alarm clock function, so that they still have only one device under their pillow. It is worth noting that the auditory impaired prefer to be alarmed by the siren just as often as the main population at night and in leisure time. Maybe this is caused by the fact that many auditory impaired can hear the siren at night, as their impairment is not very severe. More often than the main population the auditory impaired prefer to be alarmed by television and their normal phone. On TV they can possibly use subtitles-for-the-auditory-impaired or sign language.

Other comments and tips

The last question in the questionnaires and in the interviews and focus groups was if people had other comments or tips to make a success out of the SMS service. The participants said it is very important to have up to date information about what is going on on TV and radio. Another interesting tip was to let the traffic lights in the emergency area flash the red light so that people that are approaching the area know that in no case they should drive on. People want to be warned when they are in an emergency area, not just when they are in their predefined postal code regions. Most participants say that they would forward the SMS messages to relatives and loved-ones. This would again enlarge the range of the SMS service. Because the sounding of the siren will always be in a relatively small area, there will be no problems with overloading the SMS system. Furthermore, messages originating from the government get priority over private SMS messages. The participants would also appreciate a separate ringtone for the alarm messages; so that they immediately know it is an alarm SMS instead of a normal message. About the content of the message the participanst were clear; they wanted to know what is going on, how serious it is, the location, clear instructions and radio frequencies for further information. The participants also said that the SMS service should be implemented as quickly as possible. The auditory impaired added a possibility to combine the siren and the flashlights in their home, which could also be done by means of SMS. They also added the fact that a lot of people that were born deaf have a low language level and that they might need adjusted messages.

Conclusions and discussion

The current way of warning through the siren is far from perfect. The siren is often not heard, especially inside buildings. And even if the siren is heard, many people do not know how to respond to it. The proposed system with SMS messages can reduce these problems substantially by reaching more people and giving more information, so that people know what is expected of them. By these means a system like the text message service can enlarge the ability of people to rescue themselves.

The SMS service is an effective and efficient addition to the current siren. This study shows that the range of SMS is large; the number of people that are reached can be substantially increased by means of the SMS service. The siren alone would not have alarmed the people that did read the SMS message, but did not hear the siren. Besides this it is possible to send differentiated messages to different target groups. SMS messages can be heard indoors as well as outdoors and can considerably enlarge the range of warning.

Many participants mention that the SMS service should be implemented very quickly. They consider this service very valuable and would like to use the SMS service. Furthermore, many more people are alarmed with the SMS service as an addition and more information can be given. Therefore, this study shows that SMS is a good addition to the current siren. The participants were very enthusiastic prior to the start of the study and the study made the degree of acceptance even higher. This follows from several aspects of the study. First of all, people were very willing to participate in the study. Next, an average of 74 % of people sent back a reply SMS message. There was also a high response rate for the questionnaires (65 %). Then, people were more satisfied about the audibility of the siren after the test than prior to the test. Furthermore, the participants feel that the SMS service is a good addition to the current siren (95 %, see Figure 5) and they feel they are better reached with an SMS as an addition than with the siren alone (84 %, see Figure 6). Finally, at work and in leisure time people prefer to be alarmed by SMS and the siren and at night the siren scores highest.

The current siren

Participants were more satisfied about the audibility of the siren after the test than prior to the test (see Figure 2, $Z=-4.88$, $p < .001$) for the question about the audibility of the siren as well as for the number of times that people heard the siren the last three months. This may be explained by the fact that participants were primed by participating in the test, paid more attention to hearing the siren on the first Monday of the month and therefore they actually heard the siren more often, and because of that they were more contented with the audibility of the siren.

The reading of and reacting to SMS messages

The range of the SMS service is very large, 74% of the participants responded to the SMS messages; the actual range is even larger as not everybody that reads the SMS message will send a reaction. Moreover, people can forward the messages to friends and relatives. In the reply SMS messages 28% say that they did not hear the siren. This is the profit of the SMS service, these people would not have been warned if they had not read the SMS message.

Expectations and opinions

Most people very much agreed (35.5 %) or agreed (58.8 %) with the proposition that SMS is a good addition to the siren before the test (see Figure 5). Afterwards quite a lot more people very much agreed (49%) and a little less people disagreed. A cause for this difference ($Z=-2.44$, $p < .01$) can be found in the fact that people received some SMS messages when they had not heard the siren and that they had not expected this prior to the start of the test.

Most participants (91 %) feel that they will be better reached (see Figure 6) by the SMS service than with the siren alone prior to the test. Afterwards this is still true. Now fewer people (84 %) feel that they are better reached than with the siren alone ($Z=-2.96$, $p < .01$). Perhaps people have experienced that at some occasions they did hear the siren before they received the SMS message and so this may be more of a technical judgment about the SMS service, about the range of the SMS service, as people want to be reached every time and not in 90 % of the cases. This could also have been influenced by the fact that the last three SMS messages together with the siren have been sent a bit late to make sure people could have heard the siren before receiving the SMS message. This may have caused people to think that the SMS service is slower than the siren.

Positive and negative aspects of the SMS service

According to the participants in the focus groups, the SMS messages are heard where the siren is not heard, for example indoors. Besides this a lot of ambiguity and doubt is taken away by the SMS service about what is going on and about what to do, because of the possibility to give extra information. The results from the professionals were not different from the results of the citizens, that is why they are not mentioned separately in the results or the conlusions.

Recommendations

A number of things have to be taken into account when implementing the SMS service.

Professionals

The professionals now do not have any extra information available about an emergency situation. There are extra things that the professionals have to do in an emergency situation, such as the keeping inside of shop visitors. So these professionals (shopkeepers, teachers and physicians) need to be interviewed to find out what kind of extra information they need to be better able to help other people and limit the consequences of an emergency situation. Think about the specification of a spilled chemical substance for physicians and hospitals.

Information in the SMS messages

The possible information in the SMS messages is not taken into account in this study but is very important. As soon as extra information is given in the SMS messages the responsibilities for the possible consequences must be considered thoroughly. The messages sent in this study were well understood but more research is needed to determine the best possible content and form of the messages. Only one interpretation should be possible for the messages. Therefore, these messages have to be written in advance. The extra information can be very valuable, but when an incorrect or ambiguous message is sent, the consequences can be disastrous. For example, when there is a chemical spill and the message is sent that there is a fire it is possible that people will go outside to see what is happening. This implies the need for more research on the reactions of people to the content of the messages.

The design of the test

The late sending of the SMS messages together with the siren (a few minutes after the onset of the siren) has had advantages and disadvantages. The advantage is that it is sure that people were given the opportunity to hear the siren before receiving the SMS message and that they were not primed to hear the siren by the SMS message. The disadvantage is that a lot of people were not aware that the SMS messages were sent late on purpose. This may have caused participants to think that the SMS service was slow. This may have influenced peoples' opinions on the usefulness of the SMS service negatively. It could also have influenced the sending of a reply SMS message. If people think the SMS service is slow, then they may not send a reply SMS as the usefulness of the service is estimated lower.

Type of SMS warning

Many participants said they would also like to be alarmed when they are in a certain disaster area that is not in one of their predefined postal code regions. So together with Cell Broadcast, SMS could form a very successful combination in warning citizens. Than people can be informed both when they are in a disaster area, and for certain postal code regions that they have predefined, such as the school of their children.

Acknowledgements

Many thanks are owed to Citizen Alert Services B.V. in The Hague and the Risk Centre at Delft University of Technology for making this study possible.

References

Consumentenbond (2003).
  http://www.consumentenbond.nl/nieuws/persberichten/Archief/2003/320206?ticket=nietlid
De Hond, M. (2003). Personal comunication.
Jansen, H. W. (2003). *Bepaling van geluidemissieniveaus van sirenes*.[assessment of sound emission levels of sirens], Report DGT-RPT-030077a Delft: TNO TPD.
Nijdam, B. & Van Buuren, H. (1993). *Statistiek voor de Sociale Wetenschappen; Deel 3 Vergelijking Groepen en Regressie* [Statistics for Social Sciences. Part 3: comprison of Groups and Regressions]. Alphen aan de Rijn, The Netherlands: Samsom BedrijfsInformatie.
Temme, B., Bekkers, H., Geveke, H., Lemmer, L., Stuurman, M. & van Erp, J. (2003). *Stank en sirenes; crisismanagement in Vlaardingen, Evaluatie van het Vopak-incident op 16 januari 2003* [Stench and sirens; crisismanagement in Vlaardingen. Evaluation of the Vopak-incident on January 16 2003]. Den Haag: B&A Groep Beleidsonderzoek & Advies BV.
Vilella, A., Bayas, J.-M., Diaz, M.-T., Guinovart, C., Diez, C., Simo, D., Munoz, A. & Cerezo, J. (2004). The role of mobile phones in improving vaccination rates in travelers. *Preventive Medicine, 38*, 503-509.

Vos, J. & Geurtsen, F.W.M. (2003). *Een laboratoriumstudie naar de hoorbaarheid van sirenegeluiden* [A laboratory study of audibility of siren sounds]. Report TM-03-C057. Soesterberg, The Netherlands: TNO Human Factors.

Ways of representing sound

Kirsteen Aldrich, Judy Edworthy, & Elizabeth Hellier
School of Psychology
University of Plymouth, UK

Abstract

It is important to know which features of sound confer similarity and which confer difference. This paper suggests that this depends on the circumstances in which sounds are heard, and the way in which this interacts with the type of sound heard. Similarity judgements were taken across three sets of sounds, one where the sounds were acoustically similar but from different sources, one where the source was the same but the sounds were different and one where the sounds were unfamiliar. Results showed that the same source sounds tended to be grouped by existing categories such as animal sounds, the unfamiliar on the basis of their acoustic properties and the same sound/different source by a mixture of acoustic and category-based descriptors. This knowledge has implications for the design and implementation of auditory warning sounds and other sounds used in human factors and ergonomics applications. The similar sound group will be reported here.

Methodology

15 participants took part in the study (for the similar sound group) and all the participants had normal or corrected-to-normal hearing as measured by self report. The sound stimuli were presented using a purpose written grouping program (Aldrich, Oct, 2003).

The similar sound set was made up from 20 sounds (not all abstract) made up of ten pairs of sounds that sound similar but come from different sources e.g. food frying and rain falling. Two other groups were employed but not discussed in detail here.

The program presented participants with 20 icons representing the 20 sounds from the similar sound set. The task required participants to sort the icons so that they represented groups of sounds that the participants felt were similar. In addition participants were asked to provide a one sentence explanation for each of their sound groups after the grouping task was complete.

Results

Multidimensional scaling analysis using a composite matrix from all 15 participants identified 3-dimensional solutions as the most appropriate for the similar sound set. The R^2 value suggested that 96% of the variance was accounted for with a stress score (Kruskal's stress formula 1) of .069. A range of acoustic (including RMS power, pitch & length) and affective measures (including word pairs e.g. safe/dangerous) were correlated with the sounds' locations on the MDS dimensions. Different sound groups showed different features as the most salient. The similar sound results will be focussed on here.

The *similar sound* group illustrated three clear features employed in participants' grouping decisions. The MDS dimensions showed a mixture of salient features in participants grouping decisions.

- dimension 1 - two measure of RMS power (RMS < 2000hz and average RMS power)
- dimension 2 - the affective measures calming/exciting and safe/dangerous
- dimension 3 - difficult to identify.

The hierarchical cluster analysis based on the composite matrix using the furthest neighbour method for this sound group (Figure 1) showed clear pairs of sounds that reflected the original sound pairs that were chosen for inclusion in the study for example, fryfood and rain were selected for their sounds similarity. There were also a few unintentional looser clusters towards the bottom of Figure 1. An animal and a machine group showed a tendency to use category information in the grouping decisions.

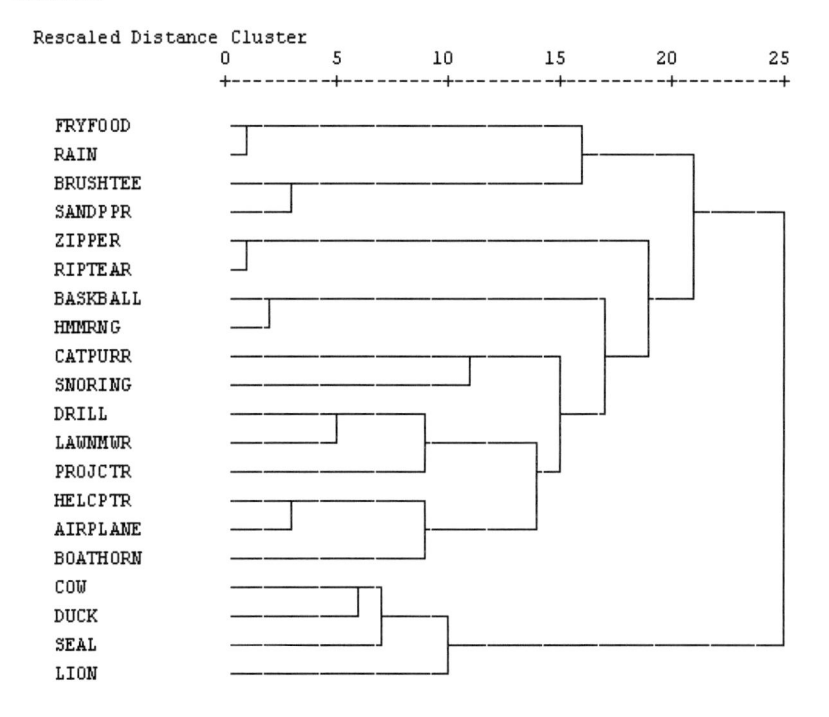

```
Rescaled Distance Cluster
                 0         5        10        15        20        25
                 +---------+---------+---------+---------+---------+
       FRYFOOD   ┐
          RAIN   ┘
      BRUSHTEE   ──┐
       SANDPPR   ──┘
        ZIPPER   ┐
       RIPTEAR   ┘
      BASKBALL   ──┐
        HMMRNG   ──┘
       CATPURR   ──────
       SNORING   ──────
         DRILL   ──┐
       LAWNMWR   ──┘
       PROJCTR   ──────
       HELCPTR   ──┐
      AIRPLANE   ──┘
      BOATHORN   ──────
           COW   ──────
          DUCK   ──────
          SEAL   ──────
          LION   ──────
```

Figure 1. Hierarchical cluster analysis for the similar sound group

The extra qualitative data highlighted that 55% of the reasons given by participants for their groups were based on categories, 25% as descriptive and 19% on acoustic explanations.

Discussion

For the similar sound group this experiment demonstrated a loose arrangement on the MDS output reflecting the sounds' original acoustically similar pairs. The hierarchical cluster analysis illustrated that the pairs were not clustered very closely but identified some categories forming. The Spearman's Rho correlations identified the following acoustic and affective features as salient in the similarity judgments for this sound group;

- dimension 1 – RMS < 2000Hz and average RMS power
- dimension 2 – calming/exciting, safe/dangerous and difficult to id
- dimension 3 – pleasing/annoying, familiar and difficult to id.

The participants' explanations provided information not provided by the statistical analyses such as the importance of category membership to the listener and rhythm which was difficult to measure in such variable stimuli.

The results suggest the relative strengths of acoustic and categorical information used to represent the sounds varies depending on the sounds used and the context in which they are used. For example, unfamiliar sounds showed little clustering on the MDS analysis compared to the other sound groups and showed a much stronger reliance on acoustic features according to participants self report.

The results demonstrate that category membership can override acoustic similarities between sounds, which has important implications for the way warning sounds and other sounds (such as telephone rings) used for human factors and ergonomics applications should be designed and implemented.

The effect of phonetic features on the perceived urgency of warning words in different languages

Mirjam van den Bos[1], Judy Edworthy[2], Elizabeth Hellier[2], & Addie Johnson[1]
[1]University of Groningen, The Netherlands
[2]Plymouth University, UK

Abstract

Previous studies have shown that acoustic and semantic features of warning signal words influence the perceived urgency of these words. The present study shows that phonetic features may influence perceived urgency as well. Experiment 1 presented listeners with eight warning signal words, each spoken in five different languages and two acoustic styles, non-urgent and urgent. The perceived urgency of the words was affected by acoustic style, even when the listeners did not understand the meaning of the words. Experiment 2 controlled for acoustic effects by using only one speaker and only non-urgently spoken words. This experiment showed that the words that were rated as least urgent tended to have a phonetic feature in common, the so-called 'sonorant'. This finding suggests that the use of warning signal words can be improved by avoiding words that contain certain phonetic features that make the word sound less urgent.

Introduction

Imagine you are at Moscow Airport and that you do not speak Russian. Suddenly, a stream of incomprehensible Russian words is spoken loudly through the speakers and the whole crowd starts running towards the exit! Obviously, something urgent is going on. But how could you have known, without observing the behaviour of the other people? An interesting question is whether people are able to estimate the urgency of a situation just by listening to the sounds, rather than the meaning of the words in a warning message. Just as there are general acoustic urgency features like high pitch and high intensity (Hellier, et al., 2002), it may be that there are also general phonetic features, that is specific vowels and consonants that make words sound urgent. The main aim in the present study is to explore the question of whether there are phonetic features of urgency and whether these are the same across different languages.

The acoustic determinants of urgency in warning systems have been thoroughly investigated over the past decades. Patterson (1982) introduced an ergonomic approach to auditory warning systems. His work elicited a stream of further research, mainly concentrated on the concept of 'perceived urgency' (e.g., Edworthy, Loxley & Dennis, 1991; Haas & Casali, 1995). Perceived urgency is considered to be inherent in the warning sound itself and to consist of acoustic features. Patterson's research concerned only the determinants of urgency in non-speech sounds. Other researchers have applied Patterson's findings about general acoustic urgency features to speech warnings (e.g., Edworthy et al., 2003; Hellier et al., 2002).

Hellier et al. (2002) compared the perceived urgency of warning signal words spoken in an urgent, non-urgent and monotone style. They found that the urgently spoken words were perceived as most urgent. They attributed this effect to the acoustic features of the urgently spoken words, which differed from the acoustic features of the words spoken in a non-urgent or monotone style. The words that were perceived as most urgent had a higher frequency, a broader pitch range and were louder than the non-urgent and monotone words. They also found that female speakers elicited a wider range of urgency judgements than did male speakers.

Edworthy et al. (2003) replicated and extended the findings of Hellier et al. (2002). They distinguished between three components of warning signal words: their physical structure, which they called acoustics; their meaning, which they called semantics; and the structure of the words at their voiced level, (i.e., the phonetics; International Phonetic Association, 1999). They investigated whether signal words contain specific urgency features that non-signal words do not contain. Therefore they compared synthetic versions of signal words with non-signal words that were similar in phonetic structure (e.g. 'deadly' compared with 'medley'). Edworthy et al. concluded that semantics had a clear effect on perceived urgency: Signal words were perceived as more urgent. Furthermore, they found that acoustics influenced the perceived urgency of both signal words and non-signal words, such that even a word like 'medley' was perceived as more urgent when spoken in an urgent style. This gives support to the conclusion that changes in speed, pitch and loudness can alter the way in which words are perceived. According to Edworthy et al., no reliable statements could be made about phonetic influences, as the phonetics of the signal words and the non-signal words were not properly matched.

In recent consumer research, a significant role has already been allocated to phonetics (Klink, 2000; Yorkston & Menon, 2004). Certain vowels and consonants have been found to make brand names more attractive. This consumer research was based on early investigations by Sapir (1929), who introduced the term 'phonetic symbolism'. Phonetic symbolism refers to the notion that the sounds of words convey meaning apart from their semantic connotation. For example, 'front vowel sounds' (such as the /i/ sound in *mill*) have been shown to connote meanings of fast, small or sharp. 'Back vowel sounds' (such as the /ɒ/ sound in *mall*) have been shown to connote meanings of slow, big or dull. The same type of effects has been noted with consonants as well. Certain vowels and consonants symbolise size (Sapir, 1929), whereas others convey degree of darkness (Newman, 1933). These findings on phonetic symbolism (Klink, 2000; Newman, 1933; Sapir, 1929; Yorkston & Menon, 2004), together with the findings on the determinants of perceived urgency in warning systems (Edworthy et al., 2003; Haas & Casali, 1995; Hellier et al., 2002; Loxley & Dennis, 1991; Patterson 1982) made us wonder whether a special role could be assigned to phonetics in warning designs. Perhaps urgency in warning messages is conveyed not only by semantics and acoustics, but also by the vowels and consonants of a word. If it can be shown that, independently of people's knowledge of the language or comprehension of the warning message, specific vowels or consonants convey urgency information, then it should be possible to improve existing warning systems and design new ones by choosing words which contain these urgency conveying vowels and consonants.

We investigated this issue in two experiments in which listeners rated the urgency of warning words from different languages. In Experiment 1, English listeners rated urgently pronounced and non-urgently pronounced words spoken in their native language and in four non-native languages. In Experiment 2, urgency ratings of words from five non-native languages, spoken in a non-urgent manner, were made and were subjected to multidimensional scaling and hierarchical clustering analyses.

Experiment 1

The main aim of Experiment 1 was to replicate previous studies showing that words with a high frequency, a broad pitch range and a high intensity are regarded as more urgent (Hellier et al., 2002) and to extend this result to languages not understood by the listeners. We also were interested in determining whether certain languages are regarded as conveying more urgency than others, presumably because they contain more 'urgent' phonetic features than others.

In this experiment, words were not adjusted for pitch and intensity but were spoken by native speakers in a non-urgent manner and an urgent manner. Post-hoc analyses were conducted to determine whether acoustic variables could account for any differences among ratings of languages or words.

Method

Participants
Thirty native English-speaking undergraduates at the University of Plymouth (9 male, mean age = 25, sd = 5.7) took part in this experiment. They received two credits towards their introductory psychology course requirement.

Stimuli
Eight warning words ('lethal', 'poison', 'deadly', 'attention', 'beware', 'danger', 'don't' and 'warning') were translated literally into Dutch, Spanish, Italian and Finnish (see Table 1). The translations were made by native speakers who were also fluent in English. Each word was spoken by a female native speaker in an urgent and non-urgent manner, resulting in 80 word stimuli (8 words x 5 languages x 2 urgencies).

Design
A 2 (urgent vs. non-urgent manner) x 8 (warning word) x 5 (language) within-subjects design was used. Each participant rated all 80 word stimuli once. The words were presented in random order for each participant.

Procedure

Participants were asked to estimate the urgency of the sounds by assigning a number to each of the words, proportionate to the urgency of that word (Engen, 1971). After they had written down their answer, the experimenter presented the next word. After the experiment, recognition of the languages in which the words were spoken was assessed by asking participants whether they spoke other languages besides English, how well they spoke these languages and whether they recognised the languages in which the words in the experiment were spoken. Four participants indicated familiarity with one or more of the languages besides English and were therefore excluded from the analysis.

Results

The urgency ratings were logarithmically transformed in order to fit the answers of each participant on the same scale. A repeated measures analysis of variance (ANOVA) with urgency (urgent or non-urgent manner), language (English, Dutch, Spanish, Italian or Finnish) and word (Lethal, Poison, Deadly, Attention, Beware, Danger, Don't or Warning) as factors was conducted on the ratings data.

A preliminary analysis including all languages revealed that all main effects and interactions were significant. The main effect of urgency ($F(1,25) = 196$, $p < .001$, MSE = .482) was such that the urgently spoken words (mean = .96) were perceived as more urgent than the non-urgently spoken words (mean = .51). The main effect of language ($F(4,100) = 4.1$, $p = .004$, MSE = .127) reflected that English words were perceived as more urgent than words spoken in the other languages (mean English = .78, mean Dutch = .70, mean Spanish = .74, mean Italian = .74, mean Finnish = .70). The Urgency x Language interaction ($F(4,100) = 22.6$, $p < .001$) was such that, presumably due to the contribution of semantics to urgency rating of non-urgent English words, the urgency effect was smaller for English than for the other languages (see Figure 1).

A second repeated measures ANOVA with urgency (urgent or non-urgent manner), language (Dutch, Spanish, Italian or Finnish) and word (Lethal, Poison, Deadly, Attention, Beware, Danger, Don't or Warning) as factors revealed that when the English words were not included in the analysis, the main effect for language ($F(3,175) = 3.0$, $p = .057$) and the interaction effect of language and word ($F(21,525) = 1.9$, $p = .110$) disappeared. However, the main effects for urgency ($F(1,25) = 178$, $p < .001$) and word ($F(7,125) = 8.6$, $p < .001$), the Urgency x Word interaction ($F(7,175) = 9.3$, $p < 0.001$) and the Urgency x Language x Word interaction ($F(21,525) = 3.8$, p < .005$) remained. Urgently spoken words received higher urgency ratings than non-urgently spoken words (.95 vs. .49 for urgently and non-urgently words, respectively). The translations of the words "attention" and "beware" received relatively high urgency ratings (.79 vs. .70 for attention and beware vs. all other words, respectively). This effect was presumably due to semantics in that the translations in some languages resembled the English word "attention". There were no discernable patterns in the three-way interaction.

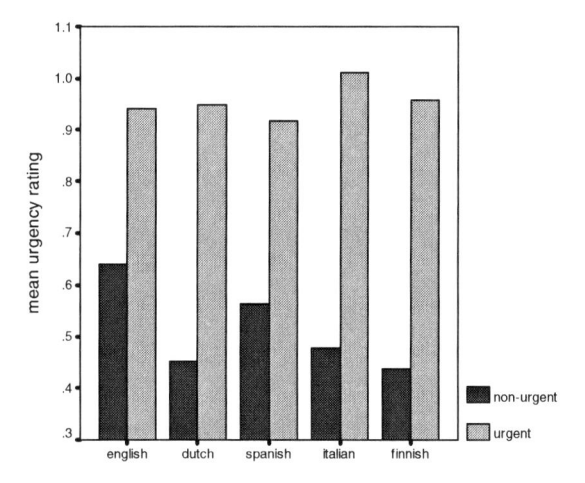

Figure 1. Mean urgency ratings as a function of urgency and language in Experiment 1

Acoustic measures

Previous research has shown that the perceived urgency of sounds increases as their mean frequency, frequency range and intensity increases (Hellier et al., 2002). Although we did not manipulate our stimuli to control for these variables, we did instruct the speakers of the 'urgent' stimuli to speak in an urgent manner. An analysis of the acoustic dimensions of the stimuli was carried out as a manipulation check. The software program PRAAT: doing Phonetics by Computer (Boersma, 2001) was used to obtain measures of average, maximum and minimum pitch, average, maximum and minimum intensity and duration. ANOVAs conducted with the acoustic measures mean, maximum and average pitch, mean, maximum and average intensity and duration as dependent variables and urgency (urgently or non-urgently spoken) as the independent variable showed that urgently pronounced words had higher values for each of the acoustic measures than the non-urgently pronounced words had (see Table 1).

Similar ANOVAs were conducted on the non-urgently spoken words with word and language as factors. Averaged across language, the words did not differ in any of the acoustic measures. The languages, however, differed in mean pitch ($F(4,33) = 3.5$, $p = .017$; see Figure 2). This difference in mean pitch could be due to speaker differences, as we used a different speaker for each of the languages.

In sum, Experiment 1 showed that words that were spoken in an urgent style were perceived as more urgent, even when they were not understood. The urgency effect was smallest for the English language, due to semantic influences on the ratings of the non-urgently spoken English words.

Table 1. Means of the acoustic variables for the urgent and non-urgent words in Experiment 1

	Urgency					
Acoustic variable	Non-urgent		Urgent			
	Mean	SD	Mean	SD	$F(1,73)$	p
Mean pitch (Hz)	193.60	20.36	288.90	54.90	100.4	<.001
Minimum of pitch	152.20	34.72	186.00	43.01	14.0	<.001
Maximum of pitch	267.40	99.92	403.80	104.53	33.4	<.001
Range of pitch	115.10	101.14	217.70	110.33	17.6	<.001
Mean intensity (dB)	72.70	3.10	76.60	2.89	31.5	<.001
Maximum of intensity	88.90	1.78	92.10	0.48	110.6	<.001
Range of intensity	31.90	2.99	35.60	1.92	40.6	<.001
Minimum of intensity	57.03	2.64	56.50	2.06	<1	NS
Duration (ms)	0.58	0.14	0.57	0.12	<1	NS

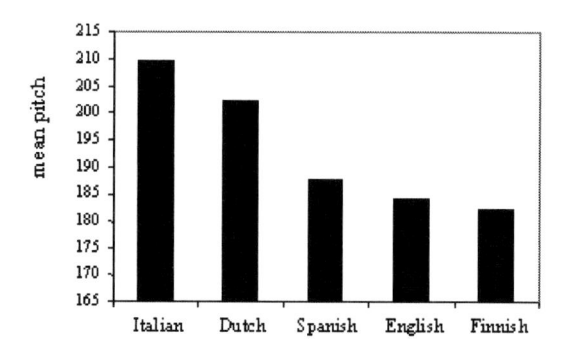

Figure 2. Mean pitch of the languages in Experiment 1

Experiment 2

In the second experiment, only non-urgently pronounced words were included, in order to minimize acoustic influences on urgency ratings. English words were replaced by German words to remove the impact of semantic influences. Another difference with Experiment 1 was that the acoustic variability was reduced by using only one speaker for all the languages. This person was trained by native speakers of each language to pronounce the words correctly. We used multidimensional scaling to investigate whether any relationship exists between the way words are sorted and the acoustic measures of the words. A hierarchical cluster analysis was performed on the same data to investigate a possible relationship between the way words are sorted and the phonetic features of the words.

Method

Participants
Thirty native English speakers (16 male, mean age = 22, sd = 4.2) took part in this experiment for credits towards an introductory psychology course requirement.

Stimuli
The same words as in Experiment 1 were used, but the English words were replaced by German words. The speaker spoke the words as if there was nothing special about the words (that is, in a non-urgent manner). The 8 warning words were 'lethal', 'poison', 'deadly', 'attention', 'beware', 'danger', 'don't' and 'warning'. These English words were translated literally into 5 languages (Dutch, Spanish, Italian, German and Finnish) as in Experiment 1. Each participant heard all 40 words once.

Design
The experiment used an 8 (warning word) x 5 (language) within-subjects design. The procedure for the urgency estimations was the same as in Experiment 1. At the end of the session a short questionnaire was administered to the participants to assess knowledge of the languages in which the words were spoken. For the multidimensional scaling, participants were asked to listen to each word as many times as they wanted and to sort the words which they regarded as belonging together into groups. They were allowed to choose their own criterion for deciding which words should be sorted together into one group (instructions adapted from Bonebright, 2001). These same data were used in the hierarchical cluster analysis.

Results

Acoustic measures
For each word, measures were taken of mean, minimum, maximum and range of pitch, mean minimum, maximum and range of intensity and the duration of each word. ANOVAs with language as a factor (Dutch, Spanish, Italian, German and Finnish) showed that the languages differed only in maximum pitch ($F(4, 34) = 2.76, p = .043$). Italian had the highest maximum pitch (mean = 411.82). A post-hoc analysis showed that Italian differed significantly from Dutch ($p = .006$) and Finnish ($p = .009$).

We also attempted to rule out semantic factors by asking participants whether they could guess the meaning of the words. A regression analysis with 'meaning guessed' as an independent variable (the value 1 was assigned when at least 90% of participants could guess the meaning and the value 0 otherwise) and estimated urgency as the dependent variable showed that the variable 'meaning guessed' was not a predictor of perceived urgency ($F(1, 38) = 2.455$, ns.).

A repeated measurement analysis with language (Dutch, Spanish, Italian, German and Finnish) and word ('lethal', 'poison', 'deadly', 'attention', 'beware', 'danger', 'don't' and 'warning') as factors conducted on the logarithmically transformed urgency ratings showed main effects of word ($F(4,100) = 4.1, p = .004$, MSE = .043) and language ($F(4,26) = 9.3, p < .001$, MSE = .04). However, no one language or word was consistently rated as conveying the most urgency, as was reflected in the significant Language x Word interaction, ($F(28,2) = 3.9, p < .001$); see Table 2).

Table 2. Mean urgency rating as a function of word and language in Experiment 2

	Language					
Word	Dutch	German	Italian	Finnish	Spanish	**Mean**
Lethal	.5012	.6188	.4628	.5929	.5056	.5439
Poison	.6353	.5458	.4513	.5798	.6058	.5530
Deadly	.5338	.5544	.4994	.4866	.6194	.5186
Attention	.6313	.6467	.6236	.6590	.7266	.6402
Beware	.6655	.6543	.5948	.6170	.6674	.6329
Danger	.4410	.5887	.6097	.5964	.6463	.5590
Don't	.5482	.6337	.7381	.5742	.6911	.6236
Warning	.4879	.5470	.4300		.7335	.4883
Other		.4471				
Mean	.5555	.5818	.5512	.5866	.6495	.5686

Multidimensional Scaling
Multidimensional Scaling is a statistical technique for determining the dimensions according to which stimuli are sorted into groups. After having obtained the results of this analysis, we correlated the obtained dimensions with estimated urgency and all our acoustic measures. This was done in order to find out whether people had sorted the words according to acoustics or estimated urgency or both. If words are grouped according to estimated urgency, we can then by means of hierarchical cluster analysis look at the separate groups and determine whether the words in each group contain the same phonetic features, which would suggest that phonetic features influence estimated urgency.

A multidimensional scaling (MDS) analysis was conducted on the word groupings and the solution with the lowest stress (.16; badness-of-fit measure for the entire MDS representation) and the highest RSQ (.76; squared correlation) with three dimensions was found. As indicated above, we looked at the correlations of the three dimensions with the acoustic measures and estimated urgency. Dimension 1 was correlated with pitch maximum ($r = .334, p = .035$) and pitch range ($r = .395, p = .012$), Dimension 2 was correlated with estimated urgency ($r = .343, p = .030$) and mean pitch ($r = .484, p = .002$), and Dimension 3 was correlated with estimated urgency ($r = -.386, p = .014$). Dimensions 1 and 3 are presented together in Figure 3. Dimension 2 is not considered because it correlated significantly with both acoustic measures and estimated urgency.

Euclidean distance model

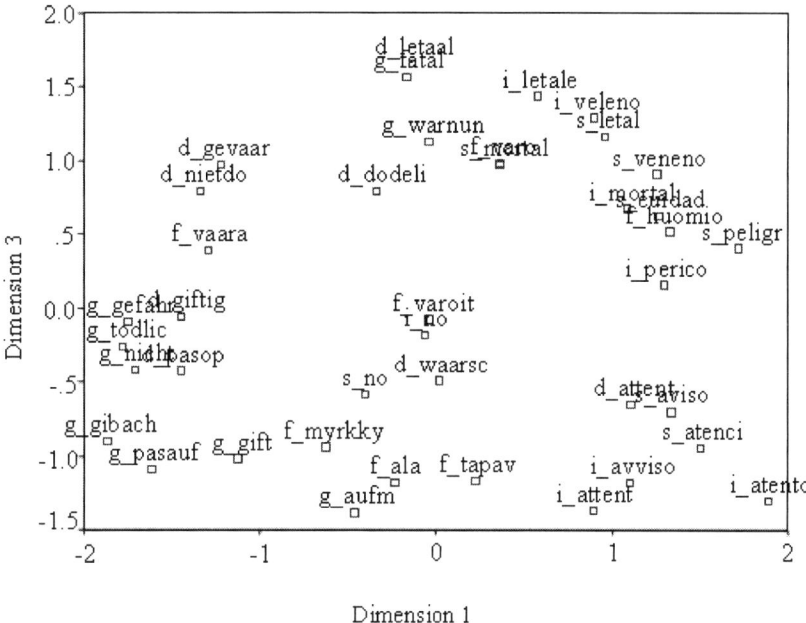

Figure 3. Pitch maximum and pitch range as a function of perceived urgency in Experiment 2

Hierarchical Cluster Analysis

Having determined that the words were at least in part grouped according to estimated urgency, we examined the separate clusters of words to determine whether the words in each cluster contained the same phonetic features. Such a finding would be an indication that phonetic features influence estimated urgency.

A hierarchical cluster analysis was performed on the same data used in the multidimensional scaling analysis. In particular, we examined whether words were grouped according to common phonetic features and whether the words in these groups were also close in distance in the urgency ranking space. The hierarchical clustering solution is shown in Figure 4.

As shown in Figure 4, people tended to group Dutch and German together, and Italian and Spanish. That is, there seemed to be a global distinction between Germanic languages and Romance languages. The Finnish words were spread between both groups rather than forming a separate group. The Romance language words had a higher mean estimated urgency (Italian + Spanish = .62) than the Germanic words (mean Dutch + German = .58), but this difference was not significant.

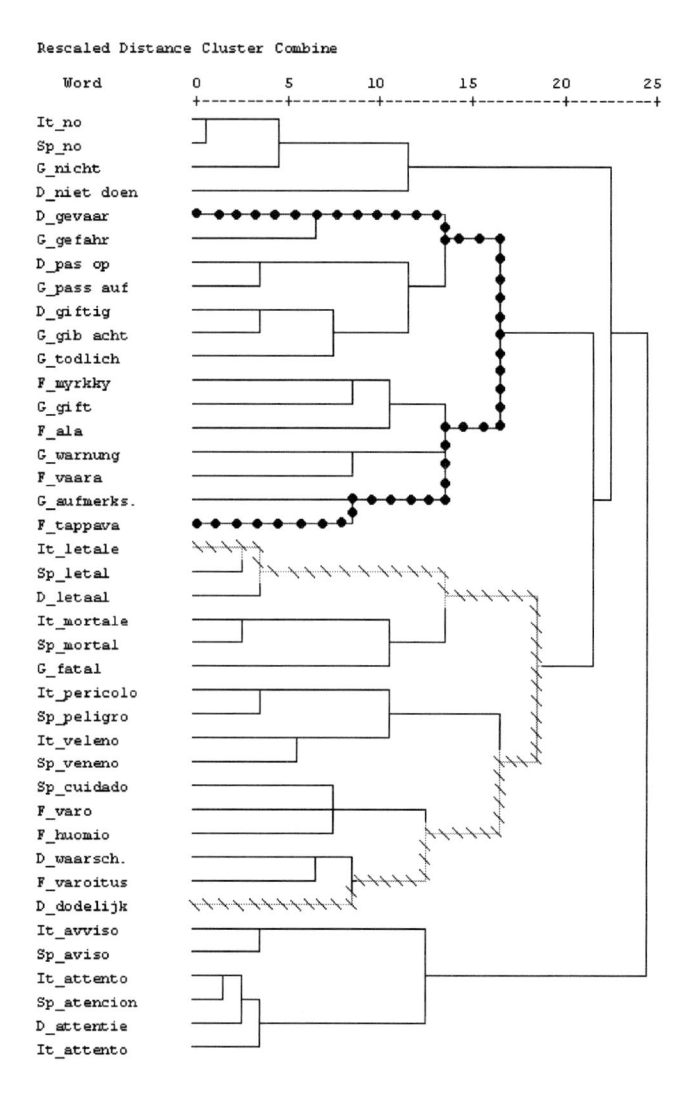

Figure 4. Hierarchical cluster analysis dendrogram (dots = Germanic languages, stripes = Romance languages)

Words that sounded similar tended to be grouped together, such as 'gefahr' and 'gevaar' and 'veneno' and 'veleno'. These words also tended to have been given similar urgency ratings. It can be asked whether this is a biasing effect or an effect that is relevant for the present study. It could be a relevant effect, because it shows that words that are similar regarding phonetic features, also obtain the same urgency ratings.

Phonetic analysis
An examination of the results of the MDS (see Figure 3) showed that words that grouped together on the low end of the 'urgency' dimension all contained a specific phonetic feature, the sonorant. Sonorants are a special class of consonants. They may be syllabic but are always voiced, that is they are produced by constricting but not closing the vocal tract, for example the 'w' in *warning*. The letters 'm', 'l' and the 'ng' combination are also examples of sonorants. In general, there are three subtypes of sonorants: liquids (lateral [l] and circumflex [r]), glides (labiovelar [w] and palatal [j]) and nasals ([m n ò ü]). Sonorants make a word sound soft and lingering and have more acoustic energy than other consonants (IPA, 1999).

We investigated the possibility of sonorants being related to urgency estimations by counting the number of sonorants in each separate word and relating this outcome to the estimated urgency ranking order. (Phonetic transcriptions of the words are presented in Appendix 1.) Sonorants were present in equal numbers in the warning words from each language, so do not seem to play a role in the distinction between Germanic and Romance languages. A correlation analysis conducted on all words for which the meaning of the word could not be guessed, showed that the words containing the most sonorants tended to be perceived as least urgent ($r = -.332$, $p = .037$). Thus, it appears that sonorants may influence the perceived urgency of warning words when phonetic features are the only clues people can rely on.

General Discussion

The key finding in the present study is that acoustic and semantic features are not the only determinants of the perceived urgency of warning words. Experiment 1 replicated results from previous studies showing that high frequency, a broad pitch range and high intensity make warning words sound more urgent (Edworthy et al., 2003; Hellier et al., 2002) and extended these results by showing that acoustic measures also influence the perception of warning words from languages not understood by the listeners. In fact, the difference in perceived urgency between urgently and non-urgently spoken words is greater when the message is not understood than when it is.

A possible shortcoming of the present study concerns the particular warning words we used, which were the English words adapted from Edworthy et al. (2003) literally translated into Dutch, German, Italian, Spanish and Finnish. These literally translated words may not have had the same urgency connotation as their English counterparts. A second shortcoming of the present study could be that in Experiment 1 we only presented the warning words once. Although this method resembles the real world, in which people may hear warning words only once, it makes it hard to compare the results of the present study with other studies in which words were presented multiple times.

In Experiment 2 we showed that when listeners cannot rely on the acoustic or semantic information in a warning word, they make use of the phonetic features of the word. The main phonetic feature that influenced perceived urgency in the present study was the sonorant. Words with sonorants were rated as being less urgent than words not containing sonorants. Future research is, however, necessary to explore the role of different consonants and vowels in more detail. In future research, it may also be better to use a computer voice to reduce the acoustic variability of the words.

We suggest that understanding the role of phonetics in perceived urgency is necessary to optimize warning systems. It may be possible to improve existing warning systems by avoiding the use of words that contain urgency reducing phonetic features and instead choosing words that contain urgency increasing phonetic features.

Another, possibly related recent development in warning designs is the use of auditory icons and earcons. Auditory icons (Gaver, 1997) are natural, everyday sounds that can be used to represent actions and objects within an interface. They rely on an analogy between the everyday world and the world being modelled. For example, copying files on a computer could be accompanied by a sound rising in pitch, indicating the receptacle was getting fuller as the copying progressed. One of the main advantages of auditory icons is the ability to communicate meanings which listeners can easily learn and remember.

Earcon systems, in contrast, use abstract sounds whose meaning is harder to learn. Earcons are non-verbal audio messages that are used in the human/computer interface to provide information to the user about computer objects, operations or interactions (Blattner et al., 1989). Unlike auditory icons, there is no intuitive link between the earcon and what it represents; this link must be learned. Earcons are composed of short, rhythmic sequences of tones, which can vary regarding level, rhythm, timbre and intensity. Earcons have shown to be an effective method for communicating information in a human/computer interface (Brewster et al., 1993; McGookin & Brewster, 2004).

Perhaps it is possible to create 'Phonicons' in warning designs. These would be word-like sounds, composed of urgency-conveying vowels and consonants. By using different kinds of vowels and consonants, urgency levels could be altered. It should be easy to learn the association between the Phonicon and the urgency level it refers to as there would be an intuitive link between the phonetics and the degree of urgency. In this way, new warning systems could be designed and existing warning systems could be extended by taking phonetics into account.

References

Blattner, M., Sumikowa, D., & Greenberg, R. (1989). Earcons and icons: Their structure and common design principles. *Human Computer Interaction, 4*, 11-44.

Boersma, P. (2001). PRAAT, a system for doing phonetics by computer. *Glot International, 5*, 341-345.

Bonebright, T. (2001). Perceptual structure of everyday sounds: A multidimensional scaling approach.
www.acoustics.hut.fi/icad2001/proceedings/papers/bonebri2.pdf

Brewster, A., Wright, P.C., & Edwards, A.D.N. (1993). An evaluation of earcons for use in auditory human computer interfaces.
http://www.dcs.gla.ac.uk/~stephen/publications.shtml

Edworthy, J., Hellier, E., Walters, K., Clift-Mathews, W., & Crowther, M. (2003). Acoustic, semantic and phonetic influences in spoken warning signal words. *Applied Cognitive Psychology, 17,* 915-933.

Engen, T. (1971). Psychophysics: 1. Discrimination and detection. In J. W. Kling & L. A. Riggs (Eds.), *Woodworth and Schlosberg's experimental psychology* (pp. 11-46). New York: Holt, Rinehart & Winston.

Esling, H. (1999) IPA Handbook. A guide to the use of the International Phonetic Alphabet. http://www2.arts.gla.ac.uk/IPA/ipa.html

Gaver, W. (1997). Auditory Interfaces. In M. Helander & T. er & P. Prabhu, Handbook of Human Computer Interaction (2nd ed., pp. 1003-1042). Amsterdam: Elsevier.

Hellier, E., Wright, D.B., Edworthy, J., & Newstead, S. (2000). On the stability of the arousal strength of warning signal words. *Applied Cognitive Psychology, 14,* 577-592.

Hellier, E., Edworthy, J., Weedon, B., Walters, K., & Adams, A. (2002). The perceived urgency of speech warnings 1: Semantics vs acoustics. *Human Factors, 44,* 1-17.

Klink, R.R. (2000). Creating brand names with meaning: The use of sound symbolism. *Marketing Letters, 11,* 5-20.

McGookin, D.K., & Brewster, S.A., (2004). Understanding concurrent earcons: Applying auditory scene analysis principles to concurrent earcon recognition. *Transactions on Applied Perception, 1,* 130-155.

Sapir, E. (1929). A study in phonetic symbolism. *Journal of Experimental Psychology, 12,* 225-239.

Yorkston, E.A., & Menon, G. (2004). A sound idea: Phonetic effects of brand names on consumer judgements. *Journal of Consumer Research, 31,* 43-51.

Appendix 1. Phonetic transcriptions of the warning words in Experiment 2.

Word	Phonetic transcription (IPA) (sonorant in bold)	Number of sonorants	Urgency ranking order
No	/n/o/	1	1
Attenzione	/a/tt/ɛ/ɲ/ʣ/i/o/n/ɛ/	2	14
Pericolo	/p/ɛ/r/i/k/ɔ/l/ɔ/	2	18
Attentoa	/a/tt/ɛ/ɲ/t/o/l/a/	1	21
Mortale	/m/ɔ/r/t/a:/l/ɛ/	2	33
Letale	/ʎ/e/t/a:/ʎ/ɛ/	2	36
Veleno	/v/e/ʎ/e/n/ɔ/	2	37
Avviso	/a/vv/i/s/o/	0	40
Aviso	/a/b/i/s/o/	0	2
Atencion	/a/t/ɛ/ɲ/T/i/o/n/	2	3

No	/n/ɔ/	1	4
Cuidado	/k/i/d/a/d/o/	0	5
Peligro	/p/e/ʎ/i/ɣ/r/ɔ/	1	10
Mortal	/m/ɔ/r/t/a/l/	2	15
Veneno	/b/ɛ/n/e/n/ɔ/	2	19
Letal	/ʎ/e/t/a/ʎ	2	31
Huomio	/h/ʊ/ɔ/m/i/ɔ	1	35
Vaaro	/v/a:/r/ɔ/	1	7
Vaara	/v/a:/r/a/	1	17
Tappava	/t/a/p/p/a/v/a/	0	22
Myrkky	/m/ə/r/k/k/i/	2	24
Älä	/e/l/e/	1	20
Varoitus	/w/a/r/oi/t/ʊ/s/	2	25
Paßauf	/p/a/s//au/f/	0	8
Gibacht	/ɣ/ɪ/p//a/χ/t	0	9
Nicht	/n/i/χ /t/	1	12
Fatal	/f/a/t/a:/l/	1	16
Gefahr	/ɣ /ɛ /f/a:/r/	1	23
Aufmerksamkeit	/au/f/m/ɛ /r/k/s/a/m/k/ai/t/	3	26
Warnung	/w/a:/r/n/u/ŋ /	4	28
Tödlich	/t/ʌ :/d/l/i/χ	1	29
Gift	/ɣ /i/f/t/	0	38
Pas op	/p/a/s//ɤ /p/	0	6
Giftig	/χ /ɪ /f/t/ɪ /χ /	0	11
Attentie	/ɐ /t/æ/n/t/i/	1	13
Niet doen	/n/i:/t//d/u:/n/	2	27
Dodelijk	/d/ou/d/ɐ /l/ɐ /k/	1	30
Letaal	/l/e/t/a:/l/	2	32
Waarschuwing	/w/a:/r/s/χ /ʊ /w/ɪ /ŋ /	4	34
Gevaar	/χ /ɐ /v/a:/r/	1	39

A contextual vision on alarms in the intensive care unit

Adinda Freudenthal[1], Marijke Melles[1], Vera Pijl[1],
Addie Bouwman[2], & Pieter Jan Stappers[1]
[1]*Faculty of Industrial Design Engineering, Delft University of Technology*
[2]*Department of Nursing Affairs, University Medical Centre Groningen*
The Netherlands

Abstract

In the intensive care unit (ICU) alarms are crucial, but the large number of auditory alarms and the lack of harmonization also cause many problems for nurses. Despite intensive research efforts, no radical improvements have resulted so far. Our aim was to find new directions to get away from the current status quo in design of alarms for the ICU. To do this the focus was on work task support in the context of the entire nursing process of the team of nurses at the ward on the one hand and on developing a congruent overarching alarm/information system on the other hand. A participatory research and design approach was used and a new vision on alarms is proposed. Related fields of relevant ergonomic research and areas for technological development are identified.

Introduction

An intensive care unit (ICU) supplies specially trained personnel and monitoring and therapy equipment. Close observation and immediate recognition of potentially life-threatening complications is crucial in treatment of the seriously ill patients (Skillman, 1975). Alarms are one of the main tools to notify nurses about a new situation in this often unpredictable work environment. In the current study the use of alarms is studied through a participatory research and design approach. New solutions tailored to support the work process are proposed.

The problems concerning alarms in the ICU are well-known and have been subject of numerous studies. First of all, there is the high proportion of alarms which are unrelated to emergencies. Chambrin et al. (1999) for example observed in an adult ICU that more than 70% of all alarms were false positives (alarms leading to no action). Lawless (1994), observing in a paediatric ICU, reported similar findings; 68% of all alarms were false alarms, and more than 94% of the alarms had no clinical significance. A consequence of this high number of false alarms is that medical staff is inclined to ignore alarms or respond more slowly (Lawless, 1994). The noise generated by this multitude of alarms leads to stress for both nurses and patients.

In D. de Waard, K.A. Brookhuis, R. van Egmond, and Th. Boersema (Eds.) (2005), *Human Factors in Design, Safety, and Management* (pp. 219 - 233). Maastricht, the Netherlands: Shaker Publishing.

The second problem is the design of the alarms: each device has its own alarm structure and its own alarm sounds. There is no standardisation across manufacturers (Meredith & Edworthy, 1995). And since devices are produced by different manufacturers, the alarm structures together do not form a consistent whole. Some devices distinguish three different levels, others distinguish four; one device might have a certain sound level meaning "medium urgent" which is equal to the sound level of another device, meaning "maximum urgent".

The third problem concerns the matching of the excessive amount of information from alarms to the cognitive abilities of the operator. Current alarm signals correspond to low-level measurements. (e.g., electrocardiogram, arterial pressure, pulse). This leaves the onus of response selection with the operator [in our case the ICU nurse] who has to interpret, prioritize and make sense of the multitude of alarms (Hollnagel & Niwa, 2001). Current alarm structures are of little help. Alarms have little intelligence on what should be considered to be really urgent and what is less. For example, when a ventilator detects a short apnoea (moment of breathlessness) it generates the same "highly urgent" alarm as when it detects a long apnoea. The first case requires intensified alertness, while the second case requires immediate action.

When the whole nursing process is considered, matters become even more complex. As Bitan puts it aptly (Bitan et al., 2004, p. 240): *Nurses combine two forms of action: they respond to events, such as alarms, and they initiate actions, often according to some more or less pre-determined schedule. (...) An excessive amount of information from alarms can possibly interfere with nurses' ability to schedule their tasks efficiently. If a nurse responds to each of the very frequent alarms she or he will not be able to perform the pre-scheduled actions.* It is therefore not surprising that all devices are equipped with a silencing knob. The nurse can temporarily shut off all alarms to avoid cognitive overload, in particular during tasks that need full attention such as reanimation.

The aforementioned problems are studied in three main fields of research. Firstly, the field of auditory perception (e.g., Edworthy, 2000): Although relevant, this type of research limits itself to improving current solutions; how to communicate hierarchy in auditory signals. Secondly, the field of computer science focusing on technical solutions to filter out superfluous alarms in the ICU (e.g., "fuzzy logic", "neural networks", "filtering by algorithms") (e.g., Becker et al., 1997, Lisboa, 2002, Schoenberg et al., 1999): The focus is on the patient's physiology only. What lacks in both these fields of research is attention to which alarms should be communicated, e.g., which hierarchy, in relation to the whole nursing process, including other tasks, such as communicating to the family, handing over of shifts and team work. Furthermore, it lacks attention to how these alarms should be designed (e.g., modality choices). A more promising approach can be found in the third field of alarm research, conducted according to methods of cognitive systems engineering (e.g., Hollnagel & Niwa, 2001 and Bitan et al., 2004). Bitan et al., for example, conclude that nurses do not respond immediately when they hear an alarm, but rather register the occurrence, evaluate the urgency of the problem that is indicated through the alarm, and eventually act on it within their ongoing flow of activity by adjusting the sequencing of their actions. The drawback of Bitan's study is that it is based only on the relation between observational data on nurses' actions and automated logging of monitoring data. There is no investigation of the how and why of nurses' reactions to alarms. Nor does it provide guidelines for the design of better alarm/information systems. They do suggest however that information to which the nurses do not have to respond immediately should probably be provided through other methods and not as alarms. Promising alternative methods to communicate time critical information are the use of various modalities (e.g., Sarter, 2000, Spence, 2002, Van Erp & Verschoor, 2004) and the ecological approach to interface design (e.g., Flach, 1995, Effken et al., 1997). Tactility is being regarded very promising as a less annoying way of alerting and is suitable for a parallel information perceiving and handling task (Sarter, 2000). Tactile information provision is currently being developed, for example in car seat belts and in the aviation domain. Sarter (2000) proposes, for cockpit automation, to use auditive signals or tactile signals, depending on the event type. In the medical field applied research on haptics is still rare, though upcoming. Recently investigations were started by the Norwegian Centre for Telemedicine (2004) on the feasibility of applying wearable computing in health care work and as one of the ideas to apply haptic signals in the nurses' uniforms. The ecological approach has a long tradition in aviation, especially in fighter jets. Interface design according to ecological principles focuses on the structure of the presented information. This approach aims to match the structure of the interface to the natural constraints of the work domain. In particular the chosen level of information should match the task of the operator in the context of his work (Flach, 1995). To support work tasks various levels of information in parallel or alternating are needed. A medical example is a display for a haemodynamic monitoring and control task developed by Effken et al. (1997) which presents anatomical constraints in pressure and flow (low-level information) nested within the relevant etiological constraints (high-level information). Effken et al. found that speed and accuracy in assessing clinical problems were improved compared to the traditional display. The principles of the ecological approach could possibly be applied to the design of alarm systems as well, providing a higher level of interpretation when appropriate.

The aim of this study (earlier published by Pijl, 2004) was to break the status quo in alarm design. Not small improvements were sought, but a radical different approach, which might open up new directions to come to solutions. A complete change of alarm structure could be considered, e.g. different modalities for the signals. The ecological design approach would serve as a source of inspiration. A participatory research and design approach was taken. Field investigations and design work involving nurses was conducted iteratively, with rapid evaluations of interim findings. Feedback from ICU nurses was used to guide and check research findings and design directions.

Methods and materials

The presented work used the Vision in Product Development (ViP) framework, developed by Hekkert and Van Dijk (2001). This is a method that prescribes a certain approach in research and design for finding creative solutions and is in particular useful for radical product design (often needed for long term product development) (Van Veelen, 2003). Van Veelen used ViP to develop a vision presented as a set of guidelines for the development of minimally invasive surgery products. ViP's step wise approach was followed (adapted for the ICU-alarm domain): (1) Restructure the context (the designer/researcher gets rid of his preconceptions): this was mainly done through ethnographic research; (2) Restructure the context: the information was restructured according to its role in the nursing process; (3) Formulate a vision on product information and user-system interaction and present this as a set of coherent guidelines. (4) In the last step material was produced to enable evaluation by nurses and experts and evaluation was carried out.

Step 1. Restructure the context

To analyze the current context (i.e. the ICU work environment) a set of research techniques was used, iteratively and in parallel. Data were gathered via three methods and analysed in four ways.

Data gathering: (a) Reviewing literature concerning the intensive care nursing process (e.g., Groen, 1995), the use of alarms in the ICU (e.g., Chambrin et al., 1999), auditory perception of alarms (e.g., Edworthy, 2000), technical solutions to filter out superfluous alarms (e.g., Becker et al., 1997), human factors (e.g., Stanton & Edworthy, 1999), interaction design (e.g., Preece et al., 2002), ecological approach to interface design (e.g., Flach, 1995), multimodal design (e.g., Spence, 2002).

(b) Conducting ethnography (Fetterman, 1998) - which can in this case also be called participatory research: conducting interviews with experts, observing and interviewing and studying work documents at ICU wards. The nursing process was studied in five ICUs (thorax ICU, internal ICU, general ICU and two neurological ICUs) in three different hospitals (the University Medical Centre Groningen, the Onze Lieve Vrouwe Gasthuis in Amsterdam and the Erasmus Medical Centre Rotterdam). In each ICU at least a day or a night was spent during several shifts and handovers of shifts. In total 50 hours was spent in the ICUs. Observations of the work and interviews with nurses, head nurses, physicians and members of the technical staff were conducted. Also regulations and advisory documents were studied concerning post surgical treatment, infection prevention, organisation and work methods at the ICU, and architectural issues. In a second round of studies the focus was on one ICU in particular, the thorax ICU in the University Medical Centre Groningen. 34 hours were spent in this ICU. The focus was on the nursing tasks related to one specific syndrome (namely clogged artery treated with bypass surgery) in one patient. The patient was followed in his path into surgery, during surgery, to ICU admission, treatment and dismissal.

(c) Observing and interviewing other critical decision making domains. The results of (a) and (b) were verified by doing observations and interviews in similar domains. Domains were selected by the similarity in the critical character of the decision making by the human operator, under great time pressure, and in case of errors with the chance of human losses. Visits were paid to the operating room of a university hospital, the fire department of a large city, and a control room of a major oil company, and a pilot was interviewed. During these observations and interviews the focus was on differences and analogies in needs for information. The aim was to identify possibly overlooked aspects.

The task analysis: The ICU nursing process was unravelled into tasks in several levels of abstraction. The various actors, i.e., single nurses and members of the team (including physicians), were considered. Rasmussen's Abstraction Hierarchy (Rasmussen et al., 1994) was used. Therefore, the process-oriented character of the levels (e.g., production flow models, information flow topology, chemical processes of components) was transformed into an activity-oriented form. The (from high to low) levels of abstraction were: Goal of the ICU nursing process (alarm related), main work tasks, main work activities, sources of information.

The analysis of information cues: These sources of information provide a range of information cues. These information cues were inventoried and put into an overview with help from the (interface) design tool by Melles et al. (2004). This tool helps categorize the information as related to the ICU work process, i.e., according to them being primarily related to the current situation, the future situation, or the past situation. Discrepancies between information cues as provided by the equipment and actually used information (e.g. built up by the nurses by combining various other cues) were identified. This way an overview of available cues (provided by the equipment or context) and missing cues (not directly provided by the equipment or context) was made.

Assessment of crucial information categories nurses use when processing alarms: These are the main information components nurses use to quickly judge an alarm signal or to process it. For this the mentioned research methods were used, but also the feedback from ICU nurses and other domain experts to the research and design work (see parallel step 4).

The assessment of main problems with alarms: Again these extended methods were used (including step 4).

Step 2. Restructure the context

The method for restructuring the alarm-related context was decided on after step 1. Restructuring was done according to the three identified categories of crucial information as identified in step 1. Evaluation of the proposed restructured information context would be postponed until step 4.

Step 3. Formulate a vision on product information and user-system interaction

Following the results from step 1 and 2 two requirement defining aspects were used to make a matrix of all possible alarm types: according to interaction phase and according to alarm type. This provided a matrix of two interaction phases by four types of events. Each cell had its own and distinct interaction style and its typical information type. Requirements, as a result of the two aforementioned steps, were defined for each cell. The design guidelines concern sensory and cognitive aspects and also requirements for type of information (content) given.

Since a design vision presented in a set of ergonomic guidelines is not very suitable to directly evaluate with end users, other materials were needed for evaluation. For this an integral design solution, including the various types of events and interaction phases, presented in scenarios, storyboards and product drawings (Preece et al., 2002) was developed. This was produced in step 4.

Step 4. Generate materials to evaluate and evaluation

Parallel to the above steps design activities were carried out. Partial solutions were constantly evaluated with ICU nurses and other domain experts, and the feedback was fed into the research work. In the last phase of design an integral set of technological solutions was developed for evaluation by nurses with the aim to assess the suitability of the product vision: Eight types of interaction styles, and information content were designed according to the requirements made. The question was whether this vision reflects the way nurses work, and whether it is sufficiently complete, and whether the new approach would improve the work process. In designing such an integral alarm system practical problems are bypassed, such as manufacturers who would not want to cooperate in practice, or ICT technology which is not yet developed sufficiently to handle the communication between the ICU devices. But that is not an issue here. The design solution does not have to be produced soon, it is meant to test the vision, from an ergonomic perspective - and the vision is meant for long term development. Additional research and technology development may be assumed. More so, an important outcome of the evaluation should be the identification of desirable further developments and research. In this last design phase, design work was done during several days inside the ICU, while nurses informally walked by and commented the design work. Furthermore, a brainstorm was held with 19 ICU nurses and one head nurse, from the University Medical Centre Groningen. See Pijl (2004) for the full method. The evaluation of the actuated design vision was done by showing the presentation materials and by explaining these to a group of nine nurses and a nursing expert, from the same hospital (some of them had been involved in earlier evaluations and idea generation rounds). This was done in a focus group session (Stewart & Shamdasani, 1990). The opinion of the participants on the new interactions and the way it would affect their work process was asked for. They discussed these issues. Next the vision was explained and that was discussed in the same way. The principal designer guided this with the aim not to interfere: she participated for clarifications but did not ask any questions. All verbal utterances were transcribed and analysed later.

Results

Step 1. The current context

Task and information analysis
Goal of the ICU nursing process: To make the patient's physiological functioning independent of ICU support.
Work tasks: Monitor the patient's physiological functioning; Support or completely take over the patient's physiological functioning with the help of specialized equipment or techniques; Prepare and maintain the medical equipment (in general, not patient specific).
Main work activities: Some examples: Estimate the current physiological condition of the patient, and remember this; Predict potential problems, and remember these; (Re)schedule mental tasks such as problem solving or making a diagnosis and actions to be performed, remember these tasks, and remember when they should be performed (how soon/ how urgent); Add new information and integrate this for an update.
Sources of information: Patient, equipment, team members, relatives.
Information cues: Some examples: Complexion and behaviour of the patient; Data from the bedside monitor; Location of team members; Presence of relatives.

The assessed crucial information categories nurses use when processing alarms
The processing of alarms should be divided into two interaction phases: immediate reaction and (if applicable) later processing.
The immediate reaction: The constant stream of alarming sounds is being processed parallel to the ongoing work. This must be done in an efficient way in order to allow the nurse to conduct the ongoing cognitive and physical work tasks. There must be some sort of rescheduling activity which is an ongoing activity; immediately nurses must know whether they need to react, whether they should reschedule and whether they should process the signal later. It became clear from the observations and discussions that nurses need *to immediately get some global information*: they need an idea of *urgency*, involved *physiological function(s)* and *needed action*. The combined and instant assessment (by the nurse) of these three topics defines whether and how the nurse will respond to the signal, e.g. whether to act or to reschedule the work. (Note that it is unlikely that every incoming auditory signal will be processed explicitly, there are simply too many of such signals; it is therefore likely that nurses raise their awareness for certain signals). The result of the immediate response is initiating immediate action (and rescheduling their other work), discarding the alarm, have it raise their awareness, or scheduling it to process later in their ongoing work. The impression was that if one of the identified crucial information components is unclear this tends to cause a negative influence on the work process.
Executing the rescheduled work: At the time that processing of the alarm is continued the same three categories of information are used (urgency, involved physiological function(s) and needed action). The continuing parallel task of new incoming alarm signals, and the continuous monitoring actions, causes alarm processing tasks to be interrupted when needed.

The assessed problems
Some of the main assessed problems with alarms: - The large number of auditory warnings, causing a noisy environment for severely ill patients and workers with high (cognitive) work loads. - The continuous task of handling parallel incoming auditory alarms. - The inappropriateness of the modality of many alarms. – The lack of integrated information.

Step 2. The restructured context

Restructuring of the context was done according to three (identified in step 1) *crucial categories of information*: urgency, involved physiological function and kind of needed action. Next the information in these categories will be described.

Alarm categorization by urgency
Currently nurses extract three types of urgency from the (alarm) cues they get:

1. "Crisis! Get here!": Immediate awareness needed, immediate understanding, immediate actions, for whole team (often help and delegation of tasks is needed). The work aims at avoiding this situation, so it is always unexpected. Sound and readings are quickly interpreted and action follows immediately. For example: An alarm indicating "asystoly" goes off. Immediately the nurse checks the correctness of the message and if correct initiates the work of several team members for reanimation. During reanimation the main responsible nurse refuses to mentally process other signals, the nurse needs all attention. Other alarms are silenced altogether.
2. "Keep an eye on this": Raised awareness needed now, no actions needed right now, understanding needed soon. Alarms in this category can be unexpected or expected, e.g., a critical change that may need intervening in the near future. Using alarms for this type of functions is usually indirectly; the nurse gets a bunch of cues and/or reactions to actions that together can be deduced as message "attention – this or that is the case!". The nurse will have to keep an eye on this. First example: - The monitor alarms that the heart rate (HR) is too high (the heart producing extra systoles). This increase in HR can be unexpected, requiring raised awareness, because it could lead to a hypovolemic shock. However, the rise can be expected as well, being a personal physical defect of the patient. Second example: The patient has been to surgery and shortly after his arrival at the ICU he coughs. As a result he puts up resistance ("contra-pressure") to the ventilator and it generates an alarm. A few minutes later the patient puts up contra-pressure again. As a consequence the nurse concludes that the patient is awakening. It is way too early for the patient to get awake and therefore the nurse supplies the patient with more sedation.
3. "Nothing wrong": *Informing messages concerning physiological readings, no action needed right now.* These "alarms" can be expected or not expected. There are two subgroups here: informing feedback and informing messages. Informing feedback gives feedback on actions carried out, for example: An "alarm" signal indicates that probes were disconnected from the patient; the nurse already knows because the patient is being washed. Second example: An "alarm" signal indicates that blood pressure is increasing. The nurse expects this to happen because a

medication to increase blood pressure was just given: the alarm gives valuable feedback on the effect of the treatment.

Informing messages are all other ''alarms'', these are often alarms which sound automatically after a value event or change event was measured. These, however, in this case are non critical (for now), (or else they would have been of one of the other two types of alarms) - so the nurse reads them as "Nothing wrong". This information can be expected or not expected. But it certainly does not need immediate processing. We found that the nurses did not want to do without the "nothing wrong" information. They apparently use it as information (not as alarm), or they might want to stay in control on choosing to use if or not. The current modality (sound) is the annoying thing about these "alarms".

Alarm categorization by involved physiological function
The second type of crucial information nurses use is the involved physiological function. Nurses think about the physiological system as a set of anatomic subsystems (e.g., heart, brains, kidneys), related to a set of physiological processes (e.g., ventilation, cardiovascular system, hypoxia). Although in alarming events multiple anatomic subsystems can be involved, almost in every event one subsystem can be indicated as "initiator". To solve the physiological problem this subsystem needs support or intervention. However, alarms currently almost only measure at an "organismic" level (e.g., arterial pressure, respiratory rate) and alarm at threshold crossings, without further intelligence. It lacks readily integrated information related to the assessment of the stability or the occurring problem in these subsystems. Interviews revealed that nurses need (and thus deduct) such information to assess the alarming event and to choose the appropriate actions.

Alarm categorization by kind of needed action
Current alarms do not directly communicate what kind of action should be taken. When an alarm goes off, the nurse has to decide on the required action. They need to do this promptly, since it needs reorganising their tasks, and in some cases reorganising the tasks of other team members as well. To assess what (re)action is needed they use a combination of various cues. Important is what medical device is involved, their own expectations, additional monitoring readings, etc.

Step 3. A vision on product information and user-system interaction

In Table 1 the vision on interaction and product (information) is presented as a table with requirements per alarm situation. All current alarms are still present in the vision. There are eight distinguished situations, depending on the interaction phase (in columns) and the type of event (in rows). There are distinct differences in requirements between the cells. The relation between the ultimately chosen solutions for the eight cells should be such that the total set becomes a congruent overarching system, which means that the relative differences and also the similarities in information support should result in a logical set communicating its hierarchy in the experience of the nurse. In the case of alarming messages global information should be communicated in the very first moment. Global information indicates urgency level, organ or function involved, and action needed. In all cases (except for immediate feedback) background information is needed (e.g., to set a diagnosis, to solve a problem) this will be done in the "later interaction phase" (which can in some cases be much later or if the event is acute can be immediately after receiving the message). Background information contains both global information and detailed information. Detailed information can be any information the nurse wishes to use for her work related to the event. In step 1 it was assessed that nurses use cues from the monitoring and supporting equipment, that they integrate such cues from one or more monitoring devices, and that they integrate these cues with other information, e.g., their experience, observations by relatives of the patient, written documents and so on. This should remain possible. The system should actively support information integration.

Table 1. Overview of the set of requirements, together forming the vision on interaction and the new alarm/information content

	First interaction moment: the signal	Later interaction moment: information support
Informing feedback *Expected*	Immediate feedback on conducted actions should be supplied. It should, however, not draw attention from anybody but the nurse.	None
Informing message *Expected or unexpected; No intervening action is needed right now.* These messages confirm expectations or predicted problems, or none critical (for now) value events or change events.	No attention or hardly any should be drawn, and certainly not from others than the nurse.	The related background information should be provided on call. It should consist of **global** information and **detailed** information to help assess the situation. This information should be provided in a way that the nurse could privately reflect on the information at his/her ease. When he/she needs help from colleagues he/she should be able to share the information.
Alerting *Unexpected; Intervention is needed in the near future.*	The burden on the cognitive capacities of the nurse should be minimised. However, the alarm should draw **attention**. Possibly some idea of what the event is about is necessary (some **global information**).	The same type of background information as in informing messages should be provided in the same way.
Alarming *Unexpected; Immediate intervention is needed.*	The alarm should draw **attention**. The nurse who is responsible should be warned, but team members also, because in such calamities often help is needed. The alarm should immediately provide **global information**. Thus, the nurse is immediately prepared and can act as soon as possible, instead of first examining the problem. The alarm should be perceived in every situation of work.	The same type of background information as in informing and alerting messages should be provided. It should be provided immediately and without having to ask for it. It should be possible to reach other information than what is shown automatically. When he/she needs help from colleagues he/she should be able to share the information.

Concluding this means that detailed information should at least contain (a) and preferably also (b) en (c):
(a) The system should present selected *factual information* related to the event (as is now the case): sensed values of parameters, crossings of critical thresholds, sensed

disconnected sensors, hindrances in supply and transport, potential hindrances in supply and support; if the nurse wants other than the selected factual information this should be possible;

(b) *Suggested interpretations*: integrated information organised by the relevant physical functions (e.g., heart, lungs), estimation of condition of the patient and of separate physical functions, sensed global changes of values, prediction of potential problems, integrated feedback on the results of the chosen support;

(c) *Optional (re)actions* to deal with an event: *monitoring-related (re)action*: e.g., intensify alertness, *supporting-related (re)action*: e.g., new settings of supporting equipment; *caring for medical equipment*: e.g., reattach power line.

Step 4a. The actuated vision

Suitability of the product vision for the ICU was to be tested, so the actuated design had to cover all eight types of interaction styles and information content; and had to follow the requirements made in these eight types of alarm handling (stages). Because this was too much for a limited set of storyboards the choice was made to give the nurses an idea of the types of interaction by choosing three representative interaction lines, one illustrating an alarming situation, one of an alerting situation and one of an informing situation.

Table 2. In storyboards (drawings and some text) the actualised vision on alarms was presented to be valuated by nurses. Here those aspects are summarized that the nurses could read from the storyboards. For the full design and its backgrounds see Pijl (2004)

The technological solutions which were presented to the nurses in the storyboards:

Crisis alarms signal with sound. Alerting is done by a tactile stimulus, in this case a 'tap' on the shoulder, and another tap if the nurse's reaction takes too long. New informing messages are communicated by a lamp on the monitor of the Patient Data Management System (PDMS) at the bedside; detailed information on call is on the PDMS monitor as well.

When the nurse decides to further process the message/alarm as a reaction to an alarming, alerting or informing signal he/she can look at the bedside monitor or at his/her portable device. There the following related information will is presented: (1) low-level measurements; (2) suggested interpretations, such as global changes of values or an entire diagnosis, and (3) suggestions for actions.

An overview per physical function is given permanently on a screen at the head side of the bed and also on the portable device: Per physical function current alarms are shown, as well as the current level of stability.

The nurse confirms or rejects the suggested interpretation and the suggested action and the event is automatically registered in the report. At the same time confirmation will make the system aware of the activity going on, so that appropriate system behaviour for the ongoing work is initiated (for example, during defibrillation the alarm will silence and the heart will be monitored auditory automatically, which is a well known way of supporting defibrillation).

In Table 2 the aspects are summarized that the nurses could read from the storyboards. The main decisions for this design were: Intelligence in the system is provided by additional software in the patient data management system (PDMS). The system assesses event type to be able to provide the appropriate modality of alarm signals – modality choices meet requirements on attention level in relation to parallel tasks, privacy needs and availability requirements. An intuitive hierarchy experience is expected (sound for alarm, tactile for alerting and visual for notifying). To assess these event types the system uses readings from devices monitoring or supporting the physiological functions, and uses data provided by the nurses, through the interface. To promote nurses to indeed enter data into the system they get useful and directly noticeable results from these actions. Information is provided by the system at the low-level (equal to the current information), at the level of assessed stability of the organ or physiological function, and suggestions for event related diagnosis and actions to be taken (both are integrated information).

Step 4b. Evaluation

- The nurses were positive overall about the proposal: They inferred from the design that less alarms would be necessary, they "would only receive the alarms that matter";
- And that the ICU would become quieter, they liked that for themselves and for the patients.
- The nurses were positive about the modality choices, auditory, haptic and visual - and to which event type they belong. However, they did question the haptic design choice: a "tap" might be too human.
- The nurses pointed out that privacy for informative alarms can be regarded an advantage, but it must be checked if this might be a disadvantage for some working situations.
- According to the nurses the control over the division between informing and alarming should be in the hands of the nurse in charge. The participating nurses explained that they are responsible for setting (and adjusting) alarming thresholds for the patient - depending on the clinical picture and medical protocols. Now there is the choice between silence and sound. In the new situation the nurse would want to be in charge for controlling thresholds between sound and haptics because this is in the critical range and defines whether the signal can be noticed by colleagues (and how intrusive it is). The division between haptics and visual (alerting/notifying) did not draw special attention in the discussion.
- The nurses expressed their concern about whether nurses would stay alert with so much support and whether they would keep on checking the equipment, and not relying too much on the alarm/information system's intelligence.

Discussion

Through a participatory research and design approach new directions to get away from the current status quo in design of alarms for the ICU were identified. A new vision on work supportive alarms to fit into the ICU context was proposed. Design guidelines for the various interaction phases and event types were presented. Technological solutions and the vision were evaluated. The next step would be to identify what of the proposed vision and which of the proposed directions in technology would be so promising that further research and development should be considered. It is important to improve those aspects of alarms in the ICU which currently cause the most adverse situation for the nurses. The impression was (from observations and interviews and supported by Bitan et al., 2004), that the situation in handling the alarms currently is especially adverse if it comes to the initial presentation modality: the ICU must become quieter and the number of 'false' auditory alarms must be reduced. This was substantiated in the last evaluation round and the nurses also indicated that the design would do just that: A main positive side of the design is that the noise at the ward will be substantially reduced. The main responsible factor for this was - as understood by the nurses - the fact that haptic and visual alarms would be used. The suitability of such alarms for the ICU was not questioned: the modalities seemed appropriate, and the type of alarms they were used for also. The three categories of alarms were chosen well (urgent/alert/notify) according to the nurses. Also in literature support is found for this division: Groen (1995) distinguishes three comparable modes of working: emergency, vigilance and routine. The nurses elaborated on design choices for the haptic signals and on the choices in intelligence of the system. From an ergonomic and technical point of view these two issues indeed need most attention. Although haptics has long been regarded as our richest and most trustworthy communication modality, designing haptic signals is still in its infancy. Currently, besides products for the blind, just a few products use haptic signally, an example is the mobile phone. The opportunities of various modalities in cockpit automation are also studied by Sarter (2000). She proposes the use of auditory cues to indicate events that require immediate action or pose an immediate threat. Tactile signals are proposed for alerting messages because they disturb the cognitive processes of ongoing work less than sound does, but also have a low chance of being missed (Sarter, 2000). The evaluated actuated design followed the same division of modalities for event types, and the conclusion is that this is an appropriate choice. However there is no solution on the shelf for exact properties of such haptic alarms. Investigations are needed to assess how to communicate alerting messages in the first phase of interaction (to be sufficient, but also minimised to avoid cognitive overload - and annoyance). Whether they should be a tap on the shoulder, a vibrating arm pad or whatever needs to be designed and tested. Secondly there is the question of how technology should choose whether an alarm is of type ''alarm'', of type ''alert'' or of type ''inform'' - and how the nurse is going to remain in control.

The envisioned design provides intensive care nurses with low level information as well as high level information (i.e., "Suggested interpretations" and "Advice to the nurses about needed actions"), following ecological principles. However, when providing high level information in a complex working environment like the ICU, prudence should be called for. There are system-related considerations, as expressed by Hollnagel and Niwa (2000, p. 369): High level alarms are less reliable, since the underlying diagnosis of the system state relies on a many-to-one mapping from measurements to alarms, which cannot be completely unique. Another important consideration is that the responsibility of interpretation should remain with the nurse/doctor. To guarantee that nurses keep on checking the equipment as they currently do, and not rely too much on the alarm/information systems intelligence, the design must be such that low level readings are indeed actively processed. Even if the decision is made to - for now - only work on intelligence for the first phase of interaction (managing the stream of incoming alarms) probably quite complex context aware ICT will be needed. There is no equivalent of this yet. It must be assessed what factors can be used to define the alarm type. Should such an assessment follow the current alarm signals, or are new signals needed? An artificial intelligence architecture must be designed which includes various context models, e.g., of the patient, which should be changeable by the user. Simple thresholds will not do, various equipment readings will probably have to be combined with some information actively provided by the nurse. The next joint research step will be to assess whether and how such context aware alarms with haptic signals can be designed to safely support the nursing work process.

Acknowledgments

We owe our thanks to John Flach who contributed to this study by providing valuable discussions and critical comments and to the nurses and other experts for their hospitality and sharing their ideas.

References

Becker, K., Thull, B., Käsmacher-Leidinger, H., Stemmer, J., Rau, G., Kalff, G., & Zimmerman, H.-J. (1997). Design and validation of an intelligent patient monitoring and alarm system based on a fuzzy logic process model. *Artificial Intelligence in Medicine, 11*, 33-53.

Bitan, Y., Meyer, J., Shinar, D., & Zmora, E. (2004). Nurses' reactions to alarms in a neonatal intensive care unit. *Cognition, Technology and Work, 6*, 239-246.

Chambrin, M.-C., Ravaux, P., Calvelo-Aros, D., Jaborska, A., Chopin, C., & Boniface, B. (1999). Multicentric study of monitoring alarms in the adult intensive care unit (ICU): a descriptive analysis. *Intensive Care Medicine, 25*, 1360-1366.

Edworthy, J. (2000). Medical device alarms: equipment- or patient-centered? In E. Haas and J. Edworthy (Eds.), *The Ergonomics of Sound: Selections from Human Factors and Ergonomics Society Annual Meetings.* (pp. 56-59). Santa Monica: Human Factors and Ergonomics Society.

Effken, J.A., Kim, N.-G., & Shaw, R.E. (1997). Making the constraints visible: testing the ecological approach to interface design. *Ergonomics, 40(1)*, 1-27.

Fetterman, D.M. (1998). *Ethnography, step by step.* Thousand Oakes, California: Sage Publications.

Flach, J. (1995). The ecology of human-machine systems: a personal history. In J. Flach, P. Hancock, J. Caird and K. Vicente (Eds), *Global Perspectives on the Ecology of Human-Machine Systems.* (pp. 54-67). Hillsdale, New Jersey: Lawrence Erlbaum.

Groen, M. (1995). *Technology, work and organisation. A study of the nursing process in intensive care units.* PhD thesis, University of Maastricht. Maastricht, the Netherlands: Maastricht Economic Institute on Innovation and Technology.

Hekkert, P. & Van Dijk, M.B. (2001). Designing from context: foundations and applications of the ViP approach. In P. Lloyd and H. Christiaans (Eds.), *Designing in Context: Proceedings of Design Thinking Research Symposium 5.* (pp. 383-394). Delft, The Netherlands: DUP Science.

Hollnagel, E. & Niwa, Y. (2001). Enhancing operator control by adaptive alarm presentation. *International Journal of Cognitive Ergonomics, 5(3)*, 367-384.

Lawless, S.T. (1994). Crying wolf: false alarms in a pediatric intensive care unit. *Critical Care Medicine, 22*, 981-985.

Lisboa, P.J.G. (2002). A review of evidence of health benefit from artificial neural networks in medical intervention. *Neural Networks, 15*, 11-39.

Melles, M., Freudenthal, A., De Ridder, H., & Snijders, C.J. (2004). Designing for enhanced interpretation, anticipation and reflection in the intensive care unit. In W. Thissen, P. Wieringa, M. Pantic and M. Ludema (Eds.), *Proceedings of the IEEE Conference on Systems, Man, & Cybernetics.* (pp. 809-815). The Hague, The Netherlands: IEEE Systems, Man & Cybernetics Society.

Meredith, C. & Edworthy, J. (1995). Are there too many alarms in the intensive care unit? An overview of the problems. *Journal of Advanced Nursing, 21(15)*, 15-20.

Norwegian Centre for Telemedicine (2004). *ICU4 - Intelligent communication uniforms for health care workers,* http://www.telemed.no/cparticle69113-4357b.html.

Pijl, V.L. (2004). *Towards supporting alarms in the intensive care unit.* Engineering thesis, no. 2846. Delft: Delft University of Technology, Department of Industrial Design. Available: http://www2.io.tudelft.nl/afstuderen/bekijken.php?afstudeernummer=2846

Preece, J., Rogers, Y., & Sharp, H. (2002). *Interaction design: beyond human-computer interaction.* New York: John Wiley & Sons.

Rasmussen, J., Pejtersen, A.M., & Goodstein, L.P. (1994). *Cognitive systems engineering.* New York: Wiley.

Sarter, N.B. (2000). The need for multisensory interfaces in support of effective attention allocation in highly dynamic event-driven domains: the case of cockpit automation. *The International Journal of Aviation Psychology, 10(3)*, 231-245.

Schoenberg, R., Sands, D.Z., Safran, C., & Deaconess, B.I. (1999). Making ICU alarms meaningful: a comparison of traditional vs. trend-based algorithms. In *Proceedings of the AMIA Annual Symposium.* (pp. 379-383).

Skillman, J.J. (1975). *Intensive care.* Boston: Little, Brown and Company.

Spence, C. (2002). Multisensory attention and tactile information-processing. *Behavioural Brain Research*, 135, 57-64.

Stanton, N.A. & Edworthy, J. (1999) Auditory warnings and displays: an overview. In N.A. Stanton and J. Edworthy (Eds.), *Human Factors in Auditory Warnings.* (pp. 129-149). Aldershot, UK: Ashgate Publishing Limited.

Stewart, D.W. & Shamdasani, P.N. (1990). *Focus groups: theory and practice.* London: Sage Publications.

Van Erp, J.B.F. & Verschoor, M.H. (2004). Cross-modal visual and vibrotactile tracking. *Applied Ergonomics, 35*, 105-112.

Van Veelen, M.A. (2003). *Human product interaction in minimally invasive surgery: a design vision for innovative products.* PhD thesis. Delft, The Netherlands: Delft University of Technology.

The influence of perceptual stability of auditory percepts on motor behaviour

René van Egmond[1], Ruud G.J. Meulenbroek[2], & Pim Franssen[2]
[1]Faculty of Industrial Design Engineering
Delft University of Technology
[2]Nijmegen Institute for Cognition and Information
Radboud University Nijmegen
The Netherlands

Abstract

In user-product interaction the way perception influences accuracy of movement or human error is important. Studies on visual illusions have shown that multi-interpretable images often lead to confusion or error. In the present study we investigated whether multiple-interpretable auditory sequences have a similar effect on motor behaviour. An auditory model accounting for the perceptual stability of tone sequences is proposed. On the basis of this model tone sequences were generated that contained multiple levels of perceptual stability. Analogous to a multi-interpretable Necker cube, only one representation of the tone sequences was presumed to be dominant at a specific moment in time. In a perceptual study, the predicted saliency of the perceptual representations was corroborated. The proposed model predicted the results adequately. While listening to the sequences in a second experiment, participants tapped on a table such that their tapping height reflected the low, middle, and high pitches that they perceived. We found that the variability of the tapping movements was higher for the unstable than for the stable percepts. Collectively, the results of the present study show that predicting the perceptual stability of auditory sequences is important to gain insights into the relationship between acoustic dimensions and performance accuracy.

Introduction

Everybody has experienced that sound influences various aspects of our motor behaviour. Music sometimes forces us to move. People tap with their feet while listening to music (often not being aware of it), they dance on music, and they run on music. In all these examples the frequency with which one moves corresponds to a perceived beat evoked by the music. But also in working environments, sound –or more specific music– has been used or is used to influence our behaviour. For example, typists are often trained by means of a metronome to achieve a certain typing speed. Alarm signals disrupt concentration and afford certain actions. In intensive cares both sound and alarm signals should not in any way disturb fine motor behaviour.

In D. de Waard, K.A. Brookhuis, R. van Egmond, and Th. Boersema (Eds.) (2005), *Human Factors in Design, Safety, and Management* (pp. 235 - 246). Maastricht, the Netherlands: Shaker Publishing.

Often people claim that they can better concentrate while listening to music. A lot of research in this context has been conducted using tapping tasks or musical-performance tasks to investigate how these tasks influence the timing aspects of people. However, there have not been (many or any) studies that have investigated the role of the perception of pitch on motor behaviour or vice versa, i.e., the effect of motor behaviour on the perception of pitch sequences. In this study we make a first step to investigate the mutual influence of sound, in particular, pitch structure, and motor behaviour by using simple melodic sequences in combination with a tapping task.

Although a melody is one of the more salient features of music, the way melodic percepts are mentally formed has not been investigated intensively. Many studies on melody perception have dealt with the perceptual invariance of melodies under different types of transformation (e.g., Dowling, 1972; Van Egmond & Povel, 1996). Deutsch (1980) and Deutsch & Feroe (1981) introduced a representational model of melodies and tested the model's melodic representation in the form of hierarchically-ordered segments. In other studies it was suggested that perceptual accents play an important role in drawing a listener's attention towards melodic events (Boltz & Jones, 1986; Jones, 1987). Perceptually-accented events can be considered as the cues by which a listener segments a melody in smaller perceptual units. These accents are evoked by inter-onset-intervals (Povel & Essens, 1982), large steps in pitch height (Jones, 1987), and contour (Jones, 1987; Thomassen, 1982)[*]. In recent studies it has been shown that often the combination and the coincidences of these factors are the main determinants in the well-formedness of tonal melodies (cf., Povel, 2002; Jansen & Povel, 2004a,b).

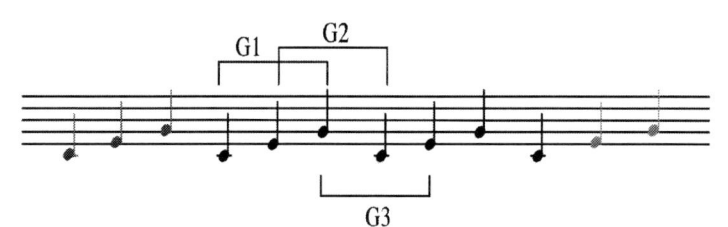

Figure 1. A cyclically repeated three-note pattern. This pattern may induce three possible perceptual groupings indicated by G1, G2, and G3.

In isochronous melodic sequences every note may –in principle– function as a starting point of a segment of that sequence. The segment in which the starting tone receives the highest perceptual accent will be the most salient. The Necker cube demonstration in analyses of visual perception has two perceptual representations even though it is one and the same physical structure. However, people can only see one representation at the same time. In auditory perception we expected that in cyclically repeated tone patterns a similar effect might be found. In Figure 1 such a cyclically repeated three-note pattern is presented. As can be readily seen in this

---

[*] Inter-onset-interval is the duration between the onsets of two tones. Contour is the directional pattern between pitches (ascending, descending, equal).

figure three possible percepts may occur. It is apparent that every grouping has a certain chance of being perceived. We propose that the relative interval size that precedes a certain tone within a sequence determines the extent of the perceptual accent for that tone (given that no rhythmic or other dynamic accents occur). The larger this relative interval size the larger the accent and as a consequence the tone will have more chance of being perceived as the first tone of the sequence. In order to be able to compare melodies with different melodic ranges (the distance between the lowest and highest pitch in the melody) the relative interval size has been chosen over the absolute interval size. The relative interval size is determined by dividing an interval size in semitones by the melodic range in semitones yielding numbers between 0 and 1. For example, the melody in Figure 1 consists of intervals consisting of +4, +3, and -7 semitones. If the interval structure would be +2, +2, and -4 (Frère Jacques) it is expected that the perceived grouping will not be different. In Figure 2 relative interval size is presented in the figure as a function of time and semitones. The stable sequences only contain one large relative-interval size, whereas the unstable sequences contain two similar large relative-interval sizes. The sequence depicted in the upper left corner of Figure 2 is the same sequence as in Figure 1. For the stable sequences only one tone receives the highest probability of being chosen as the first tone in a cyclically repeated sequence (the tone with relative accent 1.0 in Figure 2), whereas for unstable sequences two tones have an approximately similar high probability of being chosen as the first tone in the cyclically repeated sequence(the tones with relative accents .83 and 1.0 in Figure 2). Thus, it is expected that people will hear more than one group for these latter unstable sequences whereas the stable sequences will probably be heard in one way only.

In auditory-perception research it is often difficult to determine when certain perceptual representations occur or if they –just like the Necker cube– flip-flop from one percept into the other. Often judgments are asked only after the sequence has been heard entirely. It is then difficult to determine whether the percept had remained stable during the entire presentation interval or whether it had changed several times. In research on rhythmic structure tapping tasks are often used to determine how sequences are represented (e.g., Fraisse, 1980; Pöppel, Muller & Mates, 1990; Povel & Essens, 1985; Vos, Mates, & Van Kruysbergen, 1995). Research on motor transitions is more of interest to the present study. For example, Kelso (1995) modelled categorical phase transitions in bimanual task performance (such as moving the two index fingers of the left and right hand in synchrony into opposite directions at a low pace versus moving these fingers in synchrony in identical directions at a high pace) using dynamic systems theory. He showed that the dynamic system describing the fingers in terms of coupled oscillators becomes unstable just before it transfers into another coordination pattern. Inversely, a system may be considered unstable if it proves under certain conditions of a control variable to change from one coordination pattern to another (see also, Meulenbroek, Thomassen, Schillings, & Rosenbaum, 1996).

The proposed probabilistic model predicts in which tone sequences perceptual transitions may occur. To capture the transition from one perceptual categorization to another, we used a tapping task in the present study to probe into the auditory processes while people listen to a tone sequence. Similar to the abovementioned tasks it is expected that the perceptual transitions will be reflected in the movement of the hand. In particular, we expected that if perceptual transitions occurred, the variability of the hand movements would be higher than when these transitions would not occur.

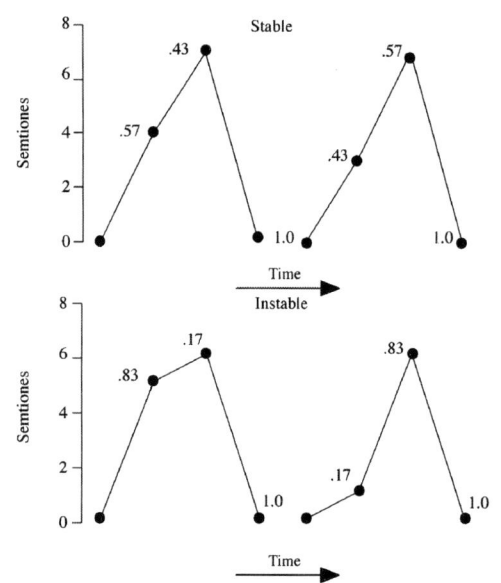

Figure 2. Two perceptually stable (C E G, C Eb G) and two perceptually unstable sequences (C F F#, C C# F#). (The y-axis represents the number of semitones that the tones are separated. The numbers next to the markers represent the relative distance between the tones (the largest distance within one sequence is chosen to be 1).

Method

We evaluated the salience of melodic sequences by means of a three forced-alternative choice task. Four three-tone melodies were used of which two were – according to our model– stable and two were unstable. The melodies were repeated cyclically. In the second task, a listener was asked to tap along with the sequence and after that sub-task was completed, (s)he was asked to indicate the heard melodic fragment.

Participants

Thirteen students (age between 20-26 years, $M=23$ years) of the Radboud University Nijmegen participated in this study. Ten of the participants were musicians and had received five or more years of musical education. All participants were volunteers and were paid. They all reported to have normal hearing. The participants were selected from a larger group of 28 participants by successfully passing a pitch-height discrimination task (see procedure).

Stimuli

Two stable and two unstable sequences were used. All sequences consisted of three tones (L(ow), M(iddle), H(igh)). These sequences will be indicated as "root sequences". The two stable triadic sequences consisted of +4 +3 semitones (major chord) and of +3 +4 semitones (minor chord). The two unstable root sequences consisted of +5 +1 semitones and of +1 +5 semitones. In Figure 2 the relative accents of these sequences are presented. It can be seen that the unstable sequences have two tones that receive a relatively large accent as a result of the preceding interval (.85 and 1.0) whereas the stable sequences have only one tone that receives a relatively large accent (1.0). The root sequences were presented cyclically. Consequently, each root sequence had three possible representatives (c.f., Figure 1). The first tone of a sequence could be the M, L, or H tone. Thus, each sequence can be represented by three pitch-height triads (i.e., L-M-H, M-H-L, H-L-M) yielding 12 different representative sequences depending on the start tone of the sequence of the repeated pattern. The timbre of the sounds was a piano sound (Roland JV-35). The loudness of each tone was calibrated to a constant loudness level to avoid accents due to perceived loudness variations (loudness perception is frequency dependent, see, e.g., Zwicker & Fastl, 1999).

Apparatus

Stimulus presentation and data collection for the perceptual experiment were controlled by means of a specially written program on a Power Macintosh 8200/120 that was connected to a Roland JV-35 synthesizer through MIDI. The sequences were presented through a headphone (Sony, MDR CD550). The loudness of each tone was calibrated using an artificial ear by adjusting the parameter speed of the MIDI code for each tone. The tapping behaviour of a participant was monitored using an Optotrak 3020 motion-tracking system consisting of three precalibrated camera units. An Infrared Light Emitting Diode (IRED) was fixed onto the nail of the index finger of the dominant hand. The 3D-position of the IRED was recorded with a sample frequency 200 Hz and an accuracy better than 0.2 mm in the X, Y and Z dimensions. The Optotrak system was connected to a personal computer (Windows-XP Operating System) for control and data storage. In order to synchronize the presentation of the melodic sequence and recordings of the hand movement, the Optotrak system and the personal computer were connected to the Macintosh that triggered these systems to start the registration.

Procedure

Pitch-height discrimination task (pretest)
In order to ensure that a participant was able to hear if a tone was low, middle of high a pitch-height discrimination task was developed. A candidate participant was asked to indicate on three horizontal lines the relative pitch of the tones of a sequence. The twelve representative sequences were employed that were also used in the main experiment. The representative sequences were presented in a random order for each participant. In order to be allowed to participate in the main experiment, a participant had to indicate 10 out of 12 stimuli correctly (sign test, $p<.05$).

General
The experiment consisted of two tasks: (1) a perceptual-decision task and (2) a tapping task followed by a perceptual-decision task. Two groups of participants were formed. One group ($n=7$) participated first in a session consisting of the perceptual-decision task and then in a session consisting of the tapping + perceptual-decision task. The other group ($n=6$) first participated in a session consisting of the tapping + perceptual-decision task and then in a session consisting of the perceptual-decision task. For all participants, the second session was at least 24 hours later than the first session.

Perceptual-decision task
A participant was seated in front of a computer and was explained the procedure of the experiment. The participant was then asked to put on the headphone and was offered one training trial. The training was exactly the same as an experimental trial, but another stimulus was used. After pressing on a Start button a participant heard a representative sequence that was repeated 15 times. After the representative sequence was heard a participant could listen to all three representatives of the sequence (including the representative sequence that was used as starting sequence of the cyclic repeated pattern) by clicking on three radio buttons. The participant was asked to activate the radio button that represented the perceived three-tone sequence. The participant could then click next and proceeded with the next trial. The order of trials was randomized for each participant. In addition, the three possible answer sequences were randomly assigned to the radio buttons in order to avoid response bias. Each of the twelve representative sequences was offered in three different speeds (IOI 650ms, 500ms, and 350ms) and each sequence was replicated twice. The different speeds were offered in separate blocks. Thus, randomization of the order took place within one block. The starting tone of a sequence was randomly selected from the range of MIDI pitches between 42 and 60 (60 corresponds to the tone of C). Each participant was offered 72 trials.

Tapping + perceptual-decision task
A participant was seated at a table on which a strip of tape (40 x 4 cm) was administered. The long side of the strip was in the sagittal plane of the body. A participant was asked to tap on this tape in such a way that lower tones were indicated closest to the body and higher tones at a further distance from the body. Before each trial started the participant heard the initial three-tone segment and had to indicate this segment in the right order on the tape. This was repeated until the participant indicated that the tapping movement was under control. A participant then heard three clicks of which the interclick intervals indicated the required tempo and then the sequence started. After a tapping trial a participant was asked to indicate the stimulus (s)he heard again using the same three-alternative forced choice task described above. However, because of the movement restrictions of the participant (being positioned in front of the Optotrak system), this time the experimentator collected the answers. The same randomization procedures as in the decision experiment were used. If the Optotrak recordings during a trial failed then this trial was repeated at the end.

Results

The perceptual data were analyzed through an implementation of the proposed perceptual model in a probabilistic model. In addition, the effect of task was tested using a nominal logistic regression analysis.

Perceptual-decision task

Two models were fitted. In the first model the factors Relative Interval size (continuous number between 0 and 1) and contour (dichotomous variable, 1 if there was a contour change, 0 if not) were used as predictor variables. In the second model, a third factor Initial Segment (dichotomous variable, 1 if the response was equal to the initial segment, 0 if the other two responses were chosen) was used. Both models were estimated using the method of maximum likelihood (CATMOD procedure, SAS). The implementation of his type of model is similar to that of a multiple-regression model. However, instead of one dependent variable there are three answer categories. In the first model, the estimates for the Relative Interval size (2.98) and Contour (-.16) differed significantly from 0, $\chi^2(1)=504.78$, $p<.0001$ and $\chi^2(1)=4.94$, $p<.05$, respectively. The proportion fit was determined by a measure comparing the empirical frequencies and the predicted frequencies. This measure was .40 for the unstable sequences and .74 for the stable sequences. In the second model, the estimates for the Relative Interval (3.53), Contour (-.17), and Initial Segment (1.24) differed significantly from 0, $\chi^2(1)=532.11$, $p<.0001$, $\chi^2(1)=4.40$, $p<.05$, and $\chi^2(1)=384.09$, $p<.0001$, respectively. The proportion fit was .79 for the unstable sequences and .96 for the stable sequences.

The choice data were then analyzed using a nominal (stepwise) regression model with "Root sequence" (4 levels), Initial Segment (3 levels) and Task (2 levels, with and without tapping) and their interactions as independent variables and Response Segment (3 levels) as dependent variable. Significant effects of Stimulus, Initial Segment, and the interaction between Initial Segment and Task were found, $\chi^2(6)=265.32$, $p<.0001$, $\chi^2(4)=493.19$, $p<.0001$, and $\chi^2(4)=20.04$, $p<.001$, respectively.

Tapping task

The Optotrak data were filtered by means of a third-order Butterworth low-pass filter with a cut-off frequency of 5 Hz. Missing data points were added using linear interpolation. The data were positioned in three dimensions (XYZ). The Z-dimension corresponded with the sagittal plane and only these positional data points were used for further analysis. Although temporal data were also available they were not used in the present analysis. In Figure 3 the normalized index finger position (with 1 as the maximum value) as a function of tone sequence (1 and 2 stable, 3 and 4 unstable) is presented. If this figure is compared with the sequences in Figure 2, it can be seen that the participants positioned their fingers relatively well according to the relative distance in pitch height. The position of the index finger for the sequences 1 and 2 should be approximately the same. The position for sequence 3 should be relatively closer to the body, whereas the position for sequence 4 should be the most distant from the body.

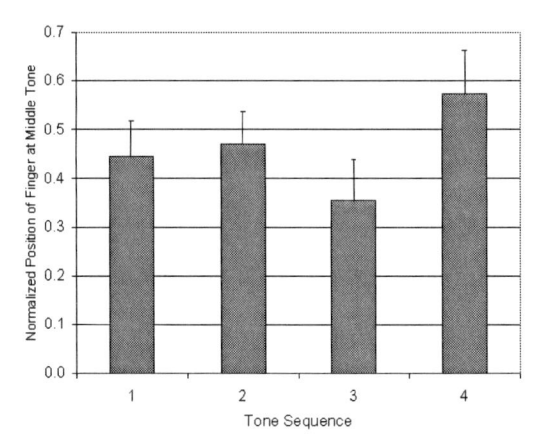

Figure 3. Normalized index-finger positions as a function of tone sequence. Normalization consisted of expressing the mean index-finger position per subject and tone sequence as fraction of the distance between low and high position. Subsequently, data were averaged across subjects. Error bars reflect between-subject variability (standard deviations).

In Figure 4 the finger position (in cm) as a function of repetition number and relative tone height (L, M, H) for a single trial is presented. In this trial a perceptual transition occurred, that is, the initial sequence was different from the sequence chosen. It is suggested that the perceptual transition occurred between the third and fourth repetition as reflected by the jump in the position of the index finger for the middle tone.

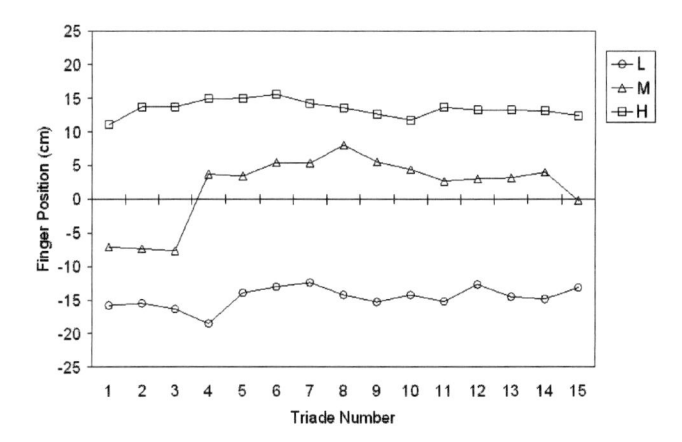

Figure 4. Example of finger positions (in cm) as a function of repetition number adopted during a single trial in which a perceptual transition was reported.

The mean middle within-subject variability was calculated to determine whether or not the perceptual stability of the sequence would be reflected in the variability of the positioning of the index finger. In Figure 5 this variability is presented as a function of sequence and of perceptual transition. It can be seen that when the perceptual transition was absent the variability was lower. In addition, it can be readily seen that the variability is systematically higher for the unstable sequences than for the stable sequences.

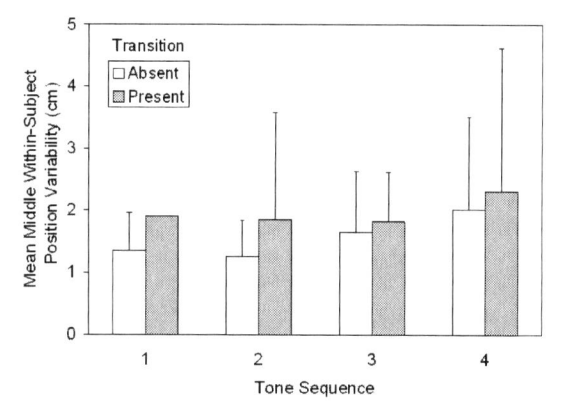

Figure 5. Mean within-subject middle position variability as a function of tone sequence when the response was identical to the initial sequence (white bars) and when the response differed from the initial sequence (grey bars). Error bars represent between-subject variability (standard deviations).

Discussion

Two main findings resulted from the present study. First, the proposed perceptual chance model sketched in the introduction section correctly predicted the perceptual stability of isochronous and iso-loud pitch sequences. Second, the perceptual stability of the sequences was reflected in the variability of the position of the index finger with which participants indicated perceived pitch height of cyclically presented three-tone sequences. These findings will be discussed in more detail.

The main parameter in the perceptual model is the relative interval size within a sequence. This relative interval size produces a perceptual accent on the tone following the largest relative interval. In contrast to earlier models (e.g., Thomassen, 1982) contour is not that important. This was reflected in the relative small contribution of contour in our regression analyses and the negative sign of the estimated weight associated with this factor. This negative sign indicates that a contour change decreases the probability of a "representative sequence" to be chosen. In other words, instead of contributing to the perceptual accent of a tone it makes the perceptual accent weaker. In addition, the initial segment plays a more important role in the response choice for the unstable sequences than for the stable sequences. It has to be noted, however, that in the present study the initial segment was always identical to the last segment. Therefore, it is difficult to determine if the influence of the initial segment relates to a primacy or a recency effect.

The perceptual instability of the mental representation of the melodic sequences was clearly reflected in the variability of the finger position in the tapping task under study. This variability was higher for the theoretically unstable sequences and it was higher for both the stable and unstable sequences in which a perceptual transition occurred. Both these effects make sense. The unstable sequences prove inherently more difficult to process and to represent mentally. This will result in larger performance variability. When a perceptual transition occurred this resulted in a changed mental representation on the basis of which the tapping behaviour was adjusted. Consequently, the movement variability was higher for the sequences in which a transition occurred than for the sequences in which no such transition was reported.

The results of the perceptual-decision task with and without tapping show very similar results indicating that there are no consequences in comparing the data from these tasks. Only a very small effect of task was found. Interestingly, the tapping task that we presently exploited also exerted an influence on the choice behaviour in the perceptual-decision task that immediately followed the tapping task. The initial segment was chosen more often after the tapping task than it was before this task. This may indicate that tapping along with a sequence supports the memory of the initial segment. However, it must be noted that before starting with the tapping task the listeners also heard the initial segment (more than once) to practice the movement pattern. This may also have been the reason for the increase in choice of the initial segment after the tapping task was completed. Consequently, the possible influence of motor behaviour on the perception seems a promising finding that warrants further investigation in future research.

In sum, the present study has shown that motor behaviour can be used to gain online insight into the process of auditory perception. Most experimental paradigms used in auditory perception let listeners only respond after the entire sequence has been heard. This makes it very difficult to predict how a percept emerges from a series of sequential auditory events. It is our aim to exploit the tapping-along paradigm in future studies to gain further insights into the process of auditory perception. In addition, the present study shows that perceptually unstable auditory sequences clearly give rise to a higher variability in motor behaviour. It is our aim to exploit in future research the relevance of these findings and the employed paradigm for actual usage situations.

Acknowledgements

We thank Chris Bouwhuizen en Hubert Voogd for the development of the computer programs and their technical support. We thank an anonymous reviewer for the useful comments on an earlier version of this manuscript.

References

Boltz, M., & Jones, M.R. (1986). Does rule recursion make melodies easier to reproduce? If not, what does? *Cognitive Psychology, 18*, 389-431.

Deutsch, D. (1980). The processing of structured and unstructured tonal sequences. *Perception & Psychophysics, 28*, 381-389.

Deutsch, D., & Feroe, J. (1981). The internal representation of pitch sequences in tonal music. *Psychological Review, 88*, 503-522.

Dowling, W.J. (1972). Recognition of melodic transformations: inversion, retrograde, and retrograde inversion. *Perception & Psychophysics, 50*, 305-313.

Fraisse, P. (1980). Les synchronisations sensori-motrices aux rythmes [Sensori-motor synchronization to rhythms]. In J. Requin (Ed.), *Anticipation et comportement* (pp. 233-257). Paris: CNRS.

Jansen, E., & Povel, D.J. (2004a). Perception of arpeggiated chord progressions. *Musicae Scientiae, 7*, 7 - 52.

Jansen, E., & Povel, D.J. (2004b). The processing of chords in tonal melodic sequences. *Journal of New Music Research, 33*, 31 - 48.

Jones, M.R. (1987). Dynamic pattern structure in music: Recent theory and research. *Perception & Psychophysics, 41*, 621-634.

Kelso, J.A.S. (1995). *Dynamic Patterns*. Cambridge: MIT Press.

Meulenbroek, R.G.J., Thomassen, A.J.W.M., Schillings, J.J., & Rosenbaum, D.A. (1996). Synergies and sequencing in in copying L-shaped patterns. In M.L. Simner, C.G. Leedham, and A.J.W.M. Thomassen (Eds.), *Handwriting and drawing research: Basic and applied issues.* (pp. 41-55). Amsterdam: IOS Press.

Pöppel, E., Müller, U., & Mates, J. (1990). *Temporal constraints in synchronizations of motor responses to a regular sequence of stimuli.* Paper presented at the Annual Meeting of the Society for Neuroscience, St. Louis, Missouri.

Povel, D. J. (2002) A model for the perception of tonal melodies. In: C. Anagnostopoulou, M. Ferrand, A. Smaill (Eds.). *Music and artificial intelligence.* (LNAI 2445). Heidelberg: Springer Verlag.

Povel, D.J., & Essens, P. (1985). Perception of temporal patterns. *Music Perception, 2*, 411-440.

Thomassen, J.M. (1982). Melodic accent: Experiments and a tentative model. *Journal of the Acoustical Society of America, 71*, 1596-1605.

Van Egmond, R., & Povel, D.J. (1996). Percieved similarity of exact and inexact transpositions. *Acta Psychologica, 92*, 283-295.

Vos, P.G., Mates, J., & Van Kruysbergen, N.W. (1995). The perceptual centre of a stimulus as the cue for synchronization to a metronome: Evidence from asynchronies. *The Quarterly Journal of Experimental Psychology, 48A*, 1024-1040.

Zwicker, E., & Fastl, H. (1990). *Psychoacoustics: Facts and Models*. Berlin: Springer-Verlag.

Working postures and physical demands on a utility vehicle assembly line

Laura K. Thompson[1,2], Thorsten Franz[1], & Heinzpeter Rühmann[2]
[1] MAN Nutzfahrzeuge AG, Munich, Germany
[2] Lehrstuhl für Ergonomie, Technische Universität München, Munich, Germany

Abstract

The working postures, physiological workload and manual materials handling demands of male workers on a utility vehicle assembly line were measured using methods that could be implemented in industry. For an analysis of working posture, ten working postures and workplace-specific tasks were defined. During observation, the elapsed time per working posture and task was recorded. The observed activities included the mounting and fastening of the drive shaft and steering rod, and the fastening of the front and rear axles. Generally the working postures were found not to be harmful, except for the severely bent posture during the mounting and fastening of the connecting drive shaft. The physiological workload was measured in terms of the energy expenditure required to complete the assembly work and was acceptable for the male workers. The manual materials handling demands were assessed using both the NIOSH equation and the VDI method. Both methods indicated that the lifting of the steering rod was too demanding. Both the NIOSH equation and the VDI method could be easily implemented in industry, but the working posture analysis was too time-consuming to be used as a general analysis method.

Introduction

Musculoskeletal disorders in the German manufacturing industry accounted for most (29%) of the lost days due to illnesses and injuries in 2002, with an average of 15.5 days per worker. This corresponds to a production shortfall of approximately 4.16 billion Euro (BAuA, 2002). Corrective measures are needed to reduce these costs and injury rates. At the MAN Nutzfahrzeug plant in Munich, a goal is to reduce the relatively high number of workers on sick leave. One of the possible measures is to improve the ergonomic conditions on the assembly line. Therefore this study was undertaken as a pilot project to analyse the ergonomic conditions on the assembly line at MAN and determine corrective measures. Three sample workplaces were chosen. The observed activities included the mounting and fastening of the drive shaft and steering rod, and the fastening of the front and rear axles on different utility vehicle models.

Working posture analysis

For an analysis of working posture, ten working postures were defined: upright standing, bent standing, stooping, twisted standing, bent and twisted standing, walking, squatting, kneeling, sitting and overhead. During observation, the elapsed time per working posture and workplace-specific task was recorded. The percentage distribution of postures for each vehicle model and workplace was overlaid on a simplified version of the Ovako Working posture Analysing System (OWAS) recommendations for corrective measures (Stoffert, 1985). Generally the working postures were not considered harmful, except for the working posture during the mounting and fastening of the connecting drive shaft. In a severely bent posture, the worker must hold the 15 kg drive shaft with one hand while fastening 4 bolts with the other hand and subsequently fasten the bolts with a 5 kg handheld screwdriver. Corrective measures to improve this task should be considered in future planning. The working body postures for the other tasks could also be improved by implementing pedestal desks and risers as well as through worker training.

Physiological workload analysis

The physiological workload was measured in terms of the energy expenditure required to complete the assembly work, which was based on the working body posture, the duration of the task and the type of work (Spitzer et al., 1982). The workload at the rear and front axel fastening workplaces was between 137 and 209 Watt and thus acceptable for men (280 Watt limit) and conditionally acceptable for women (200 Watt limit). If the cycle time is decreased from 8.1 to 7.1 minutes, the 225 Watt workload becomes only acceptable for men.

Manual materials handling

To determine the manual materials handing demands, both the NIOSH equation developed by the National Institute of Safety and Health in the USA (Waters et al., 1993) and the Verein Deutscher Ingenieure (VDI) method (VDI, 1980) were applied to the workplaces. The fetching of the bumper holders and the steering rods, as well as the operation of the side power screwdriver were analysed. The fetching of the bumper holders was acceptable; the weight was below the weight limit calculated using the VDI and NIOSH equations and the lifting index was correspondingly below 1.0. According to the VDI method, the lifting of the steering rod was too demanding for weaker and/or older workers. The results of the NIOSH equation were similar. With a lifting index as high as 2.1, there was an increased risk of low-back pain for some workers. For the operation of the side power screwdriver, the VDI method indicated that the weaker workers should always operate the machine with both hands. This task could not be evaluated with the NIOSH equation.

Discussion

In general, the working posture and physical demands are acceptable for the male workers, except for two cases: the working posture during the mounting of the connecting drive shaft and the lifting of the steering rods. These situations can be improved through the implementation of a supporting crane for the drive shaft and through an improved storage location of the steering rods. The working body postures for the other tasks can also be improved by installing pedestal desks and risers as well as through worker training.

The working posture analysis proved to be useful for classifying working postures and determining corrective measures, but it is very time-consuming. It is better suited to use as a tool to analyse certain critical situations. The physiological workload analysis gives a useful overall measure of the workload during the entire shift, but unfortunately it cannot be generally implemented since a complete workload analysis can only be completed concurrently with a working posture analysis. Both the NIOSH and VDI methods are fast and easy to use and can therefore be used to evaluate all workplaces in the plant. The VDI method additionally considers the characteristics of the worker and the static holding time and can be extended beyond lifting tasks. However, the NIOSH equation is preferred since it gives more realistic weight limits in combination with the lifting index.

References

Bundesanstalt für Arbeitsschutz und Arbeitsmedizin (BAuA) (2002) *Sicherheit und Gesundheit bei der Arbeit 2002 (Safety and Health at Work 2002)* Berlin: Bundesministerium für Arbeit und Sozialordung.

Stoffert, G. (1985). Analyse und Einstufung von Körperhaltungen bei der Arbeit nach der OWAS-Methode (Analysis and evaluation of working body postures using the OWAS method). *Zeitschrift für Arbeitswissenschaft, 39*, 31-38.

Spitzer, H., Hettinger, T., & Kaminsky, G. (1982). *Tafeln für den Energieumsatz bei körperlicher Arbeit. (Tables of energy expenditure during physical work)*. 6th Edition. Berlin: Beuth Verlag GmbH.

Verein Deutscher Ingenieure (VDI) (Ed.). (1980). *Handbuch der Arbeitsgestaltung und Arbeitsorganisation (Handbook of work design and organisation)*. Düsseldorf: VDI-Verlag.

Waters, T.R., Putz-Anderson, V., Garg, A., & Fine, L.J. (1993). Revised NIOSH equation for the design and evaluation of manual lifting tasks. *Ergonomics, 36*, 749-776.

Ergonomic rucksack design for elementary school students in Indonesia to minimise low back pain

Johanna Renny Octavia Hariandja, Bagus Arthaya, & Nana Suryani
Department of Industrial Engineering
Parahyangan Catholic University
Bandung, Indonesia

Abstract

Curriculum development in Indonesia had brought some new issues in the daily school life of students. In elementary school, the course subjects became so much extended that every subject needed two to three reference books. The students were ought to bring more books to school every day, with the result that youngsters between 6 and 12 years old had to carry heavy loads. Among many types of school bags, the rucksack was the most preferred in usage among the elementary school students because it could hold the heavy loads and gave a more symmetrical load distribution to the human body. The preference for rucksacks led to competition amongst school bag designers in Indonesia to produce rucksacks in many models. Unfortunately, the designers did not take ergonomics into consideration seriously in the design of the rucksacks. Low back pain was one of the effects associated most often with the use of rucksacks that were not designed ergonomically. This was even more risky for the elementary school students because their bones were still growing. This study focused on creating a prototype of an ergonomic rucksack for elementary school students in Indonesia. The goal was to make the rucksack comfortable to the users and minimise the chances to get low back pains.

Introduction

In elementary schools in Indonesia, school curriculum development impacts greatly on many practices in daily school life. One development is that the course subjects become much extended. Every student has to bring two to three reference books for each subject. This results in the obligation to bring a large amount of books every schoolday, which means that they have to cope daily with carrying a heavy load on their back for a long time. Facing this problem, the students tend to choose a large school bag which can enable them to carry the heavy load.

Nowadays in Indonesia, the rucksack is the most popular school bag used by elementary school students. Double-strap rucksacks give a more symmetrical load distribution to the human body and make the load much lighter to carry. For elementary school students, their consideration in choosing a rucksack is likely based on the brand and physical attributes such as shape, size, colour and accessories. This leads to competition among school bag designers in Indonesia to produce rucksacks in many models with a focus on those attributes only. Ergonomics is not taken into consideration seriously in the designing process. Many students also use rucksacks with Western brands, which mostly are designed without taking into account the cultural differences in their physical characteristics that are related to race and nationality.

In the last few years, there has been concern about musculoskeletal discomfort and increasing low back pain amongst school students throughout the Western world. Whittfield et al. (2001) investigated the weight and use of schoolbags amongst 140 students in New Zealand secondary schools. This study suggests that one group of students may be at a higher risk of developing musculoskeletal symptoms than others because of heavy schoolbags, long carriage durations and lack to access to lockers amongst them. Some studies of the paediatric population give the fact that many school children experience back pain. Troussler (1994) conducted a study among Scandinavian countries that identified the prevalence of back pain in a group of 1,178 school children at 51 percent. A study in USA shows that the use of rucksacks resulted in more than 6000 students have injuries related to rucksacks. Roth (2001) gives a figure of approximately 23% of elementary school youths and about 33% of secondary school youths complain of low back pain. The report made by archangelhealth.com (2002) informs the use of rucksack that leads to low back pain mostly occurs to students less than 14 years of age because their bones are still growing.

The main purpose of this study is to create an ergonomic rucksack design exclusively for elementary school students in Indonesia. The design is ergonomic when it is comfortable to its users and low back pain is minimised. It becomes important to investigate the current rucksack designs used amongst the elementary school students in Indonesia now, to study the complaints from their usage and to figure out how to solve them within the new design. It is thought that the dimensions of current design do not fit the physical characteristics of Indonesian elementary school students and can smooth the progress of low back pain occurrence. As a response to this, an ergonomic rucksack is designed based on their anthropometric data.

The objectives of this study are to answer the following questions: (1) what complaints elementary school students in Indonesia have related to current rucksack usage, (2) what modifications should be made to the rucksack design based on the complaints and anthropometric data, (3) what is an ergonomic rucksack design for elementary school students in Indonesia, and (4) how the performance of the ergonomic rucksack compared to current designs based on ergonomic aspects. The aim of this study is to create a new design of an ergonomic rucksack for elementary school students in Indonesia so it becomes comfortable to the users and minimises low back pain.

Methods

In the beginning of this study, a total of 106 elementary school students were surveyed to gather preliminary information about the current situation on rucksacks usage among them. A questionnaire was used to find out the following information of: (1) what kind of rucksack they use, and (2) on which body regions and rucksack parts they feel discomforts and have complaints. Based on this information and on the results of literature study, the design requirements for an ergonomic rucksack were stipulated.

After the identification of design requirements, the data collection from 106 students was conducted to gather the anthropometric data. The relevant anthropometric data that should be gathered were determined by looking at their relation with the rucksack dimensions. The raw data were processed through statistical tests first before applying them to the dimensions of the new rucksack design.

Based on the analysis of the preliminary information and anthropometric data, an ergonomic rucksack was designed and completed with a prototype. To know the performance of the new design, a comparison was made between the prototype and the current rucksacks based on three criteria: physiological (energy expenditure), biomechanical (compression and shear force), and psycho-physical (level of comfort).

Results

Questionnaire

The result from the questionnaire given to 106 elementary school students provided some preliminary information about the current situation on rucksacks usage among them. Based on the survey, it was found that there were three main brands of rucksacks used by the elementary school students in Indonesia. These rucksacks were used as current designs and later on were compared with the new design.

Several discomforts were identified from the use of current rucksack by the students. Table 1 shows the students complaints which resulted from using their current rucksacks. From those complaints, the percentages of comfort felt by the students were calculated. Table 2 and Table 3 show the percentage of comfort on body regions and rucksack parts when using current rucksacks.

In this study, the target percentage of comfort both on body region and rucksack part for the new design was aimed at 80%; maximal 20% of the users should feel discomfort when using the new rucksack design. Woodson et al. (1992) suggested that an ergonomic design based on the subjective opinions of its users should meet at least 80% of their opinion.

Table 1. Complaints of discomforts from the use of current rucksacks

Complaints	Frequency of Occurrence		
	Often	Sometimes	Never
A. Body Region			
Discomfort on neck	7	66	33
Discomfort on shoulder	26	60	20
Discomfort on upper back	7	42	57
Discomfort on lower back	3	32	71
Discomfort on waist	8	35	63
B. Rucksack Part			
Discomfort on shoulder straps	17	46	43
Discomfort on back of the body	9	53	44
Rucksack is not soft enough	7	36	63

Table 2. Percentage of comfort on body regions

Body Region	Comfort (%)
Shoulder	47.2
Neck	62.3
Upper back	73.6
Waist	75.9
Lower back	82.1

Table 3. Percentage of comfort on rucksack parts

Rucksack Part	Comfort (%)
Shoulder straps	62.3
Back of the body	66.5
Softness	76.4

The analysis of the overall questionnaire results showed some facts that in the end were taken into the design consideration as follows:

- The body region that has the most discomforts was the shoulder, followed by the neck. It also showed that the shoulder straps were the most discomfortable part of the rucksack. Therefore, the shoulder straps became the most important considered rucksack component in the new design, with its relation to overcome the discomforts on the shoulder and neck.
- The upper back was the next body region that has the percentage of comfort below target. The discomfort in the upper back was related to the discomforts felt on the back of the rucksack body and the softness factor. So, the back of the rucksack body and its softness needed to be considered as well.
- The percentage of comfort on the waist was also below target, this should be overcome by including a hip belt with an adjustable part around the waist in the new design.
- Although the percentage of comfort of the lower back was still above the target, it would be still taken into account in the new design in line with our aim to minimise low back pain for the users.

Anthropometric Data

The design of an ergonomic rucksack prototype in this study required several anthropometric data gathered from 106 elementary school students in Indonesia. The determination of the anthropometric data needed was based on their relevance for the rucksack dimensions. Most of the new rucksack dimensions were based on these anthropometric data. The relevant anthropometric data are shown in Table 4 and Figure 1, together with three percentiles: P_5, P_{50} and P_{95}.

Table 4. Anthropometric data of Indonesian elementary school students, relevant for rucksack design

Anthropometric Data (Figure 1)		P_5 (cm)	P_{50} (cm)	P_{95} (cm)
A	One shoulder breadth (biacromial)	10.1	12.1	14.1
B	Shoulder breadth (biacromial)	30.1	35.2	40.3
C	Shoulder – hip circumference	46.3	50.1	54.0
D	Nape of neck – middle point shoulder distance	3.4	5.0	6.5
E	Middle point shoulder – buttock distance	32.9	37.6	42.3
F	Back height	38.9	43.6	48.3
G	Lumbar concavity height	10.8	13.2	15.7
H	Waist circumference	58.4	69.3	80.3
I	Neck breadth	9.7	11.3	12.8
J	Waist breadth	20.2	23.1	26.0
K	Chest depth	12.0	13.5	14.9
L	Lumbar concavity depth	0.7	1.6	2.6
M	Shoulder elevation (°)	11.4	14.9	18.4

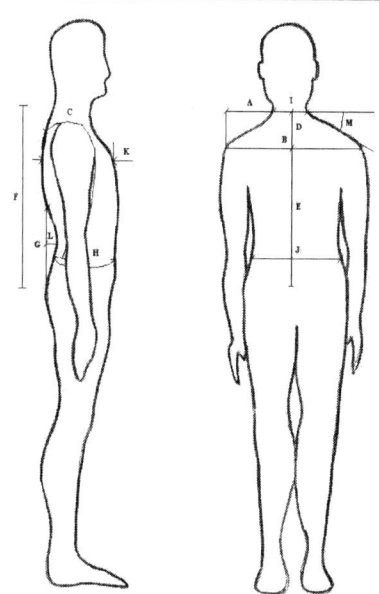

Figure 1. Relevant anthropometric data of Indonesian elementary school students for rucksack design (see Table 4). See also http://extras.hfes-europe.org

Design of new rucksack

The process of designing a new ergonomic rucksack was based on the analysis of the questionnaire and the anthropometric data. The questionnaire analysis was used as inputs for the improvement of the current rucksack design in order to give more comfort and safety to the elementary school students in Indonesia as its users.

Improvements were made on the shoulder straps, main body dimensions and additional cushions as covers of the rucksack parts that compress the user body. There were two new components in this new design, a belt and a side band, to give more comfort to the users. The anthropometric data were used as a basis to determine the dimensions of the new rucksack design. Table 5 shows the dimensions of the new rucksack design. The sketch drawings and pictures of the prototype are shown in Figure 2 and Figure 3.

Table 5. Dimensions of the new rucksack design

No.	Rucksack Dimensions	Dimension (cm)
1	Body height	38
2	Body width	26
3	Body depth	16
4	Shoulder straps length	40
5	Adjustable straps length	37.5
6	Shoulder straps width	7
7	Shoulder straps depth	0.8
8	Angle between shoulder straps and lower body (°)	25
9	Angle between shoulder straps and upper body (°)	45
10	Cushion height on lumbar region	16
11	Cushion depth on lumbar region	2
12	Cushion depth on back region besides lumbar	0.5
13	Hip belt length (one side)	12.5
14	Adjustable belt length (one side)	32
15	Hip belt width	7
16	Cushion depth on hip belt	0.8
17	First side band position (from body base)	10
18	Second side band position (from body base)	20
19	Side band length (hooked position)	16

Comparison of current and new rucksack design

The performance of the new rucksack design had to be tested whether it performs better compared to the current rucksack. Therefore, a comparison was made between the prototype and current rucksack based on three assessment criteria: (1) physiological, (2) biomechanical, and (3) psycho-physical.

In this test, 30 students were chosen as subjects and divided into 3 groups based on their grade in school: Grade 4 (age 10), Grade 5 (age 11) and Grade 6 (age 12). The subjects were asked to carry a particular load and walk normally within a certain distance using the three current rucksacks (A, B and C) and the prototype of the new design.

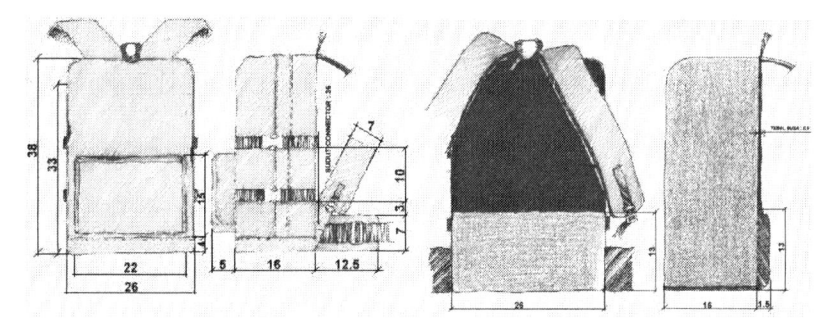

Figure 2. Sketch drawings of the new rucksack design (in colour on http://extras.hfes-europe.org)

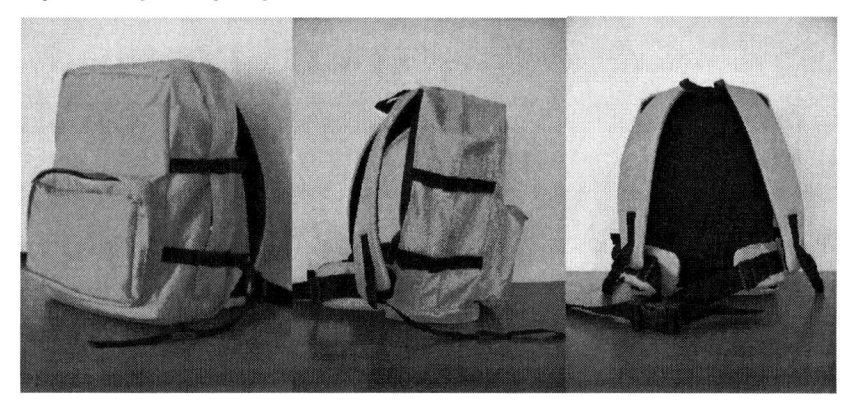

Figure 3 Prototype of the new rucksack design (see http://extras.hfes-europe.org for colour pictures).

Physiological Approach

In the physiological approach, the amount of energy expenditure needed by the subject to carry the load using the rucksack was assessed. Hence, the heart rate for each subject was measured before and after walking using the loaded rucksacks. Measurement of heart rate had been used to estimate energy expenditure over longer periods. There was a linear relationship between the physiological factors of heart rate, oxygen consumption and energy expenditure. Sutalaksana et al. (1979) proposed the conversion equation to calculate the energy needed for a job from the heart rate data as shown in Equation 1.

$$Y = 1.84011 - 0.022903 \cdot X + 4.71733 \cdot 10^{-4} \cdot X^2 \qquad (Eq.\ 1)$$

Where: Y = energy needed (Kcal/min)
X = heart rate per minute

The amount of energy expenditure was determined by calculating the difference between the energy before and after walking. The average energy expenditure in using the current rucksacks and the prototype are shown in Table 6. From the table, it shows that the average energy expenditure in using the prototype rucksack is minimal compared to the current rucksacks.

Table 6. Comparison of energy expenditure among rucksacks

Group	Energy Expenditure (Kcal/min)			
	A	B	C	Prototype
1	4.102	3.987	3.837	3.781
2	4.108	4.047	3.900	3.836
3	4.022	3.992	4.016	3.857

Biomechanical approach

Heavy loadings in using rucksacks are able to cause pains in several body regions such as shoulder, neck, back and in the end lead to injuries in low back area. Chaffin and Anderson (1991) stated that the Lumbar 5/Spinal 1 (abbreviated L5/S1) area located in the low back bone is the most critical area where pain mostly occurs because of loading forces. Low back pain happens when the disc between the low back bones can not restrain the compression force and shear force because of the loadings.

From the biomechanical point of view, the comparison factors were the compression force and shear force occurring at the L5/S1 area because of the weight of the carried rucksack. To determine the forces, several data were required from the subjects and the rucksacks.

The weight and height of subjects and the back bone elevation of the subjects when using rucksacks were measured. From the rucksacks, the following data were gathered: the weight of rucksack load, the upper and lower straps elevation, the body height, and the distance between centre point of upper and lower straps. The back bone, upper and lower straps elevations were determined with the help of Motion Analysis Tools (MAT) software.

The compression force (Fc) and the shear force (Fs) were calculated first by determining the back muscle force (Fm) that applies while the subjects use the loaded rucksacks. The free body diagram in Figure 4 illustrates all forces that apply at the back of the subjects during the carrying of a loaded rucksack.

Based on the free body diagram in Figure 4 and combining several relevant equations (Chaffin & Anderson, 1991; Chandler, 2000), three equations were established to calculate the compression force (Fc) and shear force (Fs) as shown in Equation 2.

$$\sum M_A = 0 \qquad \text{(Eq. 2)}$$

$W_{tas}\sin\theta.(a) + W_B\sin\theta.(\overline{AB}) - F_m\sin 13°.(\overline{AC}) + T_1\sin\alpha.(b) - (T_1'\sin\alpha + T_2'\sin\beta).(\overline{AD}) + W_E\sin\theta.(\overline{AE}) = 0$

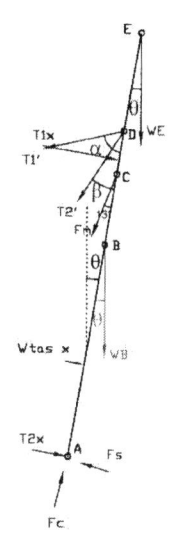

Legend:

A = L5/S1 position

B = centre of gravity of breast and stomach

C = back muscles position

D = midpoint of shoulder

Figure 4. Free body diagram of applied forces at the back during the use of rucksacks

$\Sigma F_{comp} = 0$ (Eq. 3)

$F_c - W_B \cos\theta - F_m \cos 13° - T_1' \cos\alpha - T_2' \cos\beta - W_E \cos\theta = 0$

$\Sigma F_{shear} = 0$ (Eq. 4)

$F_s - T_2 \sin\beta - W_{tas} \sin\theta - W_B \sin\theta + F_m \sin 13° - T_1 \sin\alpha + T_1' \sin\alpha + T_2' \sin\beta - W_E \sin\theta = 0$

Where:
M_A = momen at L5/S1
$\overline{AB}$ = distance between L5/S1 and centre of gravity of breast and stomach (cm)
$\overline{AD}$ = distance between L5/S1 and centre of gravity of head, neck and arm (cm)
$\overline{AC}$ = distance between L5/S1 and back muscle position (cm)
W_B = breast and stomach weight force (N)
W_E = head, neck and arm weight force (N)
m_b = body weight (kg)
H_b = body height (cm)
θ = angle between back bone and horizontal plane, where centre point is at L5/S1 (°)
α = angle between back bone and back muscle force, where centre point is at back muscles position (°)
F_m = back muscle force (N)
b = distance between centre point of upper and lower straps

a = distance between load centroid at axis y minus distance between centre point T_2 to rucksack base

In this study, the compression force of stomach onto the centre of diaphragm (F_A) was assumed not to exist during the carrying of loads with the rucksacks. Kroemer & Kroemer (1994) acknowledged that the compression force of stomach onto the centre of diaphragm (F_A) does not reduce the compression on the back bone during loadings.

The result of the average compression and shear force calculation for all rucksacks are shown in Table 7. From the table, it shows that the average compression and shear force is the least when using the prototype rucksack, compared to the current rucksacks.

Table 7. Comparison of calculated compression (Fc) and shear forces (Fs) caused by using the different rucksacks

Group	Compression and Shear Force (N)							
	A		B		C		Prototype	
	Fc	Fs	Fc	Fs	Fc	Fs	Fc	Fs
1	526.2	39.4	492.4	38.9	464.1	38.9	450.0	38.1
2	469.6	43.0	458.0	43.1	453.2	42.3	438.8	42.2
3	542.5	46.5	479.2	45.9	509.6	46.2	468.7	45.8

Psychophysical approach

The assessment of the psychophysical approach was conducted by giving a questionnaire to the subjects to know their opinion of the new rucksack design about the level of comfort they feel when using it. The questionnaire was filled each time after the subject finished using one type of rucksack. The average percentages of comfort for all rucksacks are shown in Table 8. From the table, it shows that the average percentage of comfort in using the prototype rucksack is maximal compared to the current rucksacks. It was also above the aimed target percentage of comfort (80%).

Table 8. Comparison of comfort percentage among rucksacks

Group	Comfort (%)			
	A	B	C	Prototype
1	40.0	54.4	62.5	81.9
2	44.4	46.9	68.8	83.1
3	44.4	63.1	49.4	83.8

Conclusions

The aim of this study is to design an ergonomic rucksack for elementary school students in Indonesia to minimise low back pain. In practice, ergonomics is often not taken into consideration seriously in the process of designing rucksacks, especially in

Indonesia. Low back pain is one of the mostly recognised effects from the use of not designed ergonomically rucksacks. There are increasing numbers of low back pain occurrence amongst school students and this is even more risky for elementary school students because their bones are still growing.

This study finds that there are quite high numbers of complaints from elementary school students in Indonesia related to current rucksack usage. They feel mostly discomfort on their shoulders, followed by their neck, upper back, waist and lower back. Surprisingly, the lower back is the body region they feel least discomfortable. This means that low back pain symptoms have not occurred yet until this moment. Nevertheless, they are still running a risk to have low back pain in the end if they keep on using rucksacks that are not ergonomic for them. Therefore, the effort to minimise low back pain still has to be taken into account in the new design.

Based on the intentions to overcome the students' complaints, some modifications have been made to the rucksack design such as improvements on the shoulder straps, main body dimensions and some cushions are added. Besides that, a belt and a side band are included in the new design to give more comfort to the users. The anthropometric data of elementary school students in Indonesia are used as a basis to determine the dimensions of the new rucksack design. In the end, a prototype of an ergonomic rucksack design for elementary school students in Indonesia is completed.

As a final stage in this study, the performance of the prototype of a new ergonomic rucksack design is measured based on physiological, biomechanical, and psychophysical measures. It is shown that the new design performs better compared to the current rucksacks.

References

Chaffin, D. B. & Anderson, G. B. J. (1991). *Occupational Biomechanics.* New York: John Wiley & Sons.
Chandler, A. P. (2000). *Human Factors Engineering.* New York: John Wiley & Sons.
Kroemer, K. H. E & Kroemer, H. B. (1994). *Ergonomics: How to Design for Ease and Efficiency.* London, UK: Prentice Hall.
Roth C. (2001). Parents and teachers need a lesson in ergonomics. *Industrial Safety and Hygiene News, 35,* 62.
Sutalaksana. I. Z., Anggawisastra. R., & Tjakraatmadja. J. H. (1979). *Teknik Tata Cara Kerja.* Bandung, Indonesia: Institut Teknologi Bandung.
Troussler B. (1994). Back pain in school children: A study among 1178 pupils. *Scandinavian Journal of Rehabilitative Medicine, 26,* 143-146.
Whittfield, J., Legg, S. J., & Hedderley, D. I. (2001). The weight and use of schoolbags in New Zealand secondary schools. *Ergonomics, 44,* 819-824.
Woodson, W. E., Tilman, B., & Tilman, P. (1992). *Human Factors Design Handbook.* New York: McGraw-Hill.
archangelhealth.com (accessed 11 March 2003) Backpacks Weigh Kids Down, www.archangelhealth.com/newsletters/nl060702.htm

Comparison of genetic and general algorithms of MLPs in posture prediction based on 3D scanned landmarks

Bing Zhang, Imre Horváth, Johan F.M. Molenbroek, & Chris Snijders
Faculty of Industrial Design Engineering, Delft University of Technology
Delft, The Netherlands

Abstract

Since the introduction of the 3D anthropometric techniques, engineers and ergonomists have sought to exploit the potential of this exciting technology. The added components of human body shape provided by 3D measurements offer a more detailed description of human variation compared with traditional manual 1D or 2D data. With the 3D anthropometry, it is possible to work on the landmark coordinates in 3D space directly. This paper is focused on comparison of two different algorithms of artificial neural networks in predicting of human body posture. The input is a set of demographic data and the coordinates of the landmarks characterizing a given posture. The output is another set of landmarks characterizing the transformed postures. The artificial neural networks are based on the principles of back-propagation feed forward network. With the simulating of trained networks, the users can predict the landmark coordinates in 3D space from sitting posture to standing posture, vice versa. In workspace design and automobile interior design, this technique will help the designer and ergonomists to solve anthropometric problems more effectively compared with using traditional methods. Our conclusion has been that the genetic algorithm is computationally more efficient in predicting of human body postures than general algorithm, but it needs much more computer time and cost which leads to very slow training. General algorithm is better for experienced network designer since it saves a lot of time in manually optimization and still can have the good result as using genetic algorithm.

Introduction

With the development of the IT-knowledge, many opportunities have emerged to integrate ergonomics modules into CAD environments. Computer-aided ergonomic design (CAED) supports: (1) visualization of the exact workspace of human limbs, (2) defining and planning trajectories in the workspace, (3) designing ergonomic workplaces subject to specified cost functions, (4) facilitating the design of layouts and packaging, (5) verifying measured data and validate human models, (6) predicting realistic postures, and (7) optimizing designs based on specified cost functions (Karwowski et al., 1990). Therefore, CAED is essential for industrial design engineering (Wier, 1989, Geuss, 1998). In the past couple of decades, Digital human modeling (DHM) has been becoming more and more versatile and convenient to use in ergonomics and design procedures (Verriest et al., 1991). They offer powerful tools when coupled with the knowledge of anthropometrics, ergonomics and human factors. RAMSIS, Jack and Safework are popular ergonomic software among those DHMs (Geuss & Bubb, 1994).

3D anthropometry that is based on, for example, laser scanning and stereo-photogrammetry is in the focus of the current research. 3D surface scanning makes it possible to measure the shape and offers the capability to capture the relationships between products and persons (Jones & Rioux, 1997). 3D anthropometric data processing can handle landmarks coordinates as well as other surface point coordinates, which are in between the landmarks. Several computer programs have been developed to extract landmarks and model of the shape of the human body with the involvement of known landmarks (Yavatkar, 1993, Lee, 2002, Luximon et al., 2003). The goal of research in 3D surface anthropometry is to obtain data for digital databases that can be up-dated according to the need of digital human modelling (Robinette, 1992). A common feature of current DHM software, is posture simulation. Safework, Anthropos, and Jack, all have such a feature. With the development of computer-aided ergonomics, the posture regeneration problem received more attention in connection with digital human modeling (Molenbroek, 1994). Various methodologies have been suggested by many researchers to measure postures.

Artificial Neural Networks (ANNs) are used in various applications in the field of ergonomics (Spelt, 1992, Murakami & Taguchi, 1991). A multi-layered perceptron networks (MLPs) are capable of performing just about any linear or nonlinear computation, and can approximate any reasonable function arbitrarily well. Properly trained back-propagation networks tend to give reasonable answers when presented with inputs that they have never seen (Venno & Ogawa, 1993). Typically, a new input leads to an output similar to the correct output for input vectors used in training that are similar to the new input being presented. This generalization property makes it possible to train a network on a representative set of input/target pairs and get good results without training the network on all possible input/output pairs (Simpson, 1990). A constrained generative ANN model was applied to hand posture recognition for the purpose of real-time computer visualization. The research results show that ANN can effectively recognize these hand postures. Various methods artificial intelligence based techniques have also been proposed as alternatives to statistical methods, in particular, to model non-linear functional relationships.

With the 3D anthropometry, it is possible to work on the landmark coordinates in 3D space directly instead of work on traditional statistics in 1D or 2D which has neglected many space information of human body (Zhang et al., 2004). This paper is focused on comparison of two different algorithms of artificial neural networks in predicting of human body posture. The purpose of a comparison research is to search for the better algorithm to minimize the neurons and cost of ANNs for predicting the landmarks coordinates of whole human body in 3D space. This predicting technique will save a lot of cost in time and financial aspect for ergonomists to help them acquire unknown 3D anthropometric data in their design.

Approaches of the comparison research

Samples and training data

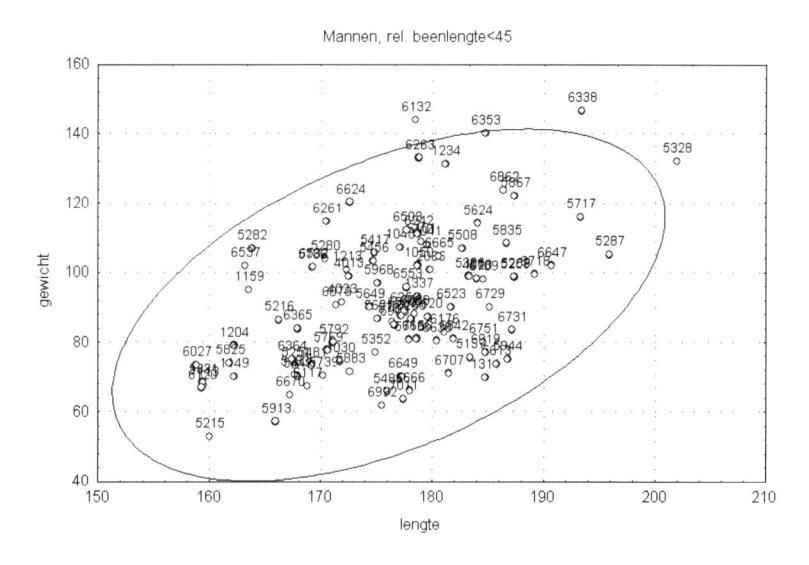

Figure 1. Samples of subjects who have ratio of leg length verse height <45%.

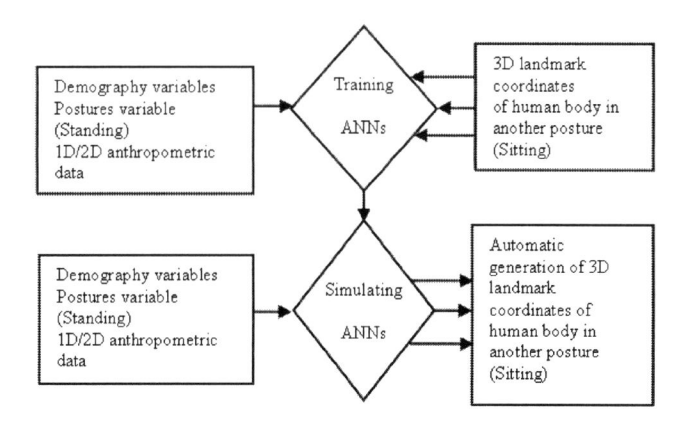

Figure 2. Workflow of training and simulating procedure in algorithm comparison.

The scanned 3D surfaces of human bodies are from the CAESAR project (Civilian American and European Surface Anthropometry Resource, Figure 1). The scanning measurements were made on people aged between 18 and 65, in three countries, namely, the United States, the Netherlands, and Italy. The raw body surface data were obtained in STL format, which is actually a polygonal (triangular) mesh. The total number of the scanned landmark data sets in our research was 40, from which 32 scans were used to train the neural network, and 8 scans were used to test the performance of the neural network.

When ANNs are used as the means of posture transformation, there are two main actions: (i) to teach the ANNs in order to get the transfer rules and (ii) to store these rules for future simulation. For teaching the neural network 25 landmarks were used from the total of 73 landmarks describing the whole body. With the learnt rule the system is able to generate the anthropometric data for the posture based on new input data, which is needed for product design. In addition to the landmark data, various demographic variables have also been used as input variables to teach the network (Figure 2). In our training experiments, these variables were gender, age, weight and height. The output is the expected coordinates of the respective landmarks. Table 1 shows the input data of neural network with relevant demography variables and 1D/2D anthropometric data, where 1 of gender means male, 2 of gender means female. When explicit posture data are used as input to ANN, the landmarks coordinates can be predicted automatically in different postures. The coordinates of landmarks can also be used to reconstruct the 3D geometric model of the human body.

Table 1. Input data of neural networks

Subject Number	Gender	Waist (mm)	Height (mm)	Weight (kg)	Sitting height (mm)	Subject Number	Gender	Waist (mm)	Height (mm)	Weight (kg)	Sitting height (mm)
1248	1	824.836	1900	65	886.4317	6023	2	677.164	1740	60	902.177
1251	1	937.733	1940	85	955.0853	6027	1	933.21	1600	74	843.3768
1329	2	766.959	1800	73	928.4067	6114	1	863.753	2050	100	994.6603
1449	2	723.918	1430	52	797.3711	6157	2	898.326	1700	70	863.9516
4042	1	1064.24	1785	87	822.1737	6268	2	879.785	1550	67	775.4662
5208	2	941.06	1800	88	897.8714	6353	1	1348.977	1870	135	924.6638
5239	2	954.388	1650	78	875.5234	6365	1	1050.28	1680	80	860.0975
5282	1	1188.33	1670	105	860.2009	6486	2	681.81	1690	50	841.0864
5287	1	1087.952	1960	100	993.1166	6550	2	1296.69	1750	126	897.8541
5317	1	1005.97	1850	91	905.0672	6551	1	872.158	2030	80	1002.6751
5440	2	1290.527	1660	122	884.1971	6562	2	1141.341	1500	92	823.5917
5514	2	710.853	1860	70	918.4264	6624	1	1248.124	1720	110	891.7229
5525	2	1140.118	1760	105	914.3433	6701	2	692.092	1820	57	871.0217
5590	2	1081.585	1680	92	817.4992	6738	1	724.771	1740	65	857.9713
5633	1	824.282	1850	70	905.7121	6754	2	668.706	1590	49	867.6571
5649	1	992.828	1780	87	936.4591	6803	1	834.493	1920	80	948.6668
5903	2	733.474	1950	81	986.4475	6870	2	906.305	1590	70	850.7872
5913	1	807.537	1700	55	859.2989	6950	1	714.688	1760	63	891.6316
5953	2	949.429	1680	78	881.6494	6992	1	712.303	1800	65	935.9212
5968	1	1095.017	1830	96	881.7302	7056	1	845.147	1960	80	965.2686

Full body transformation with general algorithm

In the latest research, the measured human body has been substituted by a proper set of landmarks, which is used as a basis of transforming the data, as they are needed to describe specific body postures. Artificial neural networks have been used for the actual conversion of data. The input variables are a set of demographic data and the coordinates of the landmarks characterizing a given posture, and the output is another set of landmarks characterizing the transformed posture.

Before designing of BP ANN, the raw data need to be pre-processed. This pre-processing included analysis of landmarks, calibration based on global coordinate system and local coordinate system respectively. Meanwhile, because of the drawbacks of scanning technique itself, some landmarks could not be acquired directly from the scanning data because laser rays could not reach. The missed landmarks have to be estimated in CAD software based on anatomical definition and the related 1D or 2D measurements have to be estimated as well in CAD software.

In the design of BP ANN, two multiple BP ANNs architectures were experimented and compared, which are two layers BP MLPs. We experiment not only different hidden layers, different number of neurons, but also different momentum and step size on different layers of MLPs.

The posture prediction was based on 25 landmarks, selected from a total of 73 landmarks identified on the measured data of the human body scanned by laser scanning technique. The steps of posture prediction for whole body are as follow:

1. Obtain point cloud of the whole body in the source posture
2. Find the landmarks on the source posture and on the target posture
3. Generate a set of descriptive input variables
4. Use the descriptive parameters, the source and target posture landmarks to teach the ANNs in general algorithm
5. Regenerate the landmarks representing the target posture based on the transformed landmarks and proportionality.

We performed an experiment with 40 subjects (20 male and 20 female) in total in 4 groups that were formed according to the ratio of leg and height of the subjects. The input of ANN were whole body landmark coordinates of 32 subjects in standing posture, with some demography information, such as gender, weight, height, head width, shoulder width and waist. The ANN is complied MLPs, which formed layered-feed forward network, typically trained with static back propagation with general algorithm. The desired/target value were landmark coordinates of the whole body of 32 subjects in sitting posture in 3D space. In the method of general algorithm, we searched for the optimal MLPs manually. It is important to find the network with the minimal number of free weights that can still learn the problem. The minimal network is likely to generalize well to new input data.

Full body transformation with genetic algorithm

Genetic algorithms are general-purpose search algorithms based upon the principles of evolution observed in nature. Genetic algorithms combine selection, crossover, and mutation operators with the goal of finding the best solution to a problem. Crossover is a genetic operator that combines (mates) two chromosomes (parents) to produce a new chromosome (offspring). The idea behind crossover is that the new chromosome may be better than both of the parents if it takes the best characteristics from each of parents. Mutation is a genetic operator that alters one or more gene values in a chromosome from its initial state. This can result in entirely new gene values being added to the gene pool. Genetic algorithms search for this optimal solution until a specified termination criterion is met.

The solution to a problem is called a chromosome. A chromosome is made up of a collection of genes that are simply the parameters to be optimised. A genetic algorithm creates an initial population (a collection of chromosomes), evaluates this population, then evolves the population through multiple generations in the search for a good solution for the problem. Therefore, genetic optimisation can be beneficial any time the network designer is unsure of optimal parameter settings.

In the design of genetic algorithm, some component configuration need to be set up, such as the maximum generations which is cell specifies the maximum number of generations that will be run until the simulation is stopped and the population size which is the number of chromosomes to use in a population. This determines the number of times that the network will be trained for each generation.

There are two types of genetic algorithm, one is generational and the other one is steady-state. Generational genetic algorithm replaces the entire population with each iteration. This is the traditional method of progression for a genetic algorithm and has been proven to work well for a wide variety of problems. It tends to be a little slower than steady-state progression, but it tends to do a better job avoiding the local minima. Steady-state genetic algorithm only replaces the worst member of the population with each iteration. This method of progression tends to arrive at a good solution faster than generational progression. However, this increased performance also in creases the chance of getting trapped in local minima. In our experiments, we choose generational genetic algorithm.

In order to make the comparison under the similar situations, we experimented with 40 subjects (20 male and 20 female) in total in 4 groups as we experimented in general algorithm. The steps of posture prediction for whole body are as follow:

1. Obtain point cloud of the whole body in the source and target posture
2. Find the landmarks on the source posture and on the target posture
3. Generate a set of descriptive input variables
4. Configure the component of genetic algorithm, use the source and target posture landmarks to train the ANNs.
5. Regenerate the landmarks representing the target posture based on the transformed landmarks.

Results and Discussion

Results of whole body-based posture transformation with general algorithm

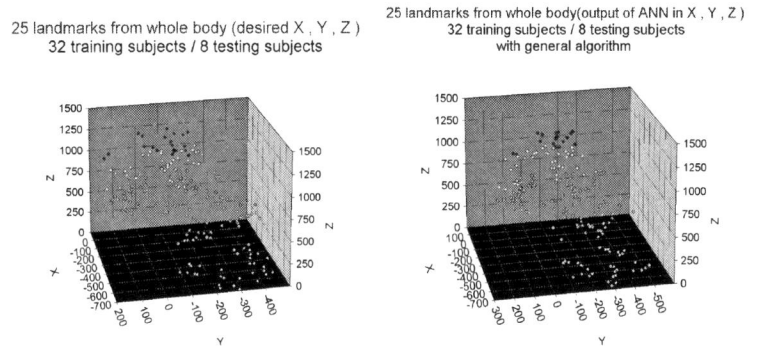

Figure 3. Visualization of the results of whole body posture prediction with general algorithm: (left) Desired (x,y,z) value of 25 landmarks of 8 testing subjects; (right) ANNs output (x, y, z) value of 25 landmarks of 8 testing subjects.

Table 2. General algorithm ANNs learning error

All Runs	Training Minimum	Training Standard Deviation
Average of Minimum MSEs	0,00736	0,00092
Average of Final MSEs	0,00761	0,00082

Best Network	Training
Run #	1
Epoch #	781
Minimum MSE	0,00626
Final MSE	0,00672

Table 3. General algorithm ANNs prediction error

MSE	0.22563
NMAE	1.23754
r	0.46944
%Error	48.969887

After many times experiments of training, we found the best weights based on minimum neurons on 2 hidden layers of MLPs. In Table 2, the MSE is the mean squared error of training progress with 32 training subjects. The best network has the minimum MSE, which is 0.00626. In Table 3, the MSE is the mean squared error of testing progress with 8 test subjects. NMSE is the normalized mean squared error of test, and r is the correlation coefficient between desired values and actual output values of 8 test data. The size of the MSE can be used to determine how well the network output fits the desired output, but it doesn't necessarily reflect whether the two sets of data move in the same direction. The correlation coefficient (r) between a network and a desired output solves this problem. A correlation coefficient of 0.88

means that the fit of the model to the data is reasonably good. In our case, the training was stopped when MSE < %Error/2 which means the configured MLPs has been trained good enough although the total r of 25 landmarks from whole body is only 0.46944. Figure 3 visualized the results of whole body posture prediction with general algorithm: (a) Desired (x,y,z) value of 25 landmarks of 8 testing subjects; (b) ANNs output (x, y, z) value of 25 landmarks of 8 testing subjects.

Results of whole body-based posture transformation with genetic algorithm

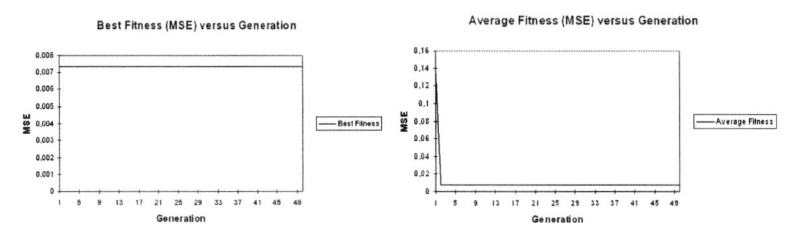

Figure 4. Plots of best fitness versus generation (left) and average fitness versus generation in genetic algorithm (right)

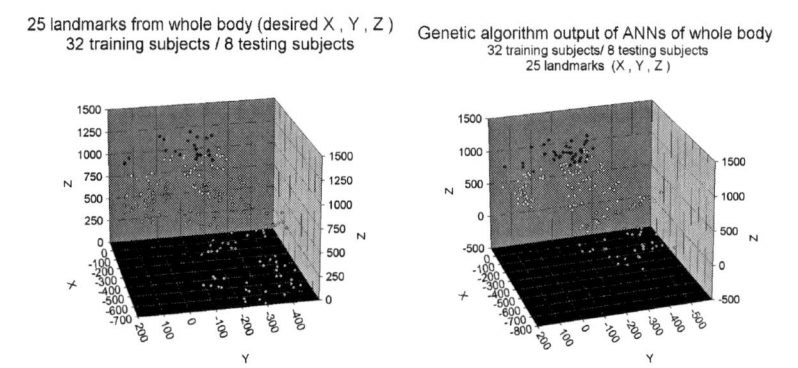

Figure 5. Visualization of 25 landmarks from whole body in 3D space: (left) desired landmarks in sitting posture; (right) actual genetic algorithm output in sitting posture

Table 4. Results of training with genetic algorithm

Optimization Summary	Best Fitness	Average Fitness
Generation #	1	2
Minimum MSE	0,00731	0,00731
Final MSE	0,00731	0,00731

Table 5. Results of testing

MSE of testing	0.19511
NMSE of testing	1.07011
r between desired and output	0.47926
% Test Error	37.14235

The experiment which finds the best fitness (low cost) of genetic algorithm is configured with the following items: number of epochs is 800; population size is 40; maximum generations are 50; maximum evolution time is 60 minutes; the bound of step size optimization is [0, 1]; the bound of momentum optimization is [0, 1]; the bound of processing element optimization is [15, 67] (Since the best network of general algorithm is based on 67 neurons, we chose 67 processing elements as the maximum value in genetic algorithm for the purpose of comparison.); Crossover is set up as one point (which means randomly selects a crossover point within a chromosome then interchanges the two parent chromosomes at this point to produce two new offspring.); mutation is set up as uniform. The best fitness (minimum MSE) is found at the first generation of chromosomes with the value of 0.00731 (Figure 4 and Table 4). Figure 5 plots and visualizes the desired output and actual network output in genetic algorithm with 25 landmarks from whole body. It shows that the prediction in Z coordinate is very precise than in X and Y coordinates. The possible reason is that the input variables related in height are efficient, but variables related in width and depth is not efficient. The testing MSE with 8 testing subjects of genetic algorithm is 0.19511. The correlation coefficient (r) between desired and actually ANNs outputs of genetic algorithm is 0.47926 which is lower than the final experiment of general algorithm (Table 5 and Table 6). However, genetic algorithm is much time consuming than general algorithm which impacts the final prediction effectiveness (Table 6).

Table 6. Comparison of general algorithm and genetic algorithm in Sum, Mean of testing error of 25 landmarks of 8 testing subjects in 3D coordinates and testing correlation coefficient (R)

	General Algorithm (mm)			Genetic Algorithm (mm)			Difference between General and Genetic Algorithm		
	X	Y	Z	X	Y	Z	X	Y	Z
Sum of testing error	-1585,93	-797,48	-381,73	-305,88	-2343,68	934,18	-1497,49	1580,33	-1324,79
Mean of testing error	-7,93	-3,99	-1,91	-1,53	-11,72	4,67	-7,49	7,90	-6,62
R	0.46944			0.47926			0.00982		
Time	50-120 minutes			5-15 minutes			45-115 minutes		

Conclusion

According to the comparison experiments between general algorithm and genetic algorithm of posture prediction of 25 landmarks from whole body, our conclusion is that genetic algorithm can help networks search for optimal design with low cost automatically, but it need a lot of time consuming and computer consuming compared with general algorithm. Because of the optimal searching periods, the genetic algorithm not only trains the good networks but also has to train the bad networks in order to get rid of them. General algorithm has result of lower R (correlation between desired landmark coordination and actual MLPs ANN output coordination) in testing procedure than genetic algorithm, but it saves a lot of time of training and manually optimal searching if the designer has good experience in training and in data pre-processing.

References

Geuss, H. (1998). Optimizing the Product Design Process by Computer Aided Ergonomics. *Digital Human Modeling for Design and Engineering.* Society of Automotive Engineers, Warrendale, Pennsylvania, USA.

Geuss, H. & Bubb, H. (1994). RAMSIS: A Newly Developed Anthropometric Measuring and Analysing Technique. Proceedings of the 12th Triennial Congress of the International Ergonomics Association, Vol.4: *Ergonomics and Design.* Toronto, Canada: Human Factor Association Canada.

Jones, P.R.M. & Rioux M. (1997). Three-dimensional Surface Anthropometry: Application to the Human Body, *Optics and Lasers in Engineering* 28, 89-117.

Karwowski, W., Genaidy, A., & Asfour, S. (1990). *Computer-aided ergonomics*, London: Taylor and Francis.

Lee, Y., (2002). *Recent Advances in Korean Anthropometry.* CARS 2000, 577-581. CARS. Springer.

Luximon, A., Goonetilleke, R.S., & Tsui, K.L. (2003). Foot Landmarking for Footwear Customization, *Ergonomics, 46*, 364-383.

Murakami, K. & Taguchi, H. (1991). *Gesture Recognition Using Recurrent Neural Networks. CHI'91-Reaching through Technology* (pp. 237-242). New Orleans, Louisiana..

Molenbroek, J.F.M., 1994, Statistische Bewerking van antropometrische data, Book: *Op maat gemaakt: Menselijke maten voor het ontwerpen en beoordelen van gebruiksgoederen* (pp. 211-218).

Robinette, K. M. (1992). Anthropometry for HMD Design. In: SPIE Proceedings 1695. *Helmet Mounted Displays III* (pp. 138-145). The International Society for Optical Engineering, Bellingham WA.

Simpson, P.K. (1990). *Artificial neural systems: foundations, paradigms, applications and implementations*, Pergamon Press, New York.

Spelt, P.F., (1992). Introduction to Artifical Neural Networks for Human Factors, *Human Factors Society Bulletin, 33*, 4-6.

Venno, K. and Ogawa K. (1993). A Design Guideline Search Method that Uses a Neural Network. In G.Salvendy and M.J.Smith (Eds.) *Human-Computer Interaction: Software and Hardware Interfaces.* (pp. 27-32). Elsevier, Amsterdam.

Verriest, J.P., Trasbot, J., & Rebiffe, R. (1991). MAN3D: A Functional and Geometrical Model of the Human Operator for Computer Aided Ergonomic Design. *Advances in Industrial Ergonomics and Safety III.* Edited by W.Karwowski and J.W.Yates.

Wier, A., (1989). Computer-aided Anthropometric Design and Assessment for Industrial Designers. In Proceedings of Conference on Design Education: Educating the 90* (pp. 206-211).

Yavatkar A., (1993). Anthropometric Shape Analysis Strategy for Design of Personal Wear. In Human Factors and Ergonomics Society 37th Annual Meeting, (pp. 411-415). Santa Monica, CA: HFES

Zhang, B., Molenbroek, J. F. M., Horvath, I., & Snijders, C. J. (2004). Automatic landmarks prediction using the artificial neural-network-based technique on 3D anthropometric data. D.Marjanovic. 2, 817-826. Zagreb, Croatia, Faculty of Mechanical Engineering and Naval Aechitecture, Zagreb. The Design Society, Glasgow. In: Proceeding of 8th International Design Conference DESIGN2004, 18-20th, May, Dubrovnic, Croatia.

The integration of anthropometry into computer aided design to manufacture and evaluate protective handwear

Gavin Williams, Simon Hodder, George Torrens, & Tony Hodgson
Department of Design and Technology
Loughborough University
Loughborough, UK

Abstract

It is important for high performance clothing and body worn products to contain enhanced fit for the wearer. This is to ensure that they provide protection while maintaining comfort, mobility and good interaction with the surrounding environment and task. This paper describes how anthropometry and computer aided design have been used to manufacture clothing that accurately fits the human body. This is demonstrated by the design of a mould tool for the manufacture of handwear, created using data taken during an anthropometric survey of undergraduate students at Loughborough University. Hand data from this survey have been processed into a series of glove sizes with CAD models of each size generated to represent the data in a 3D format. The mould tool used for the handwear manufacture is a physical manikin of one size of CAD model, created using various techniques of computer aided manufacture and rapid prototyping. Gloves produced using the mould tool have been evaluated to determine the accuracy of their fit by conducting user trials testing finger dexterity and tactile perception using subjects with appropriate sized hands.

Background

For all types of clothing and body worn products it is important to consider how they fit and integrate with the natural form of the human body. This is especially critical for high performance body worn products that attempt to protect the wearer, while simultaneously maintaining good mobility and interaction within the surrounding environment. To ensure that these products contain the high degree of fit that they require, it is necessary to use the size and shape of the end user at the beginning of the design and manufacturing process. This paper demonstrates the benefits of studying anthropometry and incorporating the data gathered into these products. It is necessary, however, to determine what anthropometric data is appropriate to use, how best to collect it and the most effective way to apply it. Consideration of how the anthropometric data might be used prior to and during its collection will greatly effect the process used to design and manufacture the desired end product. Computer Aided Design (CAD) and Computer Aided Manufacture (CAM) can also be

In D. de Waard, K.A. Brookhuis, R. van Egmond, and Th. Boersema (Eds.) (2005), *Human Factors in Design, Safety, and Management* (pp. 275 - 289). Maastricht, the Netherlands: Shaker Publishing.

incorporated into the design and manufacture of body worn products. It helps to ensure that the detail and accuracy of the design, made possible using anthropometric data, is transferred to the final product. Consideration must be made to the CAD model generation methods and prototyping techniques used and which combination is most effective, to ensure that this accuracy is transferred correctly and within an acceptable tolerance.

Integrating these two disciplines makes it possible to manufacture customised body worn technologies that enhance the fit for the user and improve their overall task performance. The anthropometric data of end users can be used directly to generate CAD models that provide a three-dimensional (3D) representation of that data. Once in a CAD model format the data can be produced as a physical manikin to be used as an aid for the design, manufacture and evaluation of any product that the original data to which it is related.

Integrating anthropometry and computer aided design

The following study demonstrates the use of anthropometric data to generate CAD models that have been used to produce mould tools for the manufacture of handwear. The data was taken from undergraduate students at Loughborough University and used to attempt to improve the fit of gloves manufactured using non-woven manufacturing techniques, dip moulding. Currently the mould tools used to manufacture gloves in this way have limited properties within them that relate to the size and shape of the wearer for which they are intended.

Anthropometry collection and processing

Since 1998 an annual anthropometric survey has been completed within the Department of Design and Technology, at Loughborough University. Each year anthropometric data from first year undergraduate students are collected. Approximately 100 students are measured during each survey, creating a database currently containing a total of over 600 records. Measurements taken in the survey include weight, stature, shoulder breadth, waist breadth and detailed dimensions of the hand, wrist and foot. The main aim of the survey is to compile a comprehensive anthropometric database of the student population. This can be used for future design projects which require anthropometry information to be used as part of the design methodology. In addition, the survey also aims to investigate the accuracy of the data collection process, determining intra and inter-observer error and generating measurement procedures for recording anthropometric data not covered in standard texts.

For this study, data of the hand have been used from the survey to create an auxiliary database with the aim to generate a size range as a basis for designing the shape of a new glove with improved fit. The data required reviewing and validating before this process could begin. Microsoft Excel was used to calculate descriptive statistics for each dimension (mean, standard deviation (SD), minimum and maximum) and correlation tables were generated to clarify the relationship between different dimensions. Each data point was inspected and accepted or deleted according to whether it was logically likely for the data to be correct. For example a subject with a large palm length and digit 3 length would also be expected to have a long hand length, see Figure 1. The processing of the raw data allowed the creation of percentiles which could represent the glove sizes. The percentiles were split into six groups with each group representing one size, i.e. six gloves sizes (6 to 11). The dimensions from the largest percentile in each group represented that respective size. Therefore within each glove size there was a range of percentiles that were accommodated.

This transformation from percentiles to glove sizes was not simply achieved by converting the percentiles into a relevant number, a consideration of glove design needed to be included. By means of a user panel, consisting of experienced experts in each relevant discipline, certain dimensions were deemed more critical than others to improve the fit of a glove (Noro, & Imanda, 1991; Krueger, & Casey, 2000). These dimensions were hand length, hand breadth, digit 2 (index finger) length and digit 3 (middle finger) length. Clearly other dimensions are also important to the fit of a glove; however, these dimensions are crucial (Bradley, 1969; Bishu, & Klute, 1995; Torrens, Williams, & M^{C}Allister, 2001). A full list of the dimensions used can be seen in Table 1 and Figure 1. Conventional glove sizing methods use circumferences as a guide to fit, primarily using the circumference around the metacarpal joints as a main criterion (BSEN 420:2003). However, length and width dimensions play a more important role in maintaining finger dexterity. Inaccuracies in circumference dimensions are less influential in providing an accurate fit and will not result in the same impairment of dexterity. Therefore, ensuring length and width dimensions are accurate means that complex tasks, such as using keyboards, donning/doffing of clothing and manipulating small objects, are not impaired when wearing a glove.

Table 1. Dimensions used to generate the CAD model

Hand Length[+]	Palm Length	Hand Breadth[+]	Digit 1 Length
Digit 2 Length[+]	Digit 3 Length[+]	Digit 4 Length	Digit 5 Length
Crotch Height 1	Crotch Height 2	Crotch Height 3	Crotch Height 4
Digit 1 PIP Breadth	Digit 2 PIP breadth	Digit 2 DIP breadth	Digit 3 PIP breadth
Digit 3 DIP breadth	Digit 4 PIP breadth	Digit 4 DIP breadth	Digit 5 PIP breadth
Digit 5 DIP Breadth	Digit 2 to Thumb Crotch Length	Wrist Breadth	Wrist Depth
Depth at Thenar Pad	Depth at MCP	[+] Dimensions critical for accurate fit	

The main factor effecting fingertip dexterity is the occurrence of excess material around or at the fingertip area. Excess material in this region of a glove occurs when it reaches the crotch of the finger before the tip, creating a gap between the finger and the glove. To overcome this, digit crotch lengths can be extended slightly to decrease digit lengths and create smaller digits, see Figure 1. This increases the probability that the tips of the digits come into contact with the glove first, eliminating the occurrence of excess material in this area. The outcome means that any gap occurring at the tips of the digits has been transferred to the crotches which can cause restrictions when attempting to span the hand. However, through user trials and practical assessments (McDonagh-Philp & Torrens, 2000; Torrens & Newman, 2000) it was found that this was less important than enhancing finger dexterity and so a restricted hand span was compromised to ensure good dexterity at the fingertips.

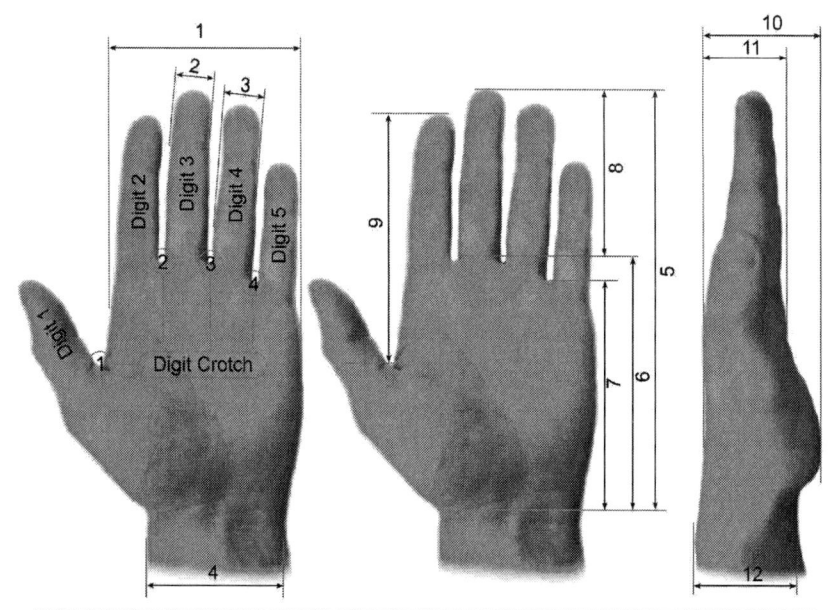

Dimension	Description	Dimension	Description
1	Hand Length	7	Digit Crotch Length
2	DIP Breadth	8	Digit Length
3	PIP Breadth	9	Digit 2 to Digit 1 Crotch - Length
4	Wrist Breadth	10	Depth at Thenar Pad
5	Hand Length	11	Depth at MCP
6	Palm Length	12	Wrist Depth

Figure 1. Key to hand dimensions

Generation of an initial CAD model

The CAD model was constructed by generating geometric parts that represented each part of the hand, e.g. the fingers, the thumb and the palm. The size and shape of the parts were determined by the corresponding dimension(s) within the size data. After each part had been generated the hand was constructed by assembling all the parts together, again using dimensions within the size data to determine their position relative to one another. Previous attempts of using anthropometry to build a CAD model (Hidson, 1991, 1992), have concluded that the collection and processing of the anthropometric data is one of the most important factors when integrating it into the CAD model. This same conclusion was also found when attempting to generate this initial CAD model, as it soon became apparent that there was insufficient data and some dimensions were not in the correct format to produce a sufficiently accurate CAD model. For example, the anthropometric survey only took one dimension to determine the depth of the palm. This part of the hand is perhaps the most complicated and so more detailed dimensions were required. Many of the dimensions taken in the survey were circumferences. This proved difficult to generate the parts within the CAD program accurately as it was necessary to have lengths and breadths to construct the appropriate geometry.

The problems with the original data used meant that new data needed to be gathered and some of the size data required re-processing. The additional data was again collected from students in the Department of Design and Technology, Loughborough University, with volunteers chosen based on how well they matched the original size data. The additional dimensions taken were wrist depth, wrist breadth, the depth at the knuckles and the depth at the thenar pad (the fleshy area of the palm located at the base of the thumb), see Table 1 and Figure 1. These dimensions would give a more detailed representation of the palm of the hand and therefore create a more accurate CAD model. To calculate breadths and widths from the circumferences, some of the raw data required re-processing; this then generated extra dimensions for detailing the fingers and thumb. The new dimensions from the re-processing and from the additional anthropometric surveys were then used to generate a revised size range and update the CAD model.

CAD model development

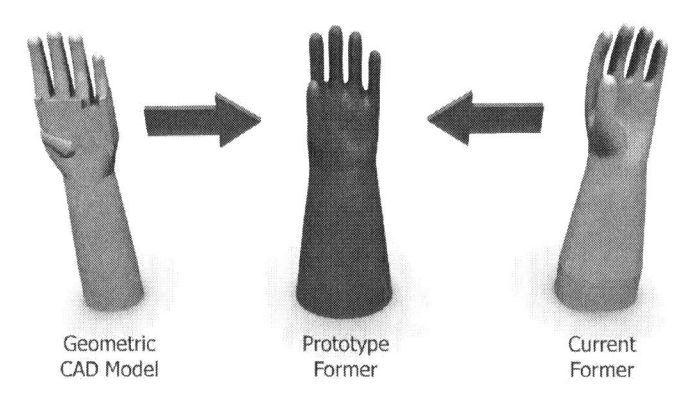

Figure 2. Integrating current glove former properties into the initial geometric CAD model to produce the prototype former

The additional data collected enabled the initial CAD model to become an improved representation of the hand, ensuring a higher degree of accuracy and detail. However, it was still unsuitable to be used as a glove former for the dip moulding manufacturing process. This method requires the former to be smooth and uniform as it produces a glove that is exactly to the size and shape of the mould former. The initial CAD model was geometric and would not create a suitable glove to wear; this meant that additional detail needed to be integrated. This came from a current glove former tool. Collaborating closely with a glove dip moulding manufacturer, the initial CAD model was successfully developed into an appropriately shaped former.

Detail from the palm area of the former and the finger crotches, as well as necessary properties for the dip moulding process, were integrated by generating extra surfaces within the CAD model, Figure 2. The integration of this detail meant that the developed CAD model contained properties for a suitable glove former as well as anthropometric data of a known sample population.

Manufacturing the prototype former

To evaluate the prototype former the CAD model needed to be produced into a physical glove former that could be used to manufacture gloves. Current formers are made from aluminium or ceramic and their manufacturing methods are very labour and time intensive. This timescale was unsuitable and so methods of Rapid Prototyping (RP) and Rapid Manufacturing (RM) were used. A CAD model generated from size 9 data was produced for a prototype former as this was deemed a popular size and the gloves manufactured using it could be evaluated using a larger number of subjects. To produce a RP model of the CAD design a 3D Systems InVision™ 3D printer was used (3D Systems, 2003), which accurately constructed a 3D representation of the CAD model in an acrylic photopolymer. The RP model was used as a master part to produce the prototype dip moulding formers for manufacturing the gloves. Due to the solvents and high temperatures involved in the dip moulding process, the prototype former needed to be produced from an inert, resistant material. Therefore, the material chosen was a polyurethane tooling resin that had good thermal resistance and did not react with the solvents it would be exposed to.

To accurately recreate the RP model in the polyurethane resin, silicone moulds were made from the master part. Silicone rubber was poured around the model and left to cure. Once cured it could be cut open to remove the RP model and leave a cavity that exactly matched the models size and shape. The tooling resin is a two part epoxy resin that is in a viscous liquid state before it hardens. This allowed it to be poured into the silicone mould and take the shape of the cavity left by the RP model. Once set, the silicone mould was opened to remove the finished resin tool. Some minor hand finishing was required to remove excess fragments of resin before it could be used as a glove former to manufacture the prototype gloves.

Accuracy of the prototype former

Before using the prototype former to manufacture gloves it was necessary to determine its accuracy. This was to ensure that the processes used to manufacture its size and shape did not differ dramatically from the CAD model. The resin former would not be an exact replica of the CAD model, but the tolerances between the two needed to be known and changes made if they were considered unacceptable. The process of was achieved 3D scanning the prototype former and comparing it to the CAD model using Computer Aided Inspection (CAI) software. Figure 3 shows a 3D comparison of the RP model and CAD model. The variation in colour indicates variations in the size and shape of the two models.

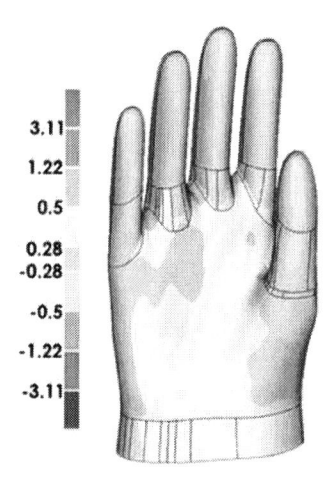

Figure 3 – 3D comparison of CAD model and RP model using CAI software, scale in mm

Manufacturing prototype gloves

Having assessed the accuracy of the former and determined that the tolerances introduced during its production were acceptable (±0.36mm), the resin prototype former was given to a handwear manufacturer to produce a batch quantity of prototype gloves. The material chosen was a lightweight Flexigum natural rubber, with a smooth finish. This was selected because it had no significant shrinkage after being stripped from the former, resulting in the glove representing its exact shape. To evaluate the prototype glove against current dip moulded gloves, a standard size 9 glove former was used to produce gloves using the same material and dip moulding process as the prototype gloves. This meant that the two gloves being tested had the same material properties and were the same thickness (table 2).

Glove evaluation

The gloves manufactured using the prototype former were then assessed through user trials to investigate the difference in performance between the prototype glove and a standard glove. This was done using subjects with relevant sized hands to undertake a series of tests that evaluated dexterity and haptic feedback.

Method

Eight healthy male subjects, 25.25 ± 6.7 years, participated in the glove evaluation. Each subject was selected by how accurately their hands matched into the size 9 data. The dimensions used to select participants were those deemed critical during the size generation process; hand length, hand breadth, digit 2 length and digit 3 length. Each subject was tested wearing the standard glove, the prototype glove and with bare hands. A familiarisation condition was also used to allow each subject to become accustomed with the various tests and reduce learning effects due to the repeated exposure to the test programme. The gloves worn for this condition were Nitrile™ chemical protective gloves.

Table 2. Glove thickness of each glove used for the evaluation testing

Glove Type	Thickness (mm)	
	Palm	Fingertip
Standard Glove	0.67	0.65
Prototype Glove	0.69	0.67
Familiarisation Glove	0.52	0.55

Four different tests were performed; a pegboard test, a nut and bolt assembly, a pin pick-up test and a keyboard typing test. Each subject was also tested for grip strength and pinch strength for each condition. A psuedo-balanced test design was used to reduce the effects of presentation order of tests and gloves. Before each trial the subject was informed of what it entailed and understood that they could stop at anytime with no obligation to give reasons for their withdrawal. They were asked to don the appropriate glove (if applicable) and carry out the test as directed by the experimenter. The subject was instructed when to start and stop each task to facilitate the timing and recording. All times recorded were taken with a digital stopwatch. The subjects were tested individually and each trial lasted approximately 30 minutes, inclusive of rest breaks between each test.

Grip strength
This was tested using a Baseline® hydraulic hand dynamometer which recorded the maximum value in kilograms. All measurements were recorded using the same apparatus. The subject was seated upright in a chair leaning against the back with their feet supported. The dynamometer handle was gripped by the subject at arms length by their side. Continuing to grip the handle, the subject was then instructed to lift their hand and forearm up, to form a right angle with their upper arm, and apply the maximum load that they could, (Mathiowetz, Weber, Volland, & Kashman, 1984; Spijkerman, Snijders, Stijnen, & Lankhorst, 1991). This value was then recorded by the experimenter. The test was performed twice, prior to and after the dexterity tests, with a mean value calculated.

Pinch strength.
This was tested using a Baseline® hydraulic pinch gauge which recorded the maximum load in kilograms. All measurements were recorded using the same apparatus. The subject was seated in the same position as when they performed the grip strength and, again, measurements were taken prior to and after the dexterity tests. The experimenter instructed the subject to extend their index finger and thumb, placing their finger pad on the pinch meter and the thumb below. They then pressed as hard as they could with their fingertip while the experimenter supported the weight of the pinch gauge. The maximum load was then recorded, (Mathiowetz et al., 1984).

Evaluating grip and pinch strengths of the subjects gives an indication to the influence the gloves have on impeding their grip and pinch patterns and therefore their ability to grip and manipulate objects effectively. Since both the standard glove and prototype were manufactured from the same material and had the same thickness, any differences in grip and pinch strength can be attributed to the fit of the glove.

Pegboard Test
Using a modified Purdue Pegboard, subjects were instructed to remove a small peg (40mm in length and 8mm in diameter) with their dominant hand from a board and place it in a corresponding cup, (Tiffin & Asher, 1948). A total of 20 pegs were removed individually and placed into separate cups. After all pegs have been removed from the board they were replaced back, one at a time. The total time taken to remove and replace all the pegs successfully was recorded by the experimenter. Dexterity at the fingertips plays a key role when manipulating and assembling small objects, however it is often hampered when wearing gloves. The Pegboard test is a multiple-operational manual test of gross- and fine-motor movements of hands, fingers, arms and the fingertips (Sweetland & Keyser, 1991), and was used to analyse the effect of gloves on the subject's capability of handling small objects. This would highlight any difference in fit occurring between the standard glove and prototype glove and the influence this has on dexterity.

Nut and bolt assembly
Subjects were asked to unscrew two nut and bolts from a vertical plane and place them into separate cups in front of them. Once both nut and bolt sets had been removed and placed into their respective cups, the subject then had to replace the bolt in the hole that it had been taken out of and attach the nut again, until it could not be turned any further. The entire task is completed with the subject facing a vertical plane perpendicular and attached to the plane which the nut and bolt is screwed into. This ensures that the subject is unable to view unscrewing and screwing the nut and bolt. The time taken to successfully disassemble and assemble the two bolts was recorded by the experimenter. This test evaluated the manipulative skills of the subject. It gives a more accurate measurement of how the gloved conditions impair their haptic feedback and dexterity as no visual input can be used to complete the task successfully.

Pin pick up test
This test has been adapted from a British Standard (BS EN 420:2003), and is a method for determining gloved finger dexterity. Four solid, steel pins each 40 mm long and with a diameter respectively of 1.5mm, 3mm, 4mm and 6mm were used. The pins were placed on a flat, smooth surface and each subject was asked to pick up each pin, with their dominant hand, by its circumference between their forefinger and thumb. The subject had three attempts to pick up each pin, without undue fumbling and within 30 seconds. The results of this test give an indication to the level of dexterity afforded to the wearer by the glove. The smaller the pin picked up the greater the level of dexterity.

Keyboard typing test
The subject was asked to type a short paragraph of text using a standard keyboard. Each subject was instructed to type using their own technique that was familiar to

themselves and to not correct any errors that they were aware of during the test. The time taken to complete the task and the number of errors within the paragraph were recorded. This test was added to include a task that the subjects were more accustomed to. Most of the subjects were unfamiliar with the battery of dexterity tests and although a familiarisation condition was included in the trial, it was necessary to analyse the affect the gloves had on an activity of daily living (ADL), where each subject would have there own individual technique.

Results

Overall all of the subjects had a reduced ability to perform the tests when wearing gloves (table 5), due to the lack of finger dexterity, haptic feedback and tactile perception. However, wearing the prototype glove their ability to perform the tasks was significantly better than when wearing the standard glove, Figure 4. Participants took, on average, 11 seconds longer to complete the pegboard test whilst wearing the standard glove when compared to bare hands, (Figure 4a) whereas when wearing the prototype glove the difference was only 4.5 seconds. Similar results are shown in the nut and bolt assembly, (Figure 4b) where the average time taken to complete the task when wearing the standard gloves is 14.5 seconds longer than it is when wearing the prototype glove. The keyboard typing test (Figure 4c and Table 3) reveals that wearing gloves significantly affects a persons individual typing style, increasing the amount of time required and introducing a high number of errors. When wearing the prototype glove there was a dramatic difference in the number of errors made, however, this difference was greater still when wearing the standard glove.

Table 3. Total number of errors recorded during the keyboard typing test

Condition	Keyboard Typing - Total Number of Errors
Bare Hands	25
Standard Glove	78
Prototype Glove	46

Table 4. Grip and pinch strength results

Condition	Grip Strength (Newtons)	Pinch Strength (Newtons)
Bare Hands	505.2	60.2
Standard Glove	464.7	60.7
Prototype Glove	473.3	60.7

The pin pick up test, (Figure 4d) is the only test where the prototype glove performed better than the bare handed condition. A possible explanation for this could be the added grip that is introduced when wearing a glove that fits correctly enabled some subjects to pick up all pins when their bare hands could not. This would also explain why when wearing the standard gloves only 65% of the subjects could successfully pick up all four pins as the fit of the glove was inadequate, experiencing too much excess material at the fingertips.

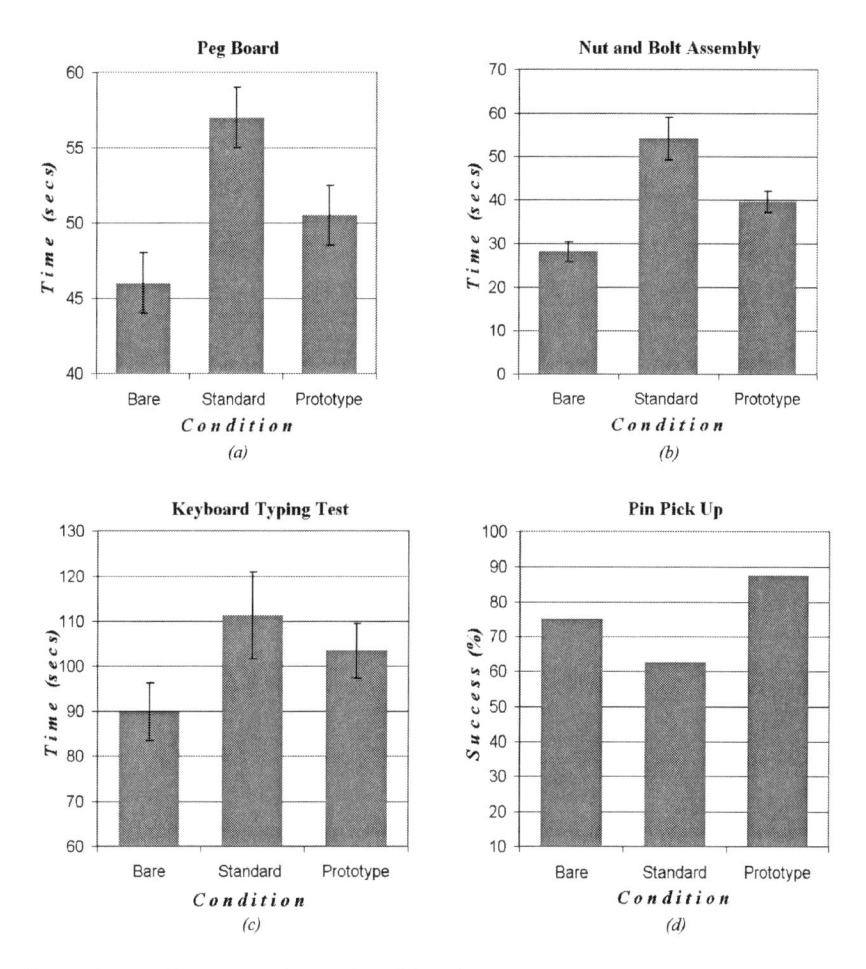

Figure 4 – Bar charts presenting results of dexterity tests

As with finger dexterity, grip strength is significantly affected when wearing gloves (table 4). Both the prototype glove and standard glove have a reduced grip strength compared to the bare handed condition; however, participants wearing the prototype glove had a stronger grip than when wearing the standard glove. This difference must be due to the variation in fit, as the material type and thickness were the same on both pairs of gloves. Results from the pinch strength show that there is only a slight difference between a gloved and ungloved hand and no difference between the different gloved conditions. Possible conclusions made from these results are that that the fit of a glove has little effect on a persons pinch, however more gloved conditions would need to be tested to fully determine this.

Table 5. Each condition ranked for each test

Test	Bare Hands	Prototype Glove	Standard Glove
Grip Strength	1	2	3
Pinch Strength	3	1.5	1.5
Peg Board test	1	2	3
Nut & Bolt Assembly	1	2	3
Pin Pick Up Test	2	1	3
Keyboard Typing Test - Time	1	2	3
Keyboard Typing Test - Error	1	2	3
Sum of Ranks	10	12.5	19.5

Discussion

Although the glove evaluation was completed using a small number of subjects, a larger trial is intended to be undertaken, the results reveal that wearing gloves made using the prototype former significantly increase dexterity and tactile perception for the wearer when compared to gloves made from a current former. The lack of anthropometric detail related to the end user within a current former means that the resulting glove does not have a high degree of fit and there is a large amount excess material, particularly around the fingers. This leads to a loss in sensitivity in this area and in some cases gives false senses to the wearer. The resulting effect is that the manipulation of small objects and recognition of different textures or surfaces is deficient, cumbersome and time consuming.

The CAD models created for the prototype glove former have been designed using a technique that allows anthropometry from different surveys to be used for their generation. This means that any set of appropriate hand anthropometric data can represented in a 3D format. This makes the process ideal for designing handwear to accommodate groups of people ranging from a single person to large populations. For example a survey of a specific workforce would generate a size range that gives the opportunity to manufacture mass customised handwear with enhanced fit, specific to their needs and requirements.

The CAD models can be used for other handwear applications. The initial CAD model, prior to development into a glove former, is been developed as a size gauge for use within a manufacturing environment as part of a quality control test procedure. Manually or machine stitched gloves can be checked against a 3D template to determine the accuracy of each glove as it is manufactured. This allows the manufacturer to monitor and control any errors introduced during the manufacturing processes, enhancing their fit by ensuring the sizes of the gloves are kept within known tolerances. This not only enables a single manufacturer to produce more accurate handwear. By installing the same range of size gauges in all manufactures of the same glove, it allows them to produce gloves to the same high degree of fit. This will therefore eliminate any dimensional variation between gloves of the same size and give a consistent, accurate size range.

The generation of CAD model using anthropometry is not limited to the hand; the same principles can be applied to design CAD models of various other parts of the body. For example, anthropometry of the foot, back, face and head can all be used to design and manufacture better fitting footwear, load carriage systems, face masks or helmets. As has been shown in this example of handwear manufacture, it is important to consider how the anthropometry will be used before it is collected. Obviously the more data gathered the more detail can be integrated into a CAD model and so generate a more accurate product. This accuracy however, must be relevant to the end product. It is unnecessary to generate a highly detailed, accurate CAD model with a large amount of anthropometric data if the product it is being used to design for does not require this level of precision. The process of integrating anthropometry into a CAD model can be time consuming and complex; therefore, integrating redundant data would create an inefficient process in terms of cost and time. Future research aims to develop recommendations for using anthropometry within different products and determining how much data is required for suitable accuracy. The methods of CAD model generation and rapid manufacturing will also be analysed for their efficiency and accuracy. This means for a given product and level of desired accuracy, the amount and type of anthropometric data, as well as the combination of CAD generation methods and rapid manufacturing techniques can be determined to ensure the entire design and manufacturing process is as efficient as possible.

Acknowledgments

The authors would like to thank Mr. Mike Reid and Mr. David Bryant from Comasec Yate Limited for their professional help and advice throughout this project.

References

3D Systems (2003). *Design It, Print It, Use It...* Retrieved February 2004 from http://www.3dsystems.com/products/multijet/invision/index.asp
Bishu, R.R. & Klute, G. (1995). The effects of extra vehicular activity (EVA) gloves on human performance. *International Journal of Industrial Ergonomics*, 16, 165-174.
Bradley, J.V. (1969). Glove characteristics influencing control manipulability. *Human Factors*, 11, 21-36.
Hidson, D. (1991). New concepts in the development of a CB protective glove using CAD/CAM. Ottawa, Defence Research and Development Canada.
Hidson, D. (1992). *Anthropometric considerations in CB (chemical biological) glove research using computer-aided design*. Ottawa, Canada: Defence Research and Development Canada.
Krueger, R. & Casey, M.A. (2000). *Focus Groups: a practical guide for applied research*. 3rd ed. London: Sage Publications.
Mathiowetz, V., Weber, K., Volland, G., & Kashman, N. (1984). Reliability and validation of grip and pinch strength evaluations. *Journal of Hand Surgery [Am]*, 9, 222-226.

McDonagh-Philp, D. & Torrens, G.E. (2000). What do British Soldiers want from their gloves? In P.T. M^CCABE, M.A. Hanson, and S.A. Robertson, (Eds.), *Contemporary Ergonomics 2000, The Ergonomics Society 2000 Annual Conference, Lincolnshire, April, 2000* (pp 349-353).. London: Taylor & Francis.

Noro, K. & Imanda, A.S., eds. (1991). *Participatory Ergonomics*. London: Taylor & Francis.

Sweetland, R.C. & Keyser, D.J. (Eds.) (1991). *Tests: a comprehensive reference for assessments in psychology, education, and business*. Austin: Pro-Ed.

Spijkerman, D.C., Snijders C.J., Stijnen, T., & Lankhorst, G.J. (1991). Standardization of grip strength measurements. Effects on repeatability and peak force. *Scandinavian Journal Of Rehabilitation Medicine, 23*, 203-206.

Tiffin, J. & Asher, E.J. (1948). The Purdue Pegboard: norms and studies of reliability. *Journal of Applied Psychology, 32*, 234-247.

Torrens, G.E. & Newman, A. (2000). The evaluation of gloved and ungloved hands. *In:* P.T. McCabe, M.A. Hanson & S.A. Robertson, (Eds.), *Contemporary Ergonomics 2000, The Ergonomics Society 2000 Annual Conference, Lincolnshire, April, 2000* (pp. 301-305). London: Taylor & Francis.

Torrens, G.E., Williams, G.L., & McAllister, C., 2001. What makes a good military glove? A discussion of the criteria used in the specification of military handwear and appropriate methods of assessment. In ANON, (ed.), *International Soldier Systems Conference 2001, Bath, 28th-30th November, 2001*.

Enhancing the use of anthropometric data

Johan Molenbroek[1] & Renate de Bruin[1,2]
[1]Faculty Industrial Design Engineering
Delft University of Technology
[2]Erin, Ergonomics Consultancy, Nijmegen
The Netherlands

Abstract

Anthropometric knowledge is most frequently used by designers and product evaluators in the form of one-dimensional data to verify whether the product dimension fits the human dimension. There are several ways in which anthropometric data are used:

- Ego design: your own body dimensions are used as a guideline;
- Average design: the body dimensions of the average person are the guideline;
- Design for P5: the body dimensions of the smallest person are the guideline;
- Design for P95: the body dimensions of the largest person are the guideline;
- Design for P5-P95: the body dimensions of the smallest and largest person are the guideline. This type is used most commonly and means that excluding 10% of the population is acceptable.
- Design for All: This implies a continuous effort throughout the design process to exclude as few persons as possible.

In this paper, two tools are discussed to make this anthropometric world easier to understand. The tool 'Ellipse' demonstrates how easy it is to analyse a fit problem with multiple 2D views. The tool 'Persona' visualises the geometrical problems in the human-product-interaction with living persons or with digital models.

Introduction

Many ergonomists are not aware of the fact that the anthropometry they use is mostly 1D. This does not mean it is of less value, but this paper explains why it is important to realise this fact and shows how information can be extended to 2D and 3D or maybe even 4D information, which may be more appropriate and valuable for daily use in a design or evaluator's environment.

There are eight design types, most of which were described earlier in Dutch literature by Molenbroek (1994) and Dirken (1997-2004).

1. Design like Procrustus
The name is taken from Greek Mythology; here the user is fitted to the product.

2. Design for the Ego
This means the designer only looks at his own size and assumes his designs will fit everyone else.

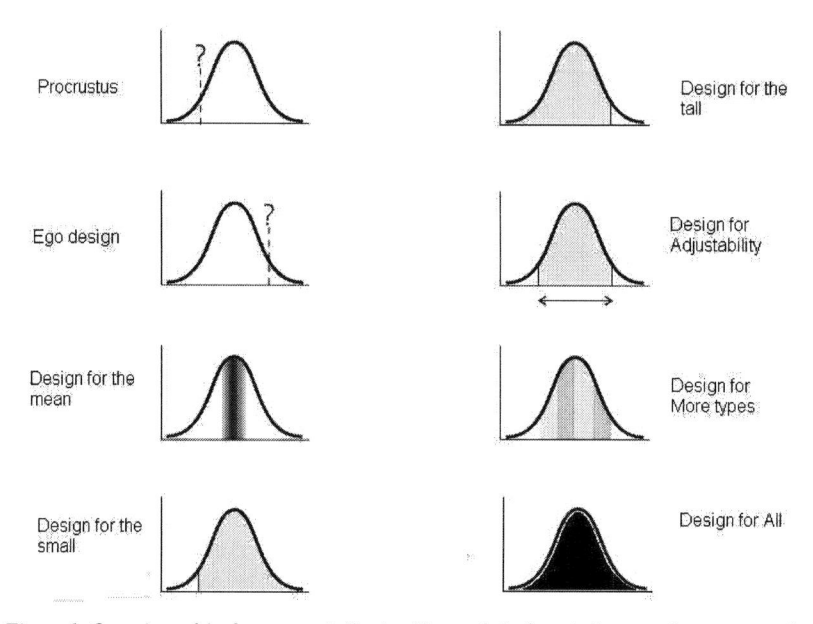

Figure 1. Overview of Anthropometric Design Types. Axis description: y = frequency and x = percentile 1 to 100

3. Design for the Average
The designer thought about the variation in sizes and decided to take the average to achieve comfort for all and to minimize the discomfort for tall and small people. Practically everyone outside the average is excluded.

4. Design for the Small
The designer is aware of the fact that in using the average values, small people will have problems. He takes care that everything fits at least the weak and the small people. This means, using the example of a nutcracker, that weak and small people can crack the hardest nut, but that the strongest person will probably crack the nutcracker.

5. Design for the Tall
The designer is very tall himself, frequently bumps his head and is therefore strongly motivated to fit at least the tall and strong users and might forget the small and the weak.

6. Design for More Types
When a product has several types to fit the variation of users, like in shoes, clothing or personal equipment, the most simple system consists of small, medium and large

sizes. Currently extra large (XL) or even extra extra large (XXL) sizes are included in this simplified size system, because of the increasing number of overweight people in our societies.

The anthropometric analysis of the data underlying the decisions of which part of the population should fit in which type of product is not extensively described. Exceptions are Roebuck (1997) and HFES (2004). Later in this paper it will be shown that the tool 'Ellipse' is meant to contribute in this field.

7. Design for Adjustability

For the users of office chairs it was a great improvement in comfort when various elements were made adjustable, starting in the seventies with sitting height, then arm rests, seat depths and other features. The disadvantages of this type of solution are how to determine what the limits are, and how to ensure that the user does not forget to adjust to his/her size.

8. Design for All

This does not mean that a designer has to design for all the 6 billion people on earth, but it means that throughout the design process it has to be taken into consideration that as few people as possible are excluded. Design for All is also known as Inclusive Design or described as Design for the widest possible audience (Include, 2005).

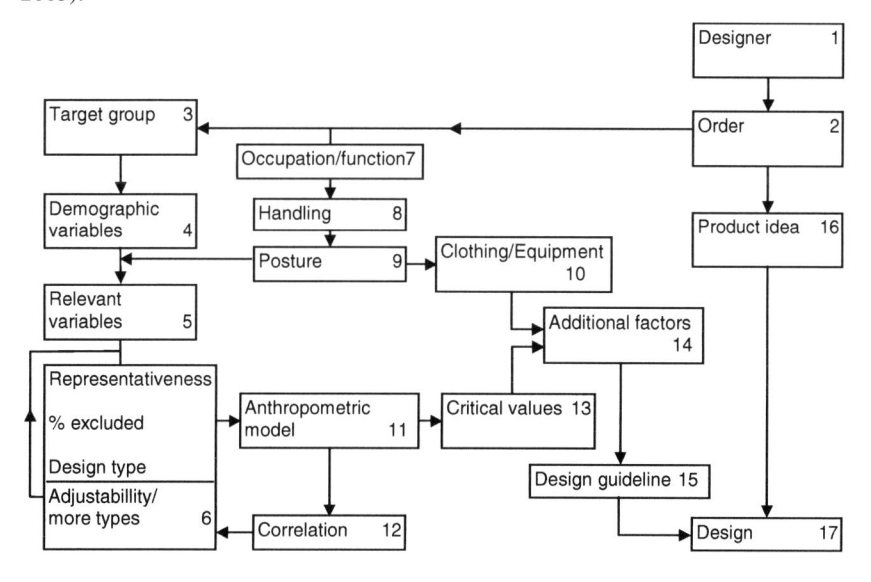

Figure 2. The Anthropometric Design Process (Molenbroek, 1993)

To understand which steps are important, Figure 2 shows the anthropometric design process, consisting of a series of sequential and parallel subprocesses.

According to this model the facts about the following elements are important:

Box 1: A designer accepts an assignment to analyse and redesign schoolfurniture.
Box 2: The object of the study (i.e. chair and table)
Box 3: The target population (i.e. European children)
Box 4: The demographic variables (i.e. 4-20 years of age, students)
Box 5: Relevant anthropometric variables (see Table 1); a key dimension is the popliteal height and not the stature as was assumed in the past (Molenbroek, 2003).
Box 6: The criteria for anthropometric data.
- Relevance of dimensions (see box 5)
- Representative of population: The population for which the objects are designed should also give the anthropometric data.
- Precision: Statistical considerations have to be applied to determine the accuracy of sample results.
- Design type: see Figure 1.
    o Adjustability: To fit a product to a range of users in various dimensions, at least three solutions are possible:
    o adapt the product to a specific user
    o make the product adjustable
    o create different sizes of the same product. For school furniture this is the most widely-used solution; it is a compromise between costs and anthropometric fit.

Box 7: The function of the user (i.e. a student sitting in a classroom)
Box 8, 9, 10 and 14: These factors result from clothing, posture etc. and also from the extra space the designer allows the user.
Box 11: Depending on the facilities of the designer, an anthropometric model will be available as a table, or as a 2D or 3D model.
Box 12: Correlation between the relevant variables should be considered to avoid exclusion of part of the user population; the tool 'Ellipse' takes care of this.
Box 13: With knowledge of correlation coefficients and scatter diagrams about body dimensions of the target population, the design and evaluation of a chair can be improved.
Box 15: Guidelines for the specific dimensions of each set (chair and table) are defined in this stage of the design process and, in the present study, will result in a proposal for a new European standard.

Table 1. The concept of product dimensions versus user dimensions. Dimensions in mm. Results are from the elderly in the GDVV project (Molenbroek, 1987)

P1	P99	User	additional	Product
429	583	Buttock-popliteal depth	-5 cm	Seat depth
344	516	Popliteal height	+ heel height	Seat height
316	454	Hip width	+ clothing + comfort	Seat width
177	299	Elbow-rest height		Armrest height
362	570	Bi-acromial breadth		Width backrest
798	1015	Sitting height		Height head support
457	651	Acromial height		Height back support

Another important aspect is the understanding that product dimensions mostly differ from user dimensions. Table 1 shows the relation between the dimension of the user and that of the product. The ergonomist has the task to determine what the content of the terms in Figure 1 and 2 and Table 1 are.

Until now this paper described 1D anthropometry, where no relations between the dimensions were made. The authors of this paper assume that over 50% of anthropometric applications are (unfortunately) only used as 1D application. This number can decrease in future if proper education tools are developed to show designers that 1D applications only show one window of the 3D real world. Useful free sources for 1D, 2D and 3D anthropometry are listed on the website of WEAR (Wear, 2000). The following paragraph elaborates more on 2D and 3D.

'Ellipse'

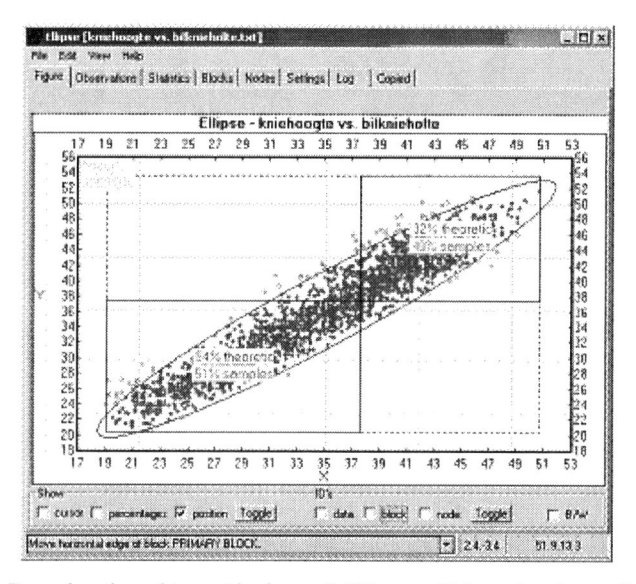

Figure 3. Example of working with the tool Ellipse, which makes it possible to shift rectangles with including % of the sample over a bivariate distribution. Axis description: x=popliteal height of Dutch children age 2-12 years, y= buttock-popliteal length of the same Kima-sample (n=2400)

Basic statistical software can be used to study the relation between two or more body dimensions, but this requires an investment in money and time which is often not made by designers and product evaluators. Therefore, a simple interactive tool called 'Ellipse' was developed, based on bivariate normal distribution algorithms from Sokal and Rohlf (1981), which draws the relation between two variables in a simple way (Molenbroek, 1994, and Molenbroek et al., 2003). 'Ellipse' gives a display of two-dimensional sample data as points in a scatter plot and finds out what percentage of the sample data are within a certain window. This window can be determined by the user, for example to see which limits to the size of a certain brand product will result in which percentage of the population.

The input data for 'Ellipse' are two columns of anthropometric data or the summative alternative: two mean values, two standard deviations and a correlation coefficient that shows the relation between the two variables.

More information about 'Ellipse' can be found on www.dined.nl/ellipse.

'Persona'

This tool is not yet available, but according to the authors it is useful to work effectively with 2D or 3D digital human models (DHM) or with test persons.

'Persona' should offer an overview of information on the widest selection of anthropometric tools in the world. It would then assist in selecting the best tool to do the job in a given context. New developments in DHM should be easy to add. Experiences of the authors with DHM resulted in earlier overviews and selection criteria for a proper DHM (Lombaers, Molenbroek and Osinga, 1986, and Molenbroek, 1994). Table 2 and Figure 4 are updated from that source.

The following criteria should be built into 'Persona'. They could be used to ask users step-by-step questions to be able to give a good recommendation to the user (Table 2), and also depend on the phase in the design process (Figure 4).

Table 2. Criteria that should be discussed in the process of setting up or using a Digital Human Model

Criterium	Description
1 Knowledge	Having basic knowledge like what is in the introduction of this paper about 1 D, 2 D and 3D data and their place in the design process, and about the concept of user dimensions distinguished from product dimensions.
2 Means	The necessary means in hard, soft- and orgware, like type of computer and operating system and which design software is used. For example, SAFEWORK works fine with CATIA on a workstation, but is only very recently available on PC and should now work with, for example, Solid Works through a Digital Mockup Unit.
3 Interpretation	If two or three people understand something else from the explanation of the DHM, then that DHM is not usable.
4 Flexibility	If the DHM is easy to modify in posture and perhaps also in design, it could be called flexible.
5 Investment in time	If a DHM takes only a short time to learn and is intuitive in itself, or if it takes a heavy course of 1 or more weeks, makes a lot of difference for a designer or researcher.
6 Investment in money	What are the costs of purchasing the system and what are the costs per design?
7 Reality	Which populations are inside the model and which products or workplaces can be simulated? How old are the data? And is it clear where the data come from? Can one's own data be added easily? Which aspects are simulated and how many dimensions? 2D or 3D? Does it take care of secular growth? What detail is simulated?

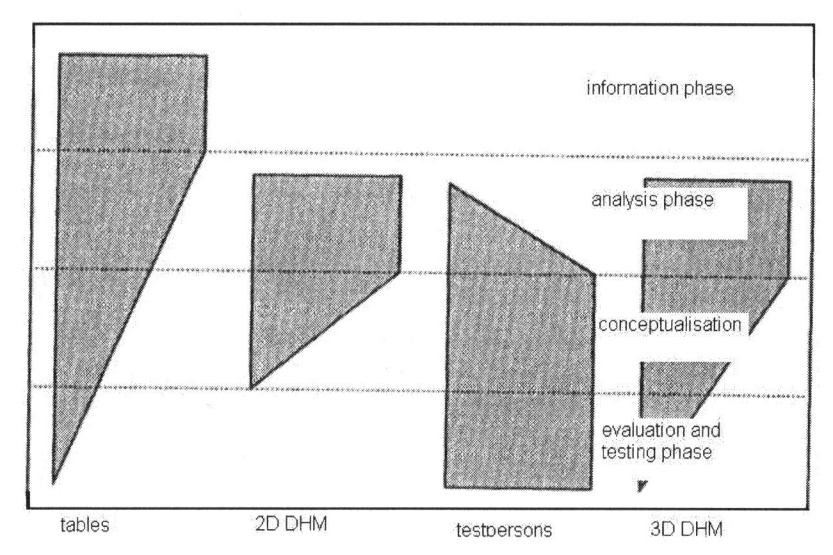

Figure 4. The usage of a (digital) human model does not only depend on its features and on the cooperation within the design, but also on the phase in the design process. Axis description: x = types of anthropometric models; -y = influence of model within the phase of the design process

For example, are the hands simple blocks or does every hand consist of 3 fingers? Is there a detection collision with own body parts or with other parts of the design? Can it take video data or 3D data as input?

'Persona': Visualisations in DHM

Furthermore, 'Persona' should have a database with short videos of usage situations with real people, like the system called HADRIAN built by Mark Porter in Loughborough. But not only data from several disabled people like Porter used, but also from the normal variety in small, large, fat and thin people, from young and old, male and female and from several cultures. In the Muybridge simulations (Muybridge,2004), a variety of usages of products in age, gender and product type can be seen.

'Persona': Testing with mock-ups

If the design is further developed and is prototype-tested, DHM should be replaced by testing with real-life test persons. These tests, but also other tests with pre-processed and not-yet-ready-to-produce product images in 2D or 3D (like in foam or wood), are extremely useful and save costs later on in the production process. Tools like 'Persona' give step-by-step advice on how to select test persons and how to instruct test persons before the test. The idea for this tool is the experience the authors of this paper gained from a European research and development project (called Friendly Rest Room, FRR, 2005).

WEAR

Perhaps the WEAR project will fulfill the demand for tools like 'Persona' and 'Ellipse' in the near future. WEAR is an international collaborative effort to create a world wide resource of anthropometric data for a wide variety of engineering applications. The WEAR group was started up in San Diego at the IEA-HFES conference (WEAR, 2000). The following countries and institutions are currently participating:

Table 3. Overview of participants in the WEAR group

Country	Institution/company	City
Australia	Sharpdummies	Adelaide
Brazil	National Institute of Technology	
South Africa	Ergotech	Pretoria
Japan	Digital Human Laboratory	
Taiwan	National Tsing Hua University	
Canada	National Research Council	
USA	Wright Patterson Air Force Base, CARD-lab	Dayton
USA	National Institute of Technology	Washington
France	Laboratoire Anthropologie Appliquée, Université René Descartes	Paris
Netherlands	TNO Human Factors Soesterberg	Soesterberg
Netherlands	Delft University of Technology, Industrial Design Engineering	Delft

Conclusion

The digital world helps to solve anthropometric problems in such a way that data are more easily accessible and can be used to improve current Digital Human Modeling. Still, this improvement is urgently needed, because misunderstandings about anthropometry are growing in number and size with the growing amount of 1D, 2D and 3D data; see for example the discussions about automatic landmarking in the 3D scanning process which, for the majority of scientists, is still an experimental process and only applicable on hard bony points, but on the other hand is used in large-scale research projects.

Another misunderstanding is that it is easy to communicate the data to the users in an understandable way. But if these users (many of them are small industrial and design companies) are only used to 1D data and only to the data related to their own products and in their own way of using them, then 2D and 3 D data are images of a completely different world and need translation and education. A shoe designer will most probably not understand how to cope with reach envelopes. Furthermore, there are almost no anthropometric tools nor good quality 3D data available for small and medium-sized companies. In the end, tools like 'Ellipse' and 'Persona' can help to use anthropometric data without a large investment in time and money to solve a size problem.

'Ellipse' is already available, but 'Persona' has to be developed in the near future.

References

Dirken, J.M. (1997-2004). *Productergonomie, Ontwerpen voor gebruik.* Delft, The Netherlands: Delft University Press.

Friendly Rest Room Project (2005). http://www.frr-consortium.org

HFES 300 committee (2004). *Guidelines for Using Anthropometric Data in Product Design.* Human Factors and Ergonomics Society.

Include 2005, http://www.hhrc.rca.ac.uk/programmes/include/index.html

Lombaers, J.H.M., Molenbroek, J.F.M., & Osinga, D.S.C. (1986). *Antropometrische modellen, Overzicht en vergelijking van modellen van mens en werkplek.* Reeks Industrieel Ontwerpen bijzondere onderwerpen, Vol. 10, Faculty of Industrial Design, University of Technology Delft, The Netherlands.

Molenbroek, J.F.M. (1987). Anthropometry of elderly in the Netherlands. Research and Applications. *Applied Ergonomics, 18,* 187-199.

Molenbroek, J.F.M. (1994). *Op Maat Gemaakt.* Delft, The Netherlands: Delft University Press.

Molenbroek, J.F.M., Kroon-Ramaekers, Y.M.T. and Snijders, C.J. (2003). Revision of the design of a standard for the dimensions of school furniture. Ergonomics, 46, 681–694.

Muybridge (2004) http://web.inter.nl.net/users/anima/chronoph/wrp/

Roebuck, J.A. (1997). *Anthropometric Methods: Designing to Fit the Human Body.* Human Factors and Ergonomics Society.

Sokal, R.R., & Rohlf, F.J. (1981). *Biometry. The principles and practice of statistics in biological research.* San Francisco, USA: Freeman.

WEAR (2000). http://ovrt.nist.gov/projects/wear/

Differential usability of paper-based and computer-based work documents for control room operators in the chemical process industry

Peter Nickel & Friedhelm Nachreiner
Carl von Ossietzky Universität Oldenburg
Work and Organisational Psychology Unit
Oldenburg, Germany

Abstract

In most process control workplaces in the chemical industry, paper-based or computer-based work documents or even both are available to the operator. It was therefore investigated whether one of these versions is preferable and what are the ergonomic recommendations for the design of each. In a repeated-measures simulation study eight operators performed process control tasks in a control room of a simulated benzene/toluene distillation plant. Task performance included typical operational states and was supported by either paper-based or computer-based work documents – identical in content but necessarily different in functional design. Differential handling and usability of the work documents were evaluated using subjective ratings of user satisfaction and workload as well as performance and observational measures. Results show that there are no categorical differences in the usability of paper based versus computer based work documents. Effects of differential usability became obvious, however, for specific tasks, operational states, and work system components in interaction with the media used for documentation. Thus, recommendations for an ergonomic design of work documents must be based on their specific context of use. The results presented were derived from the chemical process industries but should also apply to other process control settings, e.g. pharmaceutical, power, food and transportation industries.

Introduction

According to legal requirements in the European Union and its member countries (e.g. Seveso II Directive, 1996; Safety and Health Directive, 1989; Work Equipment Directive, 1989; VDU-Directive, 1990) human factors and ergonomics design strategies (e.g. task orientation) and principles (e.g. compatibility) have to be observed in the design of process control systems. This also applies to the design of work system interfaces with a view to the effectiveness, efficiency, and safety of the process under control and to the resulting workload for the operators — which in turn should improve system reliability and productivity (Nachreiner, 1998; Nachreiner et al., 2005).

In D. de Waard, K.A. Brookhuis, R. van Egmond, and Th. Boersema (Eds.) (2005), *Human Factors in Design, Safety, and Management* (pp. 299 - 314). Maastricht, the Netherlands: Shaker Publishing.

The job of process control operators can mainly be characterised as monitoring and control tasks performed at computer based work stations. These usually show the complex industrial facilities and their dynamics as a schematic representation and require the operators to continually compare information presented on the computer generated system interfaces with their mental model of the process or other information concerning the plants' actual state. In addition, operators must continually adapt to new and unforeseen changes in the technological system and environmental demands (Meshkati, 2003; Moray, 1997; Nickel & Nachreiner, 2004; Wilson & Raja, 1999). An ergonomic design of the work system (ISO 6381, 2004) is therefore of utmost importance and relevance and should cover job aids such as work documents as a systems' immanent component, serving the following purposes:

- supporting operator task performance,
- contributing to safeguarding safety and health,
- providing an indispensable part of the work equipment (or tool) within the work system,
- enabling the operator to perform usual and unusual or exceptional tasks.

According to the results of field studies in the chemical process industry, the management's objectives in providing work documents to their operators often are in agreement with these purposes (Nickel & Nachreiner, 2005). At most process control workplaces in the chemical industry, various work documents (e.g. manuals, instructions, forms) are available to the operator. These may be paper-based, computer-based or even both. Although there is a clear intention to design and to provide work documents suitable and usable for the operators and their tasks at hand, there is a lot of uncertainty about how to design such work documents and which media for presentation should ideally be used to serve these purposes.

From an ergonomic point of view, work documents and especially the operator-document interface within a work system should be designed to optimise the level of workload for the operator, to enhance operator satisfaction and to increase the reliability and productivity of the production process (Hacker, 1998). Recommendations concerning an ergonomic design of work documents have been discussed in Nickel and Nachreiner (2004). With regard to recommendations documented in standards such as EN IEC 62079 (2001) and in the relevant literature (e.g. Hartley, 1999; Hoffmann et al., 2002; Schriver, 1997) it could be concluded that broadening the perspective to a work systems orientation seems appropriate for developing recommendations for the design of work documents. Because of their intended general applicability, work document design recommendations usually cannot be given as a prescriptive set of completely specified rules to be universally applied. From an ergonomic perspective they can only be specified as general principles, rules or guidelines which need to be adapted to the specific task, user, environment, technology and topic that the work documents is addressing, i.e. a (work) systems approach will always be relevant and required.

In the chemical process industry paper as well as electronic (or even both) representations of work documents are used as job aids for operators in control rooms. The decision on the preferred media for presentation is often rather pragmatic and the rationale is related to e.g. tradition, familiarity with different media, portability, accessibility, guarantee for topicality, necessity of investments in additional computer systems and peripheral equipment, lack of a suitable software programme for easily and simultaneously preparing text, tables and figures, to name but a few. Some of these aspects are well known from discussions on the "paperless office" or on implementations of electronic process management for e.g. maintenance works (Johnson & Millans, 2000; Sellen & Harper, 2001; Swezey, 1987). Based on research findings comparing printed and electronic presentation (e.g. printed on paper versus presentation on a computer monitor) it cannot be concluded to best use both in parallel, since this option wasn't taken into account in most of the studies. However, there is some evidence to better prefer one of these versions (some studies arguing for prints and others for electronic representations) or there were no differences at all found between both versions. Findings on reading performance and accuracy results (e.g. Gould & Grischkowky, 1984; Gould et al., 1987) indicate a preference for paper-based documents while findings of e.g. Newsted (1985) support the preference for computer-based documents. However, according to e.g. Askwall (1985), Cushman (1986), Oborne and Holton (1988), and Muter and Maurutto (1991) there seem to be no differences in information acquisition via a computer monitor or paper, respectively. A closer inspection of the findings suggests that inconsistencies in these findings might partly be due to speed deficits resulting from poor image quality (e.g. resolution), the inherent characteristics of either media (e.g. paging), the differences in handling (e.g. screens' fixed position versus papers' variable position), the singularity of evaluation measures used (e.g. performance), and ignored interaction effects with other relevant variables at work (Dillon, 1992; Gould & Grischkowsky, 1985; Noyes & Garland, 2003; Oborne & Holton, 1988; Woods, 1985). Therefore it is unclear whether or when to recommend paper-based or computer-based work documents to best support the operator, particularly in contexts like control rooms in the process industry where system safety or occupational risks may be at stake.

Usability as presented in EN ISO 9241-11 (1998) is a concept to be used in the evaluation of software in office work but is in principle also appropriate for contexts other than office work (e.g. in process control; Nickel et al., 2004) and for different products (e.g. consumer products, tools, information technology; Dillon, 2004; Stanton, 1998; Woodson & Conover, 1966). When considering usability as the "extent to which a product can be used by specified users to achieve specified goals with effectiveness, efficiency and satisfaction in a specified context of use" (EN ISO 9241-11, 1998, 2) it becomes obvious that specifications of the intended users, the goals and the context of use (as work system components) are necessary before starting an evaluation based on this concept. As pointed out by Dillon (2004) the concept of usability provides a suitable framework even in comparative investigations on different versions of work documents, i.e. different versions of paper-based and / or computer-based representations.

However, no study has been found comparing work documents or different versions of them in the context of the process control industry and / or addressing other evaluations on work documents using a usability framework. Thus there seem to be no empirical basis available to decide whether or under which circumstances to recommend paper-based or computer-based work documents or even both to best support the process control operators' task performance. Therefore an experimental study was conducted in a simulated toluene/benzene distillation plant. Paper-based and computer-based work documents for process control operators were examined under comparable conditions in order to find out: (1) if the above mentioned functions of a work document (e.g. support, safety, tool) are better accomplished by paper-based or computer-based versions and (2) if there are specific design requirements with regard to the usability of work documents in both versions.

Methods

Participants

Eight students (four female, four male) with sufficient on-the-job-training and prior experience in monitoring and control operations in the experimental control room served as process control operators. All students recently attended a course on ergonomics in process control in their studies of Work and Organisational Psychology as a major subject where they all had to apply basic ergonomics concepts to the distillation plants' process control system in the laboratory. Furthermore, they all got 20 hrs training in normal operations and in handling process disturbances in preparation for another experimental study on interface design in process control operations. In order to refresh their experiences for the study presented here they were given a 30-min period, supervised by the experimenters. All participants had normal or corrected-to-normal vision and according to a questionnaire they reported neither impairments in health nor effects of preceding tasks that could have had affected their performance during the experimental sessions. The participants had been informed in general that the experiment concerned testing the suitability of work documents. Participation was voluntary and financially compensated.

Testing environment and apparatus

The study was carried out in a sound-protected and air conditioned room (7.5 x 5 x 2.4 m; Industrial Acoustics Co.) of the Work and Organisational Psychology Usability Laboratories where climatic conditions were held constant at 24°C dry temperature, 50% humidity, and < 0.1 m/s air velocity. The room was arranged as a process control room with industrial consoles equipped with a commercially used real process control system (I/A series, Invensys Systems GmbH) and an integrated full scale, real time simulator of a benzene/toluene distillation plant (see Figure 1). For the monitoring and control tasks of the operator dynamic interaction interfaces (e.g. overview, process sections, trends, and alarm manager) were available on 4 monitors. Qwerty-keyboards (ISO/IEC 9995-1, 1994), alarm-keyboards and trackballs were used as controls. Communication between the operator and others (e.g. management, outdoor operators) was provided by intercom connected to the room of the experimenter. Audio and video information and performance data of the production process were continuously recorded and the process control system was controlled by the experimenter to implement the experimental tasks.

differential usability of paper and electronic documents in process control   303

Figure 1. Simulated Control room of the benzene/toluene distillation plant in the Usability Laboratory of the Work and Organisational Psychology Unit, Carl von Ossietzky Universität Oldenburg.

Independent variables

Typical work documents used by process control operators in the chemical process industry (Nickel & Nachreiner, 2004) were selected to provide an orientation for the development of documents for the experimental study. The documents were designed considering the tasks of the operators in the distillation plant and basic ergonomic requirements for the design of instructional work documents (Nickel & Nachreiner, 2004, 2005). An operation manual contained fundamentals of the chemistry and process engineering for the benzene/toluene distillation, and information on the plant and the process control system. Instructions were drawn up e.g. for getting the distillation process started, for a shut down of the distillation column, for using the heat recovery system in the distillation process, and for arranging bypasses in different sections of the production process. Furthermore, numerical values for transmitters, sensors and actuators were compiled, set points were assigned and information was given on the localisation and function of different process control components.

Each work document was produced both in a paper-based and in a computer-based version. For the paper-based documents A4 sized paper (ISO/DIS 216, 2005) was used and the manual, the instructions and the parameter listings each were presented in a separate standing file with an appropriate indication on the spine. For the computer-based documents exactly the same content (e.g. text, figures, and tables) was redesigned for appropriate visualisation on a 21" monitor. The monitor was placed at a fixed position in the middle of the control room consoles (3rd from left in Figure 1) operated by an additional PC (233 MHz Pentium II processor, with MS-Windows NT 4 as operating system, Netscape Navigator 4.78 as hypertext markup language (html) browser and Acrobat Reader 5.0 as portable document format (pdf) viewer) and a mouse and a qwertz-keyboard as controls (the qwertz-keyboard layout is the standard in Germany according to ISO/IEC 9995-1, 1994). On the monitor an html-interface served as a navigation centre for the work documents. Textual indications on this interface were linked to the manual, each of the instructions, and the listings. The manual (similar to all other documents) was prepared as a pdf-document with e.g. each clause of the table of contents hyperlinked to the corresponding chapter, Acrobat Reader bookmarks for each chapter, and all functions of the Acrobat Reader for navigation, searching, and zooming in.

Dependent variables

According to EN ISO 9241-11 (1998) usability should be qualified and quantified by effectiveness, efficiency and user satisfaction in achieving specified goals related to the use of a product. Therefore, the operators' satisfaction with the work documents in performing the experimental tasks was collected in a concluding questionnaire by ratings on satisfaction with the documents (6 point Faces Scale; Kunin, 1955). Some of these questions referred to a whole experimental session and others referred to each sequence of a session. Qualitative and quantitative performance data yielded information on the effectiveness for the support of work documents for task performance. Therefore, audio and video taped behaviour while performing the tasks during different sequences, information on managing the sequences according to the rules, time on task, and releases of technical safety devices during task performance and performance data on the production process were gathered. Measures for efficiency in the use of work documents parameters such as activation, effort, strain, and effects of strain were collected during and after task performance. The NASA task load index (TLX) (Hart & Staveland, 1988), a subjective, post-hoc work load assessment tool, was administered after completion of each sequence within a session. The BMS questionnaire (Plath & Richter, 1984; with parallel forms of the inventory to assess effects of mental strain in a within design) was presented before and after the experimental session. Furthermore, at the end of each session the concluding questionnaire was presented asking for ratings of satisfaction with the work documents, operator preferences, and several aspects concerning the design of the work documents and the operators' experiences with the documents. After completing the second session a comparative judgement with regard to both versions of work documents was requested.

Furthermore psychophysiological data (e.g. ECG) for an assessment of mental work load were recorded, but since the results will not be reported here due to space reasons, as is the case with the results of the BMS questionnaire, no details will be given.

Experimental design

In a repeated-measures design participants completed two scenarios – with 9 task sequences – each containing both the same typical process control tasks (Nickel & Nachreiner, 2004) and different operational states of the production process, i.e. monitoring and control of the production process during normal operation and process disturbances. The second and the last task required participants to carefully look up the work documents and answering specific questions related to their contents. Each scenario (experimental session) was performed either by support of paper-based (PA) or computer-based (PC) work documents. The order of the versions of the documents used in both sessions was counterbalanced across groups, one starting with the paper based version while the other started with the computer based version (PA-PC versus PC-PA).

Procedures

All participants worked for two sessions, lasting about 4.5 h each, on subsequent days. After finishing preparatory work (e.g. fixing electrodes for the registration of psychophysiological parameters) the participants filled out a general questionnaire about demographic and health data. The participants were then given a 30-min period in order to refamiliarise themselves with the experimental environment and the equipment in the control room. After that they were handed a written instruction on their main job in the control room and their responsibility for the production process according to the rules. Next, the process control scenario started containing the following task sequences: (1) base line, (2) questions on the content of the work documents, (3) maintenance start up work, (4) documentation of system parameters, (5) maintenance finishing work, (6) process disturbance (e.g. valve or pump failure), (7) documentation of system parameters, (8) normal but complex operations (e.g. start up the heat recovery system in the distillation process), and (9) questions on the content of work documents. During sequences 3 to 8 the main task consisted of monitoring and controlling the benzene/toluene distillation process. Sequence (3) started when the operator had answered the questions on the contents of the work documents and reported that the production process and the parameters of the process control system were according to the rules (normal operation). The operator was asked giving feedback to the management (experimenter) about finishing her/his jobs in the individual sequences and controlling the process according to the rules. After having finished sequence 9 the operator was asked to fill in the BMS and subsequently the concluding questionnaire. After finishing the second session they were given the concluding questionnaire extended by comparative ratings on both versions of work documents. The sessions were audio and video taped by three cameras. Performance of the production process and psychophysiological data were recorded continuously. Ratings on the NASA-TLX were requested upon completion of each sequence.

Results

Satisfaction

The level of *overall satisfaction* for both versions of the work documents as rated on the 6 point-scale in the concluding questionnaire *for the whole scenario* was in general rather high ($M = 4.06$, $sd = 1.06$; see Figure 2). Furthermore, a significant higher level of satisfaction for the computer-based documents, has been observed according to a 2 (session) x 2 (group, i.e. order of paper-based and computer-based document versions) analysis of variance (Session x Group, $F(1, 14) = 5.65$, $p < .05$). No further effects were significant on the scenario level.

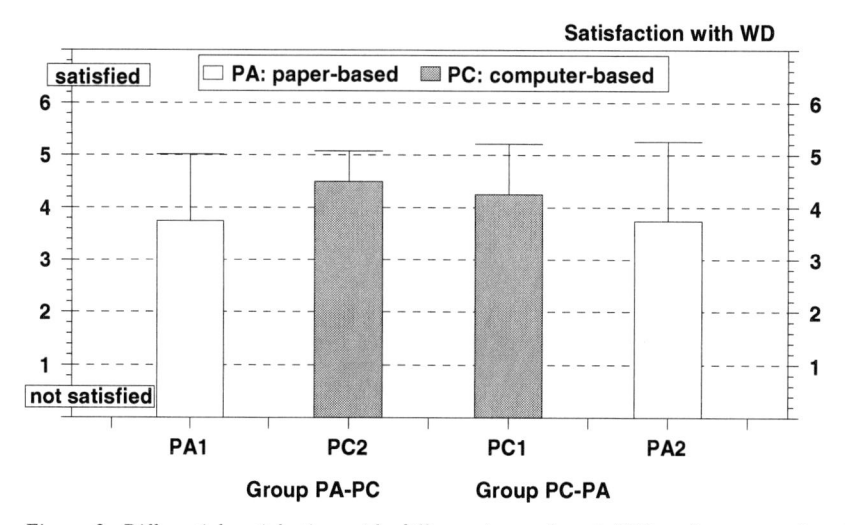

Figure 2. Differential satisfaction with different (paper-based (PA) and computer-based (PC)) work documents (WD) on scenario level

According to the ratings on *satisfaction for individual sequences*, in sequences 3 and 5 (maintenance start up and maintenance finishing) the paper-based work documents were significantly preferred (Session x Group, $F(1, 6) = 6.26$, $p < .05$ and $F(1, 6) = 6.60$, $p < .05$), whereas for all other sequences no significant effects could be identified. At first glance it seems to be rather unclear how and why the results on scenario compared to sequences levels can show these inconsistencies. However, when analysing the operators' statements for both versions of the work documents several advantages and disadvantages appeared (see Table 1 for examples). It seemed that some of the statements were related to the whole scenario (e.g. being more familiar with paper-based documents) while others were more related to specific demands in certain sequences (e.g. contents can be zoomed in). Some statements used to indicate an advantage for one media (e.g. "better orientation in all documents") were used to describe a disadvantage of the other media (e.g. "not always clear in orientation"). This sounds comprehensible. Other statements seem to refer to the same reason to express a preference for both the paper-based and the computer-based documents (see PA "easier to coordinate with work at process control system" and PC "medium is compatible with the process control system"). This might be conflicting. Looking at the interactions of the media within a work system it becomes obvious that paper-based work documents are easier to coordinate with work at process control systems because they are relative small (A4 size) and free to move or place (this doesn't apply for the computer-based work documents). However, when computer-based work documents are presented on a monitor with mouse and keyboard navigation, as is the case with process control systems, (this doesn't apply for paper-based work documents) this is superior with regard to compatibility. It could thus be concluded that the interactive component of the media with aspects within the work system affect most the operators' satisfaction and behaviour and consequently the differential evaluation of work documents.

Table 1. Examples of participants statements on advantages and disadvantages on paper-based and computer-based work documents

Work documents	Pros	Cons
PA	-familiar -easier to coordinate with work at process control system -better orientation in all documents -parallel use of different pages -finger as bookmark	-search by table of contents only -index requested -standing files bulky and unhandy -size of files and text fixed and not adaptable
PC	-flexibility of search for specific information -navigation via links, faster and easier -medium compatible with process control system -contents can be zoomed in	-not always clear in orientation -parallel use of pages not possible -turn over pages, scrolling, searching takes too much time -reading from screen is disliked -what's available in case of power failure?

Effectiveness

Since during task performance none of the technical safety devices were released and therefore no emergency shutdown occurred, there is reason to assume that task performance was effectively supported by both versions of the work documents. Looking at performance measures taken from the protocols of the experimental sessions and the audio/video tape recordings (such as time on scenario / sequence or coping with the monitoring and control tasks with the process control system in general) no statistically significant differences in the effectiveness of task performance between both versions of work documents was observed. This leads to the assumption that there was no general advantage for either version of work documents. However, parameters on quantity and quality of the production process and on handling of the document versions in specific problem solving situations have not yet been analysed in detail. This additional information on performance, however, is required before making any firm statements with regard to differential effectiveness.

Efficiency

As concerns perceived mental workload both versions of work documents were generally neutral according to the results for the NASA-TLX "overall workload" scores. Looking at single dimensions within the NASA-TLX the same indifferences for the different versions occurred for "mental demands", "physical demands", "temporal demands", and "effort" on the scenario level (for effects on the task sequences factor for "mental demands" and "effort" see below). However, operators were more satisfied with their own "performance" in all sequences using computer-based documents, as can be concluded from the marginal significant interaction effect for Session by Group, $F(1, 6) = 2.27$, $p < .10$, from the 2 (session) x 2 (between-groups of successive document versions) x 9 (task sequences) factors included in the analysis of variance for repeated measures. Furthermore, the operators perceived a lower "level of frustration" in performing the scenario supported by computer-based work documents considering the marginal significant interaction effect for Session by Group, $F(8, 48) = 3.50$, $p < .10$, and the related significant main effects for Session, $F(1, 6) = 7.78$, $p < .05$. The effects for "performance" and for "frustration level" seem to correspond to the results of the operators' differential satisfaction with work documents on the scenario level (see above).

The process disturbance sequence (no. 6) was the most prominent in being responsible for effects on the Sequence factor when analysing the NASA-TLX results on dimension levels. In this sequence "frustration level", "mental demands" and "effort" were significantly lower (according to the post-hoc comparisons) when using the computer-based work documents. Recalling that both versions of work documents are identical in content, but different e.g. in media of presentation and therefore necessarily in the functional design of the content and in the handling of the documents, the results support the hypothesis concerning differential efficiency of different work documents. However, the analyses of further available data have to be awaited to give a more comprehensive interpretation.

When analysing the concluding questionnaire ratings on certain aspects of the design of work documents differential usability was identified by the statements such as "I could take from the documentation what exactly I was expected to do" (instructions on how to do), "The documentation supported problem solving in situations of disturbance" (support for disturbances), and "The background information supports task performance in the control room" (background information). A 2 (session) x 2 (group) ANOVA for the "instructions" statement resulted in higher support ratings when computer-based work documents were used, indicated by a marginal significant interaction effect for Session by Group, $F(1, 6) = 6.18$, $p < .10$, and the related significant main effect for Session, $F(1, 6) = 13.48$, $p < .05$. On the other hand for "support for disturbances" and "background information" the paper-based work documents were preferred, indicated by a marginal significant interaction effects for Session by Group, $F(1, 6) = 5.05$, $p < .10$, and $F(1, 6) = 3.79$, $p < .10$, with no further significant effects. Therefore, it can be concluded that the usability of paper based versus computer based work documents depends on the purpose (and functional context) of use.

Discussion and conclusions

According to the results presented here, there is no overall or general advantage of either version of presenting work documents. The strengths and weaknesses identified for the paper-based and the computer-based version of the work documents, in general, point to differential usability according to different aspects. This does not seem to be dependent on the documents' content itself since both, the paper-based and computer-based work documents had exactly the same content. Differential usability rather depends on functional requirements arising from context variables (e.g. the task at hand, the operational state, amount of space in the control room and on the console) in interaction with the media of presentation. This is indicated by the results on the operators' differential satisfaction in individual sequences and their statements on the document versions' advantages and disadvantages. In addition, effects of differential efficiency were most prominent in the process disturbance sequence with different demands compared to the remaining sequences at normal operations. This is conceivable when comparing either paper-based or computer-based documents, when applying a documented process instruction in controlling the process, or when looking for specific process engineering information in a manual.

Since tasks and the resulting demands imposed on an operator vary due to the dynamics of the processes under control and the continuous adaptation processes within the work system, the design of work documents should be capable to serve these requirements. A proposal therefore might be to provide both, the paper and electronic version of documents in parallel *for different purposes*. In chemical process control in practice both versions are already available and in use, sometimes already even in parallel. Furthermore, work documents, both in paper as well as in electronic version, are mostly computer generated (sometimes, however, some parts of the documents are drawn by hand and copied). Since paper-based and computer-based work documents are different media and therefore each require their specific and suitable design one should not simply use a work document design for print versions for computer-based presentations.

From a usability perspective, however, this is not surprising, since usability can only be achieved with appropriate attention to the context of use and it can only be evaluated in a specified context of use, which is what the results would seem to suggest. However, this should be further tested and elaborated, when the effects of workload as indicated by the psychophysiological parameters have been analysed, especially during task performance at different tasks with different complexity and different requirements for the use of documents.

According to the results presented here, the design of work documents according to basic usability principles is necessary and possible, for both versions of presentation.

Moreover, differential usability seems to be important even if one stays within in the same media, since e.g. manuals or instructions should be designed according to their purpose of use, i.e. giving general or background information, improve or correct a mental model of the process under control, or giving a specific instruction for a specific task in a specified operational state. The notion of Dillon (2004, 34) on electronic documents therefore should be extended to work documents in general, i.e. when talking about work documents and their usability "it is essential that the contextual variables are made explicit and we avoid the trap of endorsing or dismissing the medium on the basis of an excellent or an inappropriate design for one specific context." Referring to Clark (1983) Swezey and Llaneras (1997) pointed out that no media, in their own right, have any influence e.g. on learning effectiveness, but are mere vehicles for more or less well designed instructions. Thus, differential usability of work documents according to their intended purpose and based on ergonomic principles and strategies is not merely desirable but also of increasing importance in providing for operational safety due to ongoing developments towards increasing complexities and dynamics in process control, e.g in the chemical, pharmaceutical, power, food, and transportation industries.

Acknowledgements

This research was initiated and supported by grant, reference number F1778 "Controlling of Industrial Process Installations by Computer Control Systems: Requirements on Working Documents for Operators", awarded to the authors by the Federal Institute of Occupational Safety and Health, Division 2 "Safety and Health of Products and Processes", Unit 2.3 "Installations and Processes, Optical Radiation", Dortmund, Germany. The process control system was financed by a HBFG grant from the FRG and the state of Lower Saxonia, FRG. The authors would like to thank the research students Tanja Brüntjen, Tim Oliver Pfingsten, Stephanie Hinnenberg, and Marko Deede for their assistance in carrying out the study. We would also like to thank two anonymous reviewers for helpful comments on an earlier draft of this paper.

References

Askwall, S. (1985). Computer supported reading vs reading text on paper: a comparison of two reading situations. *International Journal of Man-Machine Studies, 22*, 425-439.
Cushman, W. H. (1986). Reading from microfiche, a VDT, and the printed page: subjective fatigue and performance. *Human Factors, 28*, 63-73.
Dillon, A. (1992). Reading from paper versus screens: a critical review of the empirical literature. *Ergonomics, 35*, 1297-1326.
Dillon, A. (2004). *Designing usable electronic text.* Boca Raton, USA: CRC Press.
EN IEC 62079 (2001). *Erstellen von Anleitungen; Gliederung, Inhalt und Darstellung* [Preparation of instructions; table of contents, content and visualisation]. Brussels, Belgium: European Committee for Standardization.
EN ISO 9241-11 (1998). *Ergonomic requirements for office work with visual display terminals (VDTs) – Part 11: Guidance on usability.* Brussels, Belgium: European Committee for Standardization.

Gould, J.D. & Grischkowsky, N. (1984). Doing the same work with hard copy and with cathode-ray tube (CRT) computer terminals. *Human Factors, 26*, 323-337.

Gould, J.D. & Grischkowsky, N. (1985). Effects of visual angle on reading. *Proceedings of the Human Factors Society - 29th Annual Meeting* (pp. 1106-1109). Santa Monica, Ca, USA: HFES.

Gould, J.D.; Alfaro, L.; Barnes, V.; Finn, R.; Grischkowsky, N. & Minuto, A. (1987). Reading is slower from CRT displays than from paper: Attempts to isolate a single variable explanation. *Human Factors, 29*, 269-299.

Hacker, W. (1998). Allgemeine Arbeitspsychologie. Psychische Regulation von Arbeitstätigkeiten [Work psychology]. Bern, Switzerland: Huber.

Hart, S.G. & Staveland, L. (1988). Development of the NASA task load index (TLX): Results of empirical and theoretical research. In P.A. Hancock & N. Meshkati (eds.), *Human mental workload (Advances in Psychology, vol. 52)* (pp. 139-183). Amsterdam, The Netherlands: North-Holland Publishing Company.

Hartley, J. (1999). Is this chapter any use? Methods for evaluating text. In J.R. Wilson, E.N. Corlett (Eds.), *Evaluation of human work. A practical ergonomics methodology* (pp. 285-309). London, UK: Taylor & Francis.

Hoffmann, W., Hölscher, B.G., & Thiele, U. (2002). *Handbuch für technische Autoren und Redakteure. Produktinformation und Dokumentation im Multimedia-Zeitalter* [Manual for technical authors and editors]. Erlangen, Germany: Publicis Corporate Publishing.

ISO/DIS 216 (2005). Writing paper and certain classes of printed matter - Trimmed sizes - A and B series, and indication of machine direction. Geneva, Switzerland: International Organization for Standardization.

ISO 6385 (2004). *Ergonomics principles in the design of work systems.* Geneva, Switzerland: International Organization for Standardization.

ISO/IEC 9995-1 (1994). *Information technology - Keyboard layouts for text and office systems – Part 1: General principles governing keyboard layouts.* Geneva, Switzerland: International Organization for Standardization.

Johnson, W.B. & Millians, J.T. (2000). Technology based solutions for process management in aviation maintenance. In IEA/HFES (Eds.) *Ergonomics for the new millennium, Vol. 3: Complex Systems and Performance* (pp. 3/791-3/794). Santa Monica, CA, USA: HFES.

Kunin, T. (1955). The construction of a new type of attitude measure. *Personnel Psychology, 8*, 65-77.

Meshkati, N. (2003). Control rooms' design in industrial facilities. *Human Factors and Ergonomics Manufacturing, 13*, 269-277.

Moray, N. (1997). Human Factors in Process Control. In G. Salvendy (Ed.), *Handbook of Human Factors and Ergonomics* (pp. 1944-1971). New York, USA: John Wiley & Sons.

Muter, P. & Maurutto, P. (1991). Reading and skimming from computer screens and books: The paperless office revisited? *Behaviour & Information Technology, 10*, 257-266.

Nachreiner, F. (1998). Ergonomics and standardization. In J.M. Stellman (ed.), *Encyclopaedia of Occupational Health and Safety* (vol. 1) (pp. 29.11-29.14). Geneva, Switzerland: ILO.

Nachreiner, F.; Nickel, P. & Meyer, I. (2005, in press). Human Factors in Process Control Systems: the Design of Human-Machine Interfaces. *Safety Sciences*.

Nickel, P. & Nachreiner, F. (2004). Ergonomic requirements for job aids — Work documents for operators in chemical process control systems. In D. De Waard, K.A. Brookhuis & C.M. Weikert (eds.), *Human Factors in Design* (pp. 289-302). Maastricht, The Netherlands: Shaker Publishing.

Nickel, P. & Nachreiner, F. (2005, in press). *Ergonomische Anforderungen an Arbeitsunterlagen für die Prozessführung mit Prozessleitsystemen* (Schriftenreihe der Bundesanstalt für Arbeitsschutz und Arbeitsmedizin (BAuA), Fb 1778). Bremerhaven, Germany: Wirtschaftsverlag NW, Verlag für neue Wissenschaft GmbH.

Nickel, P.; Nachreiner, F. & Meyer, I. (2004). Aufgabenangemessenheit — Zur Übertragbarkeit eines ergonomischen Gestaltungsgrundsatzes von Büro- auf Prozessleitsysteme [Suitability for the task – Transfer of ergonomics design principles for office work to process control]. In W. Bungard, B. Koop & C. Liebig (eds.), *Psychologie und Wirtschaft leben. Aktuelle Themen der Wirtschaftspsychologie in Forschung und Praxis* (pp. 137-143). München, Germany: Hampp.

Noyes, J.M. & Garland, K.J. (2003). VDT versus paper-based text: reply to Mayes, Sims and Koonce. *International Journal of Industrial Ergonomics, 31*, 411-423.

Newsted, P.R. (1985). Paper versus online presentations of subjective questionnaires. *International Journal of Man-Machine Studies, 23*, 231-247.

Oborne, D. J. & Holton, D. (1988). Reading from screen versus paper: there is no difference. *International Journal of Man-Machine Studies, 28*, 1-9.

Plath, H.-E. & Richter, P. (1984). *Ermüdung-Monotonie-Sättigung-Streß. BMS. Verfahren zur skalierten Erfassung erlebter Beanspruchungsfolgen* [Mental fatigue, monotony, mental satiation, stress response; a measure of effects of mental strain]. Berlin, Germany: Sektion Psychologie der Humboldt-Universität, Psychodiagnostisches Zentrum.

Safety and Health Directive: Council Directive 89/391/EEC of 12 June 1989 on the introduction of measures to encourage improvements in the safety and health of workers at work. *Official Journal, L 183, 29/06/1989*, pp. 0001-0008.

Schriver, K.A. (1997). *Dynamics in document design*. New York, USA: Wiley.

Sellen, A.J. & Harper, R.H.R. (2001). *The Myth of the Paperless Office*. Cambridge, MA, USA: MIT Press.

Seveso II Directive: Council Directive 96/82/EC of 9 December 1996 on the control of major-accident hazards involving dangerous substances. *Official Journal, L 010, 14/01/1997*, pp. 0013-0033.

Stanton, N. (ed.) (1998). *Human Factors in Consumer Products*. London, UK: Taylor & Francis.

Swezey, R.W. (1987). Design of job aids and procedure writing. In G. Salvendy (ed.), *Handbook of Human Factors* (pp. 1039-1057). New York, USA: John Wiley & Sons.

Swezey, R.W. & Llaneras, R.E. (1997). Models in training and instruction. In G. Salvendy (ed.), *Handbook of Human Factors and Ergonomics* (pp. 514-577). New York, USA: John Wiley & Sons.

VDU-Directive: Council Directive 90/270/EEC of 29 May 1990 on the minimum safety and health requirements for work with display screen equipment. *Official Journal, L 156, 21/06/1990*, pp. 0014-0018.

Wilson, J.R. & Raja, J.A. (1999). Human-machine interfaces for systems control. In J.R. Wilson & E.N. Corlett (eds.), *Evaluation of Human Work. A Practical Ergonomics Methodology* (pp. 357-405). London, UK: Taylor & Francis.

Woods, D. (1985). Knowledge based development of graphic display systems. *Proceedings of the Human Factors Society - 29th Annual Meeting* (pp. 325-329). Santa Monica, CA, USA: HFES.

Woodson, W.E. & Conover, D.W. (1966). *Human engineering guide for equipment designers.* Berkeley, USA: University of California Press.

Work Equipment Directive: Council Directive 89/655/EEC of 30 November 1989 concerning the minimum safety and health requirements for the use of work equipment by workers at work. *Official Journal, L 393, 30/12/1989*, pp. 0013-0017.

Analogue presentation of flight parameters on a head-up display

Antoine J.C. de Reus & Harrie G.M. Bohnen
National Aerospace Laboratory NLR
Amsterdam, The Netherlands

Abstract

The present study compares tape and counter-clock formats for display of airspeed and altitude on a fighter aircraft head-up display. Flying performance measures were used side by side with objective measures of pilot visual scanning and mental workload. In the area of flying performance and workload no main effect of display format was found. However, eye fixation times on the counter-clocks were significantly shorter than on the tapes. This effect is especially prominent in flying manoeuvres where vertical velocity was to be controlled (as opposed to altitude). It is hypothesised that one of the main benefits of the counter-clock format is its spatially integrated display of trend information. In the highly demanding working environment of a fighter pilot relatively small differences may have a large effect.

Introduction

In a fighter such as the F-16, the head-up display (HUD) serves as the primary flight reference display. It presents the basic flight parameters such as, altitude, airspeed, rate of climb and descent, as well as parameters related to the on-board systems and weapons. Since the HUD plays such an important role, empirical investigations to different HUD symbologies and layouts can offer a significant contribution towards developing a more optimal pilot-vehicle interface.

A general rule for display design is that displayed parameters should be compatible with the human's internal representation of the system. In the case of an analogue physical system and an analogue internal representation, the display should be analogue too. Display compatibility is made up of two components: static aspects of orientation and dynamic properties of motion (Wickens, 1987). The flight parameters presented on a HUD, including altitude and airspeed, are physically analogue quantities. In principle, these parameters should not be presented in a digital format, since the transformation to an analogue internal representation imposes an extra information-processing step. This can lead, for example, to increased reaction times. Also, the need to perceive trend information or to estimate a rapidly changing value favours the analogue format. On the other hand, digital formats (i.e. counters) will aid reading accuracy and allow high reading speeds. The direction of the movement of an indicator on a display should be compatible with the direction of the movement of an operator's internal representation of the parameter whose change is indicated. With fixed-pointer, moving-tape indicators movement and orientation compatibility operate in opposition and so one or the other must be violated. To display the numbers in such a way that high values are at the top, the scale must move downward to indicate an increase. Another disadvantage is that, like a counter, the scale values become difficult to read when the variable is changing rapidly since the digits themselves are moving (Wickens, 1987). It is also impossible to make a check reading from the tape, as the pilot cannot have an idea about the range of potential values where the current value lies without actually reading the value. If the purpose of the display is to make a comparative check reading between several parameters, then rotating pointers are also superior.

Rotating pointers are good at conveying rate information quickly. Tapes are less good in this respect as the human is more sensitive to the angular deflection of a relatively large pointer than to the rise and fall of a tape. When vertical speed information was incorporated into a HUD format using a simple arc around the digital readout of the altitude, pilot's mean altitude capture error rates decreased (Weinstein et al., 1992). This was attributed to having rate information in addition to the rapidly changing state information that would normally be provided by a digital counter (state information conveys data concerning specific values at a given time). There are two other major arguments to investigate alternatives for moving tapes on a HUD (Weintraub & Ensing, 1992; Newman, 1995). First, tapes take up a relatively large area, which is a disadvantage on a fighter HUD where a lot of information needs to be displayed. Finally, tapes are suspected of introducing roll vection, where pilots perceive the aircraft is rolling, while actually the tapes are moving. Fixed-pointer moving-tapes with a counter, having a windowed scale format, are used to display some F-16 flight parameters. Given the information above, these parameters are perhaps not displayed in the optimum format. A clock with rotating pointers could be a better alternative. Another argument for investigating rotating pointers is that in the F-16 the head-down instrument panel contains clocks with rotating pointer for the same parameters that are displayed as tapes on the HUD. This may, for example, result in experienced difficulty in switching between the HUD and the head-down instrument panels.

In this study, the usability of 'moving tape' versus 'clock with rotating pointer' is examined, with special attention for manoeuvring and unusual attitude recovery. The standard moving tape format is not changed. However, counters are added to the clocks to aid the reading of absolute values. Moreover, an arc is added to the altitude clock to indicate the vertical velocity. The altitude and the vertical velocity may then be processed within one eye fixation. So, with only one glance (of attention), instead of two when the vertical velocity is presented in a separate scale.

Methods

Participants

Seven male participants, with an age between 31 and 35 years participated in the one-day experiment. All participants had a background as an F-16 fighter pilot (average 1143 hours). One pilot had 30 hours experience with the counter-clock format, which was obtained in the Tornado and Jaguar aircraft five years before the experiment.

Test environment

The experiment was conducted on a cockpit mock-up (Figure 1), that includes an actual F-16 throttle and stick, and a landing gear lever. A 19 inch high-resolution colour monitor in front of the participant presents a simulated out-the-window view with superimposed head-up display symbology.

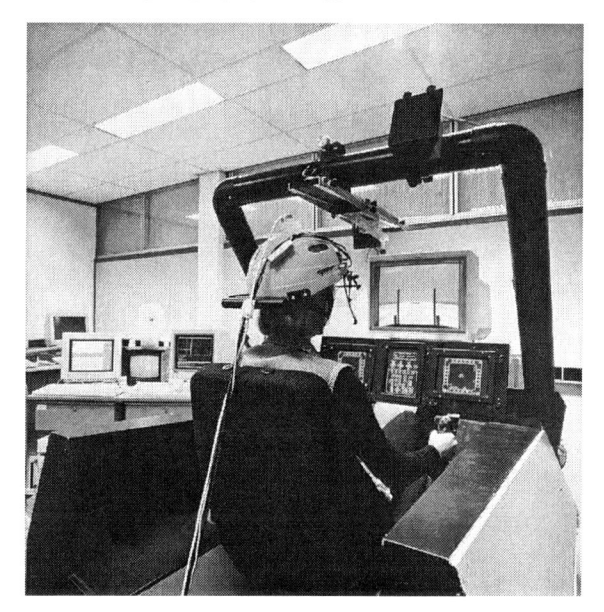

Figure 1. Fighter pilot station used in the study.

Display formats

HUD symbology was similar to that of an F-16 in navigation mode. When the landing gear was down, the landing mode of the HUD was shown, interchanging heading and bank angle scale position and with an extended altitude scale. The tape format was unaltered, but counter-clocks replaced the symbology for airspeed, altitude and vertical velocity in the alternative format.

The tape for altitude (Figure 2) indicates 500 ft per major tick in the navigation mode. Vertical velocity is shown using a separate fixed scale with a moving pointer. The middle of the scale corresponds to level flight and the pointer is at the top position when the vertical velocity is 4000 ft/min. The pointer is limited to 4000 ft/min. The counter-clock for altitude (Figure 2) indicates 100 ft per dot in the navigation mode. Vertical velocity is shown using an arc that starts at the nine o'clock position. It extends via the top of the clock and reaches the two o'clock position when the aircraft is climbing with 4000 ft/min. It extends via the bottom of the clock when the aircraft is descending. The arc is limited to 4000 ft/min. A digital vertical velocity indicator is displayed under the clock. The tape for airspeed indicates 50 Kt per major tick, while the counter-clock for airspeed indicates 10 Kt per dot (Figure 3).

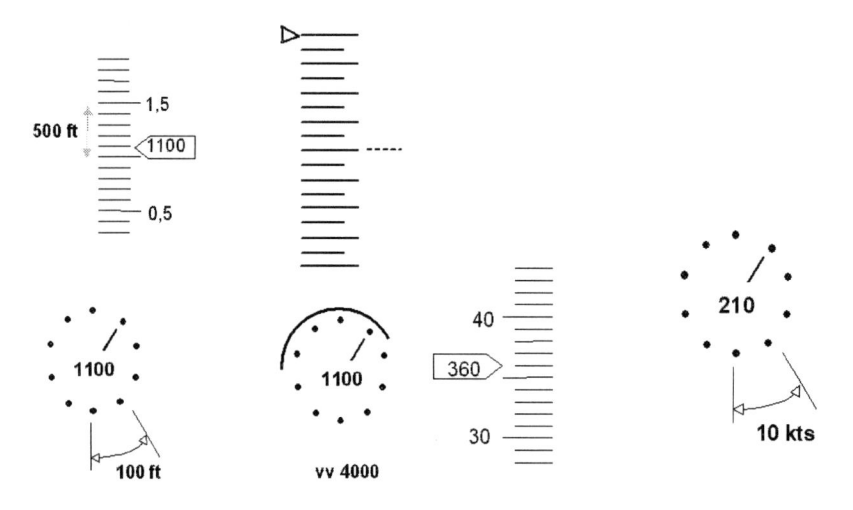

Figure 2. Representation of altitude and vertical speed in the tape and counter-clock formats. The aircraft is at 1100 ft altitude and is climbing with 4000 ft/min.

Figure 3. Representation of airspeed in the tape and counter-clock formats. Airspeed is 360 resp. 210 kts.

Experiment design

A within-subjects experimental design was used with the HUD format as independent variable with two levels: tape and counter-clock. Order of presentation of the HUD formats was balanced over participants.

Tasks

The symbology assessment involved flying a group of basic manoeuvres, where pilots had to precisely control three parameters at the same time. This included two Straight-and-Level segments (altitude and heading fixed), one vertical S-Alpha (climb followed by descent, heading and vertical velocity fixed), one vertical S-Delta (climb followed by descent, bank angle fixed), and one 180-degree Right Turn (altitude and bank angle fixed), followed by approach and landing (Figure 4). In all segments, except approach and landing, speed was fixed. Duration was approximately 30 minutes. The individual elements were taken from Weinstein et al. (1992).

In order to familiarise with the fighter pilot station and the two HUD formats, participants flew 20 minutes with each format; both periods started on the runway. Participants were encouraged to fly varying manoeuvres. Participants then flew two identical training missions, one with each HUD format. These were followed by the symbology assessment missions. The training and symbology assessment missions were shortly briefed before they were flown, and the individual mission elements were instructed at a one-at-a-time basis.

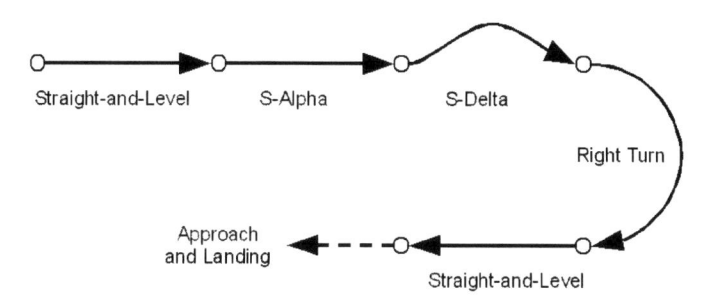

Figure 4. Overview of the symbology assessment mission.

Flight performance, visual scanning and cardiovascular data were recorded. Participants completed various questionnaires, including the Rating Scale Mental Effort (RSME) to assess experienced workload (Zijlstra, 1993). Data were analysed with a repeated measures analysis of variance (Anova) by using SPSS.

Results

Flight performance

During manoeuvring, pilots had to fly as accurately as possible and keep their aircraft within ordered limits. The types of limits depend on the specific mission phase. The percentage of time pilots exceeded the relevant limits was analysed, as well as the root mean square error (RMSE) of the relevant parameters.

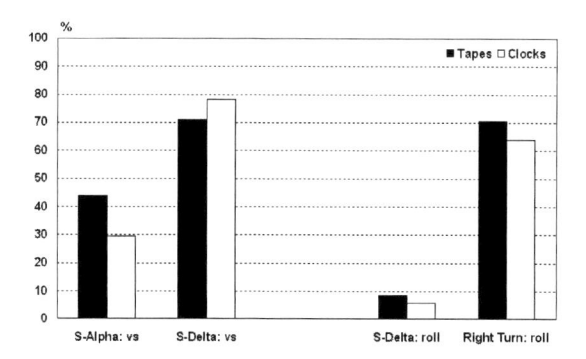

Figure 5. Percentage of time that vertical speed (vs) limits and roll limits were exceeded in the relevant flight phases.

For airspeed, altitude and heading not enough occurrences of exceeding the limits were found for statistical analysis (Figure 5). Only for the vertical speed in the S-Delta there may be an influence of HUD format (p=.095). No significant effects for HUD format were found in the RMSE measures (Figure 6 and 7). Note that the average errors are small: less than 4 Kt for airspeed and less than 25 ft for altitude.

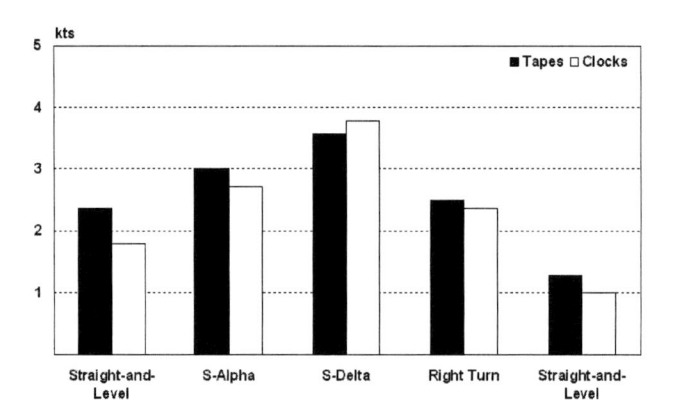

Figure 6. Root Mean Square Errors (RMSE) in airspeed in the relevant flight phases.

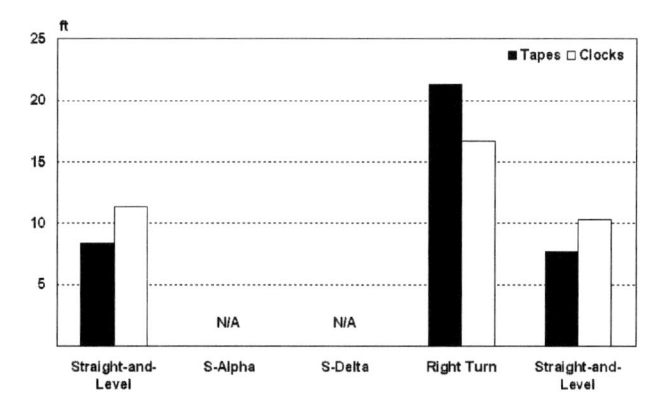

Figure 7. Root Mean Square Errors (RMSE) in altitude in the relevant flight phases. Note that S-Alpha and S-Delta are vertical manoeuvres so altitude is not fixed.

Visual scanning

Five surfaces were defined for the analysis of eye-point-of-gaze measurement: (1) the central area of the HUD, (2) the speed indicator, (3) the altitude indicator with vertical velocity indicator, (4) the heading and roll indicators, and (5) the outside view. For each of these surfaces fixation duration, fixation frequency and dwell time were calculated.

Fixation duration was defined as the duration of a single eye fixation on a specified surface. A significant effect was found for HUD format ($F(4,20)=5.034$, $p=.006$).

Significant effects were also found for mission phase and surface (Figure 8). When looking at the fixation duration on the speed indicator resp. the altitude indicator alone, significant effects were found for HUD format (F(1,5)=7.923, p=.037, resp. F(1,5)=7.373, p=.042) and for mission phase. The interaction effect HUD format x mission phase was significant for the altitude indicator, but not for the speed indicator. When looking at the specific mission phases, only for the S-Alpha and the S-Delta, significant effects were found for the type of altitude indicator (counter-clock or tape). In addition, the S-Alpha showed a significant effect in the type of airspeed indicator. Note that only these two phases required controlling the vertical velocity, as opposed to the other flight phases where the altitude had to be controlled instead.

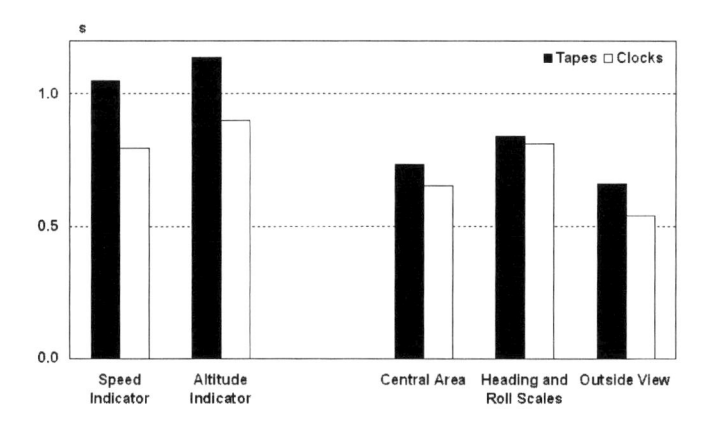

Figure 8. Fixation duration, defined as the duration of a single eye fixation on a specified surface.

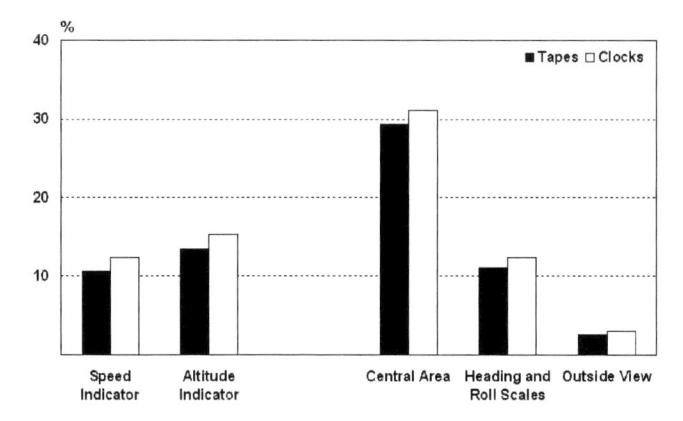

Figure 9. Fixation frequency, defined as the amount of fixations on a specified surface and expressed as a percentage of the total number of fixations.

Fixation frequency was defined as the amount of fixations on a specified surface and expressed as a percentage of the total number of fixations. It showed a potential influence of HUD format (p=.081) and significant effects for mission phase and surface. Interaction between format and surface was not significant, but the other interactions were (Figure 9). When looking at the speed indicator alone, the effects of HUD format and mission phase were not significant. However, the interaction between mission phase and HUD format was significant. When looking at the altitude indicator alone, the effects of HUD format, mission phase and the interaction between the two were significant. Again the S-Alpha and S-Delta stand out: only in these phases significant effects of HUD format were found for both speed indicator and altitude indicator. In effect, the fixation frequency was higher for the clocks.

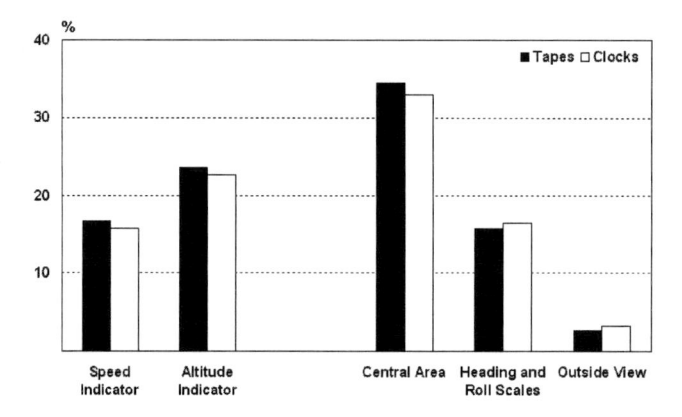

Figure 10. Dwell time, defined as the amount of time a specified surface is looked at and expressed as a percentage of the total time.

Dwell time was defined as the amount of time a specified surface is looked at and expressed as a percentage of the total time. No effect was found for HUD format. Significant effects for mission phase and surface were found. Interaction between phase and surface was significant, other interaction effects were not found (Figure 10). For the speed indicator alone and the altitude indicator alone, a significant effect on dwell time was found for mission phase, but no effect for HUD format. When looking at the separate mission phases, weak effects on dwell time for the type of speed indication were found for both Straight-and-Level segments. In the Right Turn a significant effect was found for the type of altitude indicator ($F(1,5)=15.176$, $p=.011$). In all these cases dwell time was larger for the tape scale indicator.

Summarising the visual scanning results, the conclusion can be drawn that the lower fixation duration on the counter-clocks (difference approximately 0.3 s) is compensated for by a somewhat larger fixation frequency, so that dwell times for the two representations do not differ significantly. Visual scanning patterns (re-fixations on the same surface and transitions to other surfaces) were also analysed, but no apparent effect of HUD format was found.

Cardiovascular measures

Two cardiovascular parameters were calculated: inter beat interval times, corresponding to heart rate, and the power in the mid frequency 0.1 Hz band (Mulder, 1988). The inter beat interval showed a significant effect of mission phase (F(4,20)=3.753, p=.02). This means the heart rate is significantly higher during the S-Alpha, S-Delta and Right Turn, than during the two Straight-and-Level segments (Figure 11). A higher heart rate is generally associated with a higher level of general workload.

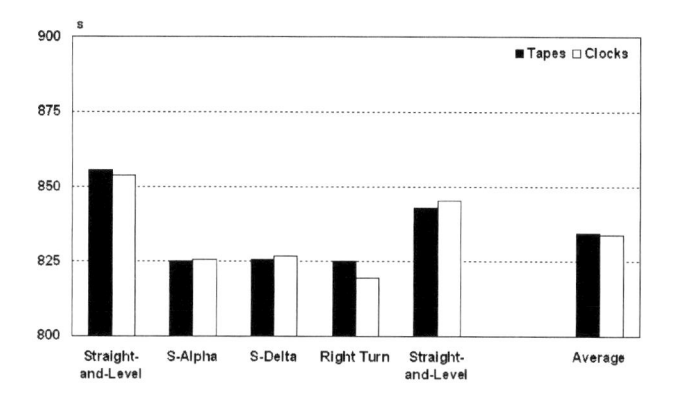

Figure 11. Inter-beat interval times

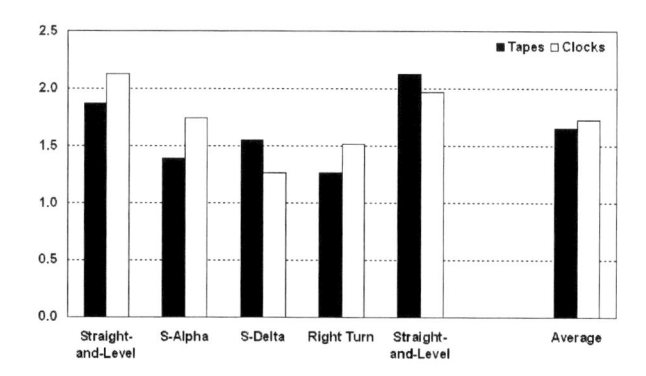

Figure 12. Power in the .10 Hz frequency band.

With respect to the power in the mid frequency band, no effect was found for the display format. Again, a significant effect was found for the mission phase (F(4,20)=5.548, p=.004). There may be an interaction between display format and mission phase, and although the present test is not powerful enough to confirm this (p=.170) it may indicate that the impact on mental workload of display format depends on the mission phase (Figure 12). Note that a lower value for the power in the mid-frequency band indicates a higher mental workload (Mulder, 1988).

Perceived effort

The perceived amount of invested effort was measured after each flight using the Rating Scale Mental Effort (RSME). No effect was found for HUD format. The basic flying task is on the average rated as a little above "somewhat effortful"

Discussion

The main purpose of the current research was to compare two forms of presentation of altitude and airspeed information: with tapes and with counter-clocks. Based on available literature, it was expected that the counter-clock format would –in some areas– show better flying performance and favourable pilot preference than the tape format.

The current study could not prove an effect of the HUD format on flight performance. Although manoeuvres similar to those used by Weinstein et al. (1992) were flown, the results obtained in that study could not be confirmed. In fact, both formats showed very similar performance. Part of the explanation may be in the differences in experimental set-up between the current study and that of Weinstein et al. The latter study used simulated turbulence during most of the mission phases (one of the Straight-and-Level segments, the S-Alpha and the S-Delta), while this was not utilised in the current study. As a consequence, it required less effort to comply with the instructed values for altitude, airspeed, roll angle and vertical velocity. This is confirmed by the lower RMSE values in the current study, and supported by the relatively low scores for perceived mental effort.

For the fixation duration, a significant effect of HUD format was found. Fixations on the counter-clocks were significantly shorter (approximately 0.3 s) than those on the tapes. This confirms one of the reasons for using the counter-clock format. Since fixation duration is related to the allocated amount of visual attention, the counter-clocks apparently require less visual attention. Especially in the S-Alpha and S-Delta, fixation times are shorter on the altitude counter-clocks in comparison to the tapes. These mission phases are the only phases that require controlling vertical velocity, which indicates that the result is caused by the way the rate information is displayed. It may be hypothesised that the counter-clock format with spatially integrated trend arc allows to use one fixation for the parameter and its first order derivative, while the tape format has a spatially separated trend scale that does not allow this. When the counter-clock with arc is indeed perceived as a single integrated object, it is likely that some parallel processing of information can take place (Wickens, 1987).

Dwell times did not differ significantly between the two HUD formats in the basic flying task. This indicates that the shorter fixations spent on the counter-clocks were compensated for by a higher fixation frequency. An analysis of the fixation frequency showed that this parameter was indeed higher on all surfaces for the counter-clock format. The effect was more profound during the S-Alpha and S-Delta. Since no visual scanning of information was required for other tasks, the current single-task setting provided the opportunity to increase the fixation frequency without loss of performance on a parallel task or an increase in mental workload.

Five out of seven pilots favoured the tape format. When looking at the rating of influence on task performance of the altitude indicator, an effect for mission phase was found. For the Straight-and-Level, S-Alpha and S-Delta segments, the counter-clocks were more favourably judged, for the approach and landing it was the other way around. In these phases the indication of vertical velocity in the counter-clocks was rated as having a slightly negative influence on task performance. A potential confounding factor for the preference scores was that participating pilots were more used to a tape format than to a counter-clock format on the head-up display. This may have resulted in more favourable judgements for the tape format, although performance and workload were on a par.

Pilots appreciated the calm and uncluttered nature of the counter-clocks. However, during approach and landing relatively large variations of the parameters occur, and under these circumstances tapes are easier to work with. The opinions on the ease of interpretation and potential flying precision varied, some pilots preferred the clocks, others the tapes. Also the opinions on the display of trend varied: some pilots liked the arc display, while others found the separate vertical velocity scale in the tape format better.

The participants in the current study all had literally hundreds of hours experience in flying with moving tapes on the HUD, while they had virtually no experience with counter-clocks. This may have biased preferences, visual scanning and flying performance in favour of the moving tapes. Even under these conditions, the counter-clock format did not perform significantly different in terms of performance and workload. In the visual scanning measures it seems to show a slight advantage over the moving tapes. This demonstrates that the counter-clock format is an interesting format for displaying altitude and speed on a head-up display.

Conclusion

The current study used objective and subjective measures to investigate the differences in performance, pilot behaviour and workload between two display formats: counter-clocks and moving tapes. In the area of performance and workload no main effect of display format was found. The majority of the pilots preferred the tape scale format. However, visual scanning data revealed some effects of HUD format, most likely caused by the way trend information is displayed. It is hypothesised that the spatially integrated display of trend information is one of the main benefits of the counter-clock format. In the highly demanding working environment of a fighter pilot this type of relatively small differences may have a large effect.

References

Mulder, L.J.M. (1988). *Assessment of cardiovascular reactivity by means of spectral analysis*. PhD thesis, Rijksuniversiteit Groningen, The Netherlands.

Weinstein, L.F., Ercoline, W.R., Evans, R.H., Bitton, D.F. (1992). Head-up display standardization and the utility of analog vertical velocity information during instrument flight. *The International Journal of Aviation Psychology, 2*, 245-260.

Weintraub, D.J.; Ensing, M. (1992). *Human factors issues in head-up display design: the book of HUD*. CSERIAC SOAR 92-2.

Wickens, C.D. (1987). Attention. In: P.E.H. Hancock (Ed), *Human Factors Psychology*. Amsterdam: North-Holland.

Zijlstra, F.R.H. (1993). *Efficiency of work behaviour. A design approach for modern tools*. PhD thesis, Delft University of Technology, The Netherlands: Delft University Press.

The effects of party line communication on flight task performance

Helen Hodgetts[1], Eric Farmer[2], Martin Joose[3], Fabrice Parmentier[4],
Dirk Schaefer[5], Piet Hoogeboom[3], Mick van Gool[5], & Dylan Jones[1]
[1] School of Psychology, Cardiff University, UK
[2] Centre for Human Sciences, Qinetiq, UK
[3] NLR, Amsterdam, Netherlands
[4] Department of Psychology, University of Plymouth, UK
[5] Eurocontrol Experimental Centre, Bretigny, France

Abstract

Cognitive streaming is an approach to human information processing that regards short-term memory as a series of cognitive processes rather than stores. The approach is used as the theoretical basis for a series of experiments, both laboratory and simulator based, that assessed the disruptive effects of the party line on flight task performance. Initial laboratory work using a computer-based visual monitoring task, a communication task and a conflict detection task demonstrated that meaningful background speech was more disruptive to performance than meaningless reversed background speech or quiet. Moreover, the negative effect of the party line was further substantiated in a more realistic flight simulator study involving eight pilots: The party line condition resulted in a greater deviation from the touchdown point on the runway, and was associated with self reports of increased distraction and workload. Furthermore, an increase in flight checklist completion time was observed when background radio/telephony (R/T) was present, and also slightly more air traffic control (ATC) calls were missed or queried in this condition. The current theme of work extends laboratory findings on the 'irrelevant sound effect' to the aviation domain, and suggests that background sound in the party line not only adds to pilot workload but may also impair flight task mental activities.

Introduction

Background sound disrupts performance on a number of laboratory-based short-term memory tasks such as serial recall (e.g., Jones et al., 1992; 1995), text comprehension (Oswald et al., 2000) and proof reading (Jones et al., 1990). The goal of the current work is to extend these findings to the aviation domain by investigating whether the presence of background speech in the party line may be disruptive to short-term memory elements of flight task performance. Modular approaches to cognition such as multiple resource theory (Wickens, 1992) would suggest that background sound should only be disruptive to those tasks that draw upon the same capacity-limited resource (i.e., a concurrent auditory-verbal task, but not a visuo-spatial task). However, cognitive streaming theory offers a different account, proposing instead that interference arises not because of a conflict due to similar *content* but due to similar *processes* (e.g., seriation, or keeping track of order).

The party line refers to the open radio channel through which all aircraft in a given airspace communicate with air traffic control (ATC), a system that allows pilots to hear both their own clearances as well as those of the other aircraft. Although anecdotal evidence suggests that this may be useful for situation awareness (Pritchett & Hansman, 1993), the need to monitor background speech for relevant information would perhaps increase pilot workload in an already demanding multitasking environment. Moreover, laboratory studies indicate that the mere presence of extraneous background sound –even when unattended– impairs performance on a range of cognitive tasks.

The most commonly used task for examining the effect of irrelevant sound in the laboratory is the serial recall task: Participants are required to recall a list of visually or auditorily presented items in serial order, and to ignore any irrelevant sound presented during the trial. Background sound incurs a cost to performance irrespective of its intensity, and regardless of whether the irrelevant stream comprises speech or non-speech (e.g., a series of changing tones; Jones & Macken, 1993). A further finding is that disruption crosses representational domains: Irrelevant sound has been found to disrupt memory for both verbal information (e.g., letters/ digits) and spatial locations (Jones et al., 1995). Traditional approaches to human information processing are unable to explain this pattern of findings. For example, an explanation in terms of the phonological similarity between to-be-remembered items and the to-be-ignored sounds (e.g., Salamé & Baddeley, 1982) does not account for the disruption caused by tones, and multiple resource theory (Wickens, 1992) would have difficulty in accommodating the finding of cross-modal interference.

An alternative approach to the irrelevant sound effect is 'cognitive streaming' (e.g., Jones, 1999), by which short-term memory is regarded as a series of cognitive processes rather than stores. Incoming information, both primary task items and unattended/ irrelevant material, is processed in similar ways and is represented in streams containing information about the order of events. For example, background sound changing in pitch (either speech or tones) comprises a series of events whose order is automatically registered in streams, just as order information is processed consciously and deliberately for to-be-remembered items in a serial recall task. Interference arises as a result of conflict between two streams of order cues; those yielded from the irrelevant sound clash with those representing the to-be-remembered items.

In the laboratory, activities that draw heavily upon short-term memory for order are particularly vulnerable to disruption by irrelevant sound. In the field, this should encompass activities that (a) involve dealing with novel information, (b) require short-term response to unpredictable events, and (c) call upon reproduction of sequences (not just spoken sequences but other sequential actions also). Cognitive streaming theory would therefore predict that performance on many types of flight task would be impaired by irrelevant sound, and not just those aspects that involve the processing of verbal information.

Experiment 1

Experiment 1 aimed to identify the range of tasks that may be prone to disruption by irrelevant sound by testing a number of tasks performed in combination, and under three different sound conditions: quiet, speech, and meaningless reversed speech. For this purpose a new task battery was developed, namely the Aviation Multi-Tasking Environment (AMTE; Figure 1), comprising four main tasks designed to represent some of those that may be undertaken on the flight deck: audio-monitoring/data entry, tracking, conflict detection and visual monitoring. Based on cognitive streaming theory, it was expected that the data entry task would be particularly disrupted by irrelevant sound since it involves the processing of order information. Conversely, background speech was expected to have little or no effect on the tracking task that involved simple psychomotor control. The conflict detection and visual monitoring tasks were more exploratory because it is unclear to what extent unattended sound will interfere with visual or auditory vigilance tasks.

Method

Participants

Twelve undergraduate students at Cardiff University were firstly trained in each task of the AMTE in isolation, and then with all in combination.

Apparatus and materials

The task was presented on a *Windows 98* PC using the *AMTE* software written in *Visual Basic 6.0*, and using sound files recorded in *Sound Forge 4.5* (Sonic Foundry Inc.). Task-relevant sounds relating to the auditory-based elements of AMTE were recorded in a female voice for presentation through the left ear of headphones, whilst task-irrelevant sound was narrative speech recorded in a male voice and presented through the right ear of headphones. The same narrative was digitally edited using the 'reverse' function of Sound Forge for use as irrelevant reversed speech. The presentation of relevant and irrelevant messages in different ears and in dissimilar voices was implemented in order to minimise the possibility of masking, whereby one message is not heard due to purely perceptual interference by a concurrent sound.

The audio-monitoring/ data entry task required participants to monitor the relevant-speech stream for their call sign and for instructions to set one of four radio ports to a six-digit frequency. The tracking task involved four main flight instruments and participants were to maintain a target heading and altitude for the duration of the trial using a joystick. For the conflict detection task participants were required to monitor party line style messages in the task-relevant stream in order to detect potential conflicts with their own aircraft due to changes in heading or altitude. Finally, the visual monitoring task required participants to check that the level of the gauges remained between 45-55%, and to take appropriate action should they deviate by pressing the corresponding key F1-F4 on the keyboard.

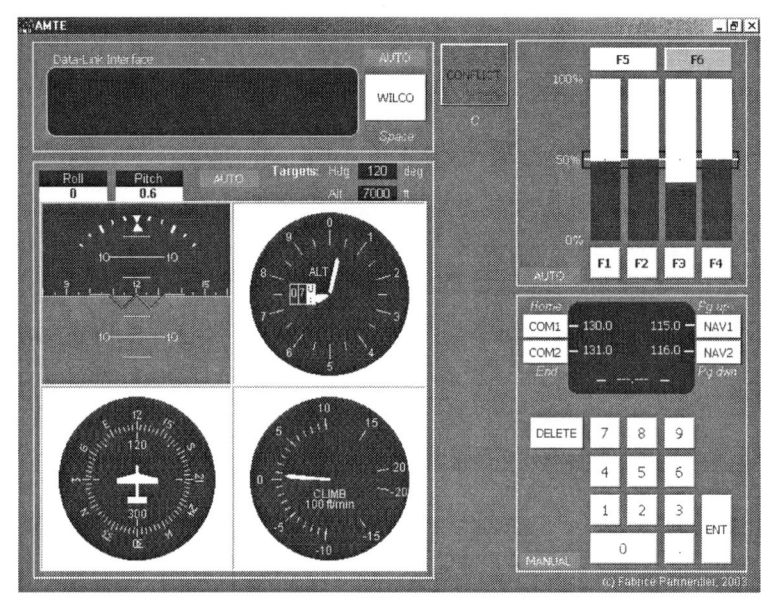

Figure 1. The Aviation Multi-tasking Environment.

Design

The four elements of the AMTE task battery were performed in combination. Each participant completed 12 trials, four in each of the three sound conditions: quiet, reversed speech and forward speech, the order of which was counterbalanced. Each trial lasted a total of 6 minutes and 40 seconds, and events were randomised within each trial.

Procedure

Participants first completed a session of training on the AMTE task. During the experimental trials, participants were told that in addition to the 'party line' messages in the left ear of their headphones, they might also hear irrelevant speech in their right ear. They were instructed to ignore this sound and to perform as well as possible on each of the elements of the AMTE task battery.

Results and discussion

Data entry task

Serial position data for the required six-digit radio frequencies are shown in Figure 2. A repeated measures analysis of variance (ANOVA) demonstrated a significant main effect of serial position ($F(5, 55) = 39.49$, MSE = .004, $p < .001$). The main effect of sound condition however, failed to reach significance (F(2, 22) = 1.69, MSE = .077, p = .21) although this could perhaps be attributed to a lack of power. Further analyses revealed quite a large effect size (Cohen's d = .78) but low power (power to detect medium effect (Cohen's d of 0.5) = .42), indicating that this non-significant result is likely due to the small sample size (n = 12).

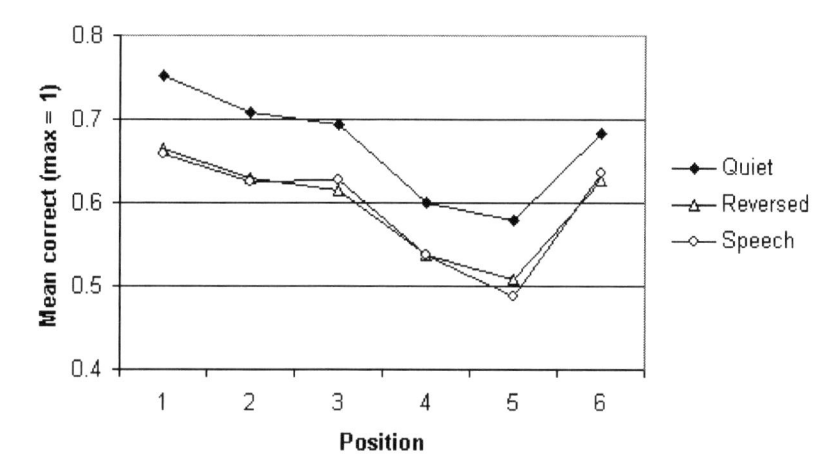

Figure 2. Serial recall performance in each of the three irrelevant sound conditions.

Figure 2 shows a clear trend for poorer performance in the two speech conditions relative to quiet. The finding that background sound –even meaningless reversed speech– disrupts performance on a task requiring seriation is in line with cognitive streaming theory. The meaning or phonological similarity of the irrelevant speech to the to-be-remembered items is not an important feature; rather, the fact that both streams yield a series of order cues is the key to disruption.

Tracking task

There was no difference between sound conditions on measures of deviation from heading or altitude. This finding is also in accordance with cognitive streaming theory because a simple psychomotor control task imposes a low cognitive load, and would not be expected to be affected by the processing of the order of the sounds in the irrelevant stream.

Conflict detection task

Repeated measures ANOVAs were conducted on hit rate, false alarms and reaction times. There was a trend for improved conflict detection in the quiet condition compared to the two speech conditions (Figure 3). This did not quite reach statistical significance (F(2, 22) = 2.63, MSE = .09, p = .09), but again this could perhaps be due to the small sample size: Further analyses indicated a large effect size (Cohen's d = 0.98) but low power (power to detect medium effect = 0.58). A Fisher's LSD post hoc analysis demonstrated that hit rate in the quiet condition was marginally better than in either of the two speech conditions (both p < .06).

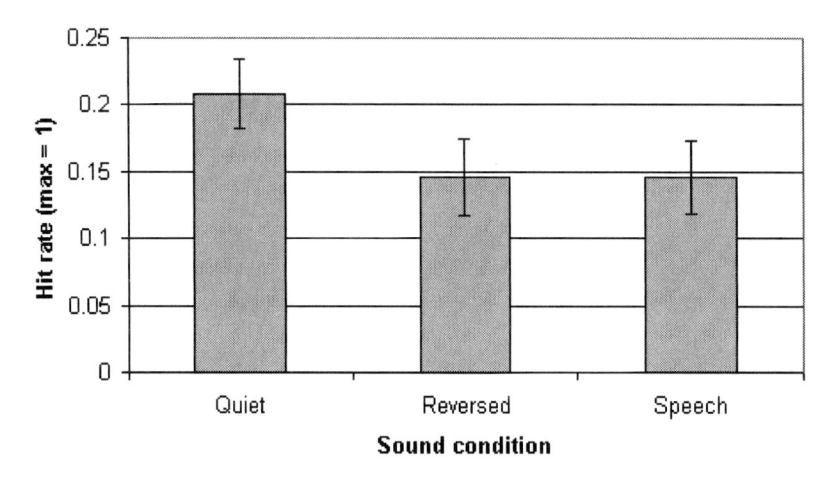

Figure 3. Mean hits in the conflict detection task. Error bars indicate standard error.

A marginally significant effect of sound condition was found in an analysis of false alarm data, (F(2, 22) = 3.15, MSE = 1.23, p < .06). A Fisher's LSD post hoc test indicated that more incorrect responses were made in the presence of forward speech than reversed speech (p < .025) or quiet (p < .08, Figure 4).

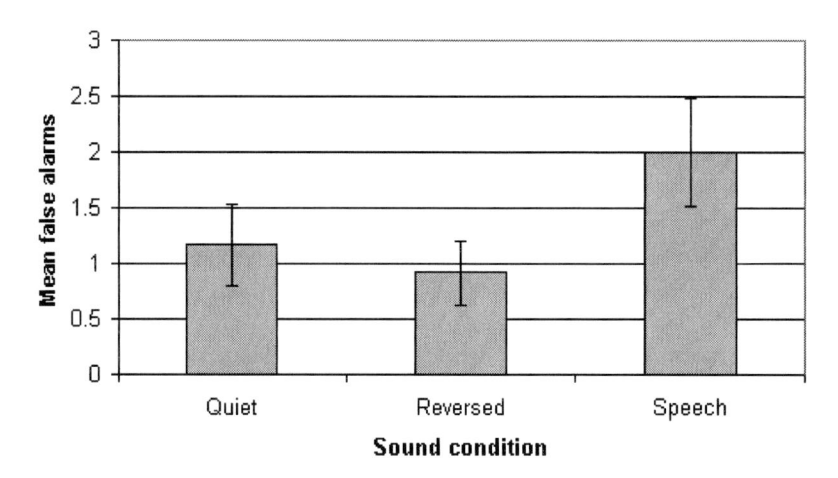

Figure 4. Mean false alarms in the conflict detection task. Error bars indicate standard error.

Reaction times were also recorded. These demonstrated a clear trend for slower conflict detection in the reversed speech condition than in quiet, and slower reaction times still in the forward speech condition. These findings show that unattended background sound can affect both accuracy and reaction times in a task of auditory vigilance.

Visual monitoring task

Both hit rate and false alarm data were collected, but neither demonstrated a significant effect of sound condition. Reaction times were also recorded (Figure 5) and demonstrated a significant effect of irrelevant sound, (F(2, 22) = 6.57, MSE = 265396, p < .01), with reaction times in the speech condition significantly longer than those in the quiet (p < .001) or reversed speech (p < .04) conditions.

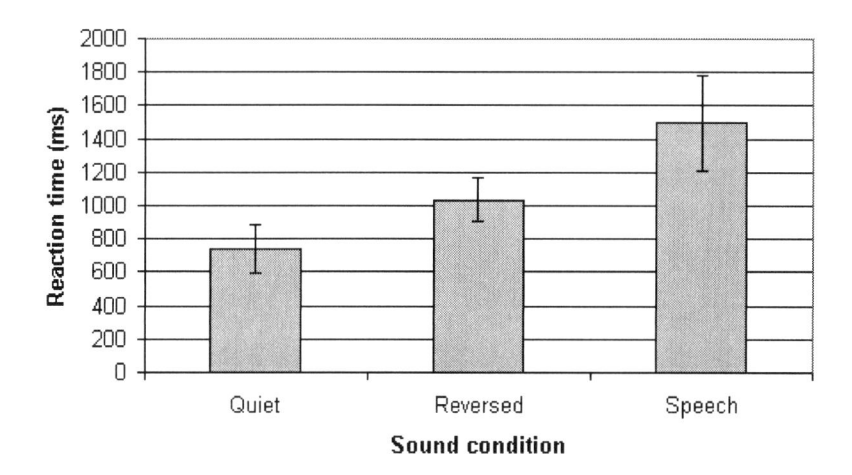

Figure 5. Mean reaction times in the visual monitoring task. Error bars indicate standard error

This result is interesting as it demonstrates that auditory-verbal irrelevant material can disrupt performance on a visual task, a finding that is difficult to accommodate within traditional modular theories of cognition (e.g., Wickens, 1992). Further research would be required to determine the exact processes involved in a visually based vigilance task and why they are susceptible to the effects of processing unattended sound.

Experiment 2

Experiment 2 examined the extent to which the effects of irrelevant sound observed in the laboratory could be demonstrated in a more realistic flying environment. Performance data, observational data, and subjective opinions were collected from four crews (eight pilots) tested under two sound conditions in a high fidelity fixed-base flight simulator. It was expected that the presence of the party line would be associated with self reports of increased frustration, distraction, pressure and workload. Moreover, it was expected that the potentially negative effects of background R/T would be evident in the objective measures of flight task performance.

Method

Participants

Eight Dutch pilots (paid volunteers) took part in the experiment. Captains had an average of around 4650 hours of flight experience and First Officers had approximately 160 hours.

Simulator

Crews were tested in the Generic Research Aircraft Cockpit Environment (GRACE) configured as a Boeing 747 with B747-400 enhanced EFIS displays and simulated Flight Management System. Pilots were instructed to disconnect the autopilot but used the flight director to help them to maintain heading and altitude.

Design

Each crew performed a total of six landings, three with the Captain as pilot flying (PF) and three with the First Officer as PF. In three of the flights the crew received 'minimal-required' R/T, and in the three remaining flights received additional 'party line' R/T. Conditions were counterbalanced between trials. The simulator collected performance data with respect to pitch angle deviation, deviation of flight director pitch angle commands, roll angle deviations, deviation of flight director roll angle commands, runway proximity (horizontal and vertical ground path approaches), and touchdown position on the runway.

Results and discussion

Questionnaire data

Post flight questionnaires (in English) were administered to both crew members immediately after each landing. An adapted version of the NASA TLX was used whereby pilots indicated on a scale of 0 – 100 how they rated each of eight items relating to the preceding flight task: mental and perceptual activity, perceived time pressure, success in accomplishing tasks, mental workload, frustration, distraction by R/T communications, perceived performance impairment by R/T, and perceived flight safety. Relative to the condition with minimal required R/T, the party line was judged to increase workload ($F(1, 7) = 62.94$, $MSE = 13.45$, $p < .01$), frustration ($F(1, 7) = 14.08$, $MSE = 104.03$, $p < .01$) and distraction ($F(1, 7) = 9.66$, $MSE = 483.14$, $p < .02$). A significant effect of party line was obtained on all items except 'success in accomplishing tasks'. Although pilots felt that task success was comparable across flights, the ratings of the other measures suggest that the crew may have had to work harder in the presence of party line to achieve this level of success.

Performance data

Data periods were analysed in which stable flight conditions occurred, but the presence of the party line had no significant effect in terms of pitch angle or roll angle deviation. When considering the landing period however, there was a main effect of background R/T on touchdown: The party line condition was associated with a higher mean standard deviation of the longitudinal position on the runway if compared to the no party line condition. This finding is surprising since almost no background R/T was present during the final approach, although perhaps the touchdown accuracy reflects a cumulative effect of background sound throughout the trial. Arguably, the landing period is the very point at which moment-to-moment correction and responsiveness to the aircraft environment is at its most pressing; one may therefore expect the effects of irrelevant sound to manifest at points when workload is greatest.

Observational data

Observations of ATC calls indicated that slightly more calls were missed, queried or incorrectly read back in the party line than the no party line condition (Table 1). However, the differences were small and there were insufficient samples to perform statistical analyses.

Table 1. Instances of ATC calls missed, queried or read back incorrectly, for Party line (P) and No Party line (NP) conditions

	Crew 1	Crew 2	Crew 3	Crew 4	Total
Calls missed (NP)	0	0	0	0	0
Calls missed (P)	0	1	0	0	1
Calls queried (NP)	1	1	0	1	3
Calls queried (P)	1	1	0	2	4
Incorrect read back (NP)	0	1	1	1	3
Incorrect read back (P)	1	3	0	1	5

Observations of checklist errors showed no apparent differences between the party line conditions: Omissions and repetitions of checklist items were perhaps more dependent upon *interruption* – the suspension and resumption of the checklist – rather than mere *distraction*. That is, a call to the specific aircraft requiring action and read back (in either R/T condition) was more disruptive than simply the presence of background R/T in the party line condition (Table 2). This is consistent with findings that interruptions are disruptive to flight deck performance (Latorella, 1999).

Table 2. Instances of checklist items omitted, repeated or resumed correctly at the next item for those occasions when an ATC call to the own aircraft caused checklist performance to be interrupted (R/T conditions combined)

	Crew 1	Crew 2	Crew 3	Crew 4	Total
Resumed at next item	2	0	1	0	3
Item omitted	1	1	0	1	3
Item repeated	2	0	0	0	2

The time taken to complete each checklist (after excluding time spent dealing with actual interruptions) was recorded and subjected to a 2 (party line condition) x 3 (checklist) repeated measures ANOVA. Although there was no main effect of party line or of checklist, there was a significant interaction, (F(2, 4) = 11.92, MSE = 0.01, p < .02). Completion of the approach and landing checklists was unaffected by background R/T, but completion of the descent checklist took significantly longer in the party line than the no party line condition (Figure 6).

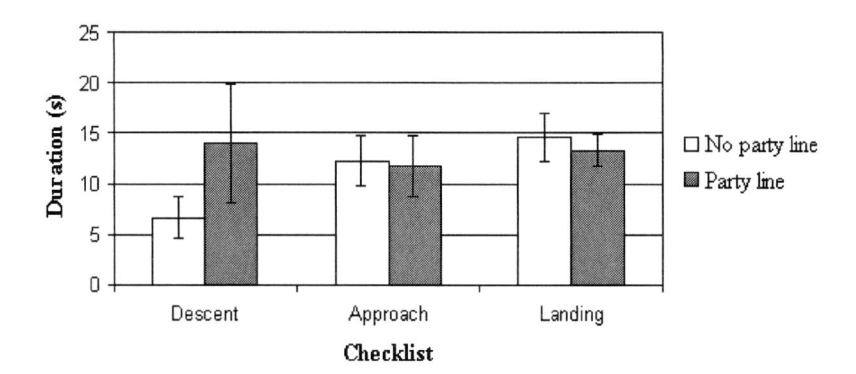

Figure 6. Checklist duration according to party line condition.

This may be because the items on the descent checklist (terrain clearance and approach preparation) involved an element of discussion between crew members, whereas the other two lists involved just quick checks and so were less susceptible to disruption from the background sound.

General Discussion

Three of the four tasks employed in AMTE showed some degree of disruption when performed concurrently and in the presence of extraneous sound. The data entry task showed typical effects of irrelevant sound on serial recall: Performance was depressed in conditions in which irrelevant material was presented, echoing the findings of stringent laboratory-based tasks (e.g., Colle & Welsh, 1976; Salamé & Baddeley, 1982). Serial recall in the reversed speech and forward speech conditions did not differ, suggesting that the physical properties of the irrelevant sounds are more critical to the degree of disruption than any meaning they may contain. In both the conflict detection task and the visual monitoring task, reactions to events slowed in the presence of irrelevant sound. This may be attributed to the nature of multi-tasking environments in which participants must divide their attention such that each task is performed to a modest level. It is likely that this process requires some form of serial order in shifting from one task to another. The conflict detection and visual monitoring tasks each required the monitoring of sources of information, one auditory and one visual. Although these tasks may not typically be disturbed by irrelevant sound when presented in isolation, the process of shifting attention from one to the other, as well as to the tracking and communication tasks, clearly utilises processes that are liable to disruption by irrelevant sound.

The findings of the simulator study were mixed. The systematic subjective reports are unequivocal in showing that the aircrew reported that the effects of party line were negative. Ratings relating to workload and distraction were higher in the presence of irrelevant R/T messages. Although the cognitive streaming model does not make firm predictions about subjective response to sound, these results are consistent with the notion that an irrelevant stream of information is difficult to ignore. In terms of the performance data, an effect of party line was obtained only in the final stage of flight. Perhaps the effects of workload and distraction were partly cumulative, so that the predicted effects were found only at the point when the demands of the task were greatest. Although differences in flight task performance measures were not observed for the main period of flight, it is interesting that a simple measure of checklist completion time showed the expected decrement in the presence of the party line. The effects of background sound in the simulator study may have been constrained by the small sample size, however, given that a few interesting findings did emerge, this is an avenue of research that warrants further study. Future research could speak more directly to the question of whether the replacement of the party line with digital data link technology may not only reduce workload, but may also circumvent the harmful effects that the party line may otherwise have on flight task mental activities.

Acknowledgement

These experiments were part of the Cognitive Streaming project, funded by EUROCONTROL in the context of the 'CARE Innovative Action'. More information on the Cognitive Streaming project can be found at www.eurocontrol.int/care/innovative/projects2002/cs.

References

Colle, H. A., & Welsh, A. (1976). Acoustic masking in primary memory. *Journal of Verbal Learning and Verbal Behaviour, 15,* 17-32.
Hoogeboom, P., Hanson, E., Joosse, M., Hodgetts, H., Jones, D., Salmoni, A., Farmer, E., & Straussberger, S. (2004). *Cognitive streaming project, report on work package 4: 'Real World' effects.* Care Innovative Action, Eurocontrol CARE-IA-CS-NLR-WP4-D4-02-C. Available: www.eurocontrol.int/care/innovative/projects2002/cs/
Jones, D.M. (1999). The cognitive psychology of auditory distraction: The 1997 BPS Broadbent Lecture. *British Journal of Psychology, 90,* 167-187.
Jones, D.M., Farrand, P.A., Stuart, G.P., & Morris, N. (1995). The functional equivalence of verbal and spatial information in short-term memory. *Journal of Experimental Psychology: Learning, Memory, and Cognition, 21,* 1008-1018.
Jones, D.M., & Macken, W.J. (1993). Irrelevant tones produce an irrelevant speech effect: Implications for phonological coding in working memory. *Journal of Experimental Psychology: Learning, Memory and Cognition, 19,* 369-381.
Jones, D.M., Madden, C., & Miles, C. (1992). Privileged access by irrelevant speech to short-term memory: The role of changing state. *Quarterly Journal of Experimental Psychology: Human Experimental Psychology, 44A,* 645-669.

Jones, D. M., Miles, C., & Page, J. (1990). Disruption of proof-reading by irrelevant speech: effects of attention, arousal or memory? *Applied Cognitive Psychology, 4,* 89-108.

Latorella K.A. (1999) *Investigating Interruptions: Implications for Flightdeck Performance,* NASA/TM-1999-209707. Washington: National Aviation and Space Administration (also published in 1996 as Doctoral Dissertation, State University of New York at Buffalo).

Oswald, C. J. P., Tremblay, S., & Jones, D. M. (2000). Disruption of comprehension by the meaning of irrelevant sound. *Memory, 8,* 345-350.

Pritchett, A., & Hansman, R.J. (1993). Preliminary analysis of pilot rankings of 'party line' information importance. In R.S. Jensen and D. Neumeister (Eds.), *Seventh International Symposium on Aviation Psychology.* Columbus, OH, USA: The Ohio State University.

Salamé, P., & Baddeley, A. D. (1989). Effects of background music on phonological short-term memory. *Quarterly Journal of Experimental Psychology, 41A,* 107-122.

Wickens, C. D. (1992). *Engineering psychology and human performance.* New York: HarperCollins.

Yellow lessens discomfort glare: physiological mechanism(s)

Frank L. Kooi[1], Johan W.A.M. Alferdinck[1], & David Post[2]
[1]*TNO Human Factors, Soesterberg, The Netherlands*
[2]*US Air Force Research Laboratory, Dayton Ohio, USA*

Abstract

A long standing mystery is the improved visibility people report with yellow glasses. Car drivers and pilots claim reduced discomfort of oncoming head lamps at night and the blue sky during the day respectively. While people clearly benefit, scientists have not been able to pin down the causal mechanism(s). Several hypotheses have been proposed regarding the physiological origin of this "blue sensitivity" including scotopic (rod) response, excessive iris contraction, blue (S) cone response, brightness perception, and circadian receptor response. These hypotheses were compared by measuring the discomfort glare to four lights differing in their spectra. The results show that more S-cone stimulation corresponds to more discomfort, while more rod stimulation does not. This is consistent but does not prove a causal relation between the S-cones and discomfort glare and it is not consistent with a major role for either the pupillary constriction hypothesis or the circadian photoreceptor. The modified spectral efficiency function $V_m(\lambda)$ and large field photopic spectral sensitivity $V_{10}(\lambda)$ do not significantly change these conclusions based on the standard $V(\lambda)$. The present study has therefore been successful in eliminating several physiological mechanisms of discomfort glare, leaving the S-cones as the primary candidate.

Introduction

There is a long standing paradox regarding the effects of yellow filtering on the sensation of glare. On the one hand scientists claim it has no objective positive effect, on the other hand users claim to benefit. The USAF and many other air forces have decided to limit the use of yellow visors because of the reduction in colour vision (Young et al., 2000). This, while scientifically defensible, leaves the user with an unsatisfied feeling because an apparently good thing (a yellow filter) is not allowed. Kooi and De Vries (2002) have shown that a yellow filter does not have a positive effect on any objective measure like visual acuity or contrast sensitivity, but does turn out to greatly reduce the so-called discomfort glare. Clearly the blue part of the spectrum is disproportionally important in causing the sensation of visual discomfort.

The spectral response of discomfort glare

Several hypotheses have been proposed that may account for the heightened discomfort glare to blue lights. These include (1) Scotopic (rod) response, (1b) Excessive iris contraction, (2) S-cone response, (3) Brightness perception and (4) Circadian receptor response. The spectral sensitivity of excessive iris pupil contraction looks like the scotopic spectral sensitivity function V'(λ) (Howarth et al., 1993; Berman et al., 1996; Vos, 2003). The iris contraction and rod hypotheses are therefore not discriminable. In fact, Howarth et al. (1993) suggests rod input to iris contraction, also during the day. Recently Adrian (2003) attributed the heightened blue sensitivity instead to the absence of the macular pigment outside the fovea. Adrian's hypothesis can be tested by comparing the 2 degree V(λ) and the 10 degree V(λ) responses. The second potential explanation, the S-cones, has received surprisingly little attention. Alferdinck (2000) for example has no treatment of it. The third hypothesis is the difference between luminance and brightness in the blue part of the spectrum. Possibly discomfort glare is directly related to brightness perception, or perhaps subjects are not able to discriminate discomfort from brightness. Berman et al. (1996) treat this hypothesis and prove it to be correct for lower light levels but not for bright lights. Fourthly the recently discovered circadian receptor provides a new visual mechanism with a high sensitivity to the blue wavelengths (Thapan et al., 2001; Brainard et al.; 2001). The discomfort matching experiment presented here is set up to discriminate between these hypotheses.

Methods

The stimulus colours

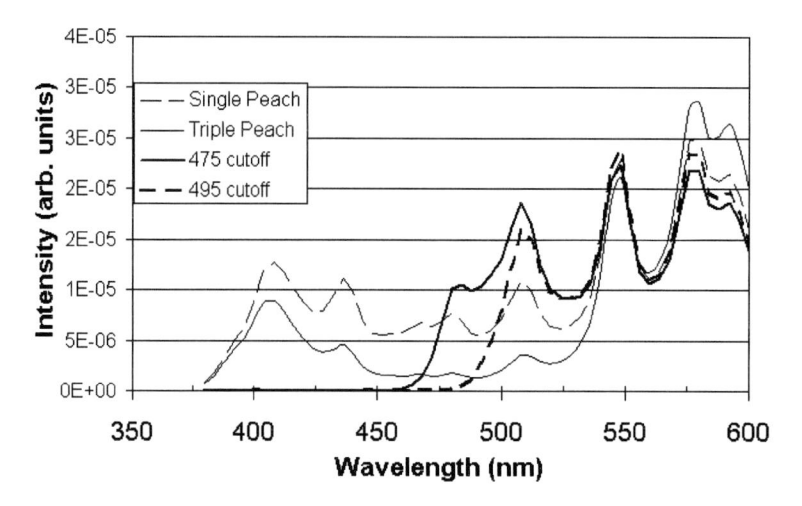

Figure 1. Spectra of the two yellow and the two peach coloured lights. The peach filter transmits much more of the short wavelength light that stimulates the S-cones. The 475 cutoff filter lets the light through that maximally stimulates the rods (500nm). The spectral spikes above 500nm are caused by the spiky light source illuminating the filters.

In order to discriminate between the rod and the S-cone hypotheses, two filter pairs had to be selected that differentially stimulate the one and not the other. Preferably the resulting colours stay close to the colours in the natural world. A combination of yellow filters (5x5cm Schott glass, type numbers GG475 and GG495) and gelatin filters with a peach colour worked out quite well. One light pair differed slightly in the S-cone response (0.14 versus 0.22, both low values) while the rod stimulation differed significantly (0.73 versus 0.38). The other light pair differed greatly in the S-cone response (0.56 versus 0.04) and marginally in the rod response (0.66 versus 0.56). for comparison, equal energy (white) light has a value of 1. Figure 1 shows the four spectra, Table 1 the relevant spectral weighting functions.

Table 1. The relevant CIE spectral weighting functions of the four lights (LEFT), normalized for photopic luminance, forcing the standard spectral sensitivity function V(λ) to be 1.0. $V_m(\lambda)$ is the "modified V(λ)", taking the 1988 CIE correction into account. $V_{10}(\lambda)$ is the large field (10 deg) V(λ), supposedly more suited to model the pupil response than the standard small field (2 deg) V(λ). The "circadian receptor" response was estimated from Thapan et al. (2001) and Brainard et al. (2001). On the RIGHT the ratios for the two filter pairs. The Yellow 475 - Triple Peach pair is designed to isolate the rods, the Yellow 455 - Single Peach pair to isolate the S-cones. The S-cone activity of the 475 cutoff filter and the Triple Peach filter are both very low. In short, both yellow filters need to be increased in luminance to match the discomfort of the two Peach filters. The bottom row shows in italics the psychophysical data from the Results section.

	Spectral filter				Filter pair ratios	
	475 cutoff	Triple Peach	Single Peach	495 cutoff	475-Triple Peach	495-Single Peach
V(λ)	1.0	1.0	1.0	1.0	1.0	1.0
Circadian receptor	1.1E-02	5.8E-03	1.7E-02	6.4E-03	1.9	0.4
V'(λ)	1.64	0.86	1.50	1.26	1.9	0.8
S-cone	0.08	0.12	0.31	0.02	0.6 (both low)	0.1
$V_m(\lambda)$	1.0001	1.002	1.004	1.0001	0.998	0.996
$V_{10}(\lambda)$	1.06	1.03	1.05	1.04	1.030	0.982
CIE x	0.41	0.47	0.39	0.45		
CIE y	0.50	0.43	0.39	0.53		
				Discomfort match:	*1.21±0.45*	*1.54±0.9*

The test procedure

The "method of adjustment" test procedure was adapted from Alferdinck and Theeuwes (1997) and Kooi and De Vries (2002). The subjects were instructed to fixate in between the two lights for 2 to 5 seconds, allowing eye movements. The subjects then indicated to the experimenter which of the two lights was more discomforting. After the experimenter adjusted the relative light level, the subject repeated the visual comparison until the two light sources are judged to be equally discomforting. After four repeated measurements the subjects were asked to rate the level of discomfort glare on the 9-point rating scale of De Boer (De Boer, 1967).

Next the same light pair was measured again four times with the left/right positions reversed. The experiment proceeded with the other light pair. In total each subject made 32 matches and four discomfort ratings, taking 30 to 45 min. The position of the coloured lights (left & right), the order of the presentations, and initial luminance mismatch were balanced within and across subjects.

Test setup and apparatus

The light stimulus consisted of two 8 cm circular lights (1.1 deg) viewed from a distance of 4.2 m. A Pritchard Spectrascan 650 spectro-photometer recorded all luminances and spectra (spectral radiance in $W.m^{-2}.sr^{-1}.nm^{-1}$ versus wavelength) from the view point of the subject. The high brightness lights are created by two Götschmann 67HT projectors aimed at a rear projection screen. The 8cm apertures are made of cardboard, attached to the rear projection screen, separated by 28cm (= 3.8 deg visual angle). At 1.9 deg besides the fovea the rod density is reported to be $30000/mm^2$, about 20 to 25% of the maximum density (Tornow & Stilling, 1998). The luminance levels are varied by widening and narrowing the light beams, guaranteeing a constant spectral composition and uniform light level. The average luminance level was $40x10^3$ cd/m^2, comparable to a weak sun reflected on a wet road. Eye safety is not an issue because of the short viewing times and the small stimulus size. The room was dimly lit (60 lx at the subject eyes), which is less than the ocular stray light caused by the two light sources.

Figure 2. The setup of the experiment. The subject looks in between the two light sources and compares the degree of discomfort. One light is yellow, the other has a peach colour. Pictures can be viewed in colour at http://extras.hfes-europe.org

Results

Pilot experiment

In a pilot experiment we asked five subjects to make discomfort and brightness matches between the light pair that differentially stimulates the S-cones at three light levels (3000 to 40000 cd/m^2). We verified that the highest light level is most suited for the main experiment. As reported by Berman et al. (1996), the sensation of visual discomfort or "visual sting" diminishes at the lower light levels, leaving the brightness sensation as the dominant sensation.

Main experiment

On the de Boer scale (1967) the average score for both light pairs was between 4 and 5, close to "just acceptable". It other words, the subjects were not overly bothered by the experiment. Figure 3 shows the discomfort matches for the two light pairs per subject. All subjects rate the Single Peach filter as more discomforting than the yellow 495 cutoff filter when matched for luminance. At first sight this result suggests a role of the S-cones in causing discomfort glare. Fourteen of the sixteen subjects rate the Triple Peach filter as more discomforting than the 475 cutoff filter. Only 21 of the total 128 measurements (16 subjects times 8 repetitions) set the Single Peach light brighter than the Yellow 495 light. The asymmetry is statistically highly significant ($Z=7.3$, $P=2.7 \times 10^{-11}$). The Triple Peach–Yellow 475 asymmetry is also highly significant ($Z=5.4$, $P=2.6 \times 10^{-7}$). Only 33 of the 128 recorded matches set the Triple Peach light brighter than the Yellow 475 light.

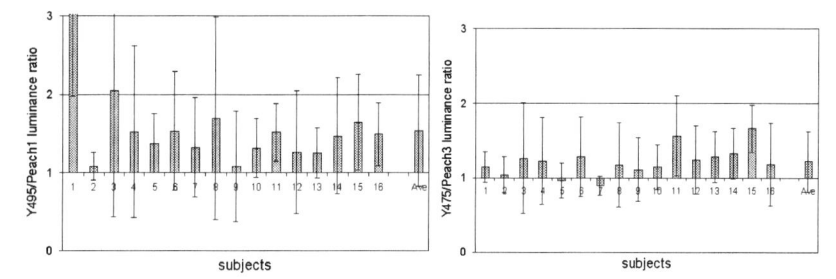

Figure 3. The discomfort glare matches for the two light pairs. The bluer light (the single peach filter) is seen as more glaring by all 16 subjects, consistent with a role for the S-cones in visual discomfort (Left). The light source with more rod stimulation (the 475 yellow filter) causes less discomfort, suggesting the absence of a causal relationship between rod stimulation and discomfort glare (Right).

Discussion

The new data summarized in Table 1 give new insight into the various theories that have been suggested to discomfort glare.

S-cone and rod role in visual discomfort.
From a statistical significance point of view it is clear that more S-cone stimulation corresponds to more discomfort glare. This does not prove a causal relationship, but does make it more likely. The data strongly suggest the absence of a causal relationship between rod stimulation and discomfort glare.

Judging discomfort versus brightness.
In the pilot experiment the matches were made on the basis of discomfort glare as well as brightness. It quickly became apparent that the distinction was not meaningful to the subjects, in particular at the medium and lower light levels. This is

an interesting finding in its own right. The judgements of visual discomfort and brightness can be hard to discern. We conclude that this type of discomfort glare experiment indeed requires high light levels with substantial discomfort to avoid settings based on brightness.

Circadian photoreceptor component in visual discomfort?
Dr. Berman of the Lawrence Berkeley Lab has given the suggestion that the supposed "circadian receptor" might be responsible for the blue dominance of discomfort glare, given its high sensitivity to blue wavelengths (Thapan et al., 2001; Brainard et al., 2001). The circadian spectral sensitivity falls in between the S-cone and rod spectral sensitivity (Figure 4). Table 2 in effect tests this hypothesis by convoluting the circadian spectral response with our four light stimuli. The result falls in between the S-cone ratio and the rod ratio for the Single Peach-495 Cutoff pair, and is virtually identical to the Rod ratio for the Triple Peach - 475 Cutoff pair. Just as the rod hypothesis can be discounted (475 Cutoff stimulates the rods more than the Triple Peach filter but the Triple Peach filter causes more discomfort), the Circadian receptor hypothesis can be discounted: 475 Cutoff stimulates the Circadian receptor more but causes *less* discomfort. The preliminary conclusion therefore is that the hypothesized Circadian receptor cannot be responsible for discomfort glare, at least not fully. It will be worthwhile to repeat the spectral calculation once the Circadian spectrum is pinned down more accurately.

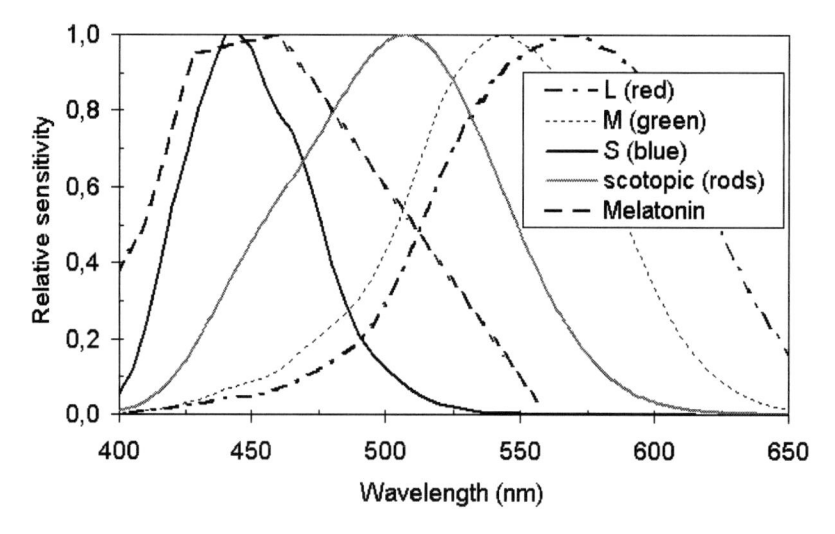

Figure 4. A comparison of the spectral sensitivity of the rods, the three cones and the hypothesized receptor responsible for the Circadian (day/night) rhythm. The spectrum of the 'Circadian receptor' falls between the S-cone and the rod spectra, making it a potential candidate to account for blue effects.

Modified spectral efficiency function $V_m(\lambda)$.
The 1990 CIE luminance correction, the *Modified* $V(\lambda)$ based on the Judd-Vos correction, models perceived brightness rather than luminance. If discomfort glare is

identical to brightness matching, this measure will show. Table 1 indeed shows a small effect in the right direction. It is too small however to impact the conclusions based on the standard spectral sensitivity function V(λ).

Large field spectral efficiency function $V_{10}(\lambda)$.
The same argument holds for the 10 degree field V(λ). The calculated ratios shift by 3%, too little to impact the conclusions.

Conclusions

S-cones
It turned out to be possible to find filters suitable to discriminate between the S-cone and rod hypotheses. The results show that more S-cone stimulation corresponds to more discomfort. This was no surprise given the well known blue glare effect.

Rods
More rod stimulation surprisingly corresponds to *less* discomfort, suggesting that the rods do not contribute. This conclusion is consistent with Berman et al. (1996) who also showed a "high scotopic light source" to be less discomforting at high light levels, undermining not only the rod but also the pupillary contraction hypothesis for discomfort glare.

Perceived brightness
Perceived brightness measured in this study appears closely related to visual discomfort. It cannot however be consistent with the brightness matching spectral sensitivity found in the literature (CIE 1978; Wyszecki & Stiles, 1982). This is one of the aspects that merits more thought and model effort.

Other factors
Other factors like the hypothesized circadian photoreceptor, the modified spectral efficiency function $V_m(\lambda)$, and the large field spectral efficiency function $V_{10}(\lambda)$ all affect the calculations but not enough to affect the conclusions based on the standard V(λ) spectral efficiency function.

Summary

The present study has successfully undermined the rod and circadian physiological mechanisms of discomfort glare. The S-cone hypothesis comes out a more likely source, but the data do not prove a causal relationship.

Remaining research issues

The present study has contributed to the understanding of the discomfort glare mechanism. One next step is to get a quantitative grip on the importance of the S-cone contribution; does it account for all or part of the yellow glare reduction? Secondly, it is intriguing that a 1.5 times luminance reduction is sufficient to compensate for a 15 times S-cone increase. A 1.5 times reduction in luminance is minimal compared to the effect of sunglasses, which typically darken by a factor of 10 to 100. We conclude that two other perceptual phenomena with yellow glasses, the increased perceived contrast and brightness (Rabin & Wiley, 1996), are worth closer investigation. Blue scattering and chromatic aberration in the eye may in principle contribute to the glare sensation through the mechanism of increased brightness. The results of this study will be useful to develop a yellow filter that simultaneously maximizes viewing comfort and colour vision.

Acknowledgement

We would like to gratefully acknowledge the USAF "Windows on Science" program and Dr. Berman for their contributions to this paper.

References

Adrian, W. (2003). Spectral sensitivity of the pupillary system. *Clinical and Experimental Optometry, 86*, 235-238.

Alferdinck, J.W.A.M. (2000). *Glare and distraction by bluish car headlamps; a literature survey* (TNO-report TM-00-C018). Soesterberg, The Netherlands: TNO Human Factors.

Alferdinck, J.W.A.M. & Theeuwes, J. (1997). The relation between discomfort glare and driving behaviour. In *Proceedings symposium PAL-congress* 1997 (pp. 24-32). Darmstadt, Germany: University of Darmstadt.

Berman, S.M., Bullimore, M.A., Bailey, I.L., & Jacobs, R.J. (1996). The influence of spectral composition on discomfort glare for large-size sources. *Journal of the Illuminating Engineering Society, Winter 1996*, 34-41.

Brainard, G.C., Hanifin, J.P., Greeson, J.M., Byrne, B., Glickman, G., Gerner, E., & Rollag, M.D. (2001). Action spectrum for melatonin regulation in humans: evidence for a novel circadian photoreceptor. *The Journal of Neuroscience, 21*, 6405-6412.

CIE (1978). *Light as a true visual quantity: principles of measurement.* (Publication CIE no 41), 1978. Vienna, Austria: Commission Internationale de l'Eclairage.

CIE (1990). *CIE 1988 2^0 spectral luminous efficiency function for photopic vision.* (CIE Publication No. 86). Vienna, Austria: Commission Internationale de l'Eclairage

De Boer, J.B. (1967). *Public lighting.* Eindhoven, The Netherlands: Philips Technical Library.

Flannagan, M., Sivak, M., & Traube, E. C. (1994). *Discomfort glare and brightness as functions of wavelength* (Report No. UMTRI-94-29). Ann Arbor, Michigan, U.S.A.: The University of Michigan, Transportation Research Institute.

Howarth P.A., G. Heron G., Greenhouse D.G., Bailey I.L., & Berman S.M. (1993). Discomfort from glare: The role of pupillary hippus. *International journal of lighting research and technology, 25*, 37-42.

Kooi, F.L. & De Vries, G. (2002). *Een subjectieve beoordeling van optische hulpmiddelen bij het autorijden in het donker [The influence of optical aids on driving comfort at night].* (TNO-report TM-02-C028, in Dutch). Soesterberg, The Netherlands: TNO Human Factors.

Rabin, J. & Wiley, R. (1996). Differences in apparent contrast in yellow and white light. *Opthalmic and Physiological Optics, 16*, 68-72.

Thapan K., Arendt J., & Skene D.J. (2001). An action spectrum for melatonin suppression: evidence for a novel non-rod, non-cone photoreceptor system in humans. *The Journal of Physiology, 535*, 261-267.

Tornow R.P. & Stilling R. (1998). Variation in Sensitivity, Absorption and Density of the Central Rod Distribution with Eccentricity. *Acta Anatomica, 162*, 163–168.

Vos J.J. (2003). Reflections on glare. *Lighting Research and Technology, 35*, 163-176.

Wyszecki G. & Stiles W.S. (1982). *Color science: concepts and methods, quantitative data and formulae.* 2nd Edition, New York: John Wiley & Sons.

Young P.A., Perez-Becerra J., & Ivan D. (2000). Aircrew visors and color vision performance: A comparative and preliminary pilot study analysis. *Aviation, Space, and Environmental Medicine, 71*, 1081-1092.

Psychophysiological predictors of task engagement and distress

Stephen H. Fairclough & Louise Venables
School of Psychology
Liverpool John Moores University
UK

Abstract

Biocybernetic systems utilise real-time changes in psychophysiology in order to adapt aspects of computer control and functionality, e.g. adaptive automation. This approach to system design is based upon an assumption that psychophysiological variations represent implicit fluctuations in the subjective state of the operator, e.g. mood, motivation, cognitions. A study was performed to investigate the convergent validity between psychophysiological measurement and changes in the subjective status of the individual. Thirty-five participants performed a demanding version of the Multi Attribute Test Battery (MATB) over four consecutive twenty-minute blocks. A range of psychophysiological data were collected (EEG, ECG, SCL, EOG, respiratory rate) and correlated with changes in subjective state as measured by the Dundee Stress State Questionnaire (DSSQ). The DSSQ was analysed in terms of three subjective meta-factors: Task Engagement, Distress and Worry. Multiple regression analyses revealed that psychophysiology predicted a significant proportion of the variance for both Task Engagement and Distress but not for the Worry meta-factor. The consequences for the development of biocybernetic systems are discussed.

Introduction

Biocybernetic systems utilise real-time changes in psychophysiology as an adaptive control input to a computer system. For example, a biocybernetic loop may control the provision of automation within an aviation environment (Byrne & Parasuraman, 1996). This loop diagnoses the psychological status of the human operator based on psychophysiological activity and relays a control signal to initiate or relinquish system automation (Pope, Bogart, & Bartolome, 1995). The affective computing concept (Picard, 1997) represents an example of the same principle where psychophysiological monitoring/diagnosis enables computer software to respond to the subjective state of the user. The concept of biocybernetic control enables a wide range of applications (Allanson & Fairclough, 2004), from adaptive automation (Scerbo et al., 2001) to health-monitoring (Gerasimov, Selker, & Bender, 2002) and biofeedback training tools (Pope & Palsson, 2001).

The goals of biocybernetic control are to promote safe and effective performance, as well as curtailing "hazardous states of awareness" (Prinzel, 2002), such as: fatigue, anxiety, boredom. Both goals are linked as hazardous states are often incompatible with reliable and adequate performance; in addition, states related to poor performance such as anxiety may have consequences for the health of operator.

The biocybernetic control loop may be designed in one of two ways: to promote positive/effective performance states or to avoid negative/ineffective states (Scerbo, 2003). Freeman et al (1999) used an EEG-based index of engagement (Pope et al, 1996) to drive a biocybernetic loop that worked on the basis of a negative control loop, i.e. system automation was only activated if participant was deemed to be engaged with the task, and the system automatically reverted to a manual mode if participants' level of automation declined. Therefore, the system was designed to maintain participants in a stable and moderate level of task engagement, thus avoiding operator complacency during system automation (Parasuraman & Riley, 1997) whilst allowing the user to experience the benefits of automation, e.g. reduced mental workload, stress, and fatigue.

The promotion of positive states such as task engagement and avoidance of negative moods such as distress and anxiety is central to biocybernetic control. However, these systems are reliant on the sensitivity and diagnosticity of psychophysiology to detect positive and negative states (Fairclough & Venables, 2004). On the one hand, the psychophysiological response appears sufficiently differentiated to discriminate broad patterns of emotional response (both positive and negative). On the other hand, it is difficult to formulate the psychophysiological signature of each subjective state with the required degree of precision, as demonstrated by inconsistent findings in this area (Cacioppo, Klein, Berntson, & Hatfield, 1993). This disparity may stem from two sources: the inclusiveness of the psychophysiological response, and the multifaceted experience of subjective states. Whenever the psychophysiological signature of a performance state is captured, it contains a non-affective content (e.g. cognitive demands, motor activity) and a contextual element triggered by the functional goals associated with that emotion (e.g. approach or avoidance) as well as an emotional signature (Stemmler, Heldmann, Pauls, & Scherer, 2001). This lack of specificity is mirrored by the experience and operationalisation of subjective states, which may involve a complex interplay between affective feelings, motivational desires and related cognitions (Matthews et al., 2002).

A partial solution to this problem is to adopt an inclusive definition of the subjective state encompassing affective, motivational and cognitive dimensions of subjective experience as well as the psychophysiological response. This was the logic underlying the development of the Dundee Stress State Questionnaire (DSSQ, Matthews et al., 2002) that attempts to integrate aspects of subjective experience within a number of meta-factors. The DSSQ was derived via factor analysis of self-report questionnaires from a large sample (e.g. Fenigstein, Scheier, & Buss, 1975; Heatherton & Polivy, 1991; Matthews & Desmond, 1998; Matthews, Jones, & Chamberlain, 1990; Sarason, Sarason, Keefe, Hayes, & Shearin, 1986). The factor analysis yielded three factors, each of which encompass at least three sub-scales:

Task Engagement (energy, concentration, motivation), Distress (tension, negative affect, confidence) and Worry (self-focus, self-esteem, cognitive interference). Task Engagement was defined as an "effortful striving towards task goals" (Matthews et al., 2002; Matthews et al., 1997), this factor increased during a demanding working memory task and declined when participants performed a sustained vigilance task (Matthews et al., 2002). The Distress meta-factor was characterised by "an overload of processing capacity" (Matthews et al., 2002; Matthews et al., 1997) and tended to increase when participants experienced a loss of control over performance quality (Matthews et al., 1997). The third Worry meta-factor was concerned with rumination and negative self-evaluation (Matthews et al., 2002; Matthews et al., 1997) and is based upon the S-REF model of anxiety (Wells & Matthews, 1996); the Worry factor was also found to increase when participants experienced a loss of control over performance (Matthews et al., 1997).

This study was performed to investigate whether psychophysiology could be used to predict positive (Task Engagement) and negative (Distress, Worry) performance states. Participants were exposed to a demanding task over a sustained time period. The high level of demand was included to provoke Task Engagement whilst the time-on-task manipulation was intended to eventually reduce engagement whilst inflating Distress and Worry.

Method

Participants

Thirty-five university students participated in the experiment, (13 female and 22 male), and all received a monetary reward. The age of participants ranged from 18-40 years, (M = 24.1 years, S.D. = 5.90). Potential participants were excluded if they were pregnant, on medication or reported any known cardiovascular problems. Participants were additionally requested not to consume large amounts of alcohol the night before, nor drink large amounts of caffeine or participate in strenuous exercise on the morning of the experiment.

Experimental task

The computer task used for the experiment was the Multi-Attribute Task Battery (MATB, Comstock & Arnegard, 1992), this is a multitasking environment containing three subtasks: tracking, system monitoring, and fuel resource management. Each subtask was pre-scripted to a high level of task demand (the parameters of which were tested and utilised in a prior experiment (Fairclough & Venables, 2004).

Psychophysiological variables

EEG was recorded with Ag/AgCl electrodes, across the four sites utilised by Pope et al. (1995) study: Cz, P3, Pz, P4, (with a ground site located midway between Cz and Pz). Each site was referenced to the left and right mastoid areas. The EEG signals were amplified (using four BIOPAC EEG100C differential, bio-electric potential modules). The high and low bandpass filters were set at 0.1Hz and 35Hz, respectively. The EEG signals were analysed via Fast Fourier Transform (FFT) in steps of 2.65 s with an overlap of 0.5 s. Epochs with total power exceeding 200% of the average for that participant were identified as outliers and removed from subsequent analysis, i.e. a pilot exercise found this criterion to be highly associated with artifacts in the EEG record identified by visual inspection. Mean % power values were obtained for: θ (4.3 – 7.8Hz), α (8.2 – 12.9Hz), and β (13.3 – 21.9Hz).

To assess vertical eye blink activity, Ag/AgCl electrodes were placed above and below the left eye, with a ground electrode positioned in the centre of the forehead. The EOG signals were filtered at 0.05-35Hz, and amplified by a BIOPAC EOG100C differential, (high gain), corneal-retinal potential amplifier. Eye-blink frequency and duration were the parameters derived from a smoothed EOG signals.

Heart rate activity was recorded using a standard Lead II configuration, and amplified using vinyl electrodes positioned on the 7th intercostal space on the right and left side of the body. A common ground electrode was placed on the sternum. ECG was measured using a BIOPAC TEL 100C differential (high gain) amplifier. The high and low bandpass filters were set at 0.5-35Hz, respectively. R peaks of the ECG were detected offline, and the interbeat interval (IBI) between successive R waves was calculated. These data were evaluated for missed and ectopic beats, the former were corrected via interpolation and the latter were discarded. HRV in mid- (0.09-0.13Hz) and high- (0.14-0.40Hz) frequency bands were calculated from the IBI data by means of an FFT analysis with Carspan software (L.J.M Mulder, van Roon, & Schweizer, 1995).

Respiration was monitored using two elasticated belts placed around the chest and diaphragm. Respiration signals were again amplified using a (differential, high gain) BIOPAC TEL100C remote monitoring module, with the filter settings at 0.05-35Hz. The waveform signals of both chest and diaphragm expansion were added together using BIOPAC AcqKnowledge software, and peaks from the combined signal were detected and used for the calculation of respiration rate (i.e. breaths per min).

Skin Conductance Level (SCL) was measured with two electrodes (which produce a continuous voltage electrode excitation of 0.5 V), attached to the side of the foot (Boucsein, 1992). These signals were amplified using a BIOPAC TEL100C remote monitoring module, and subsequently filtered (low pass) at 1Hz to rid of extraneous noise. Skin conductance values for mean and area were collected every 2secs and averaged over 4min periods. The sample rate for all channels (i.e. EEG, ECG, SCL, EOG, and Respiration) was 500Hz.

Subjective measures

Subjective state was measured using the Dundee Stress State Questionnaire (DSSQ, Matthews et al., 2002; Matthews et al., 1997). This battery of questionnaires containing Likert scales derived from earlier research which have been grouped into three fundamental meta-factors: Task Engagement, Distress and Worry.

Task Engagement is concerned with "a commitment to effort" (Matthews et al., 2002, p. 335) and contains scales for: energetical arousal (alert-tired, i.e. a mood adjective checklist, participants were asked to describe how well each adjective described how they felt at that moment on a 4-point scale from "definitely" to "definitely not", Matthews et al., 1990), motivation (8 items regarding on level of mental effort and feelings about success/failure assessed on a 9-point Likert scale) and concentration (7 items regarding the perceived efficiency of concentration assessed on a 5-point Likert scale, Matthews & Desmond, 1998). The theme of those scales grouped under the Distress meta-factor is "an overload of processing capacity" (Matthews et al., 2002, p. 336). This factor contains scales for: tense arousal (tense-relaxed) and hedonic tone (5-point Likert scale, sad-happy, both items were assessed using the mood adjective checklist described previously for energetical arousal, Matthews et al., 1990) as well as confidence/perceived control (6 items relating to positive aspects of performance and perceived control assessed via a 5-point Likert scale, Matthews & Desmond, 1998). The third meta-factor of the DSSQ is Worry and this factor is concerned with self-evaluation and self-focus; the Worry factor contains scales pertaining to: self-focus (8 items assessed via a 5-point Likert scale related to private self-consciousness, Fenigstein et al., 1975), self-esteem (a 5-point Likert scale was used to assess 6 items related to social self-esteem and 1 item relating to performance self-esteem, Heatherton & Polivy, 1991) and cognitive interference (8 items to assess the frequency of task-relevant thoughts and 8 items to assess the frequency of task-irrelevant thoughts both assessed via 5-point Likert scale from "never" to "very often", Sarason et al., 1986).

Full details regarding the factor structure of the DSSQ, population norms and state responses to different types of psychological tasks may be found in (Matthews et al., 2002).

Procedure

Upon entering the laboratory, participants were briefed about the nature of the experiment. Those who chose to participate were already fully informed as to the procedures involved in the recording of the physiological measures. Participants were prepared so their physiology could be recorded, (e.g. the location of the electrode sites, the mild abrasion of skin, the attachment of the electrodes, etc), and this was followed by a fifteen-minute baseline period for all of the physiological variables. During this baseline period, the participants were asked to lie back and relax (with their eyes open) while their physiology was measured.

Following the baseline period, participants were presented with a 5min training session to acquaint themselves with the keyboard/joystick controls. This was followed by a 20min high-demand practice block. Participants then began the formal task session of 4 x 20min (high-demand) blocks of MATB performance (80mins in total), i.e. a repeated measures design. The participants received no information about the duration of the experimental task prior to the formal task (i.e. the participants did not know how many 20min blocks must be completed); in addition, participants were asked to surrender their watches to remove anticipation of task completion. Prior to the practice block, and again after each task block, participants were presented with a computerised version of the DSSQ. The DSSQ asked participants to rate their feelings and moods as perceived *during task performance*. The DSSQ took between three and five minutes to complete. Upon completion, participants persisted with the next task block. This continued until all four blocks had been completed. The recording of the physiological measures was initiated at the same time as each task session was started.

Results

Experimental data were analysed using Statistica 6.1 (Statsoft Inc.). Outliers (defined as values lying at least three standard deviations outside the group mean) were excluded from all analyses. Significant MANOVA findings are expressed using Wilks' Lambda (Λ) and data for effect size (η^2) are also provided for additional information.

The effect of Time-On-Task on MATB performance

A MANOVA was performed on MATB performance (tracking error, accuracy on system monitoring task, mean deviation of fuel management task) over four blocks of twenty minutes. There were no significant changes in performance over time, i.e. MATB performance was stable throughout the test session. Mean values for MATB performance throughout the session were: RMS error (M = 70.31, s.d. = 3.06), target detection as percentage (M = 81.7, s.d.=2.97), and deviation from target fuel level (M = 151.48, s.d. = 36.16).

The effect of Time-On-Task on subjective states

Nine scales from the DSSQ were divided into three groups based on the factor analysis reported in (Matthews et al., 2002). This categorisation divided the DSSQ into three meta-factors: Engagement (energetical arousal, motivation, concentration), Distress (tense arousal, hedonic tone, confidence), and Worry (self-focus, self-esteem, task-irrelevant thoughts). The z-change score from each scale of the DSSQ was calculated where: *zchange = (score − group_mean from previous time period) / standard deviation of group from previous time period*. This transformation is based upon the one reported by Temple et al. (2002) and is intended to standardise change scores across all DSSQ scales.

The transformed values from those three scales associated with the Task Engagement meta-factor were analysed via a 3 x 4 MANOVA (DSSQ scale x Time-On-Task). This analysis revealed an interaction effect of marginal significance [Λ (6,29) = 0.763, p = 0.06, η^2 = 0.321]. Mean values for each component of the Task Engagement meta-factor are shown in Table 2. Post-hoc Bonferroni analyses revealed that energetical arousal showed a large decrement after 40 minutes of performance compared to other periods (p < 0.01). Similarly, motivation levels fell at a higher rate during the first half of performance compared to later periods (p < 0.01). The decrement associated with Concentration was reduced during the final period of performance compared to previous periods (p < 0.01).

The 3 x 4 MANOVA for the Distress meta-factor revealed a significant main effect for time-on-task [Λ (3,32) = 0.69, p < 0.05, η^2 = 0.253]. Both tense arousal and hedonic tone exhibited negative and positive change scores over time, but the magnitude of these changes were insignificant. Post-hoc Bonferroni testing revealed a significant effect for only the confidence factor, which declined sharply during the final twenty minutes of performance (p < 0.05).

A 3 x 4 MANOVA was conducted on the three components of the Worry meta-factor. This analysis revealed a significant interaction effect only [Λ (6,29) = 0.75, p < 0.02, η^2 = 0.382]. Post-hoc analyses of the DSSQ scales indicated that self-esteem showed a significant increase after forty minutes of performance (p < 0.01). In addition, the rate of task-irrelevant thoughts was highest after twenty and sixty minutes compared to the remaining time periods (p < 0.01).

Multiple Regression analyses

Psychophysiological data were standardised using a z-change score transformation (described in the previous section) prior to regression analyses. The z-change transformation was performed on psychophysiological data averaged over the final five minutes of each twenty minute period of task performance: this period was selected to achieve maximum coherence with the subjective self-report scales (i.e. participants were asked to report how they felt at that moment), which were administered at the end of each twenty minute period.

The transformed psychophysiological data were averaged across all four time periods and subjected to a correlation analysis. This analysis was performed to estimate the degree of redundancy between different psychophysiological measures and to identify variables for inclusion in the regression analyses. A probability level of < 0.10 was used for this analysis in order to identify both moderate as well as high levels of correlation. Based on this correlation, five psychophysiological variables with low levels of inter-item correlations were selected as independent variables for the multiple regression analyses; these variables were alpha power in the EEG (α, averaged across all four sites), inter-beat interval of the heart rate (IBI), 0.1Hz component of sinus arrhythmia (SA), respiration rate (RR) and rate of eyeblink frequency (BR).

The variables from the DSSQ were averaged into three meta-factors described by Matthews et al (2002): Task Engagement, Distress and Worry. Task Engagement was calculated by combining z-change scores from three DSSQ components: Energetical Arousal, Concentration and Motivation. To calculate the Distress factor, z-change scores for Hedonic Tone and Confidence/Control were reversed and combined with the z-change score for Tense Arousal; therefore, increased Distress was represented by rising tension in combination with negative affect and falling confidence. The Worry factor involved a combination of z-change scores for Frequency of Task-Irrelevant Thoughts and Self-Focus in conjunction with a reversed score for Self-Esteem, i.e. Worry = increased cognitive interference and self-focus in conjunction with falling self-esteem. The rationale for these formulations may be found in Matthews et al (2002).

A series of multiple regressions were performed to investigate if Task Engagement, Distress and Worry were predicted by psychophysiological variables. Four multiple regression analyses were conducted for each meta-factor using data from each period of performance. The results of the Task Engagement analysis are presented in Table 1.

Table 1. Results of the multiple regression using psychophysiological predictors of the DSSQ meta-factor Task Engagement (N=33). A summary of the regression is provided in the upper panel and significant predictors are listed in the lower panel with their Beta weights and partial correlations in brackets. Note: RR = respiration rate, BR = eye blink rate, SA = 0.1Hz component of sinus arrhythmia, α = EEG alpha power

	20min	40min	60min	80min
Regression	Adj. R^2 = 0.32 $F(5,30)$= 3.82 $p < 0.01$	Adj. R^2 = 0.43 $F(5,30)$=5.20 $p < 0.01$	Adj. R^2 = 0.41 $F(5,30)$=4.98 $p < 0.01$	Adj. R^2 = 0.53 $F(5,30)$=7.62 $p < 0.01$
Significant Predictors $p<0.05$	RR 0.57 [.59]	RR 0.69 [.69]	RR 0.32 [.40] α -0.62 [-.63] SA -0.34 [-.40]	RR 0.47 [.55] α -0.66 [-.69] SA -0.35 [-.42] BR -0.31 [-.42]

The regression analyses revealed a statistically significant relationship between Task Engagement and psychophysiological variables, which was sustained throughout the period of task performanced. Psychophysiological variables predicted between 32 and 53% of the variance associated with Task Engagement. The most consistent predictor of Task Engagement was respiration rate which had a positive relationship with Task Engagement. Mean power in the α bandwidth, the 0.1Hz component of sinus arrhythmia and eyeblink frequency exhibited a negative relationship with Task Engagement during the latter periods of the task activity.

Table 2. Results of the stepwise regression using psychophysiological predictors of the subjective meta-factor Distress (N=35). A summary of the regression is provided in the upper panel and significant predictors are listed in the lower panel with their Beta weights and partial correlations in brackets. Note: SA = 0.1Hz component of sinus arrhythmia, α = EEG alpha power

	20min	40min	60min	80min
Regression	Adj. R^2 = 0.26 $F(5,30)$=2.87 $p < 0.05$	Adj. R^2 = 0.42 $F(5,30)$=5.26 $p < 0.01$	Adj. R^2 = 0.42 $F(5,30)$=5.31 $p < 0.01$	Adj. R^2 = 0.38 $F(5,30)$=4.51 $p < 0.01$
Significant Predictors $p<0.05$	α 0.36 [.34]	α 0.38 [.46] SA 0.72 [.68]	α 0.67 [.66] SA 0.57 [.59]	α 0.64 [.63]

The results of the stepwise regressions on the Distress meta-factor are presented in Table 2. Psychophysiological variables predicted between 28 and 42% of the variance associated with Distress. It was apparent that both α activity from the EEG and the 0.1Hz component of sinus arrhythmia had a positive relationship with levels of Distress. None of the psychophysiological predictors achieved statistical significance during the multiple regression to predict the Worry meta-factor. This pattern of null findings was repeated across all four periods of task performance for the Worry meta-factor.

Discussion and conclusions

The experimental manipulations of high task demand and sustained performance had no significant affect on task performance, but caused a number of latent changes (Hockey, 1997) with respect to psychophysiology and subjective self-report. The DSSQ data were analysed as change scores to represent time-on-task trends relative to the previous period of task performance. Two components of Task Engagement, energetical arousal and motivation, showed a significant decline during the first forty minutes of performance only and falling levels of concentration accounted for the decline of Task Engagement during the latter half of the task period. The influence of time-on-task on the Distress factor was modest by comparison. The combination of high task demand and sustained performance failed to significantly increase tense arousal or induce negative affect via the hedonic tone factor; however, it was significant that confidence levels fell dramatically after the final period of performance. The sudden decline of confidence suggests that participants had reached the limits of successful coping after the fourth session (participants were not told when the task would end) and Distress may have been augmented if the task period had been extended. The Worry meta-factor was also relatively unaffected by the experimental task. The frequency of task-irrelevant thoughts increased with each period of performance; therefore participants had more difficulty focusing attention on the task as time progressed. The significant increase of self-esteem after forty minutes of performance was unexpected and is assumed to represent a perception of increased task mastery. The absence of any significant effect on self-focus was anticipated; the high temporal demands associated with multitasking MATB performance discourage rumination or a shift of attention from the task to the self (Matthews et al., 2002).

The main goal of the study was to investigate whether psychophysiological measures could predict changes in subjective states as represented by the DSSQ. The multiple regression analyses (Tables 1 and 2) provided some support of predictive validity, but with several important caveats. Psychophysiological variables predicted between one third and half of the variance associated with the Task Engagement meta-factor over the four periods of performance (Table 1). Respiration rate was a consistent, positive predictor of engagement, i.e. higher breathing rate = increased Task Engagement. A number of other variables made a significant contribution to the regression equation during the latter periods of performance (Table 1). Suppression of both α activity and the 0.1Hz component were associated with Task Engagement, both of which have been associated with increased mental effort (Gevins et al., 1998; Mulder, 1986). This finding suggests that covariation between psychophysiology and subjective self-report may be moderated by changes in sympathetic activation related to the investment of mental effort. The general pattern of the Task Engagement regression was an accumulation of psychophysiological predictors with increased time-on-task; for instance, a suppression of eye blink frequency was also associated with Task Engagement during the final period of performance (Table 1). This pattern may be indicative of increased mental effort as a compensatory strategy to counteract the influence of fatigue on performance (Hockey, 1997).

The prediction of the Distress meta-factor was modest during the initial period of performance (Table 2). This multiple regression presented a positive association between Distress and both the 0.1Hz component and α activity (Table 2). The Distress factor represents "an overload of processing capacity" (Matthews et al., 2002); in the context of the current study, any overload of capacity was induced by a failure to sustain performance over time-on-task. The effect of task activity was to suppress the level of α activity, which has been associated with mental effort investment (Gevins et al., 1998), and the Distress factor was associated with a failure to sustain α suppression. The positive association with the 0.1Hz component indicated that Distress was associated with a tendency to reduce or conserve mental effort (Hockey, 1997), i.e. the 0.1Hz component is suppressed when mental effort is invested. Therefore, the Distress meta-factor was associated with a "giving up" pattern from the psychophysiological domain.

None of the psychophysiological measures used in the study could successfully predict the Worry meta-factor. This null finding may stem from the failure of the independent variables to induce Worry in the participants. In addition, the Worry meta-factor is characterised by attentional/cognitive scales and it is possible that the psychophysiological variables used in the study failed to tap this cognitive dimension. The Worry meta-factor may have been predicted by measures of cognitive psychophysiology such as evoked-cortical potential variables, e.g. the P300 component (Prinzel et al., 2003).

The current study had at least one major weakness concerning the range of psychophysiological variables used during the study, which excluded several important measures such as blood pressure and facial EMG. The former has been used to differentiate states of challenge and threat (Blascovich & Tomaka, 1996; Tomaka, Blascovich, Kibler, & Ernst, 1997); two states that bear a resemblance to the DSSQ concepts of Task Engagement and Distress used in the current study. Facial EMG has demonstrated consistent changes in response to pleasant and unpleasant stimuli; particularly in the corrugator muscles above the eyebrow (Cacioppo, Bush, & Tassinary, 1990) and EMG activity from these sites has been used to differentiate between positive and negative affect (Bradley, Cuthbert, & Lang, 1996). The inclusion of these measures may have increased the explanatory power of the regression analyses if blood pressure and facial EMG provide a unique contribution to the variance of the subjective measures. The regression analyses indicated that psychophysiology explained between twenty-six and fifty-three percent of variance in the subjective states data (Tables 1 and 2), the average was approximately forty per cent, leaving more than half the variance unexplained. Future research could investigate how to improve the explanatory power of regression analyses by supplementing psychophysiology with other data sources, e.g.

real-time performance, cognitive models. For example, individual traits such as age and personality may play a role in the prediction of subjective states. The influence of both coping style and neuroticism on Distress from the DSSQ has been demonstrated (Matthews et al., 2002) and other more transitory variables such as sleep quality and time-of-day may also play a role.

Task Engagement and Distress are relevant dimensions of subjective state for biocybernetic control as both have implications for performance and the wellbeing of the human operator. The current study operationalised engagement and Distress using the meta-factors devised by Matthews et al (1997, 2002) which integrate mood, motivation and cognition within unitary factors. This level of specificity is sufficient to represent the subjective state of the operator as a two-dimensional space and construct a biocybernetic loop designed to counteract low levels of Task Engagement and high Distress. This characterisation should suffice for many applications where performance is important such as adaptive automation, computer games and educational software. However, this level of specificity will not suffice for those biocybernetic systems that require a more detailed level of mapping, e.g. between distinct emotional states and psychophysiology.

The rationale underlying the development of biocybernetic control is that these systems can deliver timely and intuitive system interventions. The fact that psychophysiology was capable of explaining a substantial amount of the variance associated with both Task Engagement and Distress in the current study provides momentum for the continued development of these systems. However, it is difficult to predict how this degree of convergent validity will translate into operators' perceptions of system reliability and influence related variables such as trust. A detailed understanding of how the mapping between psychophysiology and the subjective state influences user perceptions of biocybernetic control is a topic for future research.

Acknowledgements

This work was funded by the Engineering and Physical Sciences Research Council (EPSRC) under project number GR/R81077/01.

References

Allanson, J., & Fairclough, S.H. (2004). A research agenda for physiological computing. *Interacting With Computers, 16*, 857-878.
Blascovich, J., & Tomaka, J. (1996). The biopsychosocial model of arousal regulation. *Advances in Experimental Social Psychology, 28*, 1-51.
Boucsein, W. (1992). *Electrodermal Activity*. New York: Plenum Press.
Bradley, M. M., Cuthbert, B. N., & Lang, P. J. (1996). Picture media and emotion: effect of a sustained affective context. *Psychophysiology, 33*, 662-670.
Byrne, E., & Parasuraman, R. (1996). Psychophysiology and adaptive automation. *Biological Psychology, 42*, 249-268.

Cacioppo, J.T., Bush, L.K., & Tassinary, L.G. (1990). Microexpressive facial actions as a function of affective stimuli: Replication and extension. *Personality and Social Psychology Bulletin, 18*, 515-526.

Cacioppo, J.T., Klein, D.J., Berntson, G.G., & Hatfield, E. (1993). The psychophysiology of emotion. In M. Lewis & J. M. Haviland (Eds.), *Handbook of Emotions* (pp. 119-142). New York: Guilford Press.

Carver, C.S., & Scheier, M.F. (2000). On the structure of behavioural self-regulation. In M. Boekaerts, P.R. Pintrich, and M. Zeidner (Eds.), *Handbook of Self-Regulation* (pp. 41-84). San Diego: Academic Press.

Comstock, J.R.J., & Arnegard, R.J. (1992). *The Multi-Attribute Test Battery for human operator workload and strategic behaviour research* (No. 104174): National Aeronautics and Space Administration.

Fairclough, S.H., & Venables, L. (2004). Psychophysiological candidates for biocybernetic control of adaptive automation. In D. de Waard, K.A. Brookhuis, and C.M. Weikert (Eds.), *Human Factors in Design* (pp. 177-189). Maastricht, The Netherlands: Shaker Publishing.

Fenigstein, A., Scheier, M.F., & Buss, A.H. (1975). Public and private self-consciousness: Assessment and theory. *Journal of Consulting and Clinical Psychology, 43*, 522-527.

Freeman, F.G., Mikulka, P.J., Prinzel, L.J., & Scerbo, M.W. (1999). Evaluation of an adaptive automation system using the three EEG indices with a visual tracking task. *Biological Psychology, 50*, 61-76.

Gerasimov, V., Selker, T., & Bender, W. (2002). Sensing and effecting environment with extremity-computing devices. *Offspring, 1* (1), 1-9.

Gevins, A., Smith, M. E., Leong, H., McEvoy, L., Whitfield, S., Du, R., et al. (1998). Monitoring working memory load during computer-based tasks with EEG pattern recognition models. *Human Factors, 40*, 79-91.

Heatherton, T.F., & Polivy, J. (1991). Development of a scale for measuring state self-esteem. *Journal of Personality and Social Psychology, 60*, 895-910.

Hockey, G.R.J. (1997). Compensatory control in the regulation of human performance under stress and high workload: a cognitive-energetical framework. *Biological Psychology, 45*, 73-93.

Matthews, G., Campbell, S. E., Falconer, S., Joyner, L. A., Huggins, J., Gilliland, K., et al. (2002). Fundamental dimensions of subjective state in performance settings: Task engagement, distress and worry. *Emotion, 2*, 315-340.

Matthews, G., & Desmond, P. A. (1998). Personality and the multiple dimensions of task-induced fatigue: a study of simulated driving. *Personality and Individual Differences, 25*, 443-458.

Matthews, G., Jones, D.M., & Chamberlain, A.G. (1990). Refining the measurement of mood: The UWIST Mood Adjective Checklist. *British Journal of Psychology, 81*, 17-42.

Matthews, G., Joyner, L., Gilliland, K., Campbell, S., Falconer, S., & Huggins, J. (1997). Validation of a comprehensive stress state questionnaire: Towards a state 'Big Three'? In I. Mervielde, I. J. Deary, F. De Fruyt, and F. Ostendorf (Eds.), *Personality Psychology in Europe* (Vol. 7). Tilburg: Tilburg University Press.

Mulder, G. (1986). The concept and measurement of mental effort. In G.R.J. Hockey, A.W.K. Gaillard & M.G.H. Coles (Eds.), *Energetical issues in research on human information processing* (pp. 175-198). Dordrecht, The Netherlands: Martinus Nijhoff.

Mulder, L.J.M., Van Roon, A.M., & Schweizer, D.A. (1995). *Carspan: Cardiovascular Experiments Analysis Environment*: IEC ProGamma, Groningen, The Netherlands.

Parasuraman, R., & Riley, V. (1997). Humans and automation: use, misuse, disuse, abuse. *Human Factors, 39*, 230-253.

Picard, R. W. (1997). *Affective Computing*. Cambridge, Mass.: MIT Press.

Pope, A.T., Bogart, E.H., & Bartolome, D.S. (1995). Biocybernetic system evaluates indices of operator engagement in automated task. *Biological Psychology, 40*, 187-195.

Pope, A.T., & Palsson, O.S. (2001). *Helping video games "rewire our minds"*. Paper presented at the Playing by the Rules: The Cultural Challenges of Video Games (26-27th October), Chicago.

Prinzel, L.J. (2002). *Research on Hazardous States of Awareness and Physiological Factors in Aerospace Operations* (No. NASA/TM-2002-211444). Hampton, Virginia: NASA.

Prinzel, L.J., Hitt, J.M., Scerbo, M.W., & Freeman, F.G. (1995). *Feedback contingencies and bio-cybernetic regulation of operator workload.* Paper presented at the Human Factors and Ergonomics Society 39th Annual Meeting.

Prinzel, L.J., Parasuraman, R., Freeman, F.G., Scerbo, M.W., Mikulka, P.J., & Pope, A.T. (2003). *Three experiments examining the use of electroencephalogram, event-related potentials, and heart-rate variability for real-time human-centred adaptive automation design* (No. NASA/TP-2003-212442): NASA.

Sarason, I.G., Sarason, B.R., Keefe, D.E., Hayes, B.E., & Shearin, E.N. (1986). Cognitive interference: Situational determinants and traitlike characteristics. *Journal of Personality and Social Psychology, 57*, 691-706.

Scerbo, M.W., Freeman, F.G., & Mikulka, P.J. (2003). A brain-based system for adaptive automation. *Theoretical Issues in Ergonomic Science, 4*, 200-219.

Scerbo, M.W., Freeman, F.G., Mikulka, P.J., Parasuraman, R., Di Nocera, F., & Prinzel, L.J. (2001). *The Efficacy of Psychophysiological Measures for Implementing Adaptive Technology* (NASA/TP-2001-211018). Hampton, Virginia: NASA.

Stemmler, G., Heldmann, M., Pauls, C.A., & Scherer, T. (2001). Constraints for emotion specificity in fear and anger; the context counts. *Psychophysiology, 38*, 275-291.

Temple, J.G., Warm, J.S., Dember, W.N., Jones, K.S., LeGrange, C.M., & Matthews, G. (2002). The effects of signal salience and caffeine on performance, workload and stress in an abbreviated vigilance task. *Human Factors, 42*, 183-194.

Tomaka, J., Blascovich, J., Kibler, J., & Ernst, J.M. (1997). Cognitive and physiological antecedents of threat and challenge appraisal. *Journal of Personality and Social Psychology, 73*, 63-72.

Wells, A., & Matthews, G. (1996). Modelling cognition in emotional disorder: The S-REF model. *Behaviour Research and Therapy, 34*, 881-888.

Consequences of shifting from one level of automation to another: main effects and their stability

Francesco Di Nocera[1], Bernd Lorenz[2], & Raja Parasuraman[3]
[1]*Cognitive Ergonomics Laboratory, Department of Psychology,*
University of Rome "La Sapienza", Rome, Italy
[2]*Institute of Flight Guidance, German Aerospace Center (DLR), Germany*
[3]*ARCH Lab, George Mason University, Fairfax, VA, USA*

Abstract

A simulated space operations environment was used to investigate the effect of changes in the distance between levels of automation (LOA) on operator telerobotic performance. Participants were assigned to four experimental conditions corresponding to four different LOAs. During the simulation they both switched and did not switch between these LOAs. Participants performed three tasks: *i*) rover task: controlling the rover's cruise and picking up (by means of a mechanical arm) samples of rocks from the terrain; *ii*) fault detection task: detecting the occurrence of a fault; and *iii*) recovery task: remembering the location of each of the simulated operators within the station, in order to recover from the fault. No performance costs were observed when two successive trials involved the same LOA. On the contrary, upward and downward shifts in LOA led to performance costs that were modulated by stage of processing and workload. Particularly, when workload was high, performance was primarily negatively affected when shifting from decision support to manual control (LOA 2 and 0 respectively) in a detection task, whereas the same shift led to a better performance in a working memory task. Finally, moderate LOA seems not to be suited for supporting working memory tasks.

Introduction

Automation is being introduced into various systems in various domains of work and everyday life, part of the move towards "ubiquitous computing". Automated sub-systems now provide the human operator valuable support in such domains as air, ground, space, and maritime transportation, military command and control, health care, and other areas. These types of computer support can be considered to define different levels of automation (LOA) between the extremes of full manual and full automation control (Sheridan, 2002). Between these two extremes a variety of intermediate LOA can be identified, and each one could be conceptualized as a compromise between human and machine responsibilities.

A given system could be designed for a particular LOA on the basis of criteria such as system safety and efficiency, as well as human performance criteria such as the maintenance of situation awareness and balanced workload (Endsley & Kaber, 1999; Parasuraman, Sheridan, & Wickens, 2000). LOA may also be modified in real time during system operations, as in the so-called adaptive automation (Moray, Inagaki, & Itoh, 2000; Parasuraman, Molloy, & Singh, 1996; Scerbo, 1996; Scerbo et al., 2001). In adaptive systems, task allocation between the operator and the computer systems is flexible and context-dependent. Adaptive automation may reduce the human performance costs (unbalanced mental workload, reduced situation awareness, complacency, skill degradation, etc.) that are sometimes associated with high-level decision automation. Several investigators have looked at the effects of different Levels of Automation (LOA) on performance. According to Parasuraman et al. (2000) high LOA can be usefully implemented for information acquisition and analysis functions. Nevertheless, decision making functions are acknowledged to be best supported by moderate LOA. Studies by Crocoll & Coury (1990), Sarter & Schroeder (2001), and Rovira, McGarry, & Parasuraman (2002) support this view by showing that unreliable decision automation leads to greater costs than unreliable information automation.

Kaber, Onal, & Endsley (2000), Endsley & Kiris (1995), Endsley & Kaber (1999), and Kaber, Onal, & Endsley (1998) also provide support for a "moderate" LOA philosophy. The underlying rationale views moderate LOA as an optimal balance with respect to the performance trade-off resulting from the benefits of reduced workload associated with higher LOA on the one hand and with better maintenance of situation awareness associated with lower LOA on the other hand. These studies induced rare automation failure events that require operators to return to full manual control. Typically, the higher the LOA prior to this event the poorer the return-to-manual performance or -in other words- the higher the out-of-the-loop performance cost. Lorenz, Di Nocera, Röttger, & Parasuraman (2002), however, have shown that a higher LOA in a complex fault-management task does not necessarily lead to poorer return-to-manual performance under automation failure in comparison to a moderate LOA, as long as the interface supports operator information sampling to maintain situation awareness. In fact, the moderate LOA was found to be linked to a higher disengagement of sampling fault-relevant information. Apparently this LOA directed the operator attention to lower-order manual implementation of fault-recovery actions at the expense of monitoring the impact of these activities on higher-order system constraints. A mitigation of this effect could be achieved in a follow-up study that used an integrated display in support of fault state monitoring (Lorenz, Di Nocera & Parasuraman, 2004). According to these studies the LOA per se is not necessarily the crucial factor affecting the out-of-the-loop performance costs. In general, it appears that there are differential effects of LOA by stage of processing and interface type. Yet, the experimental procedure used in these studies involved LOA shifts in different blocks, making it difficult to generalize the effects found to the adaptive automation domain. Indeed, adaptive automation assumes changes in LOA within shorter time frames, e.g. even from trial to trial, and there is very little research on such dynamic shifts in LOA. Furthermore, it is unclear whether the direction of the shift (up or down the LOA continuum) affects performance.

Accordingly, the present experiment was carried out to understand whether distance (extension of the "jumps" in the LOA hierarchy) and direction (upward *vs.* downward) in the shift from one LOA to another may affect human performance. To this aim, the impact of different LOA at different stages on performance during a complex task was examined. Particularly, the task was a simulated space telerobotic operation. Subjects were assigned to four groups, each one interacting with the system most of the time at a particular level of automation, and occasionally at other levels.

Two main hypotheses can be tested with this design. The first hypothesis is that, when considering only distance "zero" (no shift between levels), the performance of the four groups should be equivalent, independently of the task that the subjects are carrying out. Indeed the groups are set to act within a particular LOA and they should show neither benefits nor costs associated with the execution of tasks within that level. The second hypothesis is that, when shifting from one LOA to another (as occurs in adaptive systems), specific distance and direction effects should be observed. Likely, a linear pattern should be expected downward (the larger the downward distance the larger the performance decrement), because of the costs associated with re-engagement of a previously dismissed process. This general pattern can be expected from the studies cited above, which address the out-of-the-loop performance problem associated with varying LOA using the return-to-manual paradigm. Conversely, upward shifts should provide less predictable results. Indeed, the opposite linear pattern (the larger the upward distance the larger the performance increment) would suggest that there is no cost associated with the disengagement of a process. Hence, operators would adopt a very passive role when interacting with the system. Considering that subjects in the present experiment did not deal with a 100% reliable system, one would expect them showing some sort of resistance to the disengagement of a process.

Experiment

Method

Participants
Sixteen Catholic University of America undergraduate students (8 males and 8 females) volunteered to participate. Their mean age was 23 years, ranging from 18 to 35 years. All of them had normal or corrected-to-normal vision. All participants reported being right handed, and were naive as to the hypotheses of the experiment. Each student received $40 for participation.

Tasks
The tasks used in this experiment was composed of two separate simulations: 1) a telerobot simulation and 2) a station management simulation. One could consider the whole task as a videogame, with its rules and scores. The aim was to immerse the participants in a simulated scenario: the exploration of an area on Mars, pretending that they were scientists who controlled the entire operation from their workstation. The supposed location of the scientist was a space station in orbit around the planet, with no communication delay between station and planet surface. There were several

differences between some actual telerobot workstations used in space research and this simulation, since no communication delay was considered, and the scientist had more responsibilities (e.g. station monitoring) than is typically the case. However, space telerobotics for scientific exploration is still in its early days, and no standard exists yet. Furthermore, the aim of this apparatus was to provide a challenging as well as general task that allowed generalization of the results. The use of such microworlds is considered a viable compromise between the amount of complexity needed to derive meaningful conclusions and the amount of experimental control needed to assess the impact of target experimental variables with sufficient reliability (Gray, 2002; Lorenz, Di Nocera & Parasuraman, 2003).

Telerobot simulation. The telerobot simulation software was developed in Virtual Reality Modelling Language (VRML 2.0). Actually, VRML is neither virtual reality nor a modelling language. Virtual reality typically implies an immersive 3D experience, and 3D input devices, while VRML does not require immersion, even if it does not preclude it. The scenario provided the participants a picture that was close enough to the true operational environment. The terrain was a reasonable simulation of the Martian soil. The rover structure was very spartan, but it was close enough to provide all the elements needed to operate. Sensors were attached to the joints in order to allow routing from a control panel (a set of sliders) to them. The manipulator and the lower arm moved only vertically, while the upper arm moved vertically, horizontally, and around an imaginary cone. The arm was dexterous enough to allow pick up and release operation, but tricky enough to be still a complex task.

Due to programming constraints it was not possible to detect collision between arm and samples, therefore a "rule" was implemented into the task: the samples were located in coloured boxes, defined as "areas of interest", and to pick up the sample the participants had to immerse the manipulator into such boxes. To provide also a rule for releasing, subjects needed to place the manipulator within a green frame prior to pressing the release button. Users' behaviour was collected using scripts associated to each element of the control panel and as the rover approached a sample. In this way it was possible to get specific data regarding:

1. time needed to approach a specific sample;
2. time needed to collect the sample;
3. number of movements needed for the operation;
4. number of samples collected.

Station management simulation. The station management interface was developed using Microsoft Visual Basic 5.0. It provided a set of indicators and controls that the participants utilized to monitor and control the station environment, as well as part of the rover functions, namely the speed and the camera view. Another important function of this interface was to provide a set of tools for scientist-operator and system-scientist communication. Everything in the simulation used in the present study was scripted; timing for each event was strictly matched for all the participants. Audio warnings, also script-driven, started one minute before the time allowed for the area exploration expired.

Six environmental support indicators were placed in the right part of the interface. Four of them (Atmosphere, Temperature, Humidity, and Pressure) had a red scale with a green central part indicating the normal values interval. Other two indicators were LEDs indicating the amount of energy available and the level of radioactivity in the station environment. Other indicators included a little window showing updates of the operators' position and a strip chart showing the solicitation from the terrain to the rover. The Intercom was a separate window invoked by the Intercom button, which was located in the lower right part of the interface. It provided three sets of mutually exclusive options corresponding to:

1. ten faults that could occur (for the first four indicators too high or too low values, while radioactivity could be only too high, and power only too low);
2. names of the three operators in charge of the different systems;
3. three areas where the operators could be located.

The Automatic Aid System was represented by a window (usually blank) showing messages from the system. These messages were scripted as everything else in the simulation and provided help according to the four LOA used. Response times and subjects' choices were collected during the mission simulation. The LOA used in this experiment were "System Action" (LOA 3), automation took control of the task from the operator, and then sent a notification message to the operator reporting what was done; "Notification + Suggestion" (LOA 2), automation notified events deviating from the normal behaviour of the system, and suggested to the operator a possible solution to the problem; "System Notification" (LOA 1), automation notified events deviating from the normal behaviour of the system; "No Automation" (LOA 0), automation was absent, the human operator carried out the task without help.

Readers should be advised that in this paper the LOA number is unrelated to Sheridan's LOA scale. However they represent a continuum of levels (0 is one step under 1, 1 is one step under 2, and 2 is one step under 3). Using numbers 0 to 3 has been considered useful for computing shifts in terms of distances from -3 (three steps down) to +3 (three steps up). See Table 1 for details.

Procedure
At the beginning of experiment, the Immersive Tendencies Questionnaire (Witmer & Singer, 1994; 1998) was administered. After they completed the questionnaire, participants sat in front of a computer monitor placed in a dark room. They were given a practice session (LOA 3 for all the participants), then randomly assigned to one of 4 experimental conditions corresponding to the four LOAs described above. After the training, they engaged in 3 experimental sessions, each 45 minutes long. Participants of each group performed two tasks in the simulation. The fault detection task consisted of detecting the occurrence of a fault, and the recovery task involved remembering the position in the station of each of the operators, in order to recover from the fault. Each group executed the Detection (of a fault) and Recovery (remembering the recovery procedure) tasks receiving help by the system at the four levels of automation: on 40% of fault occurrences they received help at the LOA characterizing the group, while on the remaining fault occurrences they received help at the other 3 LOA (20% each one). The order of the fault occurrences was

randomized across subjects. LOA distance and direction were defined as the difference between the LOA which characterized the group ($LOA_{[g]}$) and the LOA at which the system help was set on each trial ($LOA_{[t]}$). Hence, distance "0" means that on that trial the $LOA_{[t]}$ and the $LOA_{[g]}$ were identical. Distance "1" means that on that trial the $LOA_{[t]}$ was one level above the $LOA_{[g]}$. Distance "-2" means that on that trial the $LOA_{[t]}$ was two levels below the $LOA_{[g]}$, and so on. The following table shows how distances between LOA were computed (distances are indicated in cells).

Table 1. This table shows how distances between LOA were computed (distances, or automation shifts, are indicated in cells). LOA[g] stands for Group's LOA, whereas LOA[t] stands for Task's LOA

	$LOA_{[t]}=0$	$LOA_{[t]}=1$	$LOA_{[t]}=2$	$LOA_{[t]}=3$
$LOA_{[g]}=0$	0 [no shift]	+1	+2	+3
$LOA_{[g]}=1$	-1	0 [no shift]	+1	+2
$LOA_{[g]}=2$	-2	-1	0 [no shift]	+1
$LOA_{[g]}=3$	-3	-2	-1	0 [no shift]

Two workload conditions were used. In the low workload (single-task) condition subjects only monitored the station functions, whereas in the high workload (dual-task) condition they also controlled the rover. These conditions were introduced in order to test whether changes in task demand would differently affect performance in the different LOA conditions. After the three sessions participants were asked to rate their trust in the automated system using a scale ranging from 1 "I don't trust this system" to 7 "I trust this system". Finally, the Presence Questionnaire (PT) (Witmer & Singer, 1994; 1998) was administered to all participants following the conclusion of the task.

Design

Since a very small number of errors occurred (1.6 mean errors per condition), only response times were analyzed. A comparison between "no-shifts" trials (that is, trials where $LOA_{[g]}$-$LOA_{[t]}$ was 0 (i.e., 0-0, 1-1, and 2-2) was carried out using an ANOVA design $LOA_{[g]}$ (0 vs. 1 vs. 2) by Workload Condition (low vs. high) by Task (detection vs. memory). $LOA_{[g]}=3$ was not taken into consideration here, because subjects did not provide any response at this level.

Analyses on distances were carried out separately for the two tasks using a $LOA_{[g]}$ by Distance (2 vs. 1 vs. 0 vs. -1 vs. -2 vs. -3) by Workload nested design, with Distance($LOA_{[g]}$) as nested factor. Indeed, this unbalanced design allowed for the analysis of distances 0, 1, and 2 (derived from $LOA_{[g]}=0$), distances -1, 0, and 1 (derived from $LOA_{[g]}=1$), distances -2, -1, and 0 (derived from $LOA_{[g]}=2$), and distances -3, -2, -1 (derived from $LOA_{[g]}=3$).

Additionally, one-way ANOVAs by $LOA_{[g]}$ were performed on 1) a composite index for telerobotic performance (number of the telerobotic arm movements / number of samples collected), 2) the immersive tendencies and presence scores, and 3) the trust in automation as reported by participants using the scale described above.

Results

The ANOVA LOA$_{[g]}$ by Workload by Task carried out only on distance 0 trials showed no significant differences between response times associated with the three LOA$_{[g]}$ ($F_{2,8}$=.65; NS). A main effect of Task was found ($F_{1,8}$=28.01; p<.01). Mean response times were higher in the memory task than in the detection task. Additionally, a significant Task by Workload interaction was found ($F_{1,8}$=43.46; p<.01). Duncan post-hoc testing showed that, in the detection task, response times associated to the high-workload (HW) condition were significantly higher than those associated to the low-workload (LW), whereas they were significantly lower in the memory task.

No main effects were observed in the analysis of distances for the Detection task. However, a Workload by Distance(LOA$_{[g]}$) interaction was found ($F_{8,36}$=3.48; p<.01). Duncan testing showed that this effect was mainly due to two strong patterns. In the low workload condition, response times associated to distance 0 were significantly faster than the others, whereas, in the high workload condition, response times associated to distance -2 generated from LOA$_{[g]}$=2 were significantly slower than the others (Figure 1). Significant differences were also observed in the LOA$_{[g]}$(Distance) nested factor ($F_{6,78}$=4.77; p<.01) for the Memory task. In this case, a main effect of Workload was also observed ($F_{1,78}$=7.40; p<.01), whereas Distance by Workload interaction showed only a tendency toward statistical significance ($F_{5,78}$=2.32; p=.06). Overall, response times associated with the low workload condition were faster than those associated with the high workload condition. However, patterns were not clear enough to be easily interpreted (Figure 2). No significant differences between LOA groups in teleoperation performance were observed both in terms of task completion time ($F_{3,12}$=1.07; p=.40) and arm movements / samples ratio ($F_{3,12}$=1.70; p=.22). Furthermore, no differences in the trust rate was found ($F_{31,12}$=1.81; p=.20).

LOA groups showed homogeneous immersive tendencies (ITQ total score) and presence (PT total scores). However, a significant difference was found in the "Involvement/Control" subscale of the PT questionnaire ($F_{3,12}$=4.73; p<.05). Duncan testing showed that the lowest level of involvement/control was associated with LOA$_{[g]}$=0.

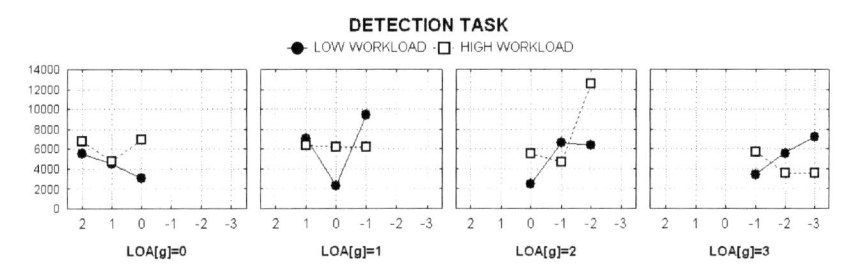

Figure 1. Mean response times (milliseconds) to the detection task at the different distances, by workload conditions and LOA[g].

MEMORY TASK
— LOW WORKLOAD   □ HIGH WORKLOAD

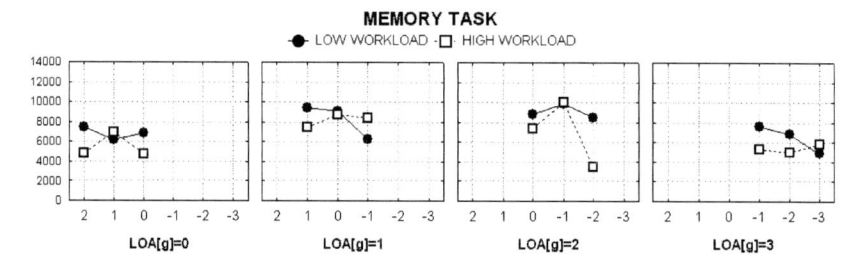

Figure 2. Mean response times (milliseconds) to the working memory task at the different distances, by workload conditions and LOA[g].

Reliability study

In order to rule out potential effects of subjects (outliers and the like), a cross-validation is needed. Resampling procedures (see Di Nocera, 2003 for applications to human factors research) can be used for testing the stability of the results obtained in a study. The Jackknife procedure (Efron & Tibshirani, 1993) was used here. This procedure is quite straightforward and is based on the replication of the analyses on samples extracted from the original one, leaving out one subject per time (*one-leave-out*). Obviously, the stability (reliability) of the results obtained with conventional analyses will be confirmed if the same results are consistently found through the replications. Sixteen Jackknife replications (each one leaving out one different subject from the original sample) were run.

Results
Detection task. LOA and Workload main effects were confirmed to be not significant 16 out of 16 times. Traditional analyses showed only a tendency toward statistical significance for Distance(LOA) main effect and the Jackknife confirmed this result, showing that this effect was statistically significant only in 5 out of 16 replications. Workload x LOA interaction was found to be not significant in 13 out of 16 replications, and Workload x Distance(LOA) was confirmed to be significant 16 out of 16 times.

Working Memory task. LOA and Workload main effects were confirmed to be significant 16 out of 16 times. Traditional analyses showed only a tendency toward statistical significance for Distance(LOA) main effect and the Jackknife confirmed this result, showing that this effect was statistically significant only in 2 out of 16 replications. Workload x LOA interaction was found to be not significant in 15 out of 16 replications, and Workload x Distance(LOA) was confirmed to be significant 16 out of 16 times.

For both tasks, the removal of two particular subjects from the original sample consistently provided opposite results respect to the conventional analyses. Overall, the Jackknife procedure provided support for the robustness of the effects found and confirmed the unreliability of those results showing only a tendency toward statistical significance.

Discussion

Adaptive automation is thought to be a key towards optimizing the benefits of automation for system performance. These systems should adapt to operators' overt and covert behaviour, but changes in the system behaviour could affect operators as well. For example, the direction of the automation shift (toward full manual or full automatic control) as well as the distance between two successive LOA (one, two, or more "jumps" in the hierarchy) could differently affect the performance of an individual. Also, as discussed by Parasuraman et al. (2000), automation can differ in type and complexity. Some forms of automation may simply organize information sources or integrate them. Such forms of "information automation" differ from automation of decision-making functions, in which decision options that best match the incoming information are provided to the user. Automation support at any or all of these stages of processing could be engaged and disengaged by the human operator. In doing so, operators (or adaptive systems) could trigger different shifts in the distance between LOA.

This study investigated this issue using a simulation of a telerobot workstation. The hypothesis was that switching from one LOA to another would affect operators' performance because of the costs associated with the engagement/disengagement process. Of course, these results are not limited to the type of tasks depicted by the simulation used, but can be extended to any other operative environment involving dynamic shifts between levels of automation. The simulation we developed provided the level of complexity needed to accomplish our goal; unfortunately, complexity in the system was also accompanied by complex patterns in the data. For example, the small occurrence of changes in the LOA likely led to a small number of errors, confining the analyses to the response times. Moreover, response times were sometimes extremely high and thus excluded from the analyses. Nevertheless, it was possible to gain important information from these data.

Indeed, the first prediction was supported so far. Groups' uniformity with respect to the no-switch condition (namely, the distance 0 trials) cannot be excluded in both the detection and memory tasks. This is a rather important result, considering that this paradigm was basically unexplored. Results showed that, even if the four LOAs (manual, notification, suggestion, action) were intrinsically different and led to differences in performance, they did not lead to any cost when involved in the no-switch condition. In other words, subjects did set their "level of performance" according to the level of automation. Costs were observed when they interacted with the system acting at another LOA, but no costs were observed when they did not switch to another LOA. As expected, mean response times were higher in the memory task than in the detection task, but the interaction between type of task and workload condition showed a puzzling pattern. Indeed, workload affected performance to distance 0 trials in the detection task, but a better performance was associated to the high workload condition in the memory task. This effect could be explained by the specific nature of the memory task. The memory task was continuous (subjects always need to keep in memory the information provided by the system) and might have gained salience. In the high workload condition subject could adopt a strategy aimed at focusing on the memory task.

The second prediction was only partially supported. In fact, when shifting from one LOA to another, it was possible to observe very peculiar patterns for distance and direction, even if they were limited to some distances. However, within and between subjects variability probably affected the results: quite often, patterns were not clear enough to be easily interpreted. Even if the Jackknife procedure provided support for the robustness of the effects found, it also showed that at least two subjects were a source of unreliability.

The two workload conditions led to two opposite patterns in the detection task (Figure 2). In the low workload condition the best performance was associated to the no shift condition (distance 0), whereas extreme jumps led to a poorer performance under the high workload condition, independently of direction. More specifically, the worst performance was associated to the distance -2 that was generated by $LOA[g]=2$, thus suggesting that the shift from a condition in which the automated system suggest a possible solution to the condition of full manual control, is the most critical. Conversely, the memory task clearly imposed a different type of load, leading to a completely different pattern in automation shifts (Figure 3). Paradoxical effects were observed in this case, such as a better performance associated with distance -2, and an overall better performance in the high workload condition. Again, this result may be due to the very nature of the task we used. The detection of a fault may be considered as a process easy to re-engage when the automation fails. On the contrary, the working memory task has to be carried out throughout the experimental session, because subjects always needed to remember the position of the operator.

Rovira et al. (2002), who also found that unreliable automation has greater costs when decision and action selection are into play, suggested that -when this is the case- operators may not longer routinely develop or explore novel alternatives apart from those provided by the automation. Overall, the commonsense consideration that only shifts toward a lower level of automation should reflect poor performance is unsupported. Upward shifts may affect performance as well, particularly when they follow a "no shift" condition and when workload is moderate. This costs best describes the disengagement problem that the present study wanted to investigate. Unfortunately, no differences were observed for teleoperation performance, probably because of the high variability of subjects' performance. Finally, the more common LOA experienced by the participants (that is the group they belonged to) weakly affected their sense of presence. Indeed, the presence questionnaire failed to show any differences between people interacting with the simulation at different levels of automation, except for the Involvement/Control dimension whose lowest values were associated with the manual condition (Figure 4). This is not surprising, because the continuous switching from teleoperation ("remote" site) to station monitoring ("actual" site) requested at this level of automation does not allow the operator to feel like "being there". Tasks imposing disproportionate amount of load may negatively affect the operator sense of presence, whereas moderately challenging tasks seems to be best suited for improving it. However, it should be noted that desktop virtual reality, like the one we used in our study, has different properties from "immersive" virtual reality studied by Witmer and Singer (1994; 1998) for developing their questionnaire.

Overall, these findings suggest that, when individuals perform a task, their cognitive systems set to a particular level (represented by the activation of a particular set of behaviours) and no costs are observed until the level (or rule) remains the same. Actually, under some circumstances, no shifts can even lead to a better performance.

Conclusions

Teleoperators and astronauts working together pose problems (Sheridan, 1993), and there is a strong need for research in this domain. The human control of remotely operated vehicles or robotic arms at different levels of automation has been recently approached by other authors (Kaber, Onal, & Endsley, 2000; Ruff, Narayanan, & Draper, 2002), but it surprising that only few studies are available on this issue.

The present study was carried out to verify whether distance and direction in LOA shifts may affect human performance when interacting with complex tasks. Results showed that specific costs are associated with the process of disengaging from one cognitive-behavioural set the operator is currently using to the engagement of another -more appropriate- set. Such effect is not only associated with variations in the difficulty of the task, and it is affected by the mental workload the operator is experiencing on the moment. Recent research results (Trafton et al., 2003) suggest that preparation may have an important role in resuming a task previously carried out, and we wonder if "preparation lag" may have a role also in adjusting to the next level of automation. It is worth noting that, albeit the experiment reported here was aimed at studying LOA shifts effects for future development of adaptive automation, a main difference exist between our experimental setup and actual adaptive systems. Indeed, in adaptive systems LOA shifts would happen according to some modification in human physiology and/or behaviour, whereas in this study they have been programmed by the experimenters. That should be taken into consideration for interpreting the results. Nevertheless, the outcome of this study may be of interest for understanding what type of response we may expect from operators when LOA shifts are inconsistent or partially unrelated to the operator functional state.

Further investigation of these phenomena is needed. Particularly, we consider the development of a model including these findings as highly desirable. As reported above, mental workload was also a main issue in the present study, and results showed different patterns according to the workload level, suggesting the need to include this factor in any future model. Results from this pilot study are far from being conclusive, but they seem to be quite relevant, and may be of use for designing adaptive systems that provide cues to re-engage the task when LOA shifts occur.

Acknowledgments

This research was supported by NASA Grant NAG 5-8761 from Goddard Space Flight Center, Greenbelt, MD.

References

Crocoll, W.M., & Coury, B.G. (1990). Status or recommendation: Selecting the type of information for decision aiding. *Proceedings of the Human Factors Society 34th Annual Meeting* (pp.1524-1528). Santa Monica, CA, USA: HFES.

Di Nocera, F. (2003). On Reliability and Stability of Psychophysiological Indicators for Assessing Operator Functional States. In G.R.J. Hockey, A.W.K. Gaillard, and O. Burov (Eds.), *Operator Functional State: The Assessment and Prediction of Human Performance Degradation in Complex Tasks* (pp. 162-173). Amsterdam: IOS Press.

Efron, B., & Tibshirani, R. (1993). *An Introduction to the Bootstrap*. London: Chapman and Hall.

Endsley, M.R., & Kaber, D.B. (1999). Level of automation effects on performance, situation awareness and workload in a dynamic control task. *Ergonomics, 42*, 462-492.

Endsley, M.R., & Kiris, E.O. (1995). The out-of-the-loop performance problem and level of control. *Human Factors, 37*, 381-394.

Gray, W.D. (2002). Simulated Task Environments: The Role of High-Fidelity Simulations, Scaled Worlds, Synthetic Environments, and Laboratory Tasks in Basic and Applied Cognitive Research. *Cognitive Science Quarterly, 2*, 205-227.

Kaber, D.B., Onal, E., & Endsley, M.R. (1998). Level of automation effects on telerobot performance and human operator situation awareness and subjective workload. In M. Scerbo and M. Mouloua (Eds.), *Automation Technology and Human Performance: Current Research and Trends* (pp. 165-170). Mahwah: Lawrence Erlbaum Associates.

Kaber, D.B., Onal, E., & Endsley, M.R. (2000). Design of Automation for Telerobots and the Effect on Performance, Operator Situation Awareness, and Subjective Workload. *Human Factors and Ergonomics in Manufacturing, 10*, 409–430.

Lorenz, B., Di Nocera, F., Röttger, S., & Parasuraman, R. (2002). Automated fault-management during simulated space flight. *Aviation, Space, and Environmental Medicine, 73*, 886-897.

Lorenz, B., Di Nocera, F., & Parasuraman, R. (2004). Examination of the proximity-compatibility principle in the design of displays in support of fault management in automated systems. In D. de Waard, K. Brookhuis, and C. Weikert (Eds.) *Human Factors in Design* (pp. 213-229). Maastricht, The Netherlands: Shaker Publishing.

Moray, N., Inagaki, T., & Itoh, M., 2000. Adaptive automation, trust, and self-confidence in fault management of time-critical tasks. *Journal of Experimental Psychology: Applied, 6*, 44-58.

Lorenz, B., Di Nocera, F., & Parasuraman, R. (2003). Cognitive performance assessment in a complex space-system microworld: on the use of generalizability theory. *Proceedings of the 12th International Symposium on Aviation Psychology*. April 14-17, Dayton (OH).

Parasuraman, R., Mouloua, M., & Molloy, R. (1996). Effects of adaptive task allocation on monitoring of automated systems. *Human Factors, 38*, 665-679.

Parasuraman, R., Sheridan, T. B., & Wickens, C.D. (2000). A model for types and levels of human interaction with automation. *IEEE Transactions on Systems, Man, and Cybernetics-Part A: Systems and Humans, 30*, 286-297.

Rovira, E., McGarry, K., & Parasuraman, R. (2002). Effects of unreliable automation on decision making in command and control. *Proceedings of the Annual Meeting of the Human Factors Society*, (pp. 428-432).). Santa Monica, CA, USA: HFES.

Ruff, H.A., Narayanan, S., & Draper, M.H. (2002). Human Interaction with Levels of Automation and Decision-Aid Fidelity in the Supervisory Control of Multiple Simulated Unmanned Air Vehicles. *Presence: Teleoperators & Virtual Environments, 11*, 335-351.

Sarter N.B., & Schroeder, B. (2001). Supporting decision making and action selection under time pressure and uncertainty: the case of in-flight icing. *Human Factors, 43*, 573-583.

Scerbo, M.S., Freeman, F.G., Mikulka, P.J., Parasuraman, R., Di Nocera, F., & Prinzel, L.J. (2001). The efficacy of psychophysiological measures for implementing adaptive technology. *Technical Paper NASA/TP-2001-211018*. Hampton: NASA Langley Research Center.

Sheridan, T. B. (2002). *Humans and Automation*. New York: Wiley.

Trafton, J.G., Altmann, E.M., Brock, D.P., & Mintz, F.E. (2003). Preparing to resume an interrupted task: effects of prospective goal encoding and retrospective rehearsal. *International Journal of Human-Computer Studies, 58*, 583-603.

Witmer, B. & Singer, M. (1994). *Measuring immersion in virtual environments* (Tech. Rep. 1014) Alexandria: U.S. Army Research Institute for the Behavioral and Social Sciences.

Witmer, B. & Singer, M. (1998). Measuring presence in virtual environments: A Presence Questionnaire. *Presence: Teleoperators and Virtual Environments, 7*, 225-240.

The Ergonomics of Attention Responsive Technology

Alastair G. Gale, Kevin Purdy, & David Wooding
Applied Vision Research Centre, ESRI
University of Loughborough, Loughborough, UK

Abstract

ART (Attention-Responsive Technology) is a new three year UK research project which will enable individuals to access technology efficiently in situations where their mobility is either impaired, as a result of disability or age, or because movement is undesirable due to environmental hazards. The system works by monitoring both the individual and the ICT (Information and Communication Technologies) devices (termed here 'objects') in his/her environment and then uses knowledge of the individual's gaze behaviour to determine to which ICT device they are attending. This information is relayed to a user-configurable control panel, which then displays as a graphical user interface (GUI) only those controls that are appropriate, both to the user and to the particular object in question. The user can then choose to operate the object. ART therefore acts as an enabling technology, with the system fully user configurable and able to cater for future developments in technology.

Background

Numerous disabilities seriously restrict mobility but leave saccadic eye movement control intact and so these movements can be used as a communication or a control aid. Systems already exist which afford physically impaired individuals the ability to interact with a computer or other devices using their saccadic eye movements. For instance an individual can 'type' by looking at keys on an on-screen keyboard, or move the computer cursor according to their point of gaze.

Unfortunately, these systems suffer from the so-called 'Midas touch' problem (Jacob, 1990). That is, a system which uses the user's point of gaze to activate controls *directly* is prone to false alarms as the point of gaze is unconsciously drawn to objects which the visual system finds 'interesting'. As a result, users of such systems generally find that such direct eye movement control can be unreliable and fatiguing, which renders the long-term utility of such systems less than ideal.

In order to control numerous objects in the environment typically a complex menu system of some kind is required which encompass all of the potential objects together with their various levels of control. This can either end up as a physically large menu selection display or else a deep menu structure. Neither is ergonomically acceptable.

In D. de Waard, K.A. Brookhuis, R. van Egmond, and Th. Boersema (Eds.) (2005), *Human Factors in Design, Safety, and Management* (pp. 377 - 380). Maastricht, the Netherlands: Shaker Publishing.

The ART system

This project overcomes the problems associated with eye-movement control of ICT devices by using eye movements to select an object for control; not necessarily to operate the object per se. In everyday tasks our point of gaze generally precedes action, i.e. we look towards an object before we reach for it (Land and Hayhoe, 2001). In this way, saccadic eye movements can therefore be used as an indicator of the intention of the individual when interacting with his/her environment (Vertegaal, 2002).

In the initial formulation, actual control of a device will be by a GUI using a touch sensitive tablet PC – although this can be replaced with numerous other potential interfaces/controls to suit a particular user. The project is concerned with the appropriate selection of objects in an environment and less with their actual operation, thus the GUI itself is not a major issue. The interface is configurable to the individual in question, ranging from a fully featured panel to a simple two-switch set up depending upon the individual's physical and cognitive abilities. Where necessary the panel could be replaced by a head- or breath-operated switch.

It is envisioned that a user, potentially in a wheelchair, is located within an environment that contains numerous ICT devices or objects. In a domestic setting such objects will include: TV, media centre, lights, curtains, air conditioning, door entry system, computer, kitchen appliances etc. Some of these objects will need to be controlled by the user and the objective of the project is to determine accurately the selection of these objects by the user through monitoring their eye gaze vector. The difference between this approach and previous eye movement control systems is that the user's point of gaze does not actually control the object, instead it pre-selects objects for operation. By presenting the user with a set of controls specific to the object, the system in effect then sets the chosen object into a state of readiness to receive a command.

Implementation

A suitable environment, which encompasses both domestic and office objects, is currently being constructed in a dedicated laboratory. The objects which the user will control are fitted, where necessary, with appropriate control systems (e.g. electrically operated curtains). Digital cameras, in fixed positions in the environment, are used to construct a computer vision system which provides a 3D environmental model. This determines the location of the ICT objects, together with the location of the user. As the user moves about in the environment, or as objects are moved, then their respective locations are monitored in real time.

Initially a head mounted eye movement system (ASL 501) is being implemented to acquire the participant's eye gaze relative to their head position. Head direction relative to the environment will be obtained via a 3D 'flock of birds' monitoring system. From these two measures the eye gaze vector relative to the environment is obtained. The system then calculates whether the user's gaze falls upon an ICT object and if so then an appropriate GUI for that object will be triggered on the tablet

PC. This obviates the need for a complex GUI to present the user with every available ICT device. Subsequent developments will replace the head mounted eye movement system with a remote eye monitoring system (Smarteye).

Discussion

The ART system provides the user with the appropriate controls for every device they might require, whilst maintaining a simple and usable control interface. It removes the need for a complex hierarchical user menu control system to handle a wide range of ICT devices. The ART system overcomes any possibility of the false operation of devices resulting from the user looking at the object 'inadvertently'. It is the user's decision whether to interact with the control panel; the control panel relies on the user to physically activate an appropriate control.

Acknowledgements

This research is supported by the ESRC PACCIT (People At the Centre of Communication and Information Technology) Programme.

References

Jacob, R.J. (1990). What You Look at is What You Get: Eye Movement-Based Interaction Techniques. In *Human Factors in Computing Systems: CHI '90 Conference Proceedings* (pp. 11–18), ACM Press.

Land, M.F. & Hayhoe, M.M. (2001). In what ways do eye movements contribute to everyday activities? *Vision Research, 41*, 3559-3565

Vertegaal, R. (2002). Designing Attentive Interfaces. In Proceedings of ACM ETRA Symposium on Eye Tracking Research and Applications 2002. New Orleans: ACM Press.

ACC effects on driving speed – a second look

Nina Dragutinovic[1,2], Karel A. Brookhuis[1,3],
Marjan P. Hagenzieker[2], & Vincent A.W.J. Marchau[1]
[1]Faculty of Technology, Policy and Management
Delft University of Technology, Delft
[2]SWOV Institute for Road Safety Research, Leidschendam
[3]Department of Psychology, University of Groningen
The Netherlands

Abstract

Advanced cruise control (ACC) systems provide assistance to the driver by automating parts of the longitudinal driving task: operational control of speed and headway. In the past decade a series of studies with respect to effects of several ACC aspects were reported. In order to integrate different results of a few studies on the ACC effects on driving speed, a meta-analysis of seven simulator studies has been performed. Although in the meta-analysis the obtained average effect on speed was near zero ($ES_{um}= 0.0956$), further analysis showed that studies clearly clustered in two opposite groups: a group that demonstrated an ACC effect of increase in the mean speed ($ES_{um\ positive} = 2.5$) and a group that demonstrated an ACC effect of decrease in the mean speed ($ES_{um\ negative} = -2.3$). The predominant difference between these two groups of studies was in the type of the ACC system they used: ACCs that support drivers on a higher level (allocate more tasks to the system in comparison to other types of ACC) and/or that support drivers within a wider speed range, have the effect of an increase in the mean driving speed.

Introduction

Advanced cruise control systems provide assistance to the driver by automating parts of the longitudinal driving task: operational control of speed and headway (cf. Hoedemaeker, 1999). In the past decade a considerable number of studies with respect to effects of ACC aspects were reported (e.g. Hoedemaeker & Brookhuis, 1999, Hoedemaeker, 1999) It is expected that general implementation of ACC would lead to the choice of "correct" speed and safer headway as well as less variation in speed and headway. Thus, safety would be enhanced, however, sufficient evidence for this projection is still lacking. Different studies show different results about the effects of ACC on driving behaviour. Therefore, a general conclusion as to what extent ACC will improve driving behaviour and traffic safety is difficult.

An approach that would help to reduce the confusion as a result of these different insights is meta-analysis of the respective outcomes of these studies. Meta-analysis involves a (statistical) procedure that integrates the results of several independent studies which are considered to be comparable (Egger, Smith & Phillips, 1997). Therefore, in the present study a meta-analysis approach has been used to integrate the different results on the effects of ACC on driving behaviour. To enhance comparability, it was decided to restrict the survey to driving simulator studies.

Method

To assure that all relevant studies are included in the analysis, various scientific literature sources were consulted. The search for relevant studies was conducted by using the International Transport Research Documentation (ITRD), the ISI web of knowledge databases, the library of the Dutch SWOV Institute for Road Safety Research, tables of contents of relevant journals, the reference lists from relevant studies, and comprehensive science-specific internet search engines (e.g. www.scirus.com). Using a suitable set of search key words produced seven relevant publications comprising nine separate studies. The studies differed in many ways, among others, by the type and sophistication of simulator, and the specifications of the ACC system.

Although significant differences were found between the operational characteristics of the ACC systems used in the nine studies, the number of studies was too small to enable the analysis for each system separately. Therefore, it was decided that the main characteristic of the used ACC systems, "support driver in controlling vehicle speed and headway", was comparable enough across the systems to allow all seven studies to be included in the analysis.

Determining the ACC effect - choice of dependent variables

Looking at the selected studies it appeared that the number and type of driving behaviour variables varied considerably between different studies. The meta-analysis did not allow all these variables to be included. Therefore, the effect of ACC on driving speed was chosen for the analysis, based on the following ground:

1. ACC's direct and foremost support to the driver is in the control of speed
2. Speed is considered one of the most contributing factors in traffic unsafety (Aarts, 2004)
3. Meta-analysis requires a sufficient amount of measurements, while speed turned out to be the one consistently used driving behaviour indicator in the present selection of studies

The choice of the specific speed indicator as the ACC effect for the meta-analysis among all different speed indicators (mean driving speed, maximum speed, percentage of time driving over the speed limit, etc.) fell on the "average driving speed". Hence, the focus of the meta-analysis is on the effect ACCs have on the average driving speed within trials or blocks of trials when driving was supported (i.e. driving speed in ACC condition) as compared to when driving was not supported (i.e. control or no-ACC condition).

The selection and computation of effect size

The selection of the effect size is one of the most important choices in meta-analysis. The type of the analysed research plays an important role in the choice of effect size but this choice is also influenced by the statistical data that are available in analysed studies (McGauw & Glass, 1980). One of the biggest problems when coding for meta-analysis is that very often researchers fail to report even results such as the mean, standard deviation or sample size per condition. The consequence of not reporting basic statistical data is that effect sizes must be estimated from incomplete information. Sometimes, studies cannot be included in the analysis because of the lack of adequate statistical information, although they fulfil the inclusion criteria in general.

The problem of incomplete data was also encountered in the present meta-analysis. A few studies did not report some of the basic statistical data at all or included two- or multi-factorial designs, so that effects of ACC on average driving speed have been reported by levels of another factor (e.g. type of the road, driving style of participants, traffic density, etc.). Occasionally, researchers did not report actual data but merely indicated that results (in this case speed difference between ACC and no-ACC condition) were found to be non-significant, leading to a choice between two (bad) options:

1. To set the effect size in this case at zero
2. To exclude these findings from meta-analysis

The first option seemed preferable, the drawback of this conservative approach being that it generally underestimates true effect size because, most likely, effect sizes for these results are not exactly zero. The problem of missing data has been solved in some cases by estimating additional results based on graphics (although this is not always the most precise way to obtain required information) or by consulting additional sources about the same experiment.

Finally, the "unstandardized" mean difference (ES_{um}) was chosen as the effect size to be used in this analysis. In order to be able to use this effect size, the following conditions have to be fulfilled:

1. The same operationalisation of the variable of interest has to be used in all research findings
2. The variable of interest has to be continuous

Both these two conditions were satisfied because all studies included did measure speed on the same scale and in the same resolution (1 km/h), using roughly the same measurement procedures. The unstandardized mean differences were calculated according the following formula:

$$ES_{um} = \overline{X}_{G1} - \overline{X}_{G2} \text{ (Lipsey \& Wilson, 2001), i.e. } ES_{um} = \overline{X}_{ACC} - \overline{X}_{noACC}$$

Because most of the studies did not report any statistics that would allow to correct obtained individual effects sizes for unreliability (e.g. measurement error) and because obtained effects sizes were based upon similar sample sizes and all studies used the same methodological approach (i.e. all were performed in a driving simulator), all experiments have been given equal weight.

Results

The magnitude of effect size is represented by the difference in average speed between the driving-with-ACC and driving-without-ACC condition. The calculated mean speed differences for each study are presented in Table 1 in the column "ES_{um}". The computed effects show both positive and negative signs. Effects with a positive signs refer to an increase in average speed when driving with ACC, as compared to driving without ACC. A negative sign refers to a decrease in average speed when driving with ACC.

Table 1. Unstandardized mean differences in speed between ACC and no-ACC condition

	Study	Sample size	Type of ACC	ES_{um} (difference in ACC and no- ACC mean speed)	Workload in ACC as compared to no-ACC condition
1	Hogema, van der Horst, Janssen, 1994.	12	ICC	-3.85	-
2	Hogema & Janssen, 1996.	12	ICC	-3.68	-
3a	Nilsson & Nåbo, 1996. (exp 1)	20	Info ICC	0.90	Lower
3b	Nilsson & Nåbo, 1996. (exp 2)	20	Automatic ICC	-1.50	Lower
4	Stanton, Young, McCaulder, 1997.	12		0.00	Better secondary task performance
5a	Hoedemaeker, 1999. (exp 1)	38	ACC & complete stop	8.00	Lower
5b	Hoedemaeker, 1999. (exp 2)	30	ACC	-0.11	Lower
6	Brook-Carter, Parkes, Burns, Kersloot, 2002.	32	ACC stop and go	1.00	Lower
7	Tornros, Nilsson, Ostlund, Kircher, 2002.	24	ACC	0.10	Lower
	Total: 200			average : ES_{um}= 0.0956	

The mean effect size is almost nil, i.e. ES_{um} = 0.0956. Because the computed effect sizes per study mean actual difference in speed, the mean effect size of 0.0956 actually means that overall ACC increased the mean driving speed with 0.0956 km/h. From a road safety point of view, an increase in average driving speed of less than 0.1 km/h is considered negligible. Hence, it could be concluded that ACC has no effect on average driving speed. However, given the sensitivity of a mean to the sample size and presence of outliers (both characteristics highly challenging for this analysis) required a more detailed analysis of this odd mean effect size. Looking in detail at the distribution of the nine individual effect sizes, then it is obvious that studies cluster in two groups: a group with negative and a group with positive effect sizes. The mean effect sizes for each of these groups are $ES_{um\ positive}$= 2.5 and $ES_{um\ negative}$= -2.3. Not surprisingly, the two means differ mainly in direction.

Discussion

In all three experiments from the studies that resulted in positive effect sizes (Hoedemaeker: 5_a, Nilsson & Nåbo: 3_b and Brook-Carter et al.), a more assisting application than just common ACC has been used. The ACC used in the first Hoedemaeker's experiment was ACC & Stop-and-Go, capable of stopping in every possible situation; Brook-Carter et al. also used a kind of "Stop-and-Go ACC". The Nilsson & Nåbo study compared automatic with only informative ICC. It seems that those ACCs that support the driver on a higher level (allocate more tasks to the system in comparison to other types of ACC) and/or that support drivers within a wider speed range than a common ACC, show an effect of increase in the mean driving speed. The bare types of speed control systems, common ACC and ICC were accompanied with no or negative effects on average speed.

The enhanced support level of the more sophisticated ACCs as compared to the common ACCs, raises the question whether a difference in workload effects was reported in these two groups of studies. The hypothesis could well be that sophisticated ACCs would be associated with lower levels of workload when compared to the common types of ACC. The ACC effects on workload as measured in the various studies are presented in Table 1. in the column "Workload in ACC condition". Unfortunately, the hypothesis about the lower workload for the atypical group of ACC could not be tested. Although a variable of workload is usually investigated /reported in the included studies (only two studies, Hogema et al. 1994. and Hogema & Janssen, 1996. did not report any result about workload) and although the studies in generally use the same kind of scales (i.e. NASA-TLX or RSME) it was not possible to test this hypothesis because the reported workload data are not precise enough to enable testing the differences. Therefore, no effect sizes for workload were computed. Only a descriptive effect of ACC on workload in the form: "workload lower in ACC condition" or "workload higher in ACC condition" was reported. Available data show that regarding ACC and workload, results across the studies are in every case in the same direction, i.e. driving with ACC was associated with lower workload than driving without ACC.

Conclusions

The effect of ACC on mean driving speed seems to be dependent on type of ACC used in a study. When driving with ACC types that take over more of the driving task and offer more support to drivers in more critical situations (e.g. stop and go function, capabilities of complete stop every situation), drivers seem to adapt their behaviour, i.e. increase their mean driving speed. Driving with ACC is experienced as less demanding than unsupported driving and this is the unanimous conclusion of all analysed studies. However, these results have to be taken as suggestive because they are based on statistical data whose quality differed across the analysed studies. Nevertheless, the results of this analysis will serve as starting hypotheses for further experimental research on the adaptation effects of ACC with respect to driving behaviour.

References

Aarts L.T. (2004). Snelheid, spreading in snelheid en de kans op verkeersongevallen. Report R-2004-9. , Leidschendam, The Netherlands: SWOV.

Brook-Carter, N., Parkes, A.M., Burns, P., & Kersloot, T. (2002). An investigation of the effect of an urban adaptive cruise control (ACC) system on driving performance. In: *Proceedings of the 9th World Congress on Intelligent Transport Systems ITS, 14-17 October 2002, Chicago. ITS America, Washington, D.C.*

Egger, M., Smith G.D., & Phillips A.N. (1997). Meta-analysis- principles and procedures. *British Medical Journal, 315,* 1533-1537.

Hoedemaeker, M., & Brookhuis, K.A. (1999). Driving with an adaptive cruise control (ACC). *Transportation Research, Part F, 3,* 95-106.

Hoedemaeker, M. (1999). *Driving with intelligent vehicles; Driving behaviour with Adaptive Cruise Control and the acceptance by individual drivers.* TRAIL thesis series nr 99/6, Delft University Press, Delft

Hogema J.H., Van der Horst, A.R.A., & Janssen, W.H. (1994). A simulator evaluation of different forms of intelligent cruise control, TNO report TNO-TM 1994 C-30, TNO Human Factors, Soesterberg, The Netherlands

Hogema, J.H. & Janssen, W.H. (1996). Effects of intelligent cruise control on driving behaviour: a simulator study, TNO report TM-96-C012, TNO Human Factors, Soesterberg, The Netherlands

Lipsey W. M. & Wilson B. D. (2001). *Practical Meta-analysis.* London: Sage.

McGauw B.& Glass G.V. (1980). Choice of Metric for Effects Size in meta-analysis. *American Educational research Journal, 17,* 325-337

Nilsson, L. & Nåbo, A. (1996). Evaluation of application 3: intelligent cruise control simulator experiment: effects of different levels of automation on driver behaviour, workload and attitudes. Reprint of Chapter 5 in 'Evaluation of Results, Deliverable No. 10, DRIVE II Project V2006 (EMMIS)' Linköping, Sweden: Swedish National Road and Transport Research Institute VTI.

Stanton, N. A., Young, M., & McCaulder, B. (1997). Drive-by-wire: the case of driver workload and reclaiming control with Adaptive Cruise Control, *Safety Science, 27,* 149-159

Tornros, J., Nilsson , L., Ostlund, J., & Kircher, A. (2002). Effects of ACC on driver behaviour, workload and acceptance in relation to minimum time headway. In: *Proceedings of the 9th World Congress on Intelligent Transport Systems ITS, 14-17 October 2002, Chicago.*

Ergonomics in surgery

Martine A. van Veelen & Richard H.M. Goossens
Faculty of Industrial Design Engineering
Delft University of Technology
Delft, The Netherlands

Abstract

To improve patient safety during laparoscopic surgery a three fold approach of the problem is conducted. Firstly, by improving the physical ergonomics of the team, secondly by improving the informational ergonomics during the operation, and thirdly by take a human factors look at the procedures. In this paper the first part improving physical ergonomics is discussed. One of the main and basic ergonomic problems during laparoscopy is the surgeon's non-neutral posture during laparoscopy. There are five main issues that influence the posture of the surgeon: the (hand-held) instrument design, the position of the monitor, the use of foot pedals to control diathermy, the poorly adjusted operating table height, and the static body posture. This paper shows an overview of the ergonomic guidelines that were developed on these five areas and shows product solutions that were developed according to these guidelines. The guidelines can be used by operating room (OR) staff to evaluate the ergonomics of their OR environment and to improve issues that do not satisfy the ergonomic guidelines. When designers use these guidelines to design new OR equipment, the new designs are an improvement in the field of human factors compared to the currently used laparoscopic products. When all these products are applied in the laparoscopic operating room, a new and ergonomic environment is created for the surgeon as well as for the assistants.

Introduction

In the past decade many studies have reported on the physical discomfort during and after the use of laparoscopic instruments (Berguer, 1997, 1998b, 1999a, Matern & Waller, 1999, van Veelen, 1999). Complaints such as pain or numbness in the neck and/or upper extremities and disturbed eye-hand coordination have been mentioned. These complaints are mainly ascribed to the non-optimal ergonomic environment of the laparoscopic surgery room (Berguer, 1998b, 1999a, Matern & Waller, 1999). Examples of ergonomic problems are the neck torsion imposed on the surgeon by an inadequate position of the monitor and extreme upper limb joint positions caused by a decrease in the measures of freedom and insufficient adjustability of the operation table (Berguer, 1998a, Matern & Waller, 1999). Since the publication of the National Academy of Sciences report 'To err is human: building a safer health system' by Kohn et al. (1999) the application of human factor analyses to prevent medical error is well accepted. Kohn et al. (1999) concluded that in the United States of America 98.000 people die annually from medical errors. To improve patient safety during laparoscopic surgery a three fold approach of the problem is conducted. Firstly, by improving the physical ergonomics of the team, secondly by improving the informational ergonomics during the operation, and thirdly by take a human factors look at the procedures. In this paper the first part, that on improving physical ergonomics, is discussed.

One of the main and basic ergonomic problems during laparoscopy is the surgeon's non-neutral posture during laparoscopy, which lead to complaints of fatigue and discomfort by surgeons (Berguer, 1998a, 1999a, Berguer et al., 1998b, 1999b, Hanna et al., 2001, Joice et al., 1998, Matern & Waller, 1999). Four main issues are associated with the non-neutral posture of the surgeon: the (hand-held) instrument design, the position of the monitor, the use of foot pedals to control diathermy, and the poorly adjusted operating table height. The fifth related issue is the limited natural changes in body posture. During laparoscopy surgeons exhibit a more static posture, which can lead to physical fatigue during surgery (Berguer et al. 1997). At the Delft University of Technology, faculty Industrial Design Engineering, these five ergonomic problem areas where studied and ergonomic guidelines were developed for the design of products in these five areas (van Veelen et al., 2001, 2002ab, 2003ab).

The amount of problems that surgeons suffer from, which is described in the literature(van Veelen & Meijer, 1999, Berguer, 1999a), indicates that the current products used by surgeons do not satisfy these new criteria (Berguer et al., 1998a, Van Veelen et al. 2003c). Only the use of currently available flat-screen monitors has proven to positively influence the posture of the surgeon and the assistant (van Veelen et al., 2002b). Therefore, new products that do satisfy the ergonomic guidelines were developed at the Faculty of Industrial Design Engineering. Papers that were recently published in the literature show the isolated results of the five ergonomic items related to the posture of surgeons and assistants (van Veelen et al., 2001, 2002ab, 2003ac). If a hospital wants to have a full benefit of ergonomic adjustments in the operating room it is important that all of these items are implemented simultaneously. For the communication to operating room staff it is important that they realise that the problem of their non-optimal posture is not solved by the implementation of only one single product solution, e.g. the implementation of flat screen monitors to realise a better monitor position.

The aim of the current article is to give an overview of the guidelines that were developed on the areas of hand-held instruments, monitor placement, foot pedals, operating table height and static body postures to improve the ergonomics of the laparoscopic operating room. It also gives an overview of the products that were developed according to these guidelines. These guidelines can be used by operating staff to evaluate the usability of their (new) OR set-up, and by designers to create new laparoscopic products.

Ergonomic guidelines that support an ergonomic posture of the laparoscopic surgeon

The development of new ergonomic guidelines and new ergonomic products was and is still done in close collaboration with medical doctors and their hospitals. In this case the collaboration mostly took place together with the medical institutes Erasmus Medical Centre (Rotterdam) and the Catharina Hospital (Eindhoven), both in The Netherlands. During the design process the design approach of Participatory Design (PD) was used. This is a design method which involves the user group throughout the whole design process. The method has shown significant achievements in collaboratively shaping technology and social environments and increases the chances that a design corresponds to real needs and will be used as intended (Muller & Kuhn, 1993).

A. Hand-held instruments

The currently used hand-held instruments cause ergonomic problems such as extreme wrist excursions during manipulation and pressure peaks on both the fingers and palm of the hand. In the literature general human factors guidelines for the design of handles can be found, and they are useful, but some of these criteria are not operational, i.e. concrete with respect to certain functions. Another problem is that instrument interfaces (e.g. handles) are designed for multi-functionality. This results in products that are not specifically designed for their main function and therefore increase the chance on ergonomic problems. Therefore, new guidelines were developed for the different functionalities of hand-held instruments. Table 1A shows guidelines for the interfaces of laparoscopic dissection forceps, grasping forceps, needle holders, and cameras.

B. Position of the monitor

It is common knowledge that the position of the monitor during laparoscopy influences the posture of the surgeon and the assistant. Former studies showed that the optimal monitor position is in front of the surgeon, on a height between the surgeons head and hands, in such a way that his head flexions are between 15-45° to the horizontal. This posture prevents fatigue and muscolo-skeletal disorders and shows an optimal task performance. It therefore is recommended as a guideline for the position of the monitor for endoscopic tasks (van Veelen et al., 2002b). Table 1B shows this guideline for the position of the monitor.

C. Foot pedals

In laparoscopy, diathermy and ultrasonic equipment is operated by means of one or more foot pedals positioned on the floor in front of the surgeon. A foot pedal is equipped with two identical switches, activated by pushing them as required. The surgeon has no direct vision on the foot pedals because they are on the floor, under the operating table and covered by the sterile sheets. This increases the risk of hitting the wrong switch, which can be dangerous (e.g. cutting instead of coagulation) (van Veelen et al., 2003b). During control of the foot pedal the surgeon's posture is

Table 1. Ergonomic Guidelines for laparoscopic products

A			
	Laparoscopic dissection forceps	1	The angle between handle and shaft must be between 14° and 50°.
		2	When the handle is manipulated with a precision grip, wrist excursions must be neutral for 70% of the manipulating time.
		3	When the handle is manipulated with a force grip, wrist excursions must be neutral for 70% of the manipulating time.
		4	The grip opening must be between 60 and 80 mm.
		5	Any disturbances (e.g. friction and spring forces) must be avoided to enable an optimal force feedback of tissue on the surgeon's hands: if the handle is manipulated in free spaces, no friction must be experienced.
		6	The handle must have a minimum width of 10 mm to prevent extreme contact area pressure.
		7	The instrument must be provided with a rotation knob to allow rotation of the instrument tip. This control switch must be manipulated with the thumb or 2nd finger and when the instrument is manipulated in free spaces, no friction must be experienced.
		8	The handle must allow left and right-handed manipulation.
		9	The dimensions of the finger rings must be: inner length minimal 30 mm, inner width minimal 24 mm.
		10	The handle of a dissection forceps has to support a precision as well as an force grip for manipulation.
	Laparoscopic grasping forceps	1	The grasping forceps must be operated by one hand of the surgeon.
		2	The handle of the grasping forceps has to support right- and left-handed manipulation.
		3	The angle between handle and shaft must be between 40° and 50°.
		4	The grip opening must be between 60 and 80 mm.
		5	For in-line handles: the length of the grip handle may not be longer than 170 mm.
		6	During manipulation, the wrist of the surgeon has to stay within the neutral zone of 20° extension, 40° flexion, 15° radial deviation and 25° ulnar deviation. During manipulation, extreme wrist excursions should not occupy more than 30% of the total manipulation time.
		7	The handle of the grasping forceps must have a minimum width of 10 mm to avoid pressure areas.
		8	If the handle contains fingerings, the dimensions of these fingerings must be: length 30 mm, width 24 mm.
		9	Forces may not be exerted on the ball of the thumb or the palm of the hand.
		10	The necessary closing force on the hand or fingers may not be more than 15 N.
		11	Specific grooves in a handle for positioning the fingers must be avoided.
		12	The handle of the grasping forceps has to support a "force-precision grip".
		13	Operation of the handle must be able for different hand postures.
		14	If the handle is manipulated in free spaces, no friction must be experienced.
		15	The instrument must be provided with a rotation knob to allow rotation of the instrument tip. This control switch and/or other control buttons must be manipulated with thumb or fore finger and manipulated in free spaces no friction must be experienced.
		16	The ratchet must be controlled without repositioning of the hand.
		17	The handle must have an option to put the ratchet function on and off.
	Laparoscopic needle holder	1	The needle holder must be operated by one hand of the surgeon.
		2	The handle of the needle holder has to support right- and left-handed suturing.
		3	The angle between handle and shaft must be between 40° and 50°.
		4	For in-line handles: the length of the grip handle may not be longer than 170 mm.
		5	If the four fingers are used to create a grip force, the length of the area where the force will be placed may not be smaller than 93 mm.
		6	During manipulation, the wrist of the surgeon has to stay within the neutral zone of 20° extension, 40° flexion, 15° radial deviation and 25° ulnar deviation. During manipulation, extreme wrist excursions should not occupy more than 30% of the total manipulation time.
		7	The handle of the needle holder must have a minimum width of 10 mm to avoid pressure areas.
		8	If the handle contains fingerings, the dimensions of these fingerings must be: length 30 mm, width 24 mm.
		9	Forces may not be exerted on the ball of the thumb or the palm of the hand.
		10	The necessary closing force on the hand or fingers may not be more than 15 N.
		11	Specific grooves in a handle for positioning the fingers must be avoided.
		12	The handle of the needle holder has to support a "force-precision grip".
		13	The ratchet must be controlled without repositioning of the hand.
		14	The handle must have an option to put the ratchet function on and off.
	Laparoscopic cameras	1	Attaching and detaching the camera adapter to the scope must allow one-handed operation through a sterile drape.
		2	During manipulation, the wrist of the user has to stay within the neutral zone of 20° extension, 40° flexion, 15° radial deviation and 25° ulnar deviation. During manipulation, extreme wrist excursions should not occupy more than 30% of the total manipulation time.
		3	The diameter of the grip entity must fall between 10-90 mm, with a preferred value of 35-65 mm.
		4	The length of the handle entity must fall between 72-140 mm, with a preferred value of 102-120 mm.
		5	The centre of mass of the entire scope assembly must be near the grasping point.
		6	Operation of any function controls on the camera head, including focus and zoom adjustment, must allow one-handed operation through a sterile drape.
		7	The camera head must allow manipulation with different hand postures.
		8	The handle of the camera has to support right- and left-handed manipulation.

Table 1 (continued)

B	Monitor	1	The optimal monitor position is in front of the subject, on a height between the surgeons head and hands, in such a way that his head flexions are between 15-45° to the horizontal.
C	Foot pedals	1	The design of the foot pedal must avoid a static standing posture.
		2	A dorsal flexion of more than 25° to control the foot switch is not allowed.
		3	The force for activation must be maximum 10 N.
		4	A frequent dorsal flexion of the foot should be avoided.
		5	The foot pedal should be controllable by clogs with the following (external) dimensions: maximum length 295 mm and minimum length 230 mm; maximum width 108 mm and minimum width 85 mm; maximum toe height 60 mm and minimum toe height 50 mm; and a total maximum clog height of 115 mm.
		6	The foot pedal must be controlled without looking at the foot pedal.
		7	The chance of accidentally activating the wrong switch function must be minimized by having a difference in control between the different functions.
		8	The foot pedal must be controllable with or without clogs.
		9	During control of the foot pedal the foot pedal may not move.
		10	The chance of losing contact with the foot pedal must be minimal.
D	Operating surface	1	The operating surface height must lay between a factor 0.7 and 0.8 of the subjects elbow height.
E	Standing support	1	During surgery the subject should use a standing support that is adjustable between 780mm and 1020mm

influenced by efforts to prevent losing contact with the foot pedal, which is achieved by keeping one foot flexed above the pedal and loading the bodyweight on the other foot. This causes a non-neutral posture which can lead to physical discomfort. Table 1C shows the guidelines that were developed to prevent these ergonomic problems with foot pedals.

D. Operating table height

The height of the operating table influences the excursions of the upper extremities of the surgeon and the assistant during laparoscopy. Studies showed that in order to prevent extreme upper limb excursions, the optimal operating surface height must lay between a factor 0.7 and 0.8 of the surgeons elbow height (van Veelen et al., 2002c). This value can be used as a guideline for the operating surface height for laparoscopic tasks and is presented in Table 1D.

E. Standing support

Operating room personnel have a static standing posture during the whole minimal invasive surgery procedure, which can be very tiring (Berguer et al., 1997). Some of the disadvantages are the obstruction of blood flow in the leg muscles and a large venostatic pressure. It would therefore be advisable to use a standing support during laparoscopic surgery, because it reduces the pressure on the leg muscles with 60%

392   Van Veelen & Goossens

and it stabilizes the pelvic and reduces the pressure on the spine (Wilson & Corlett, 1995). Table 1E shows the corresponding guideline for the use of a standing support.

New products that improve the physical ergonomics of laparoscopic surgery

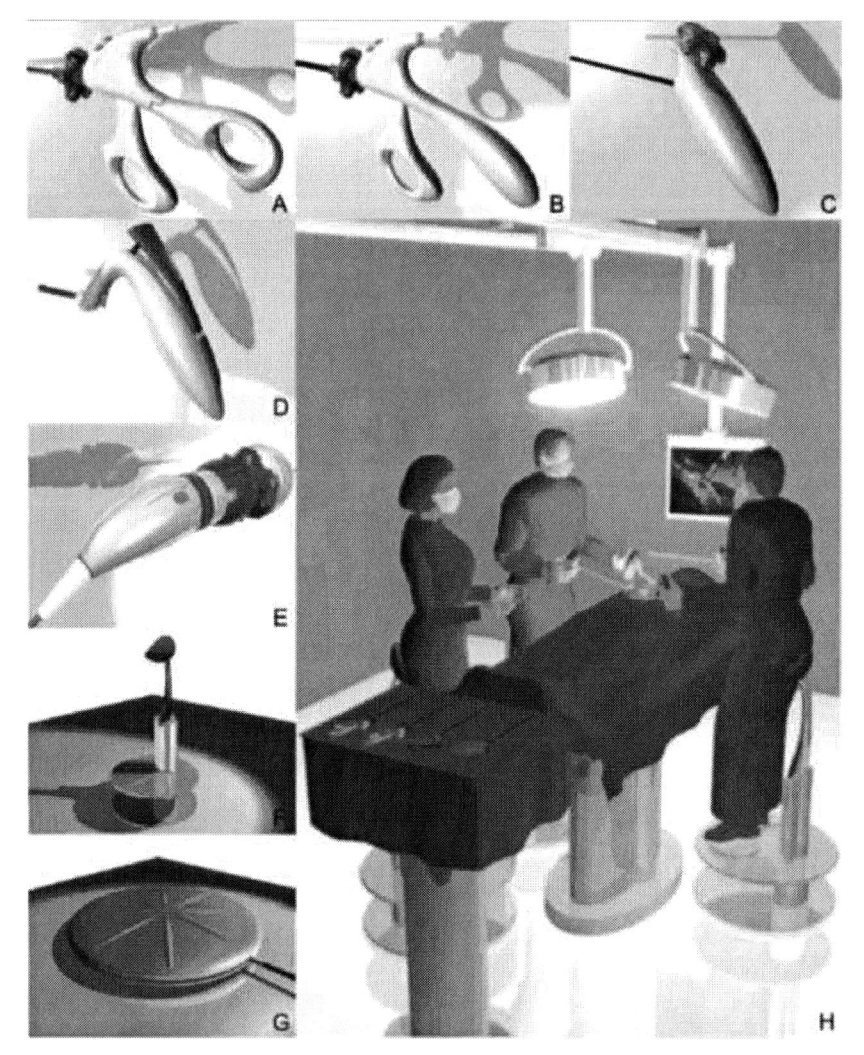

Figure 1. New designs for laparoscopic products that satisfy the new ergonomic guidelines and afford a neutral posture for surgeons and assistants.
1A: New design for a dissection forceps; 1B: New design for a grasping forceps without ratchet ; 1C: New design for a grasping forceps with ratchet; 1D: New design for a needle holder ;1E: New design for a camera head ; 1F: New design for a standing support with adjustable foot floor; 1G: New design for a foot pedal; 1H: Overview of all new products in the OR to support the neutral posture of surgeon and assistants
Figure 1 is available in colour at http://extras.hfes-europe.org

Figure 1 presents new products that satisfy the former mentioned ergonomic guidelines. Figure 1A through 1E shows respectively the new designs of a laparoscopic dissection forceps, a grasping forceps without ratchet, a grasping forceps with ratchet, a needle holder and a camera head. The new designs prevent:

- extreme wrist and shoulder excursions during manipulation
- extreme pressure peaks on the hand
- extreme force exertion

and they all are designed with an intuitive interface. Prototypes of the new instrument designs were manufactured and evaluated by at least eight surgeons in training situations. The results of these tests showed that especially extreme body postures were significantly reduced compared to the current available instruments (Van Veelen et al., 2002b, van Veelen et al., 2003b).

Figure 1F shows a design of a standing support with an adjustable foot floor. Because the height of the floor can range between 60 mm (minimum) to 460 mm (maximum), 95% of the user group will have a comfortable posture in combination with the current operating tables. The standing support is also adjustable in height allowing an ergonomic support for the whole user group.

Figure 1G shows a new design of a foot pedal, a flat round disc. The new design uses a left and right rotation of the foot to control the two diathermic functions (coagulation and cutting) instead of an identical movement of flexion and extension of the foot for both functions. Results of a user group evaluation showed that this way of controlling the foot pedal will probably reduce the errors made when activating a specific diathermic function. Physical discomfort will also be reduced because no flexion of the foot is needed and the endo- and exo-rotations needed for control are not frequent.

Figure 1H presents the whole OR arrangement, demonstrating the improved posture of the surgeon, assistant, and surgical nurse.

Discussion and conclusions

To fully solve the problem of the un-ergonomic posture of the laparoscopic surgeon, it is important to consider all the (product-) factors that influence the posture of a laparoscopic surgeon and not to focus on an isolated issue only. For example: the redesign of only the hand-held instruments will not solve the problems of extreme wrist excursions if not also the table height is adjusted to a proper ergonomic value. It is important that ergonomic guidelines and ergonomic solutions are communicated to the users that spent a significant amount of time in a laparoscopic operating room. This paper gives an overview of guidelines and solutions that should be applied all together in order to improve the physical ergonomics of laparoscopy. In the past few years the five factors that influence the posture of surgeons and assistants were subject of studies performed at the Delft University of Technology in collaboration with medical institutes. The new ergonomic guidelines for the design of hand-held instruments and foot pedals together with guidelines for monitor placement, table height and body support can be used by surgeons and other OR personnel to evaluate the user friendliness of their OR environment and to improve issues that do not satisfy the ergonomic guidelines. When designers use these guidelines to design new OR equipment, the new designs are an improvement in the field of human factors compared to the currently used laparoscopic products. When all these products are applied in the laparoscopic operating room, a new and ergonomic environment is created for the surgeon as well as for the assistants.

References

Berguer, R., Rab, G.T., Abu-Ghaida, H., Alarcon, A., & Chung, J. (1997). A comparison of surgeons' posture during laparoscopic and open surgical procedures. *Surgical Endoscopy, 11*, 139-142.

Berguer, R. (1998a). Surgical technology and the ergonomics of laparoscopic instruments. *Surgical Endoscopy, 12*, 458-462.

Berguer, R., Gerber, S., Kilpatrick, G., & Beckley, D. (1998b). An ergonomic comparison of in-line vs pistol-grip handle configuration in a laparoscopic grasper. *Surgical Endoscopy, 12*, 805-808.

Berguer, R., Forkey, D.L., & Smith, W.D. (1999a). Ergonomic problems associated with laparoscopic surgery. *Surgical Endoscopy*, 13, 466-468.

Berguer, R. (1999b). Surgery and ergonomics. *Archives of Surgery,134*, 1011-1016.

Hanna, G.B., Elamass, & M., Cuschieri, A. (2001). Ergonomics of hand-assisted laparoscic surgery. *Laparoscic Surgery, 8*, 92-95.

Joice, P., Hanna, G.B., & Cuschieri, A. (1998). Ergonomic evaluation of laparoscopic bowel suturing. *American Journal of Surgery 176*, 373-378.

Kohn, K.T., Corrigan, J.M. & Donaldson, M.S. (Eds.) (1999). *To err is human: Building a safer health system.* Washington, DC. National Academy Press.

Matern, U., & Waller, P. (1999). Instruments for minimally invasive surgery: principles of ergonomic handles. *Surgical Endoscopy, 13*, 174-182.

Muller, M.J., & Kuhn, S. (1993) Participatory design. *Communications of the AMC, 36*, 24-28.

Van Veelen, M.A., & Meijer, D.W. (1999). Ergonomics and design of laparoscopic instruments: results of a survey among laparoscopic surgeons. *Journal of Laparoendoscopic Advanced Surgical Technique A, 6*, 481-489.

Van Veelen, M.A., Meijer, D.W., Goossens, R.H.M., & Snijders, C.J. (2001). New ergonomic design criteria for handles of laparoscopic dissection forceps. *Journal of Laparoendoscopic Advanced Surgical Technique A, 11*, 17-26.

Van Veelen, M.A., Meijer, D.W., Goossens, R.H.M., Snijders, C.J., & Jakimowicz, J.J. (2002a). Improved usability of a new handle design for laparoscopic dissection forceps. *Surgical Endoscopy, 16*: 201-207.

Van Veelen, M.A., Jakimowicz, J.J., Goossens, R.H.M., Meijer, D.W., & Bussmann, H.B. (2002b). Evaluation of the usability of two types of image display systems during laparoscopy. *Surgical Endoscopy, 16*, 674-678

Van Veelen, M.A., Kazemier, G., Koopman, J., Goossens, R.H.M., & Meijer, D.W. (2002c). Assessment of the ergonomically optimal work surface height for laparoscopic surgery. *Journal of Laparoendoscopic & Advanced Surgical Technics, 1*, 47-52.

Van Veelen, M.A., Meijer, D.W., Uijttewaal, I., Goossens, R.H.M., Snijders, C.J., & Kazemier, G. (2003a). Improvement of the laparoscopic needle holder based on new ergonomic guidelines. *Surgical Endoscopy, 5*, 699-703.

Van Veelen, M.A., Snijders, C.J., Van Leeuwen, E., Goossens, R.H.M., & Kazemier G. (2003b). Improvement of foot pedals used during surgery based on new ergonomic guidelines. *Surgical Endoscopy, 7*, 1086-1091.

Van Veelen, M.A., Nederlof E.A.L., Goossens, R.H.M., Schot, C.J., & Jakimowicz, J.J. (2003c). Ergonomic problems encountered by the medical team related to products used for minimally invasive surgery. *Surgical Endoscopy, 7*, 1077-1081.

Wilson, J.R., & Corlett, E.N. (1995). *Evaluation of human work: a practical ergonomics methodology*. London: Taylor and Francis.

The use of colour on the labelling of medicines

Ruth Filik[1], Kevin Purdy[2], & Alastair Gale[2]
[1]Department of Psychology
University of Glasgow
[2]Applied Vision Research Institute
University of Derby
UK

Abstract

Medication errors occur as a result of a breakdown in the overall system of prescribing, dispensing, and administration of a drug. Problems with packaging and labelling can be thought of as being latent conditions in the system that can predispose to errors during dispensing and administration. Errors often arise through different strengths of the same product and different products from the same manufacturer having similar packaging. One possible way to aid to product differentiation is to use colour, however, whether or not colour is an appropriate aid to the correct identification of a product is a controversial issue. We present a series of visual search studies investigating the use of colour on drug labelling as a systems change to aid the identification of drug products. Participants were given the task of searching for a target drug product amongst a range of products. In some trials colour could be used as a cue to product identification. Findings provide support for the judicious, or unambiguous use of colour on the packaging and labelling of medicines; participants made fewer errors when colour was an unambiguous cue to identity, but made more errors when similar products were also the same colour.

Introduction

The systems approach to error has been widely documented (e.g., Leape et al., 1995, Reason, 2000). Problems with packaging and labelling can be thought of as being latent conditions in the system that can predispose to errors occurring in the dispensing and administration of drugs. Problems often arise through different strengths of the same product and different products from the same manufacturer having similar packaging (Department of Health, 2004). System changes or 'error traps' that can be put in place relating to packaging and labelling can include; the physical separation of drugs with similar packaging in the dispensary or where they are to be administered, and the design of systems to avoid look-alike containers and unclear labelling. One possible way to aid product differentiation is to use colour.

The appropriate use of colour on drug packaging has been the subject of much debate (see e.g., Institute for Safe Medication Practices, 2003). One perspective is

In D. de Waard, K.A. Brookhuis, R. van Egmond, and Th. Boersema (Eds.) (2005), *Human Factors in Design, Safety, and Management* (pp. 397 - 400). Maastricht, the Netherlands: Shaker Publishing.

that colour can aid the correct identification of medicines, for example, by reducing the similarity in appearance between confusable medicines, and as a further safeguard to complement the appropriate presentation of text. On the other hand, proponents of black and white labels believe that colour should not be used as a means of identification in order that more emphasis is placed upon reading the label. A key concern over the use of colour is the possibility that it may lead to too much reliance on colour to identify the product, resulting in less careful inspection of other label information, such as the name of the drug.

Although there are a number of arguments both for and against the use of colour, there is not yet much scientific evidence in support of either perspective. Two key empirical questions that emerge are: firstly, whether colour can be used to help distinguish different products (e.g., products from the same range that are different strengths). Secondly, whether this could lead to colour being used as a shortcut to identify a product. The first of these questions is addressed in Experiment 1, and the second in Experiment 2.

Experiment 1

The issue investigated in Experiment 1 was whether colour could aid the identification of a product of a particular strength within a range. Participants were given a computer-based visual search task. They were shown an image of a target drug product of a particular strength that they subsequently had to search for amongst an array of products of different strengths. The task was to indicate whether or not the target product was present in the array. For some of the trials, colour could be used as a cue. The task was designed to be analogous to the situation of someone having a mental image of the pack that they are looking for, and then searching for it amongst a range of products.

If colour does aid identification, it was predicted that participants would make fewer errors in indicating whether or not the target product was present in the array when the packs were coloured. There are two classes of error that participants could make. If the target was present in the array but participants reported it as being absent, this was a false negative error. If it was absent from the array but reported as being present, this was a false positive error. The latter would correspond to the wrong product being identified as being the target product, it was therefore the number of false positive errors that are of primary interest.

Twenty-eight staff and students from the University of Derby, who were not healthcare professionals, participated. Stimuli consisted of 64 different drug products. Mock drug packs were designed for the purpose of the experiment. Information on the packs consisted of the generic name, dosage form (e.g., tablets) and the strength (e.g., 100mg). Participants were given four seconds to view the target drug product, and then as much time as they needed to search for it in the array. The array consisted of eight different strengths of the target drug, which in the coloured trials were eight different colours. For example, the target might be 'warfarin tablets 5 mg', and the array would be 'warfarin tablets 0.5, 1, 2, 2.5, 3, 3.5,

4, and 5 mg'. Results showed that participants made significantly fewer false positive errors when they could use colour as an aid to identify a product in a range.

Experiment 2

Experiment 2 addressed the question of whether the use of colour may lead to errors being made if similar products are also the same colour. One real-life situation in which this could be a problem would be where a product with a similar name as the target product was also available in the same strength and was the same colour as the target. If, in this scenario, colour became a shortcut to the identification of the drug, then in a visual search task where the target is replaced with a drug with a similar name in the array, participants should make more errors in which they incorrectly say that the target is present in the array when the packs are coloured, than when they are not. The task and participants were the same as those used in Experiment 1. However in this study, the target product was always replaced in the array with a product with a similar name. Results from Experiment 2 showed that when colour was an ambiguous cue to identification, participants were more likely to say that the target drug was present, when in fact a similarly named drug was present instead.

Discussion

In summary, Experiment 1 provided support for colour aiding the identification of a particular strength drug product within a range, with participants making fewer false positive errors when they could use colour as a cue to identification. However, results from Experiment 2 provided evidence for colour possibly being used as a shortcut to identify the medicine, leading to incorrect identification. That is, when a product had a similar name and the same strength as the target product, participants were more likely to incorrectly identify it as being the target drug when it was also the same colour. It must be noted that our findings should be verified with a range of stimuli and methods before more general conclusions regarding the use of colour can be made. It is concluded that colour should be used carefully, with further research being required to identify situations in which colour can aid product identification.

References

Department of Health (2004). *Building a safer NHS for patients: Improving medication safety*. London: HMSO.
Institute for Safe Medication Practices (2003). A spectrum of problems with using color. *ISMP Medication Safety Alert*, November 13 Issue.
Leape, L., Bates, D. W., Cullen, D. J., Cooper, J., Demonaco, H. J., Gallivan, T., et al. (1995). Systems analysis of adverse drug events. *Journal of the American Medical Association, 274*, 35-43.
Reason, J. (2000). Human error: Models and management. *British Medical Journal, 320*, 768-770.

Data display in the operating room

Noemi Bitterman[1], Gideon Uretzky[2], & Daniel Gopher[3]
[1]Industrial Design, Faculty of Architecture & Town Planning Technion,
[2]Faculty of Medicine,
[3]Faculty of Industrial Engineering & Management,
Technion, Haifa, Israel

Abstract

A major problem in cardiothoracic surgery is to continuously follow physiological and clinical data of the patient in the operating room while simultaneously performing highly skilled manual manipulations, focussing on different visual areas. Goal of the present study was to identify the surgeons' habits in monitoring, and to assess the problems of the on-line use of physiological data, the essential physiological parameters and the use of auditory information at critical stages during cardiothoracic surgery.

Data were collected from continuous video recordings with manual follow up of cardiac operations (n=10), and, questionnaires followed by personal interviewing (n=20). These data revealed significant variability in the use of hemodynamic monitors during various stages of CPB assisted surgery. The highest frequency of monitor scans was measured during "coming off bypass" and the lowest during the "on bypass" period. 75% of the surgeons felt they are not watching enough the monitors during routine surgery and 28% on acute situation. Adding auditory systems to allow optimal follow up of physiological parameters of the patient without interference with the surgical team performance, at the most critical period of cardiac surgery are suggested.

Introduction

Recent technological developments in surgery and especially in cardiac surgery confront the surgeon with visual, manual and perceptual constraints calling for the need to define technological and design solutions for improving the performance of the surgical team (Berguer, 1996; Berguer, 1997; Stone & McCloy, 2004). One problem in cardiothoracic surgery is the need for continuous monitoring of physiological and clinical data of the patient, simultaneously with performing highly skilled manual manipulations, functioning between numerous visual fields and a coordinated teamwork. Physiological data of the patient are of major importance during cardiothoracic surgery, as they serve as a base for operating decisions which have to be taken rapidly by the surgical team (Cook & Woods, 1996ab; Gaba, Howard, & Small, 1995)

The present study examined the "on line" use of patient's physiological data by surgeons during cardiothoracic surgery. The surgeons' monitoring behaviour and the problems associated with frequent monitor scanning were assessed, as were critical operation stages. Cardiopulmonary bypass (CPB) assisted surgery was selected as representing a demanding surgical task.

Results & discussion

The events of monitor inspection are not scattered uniformly throughout the entire surgery period. Clusters of short time intervals between monitor observations are found near the critical points of connecting and disconnecting the CPB (Cardiopulmonary Bypass) system ("coming on" and "coming off" bypass), and during incidents such as using the defibrillator. Prolonged time intervals between monitor inspections are common during the "on bypass" period. The frequency of monitor inspections was significantly higher during "coming off bypass" compared to "on bypass" and "pre bypass" phases (Figure 1, p value = 0.021 using F test in repeated measures ANOVA. Differences of 0.84 and 0.91 respectively, $p< 0.05$ using Tukey's test).

The surgeons (n=20) ranked the "coming off" bypass phase as the most intensive period for monitor use (p value = 0.0001 using F test in repeated measures ANOVA. Differences of 1.0 and 1.31 compared to pre and on pump respectively, $p< 0.001$ using Tukey's test). There were no significant differences between residents and senior surgeons in their ranking of monitor use during the three surgery phases (p=0.7 (pre pump), p= 0.19 (on pump), p=0.9 (off pump), Mann Whitney Test).

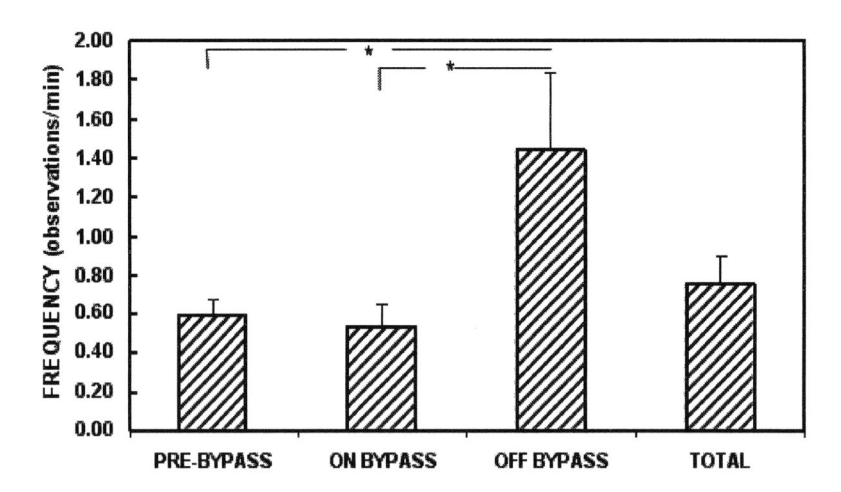

*Figure 1. Frequency of monitor inspections (observations/min) by surgeons during various stages of cardiac surgery (mean + SE), * p<0.05 in Tukey test (n=10).*

Seventy-five percent of the surgeons indicated that they do not watch the hemodynamic monitor enough during routine cardiac surgery. This number was much lower for acute surgery, when only 28% stated they do not observe the monitor screens enough (p=0.0135, Fisher's Exact Test). No correlation was found between level of seniority (resident or senior surgeon) and subjective estimate of monitor use (r=0.34, Spearman Rank Correlation).The greatest significant difference between residents and senior surgeons (p= 0.019 Mann Whitney Test) was recorded for the problem of "relocation of hands and surgical tools" resulting from frequent head movements between the monitor screen and the surgical field. 90% of the surgeons reported that they require auditory information on the patient's physiological parameters in addition to monitor inspection.

Adding auditory display of several physiological data especially during bypass, running concomitantly with existing monitors, will preclude the frequent need to raise the eyes from the surgical field. The supplemental auditory information system will divide the attention more efficiently between two modalities (visual and auditory), and increase the level of alertness and team feeling.

References

Berguer, R. (1996). Ergonomics in the operating room. *American Journal of Surgery, 171*, 385-386.

Berguer, R. (1997). The application of ergonomics in the work environment of general surgeons. *Review of Environmental Medicine, 12,* 99-106.

Stone, R., & McCloy, R. (2004). Ergonomics in medicine and surgery. *British Medical Journal, 328*, 1115-1118.

Cook, R.I., & Woods, D.D. (1996). Adapting to the new technology in the operating room. *Human Factors, 38,* 593-613.

Cook, R.I., & Woods, D.D. (1996). Implications of automation surprises in aviation for the future of total intravenous anesthesia (TIVA). *Journal of Clinical Anesthesia, 8,* 29S-37S.

Gaba, D.M., Howard, S.T., & Small, S.D. (1995) Situation Awareness in Anesthesiology. *Human Factors, 37*, 20-31.

Simulation and assessment of a North Sea rescue vessel

Alistair Furnell, Matthew Mills, & Paul Crossland
QinetiQ, Centre for Human Sciences
Farnborough, UK

Abstract

BP Exploration has developed a new concept for providing safety cover for personnel working on offshore installations. The concept involves the use of a new design of rigid inflatable boat, which can be deployed in emergencies to recover casualties from the water, sustain their life and then return them to a surgical facility for primary care. This paper describes the innovative trials run by QinetiQ in order to simulate elements of the vessel and to build a more complete picture of its usability and operational capability from a human factors point of view. The trials encompass the retrieval of casualties from a wave tank onto an afterdeck mock-up, the flow of casualties within a mock-up of the vessel and treatment under conditions of motion in a large displacement motion simulator. The paper describes how the data gathered from these discrete trials were then integrated to allow an appreciation of likely casualty handling capability of the vessel.

Introduction

A new concept for providing safety cover for personnel working on offshore installations is currently under investigation by BP Exploration. This new concept involves the use of a new design of Rigid Inflatable Boat (RIB), 18.6 m in length (the largest RIB yet designed) with the capacity for six crew and room for up to 21 survivors, including six stretcher cases. It is proposed that this vessel will provide treatment facilities to enable a team of paramedics to administer medical assistance to casualties following an offshore incident. The vessel should provide the means to stabilise casualties until they can be returned to a surgical facility for primary care. Clearly, such a concept requires a robust approach to de-risk the design, and that the concept must be demonstrated to be a true place of safety in the harshest of environments. Consequently, a thorough investigation was required into the design, functionality and suitability of the afterdeck arrangement for casualty retrieval from the water, the assessment of casualty handling (triage), and the medical treatment area and accommodation facilities.

Initial studies were undertaken and reported by Bridger *et al.*, 2002 to the internal design of the treatment facility. However, the central issue was that the customer had a largely unproven concept for which no objective performance data existed. Furthermore, the customer would not wish to go to the expense of manufacturing a vessel that did not meet the specified performance capability. Therefore, QinetiQ was approached with the aim of running several discrete trials which could gather data about the performance of the vessel prior to building one. The performance metrics would primarily be quantitative time measures, since speed is of the essence in recovery from hostile environments, and secondly qualitative, taking into account issues such as health and safety and the general ergonomics crucial to effective system performance. During the planning stages of the project, three key processes were identified that required investigation:

1. Retrieval of casualties from the water. The vessel must be able provide an effective means of allowing casualties to be retrieved from the water onto the afterdeck.
2. Man movement through the vessel. With space at a premium, the proposed layout should allow effective casualty handling and flow from the afterdeck through to the cabin containing the seating and primary treatment areas.
3. Surgical procedures under conditions of motion. Paramedics must be capable of stabilising any serious injuries whilst the vessel is at sea.

Aims

Due to restrictions in space, it is not possible to detail the full extent of each discrete trial. This paper gives a higher level overview of the trials, focussing on the trial methodologies rather than the results per se. The aims of this paper are twofold, firstly to give the interested reader an appreciation of the issues involved in the simulation and assessment of the various capabilities of a prototype rescue vessel; and secondly, to show how the data were used to form a more complete picture of the vessel's capability. For the purposes of this paper the trials are presented in the order detailed above.

Retrieval from water

The Issue

This trial was conducted to investigate whether the proposed vessel is able to provide an effective means of allowing casualties to be retrieved from the water onto the afterdeck. In addition, the timing data for recovery times could be used to replace the estimated times for recovery that had been assumed in the man-movement study.

The Apparatus

MTMC built a mock-up of the vessel's afterdeck that was then suspended from the side of the Ocean Basin at QinetiQ Haslar. The aim was to create the "roughest" seas possible in the Ocean Basin – given that this is effectively a full scale experiment. The wavemakers were tuned to give, simply, a modulated repeating sine wave with amplitude as high a possible. The direction of waves travelled was along the ocean basin to simulate beam seas. This was deemed to be the zero speed heading that the vessel would adopt in rough seas when recovering the casualties

The platform was stationary and no attempt was made to reproduce the motion of the vessel itself. Waves were approximately representative of the relative waves that would be experienced aboard the vessel, so that (for example) a camera recording the waves passing the platform in the wave pool would produce a picture very similar to that produced by a camera bolted down to the deck of a real vessel.

Two different designs for the aft end were tested, the first of these, termed Mock-up A, allowed a section of the vessel's inflatable hull to collapse into the water with the intention that this should be used as a kind of ramp to aid casualty survivor retrieval. A second, improved design, Mock-up B, was then tested in the same manner to investigate whether the changes aided the ergonomics and time duration of recovery. This second design had two collapsible sections that fell either side of the recovery area out of the way and then a ramp was provided to aid survivor recovery to the deck.

Figure 1. Crew retrieving a survivor using Mock-up B

Method

Three scenarios were created detailing worse case scenarios of survivors in the water. In each scenario the survivors would be slightly dispersed over the sea area in small or larger groups, with a mixture of six seriously injured and 15 uninjured/minor injury survivors. For the purposes of this trial the periods taken for the vessel to move between groups were allotted time durations based on the vessel's capability. The survivors were dressed in the appropriate survival suits with lifejackets. They entered the wavetank ahead of the afterdeck and then floated down towards it for recovery at appropriate intervals. The total time taken to recover all 21 subjects in each scenario was measured; in addition the time taken to recover each survivor was recorded. The time taken for the vessel to manoeuvre between groups was included in real time and no survivors were recovered during this period. This was to allow the physical workload durations of the retrievers to be realistically modelled. Similarly, the retrievers were swapped between the three scenarios as these represent three distinct rescue situations.

Two crewmen were selected to carry out survivor retrieval from the water. A third crew person was positioned behind the two retrievers and helped with stabilizing them as they lent forward and with pulling the subject aboard. This third crew person was also responsible for leading the retrieved survivor away from the retrieval area. The trial was covered by video cameras from several angles, which would allow for later ergonomic analysis. As in previous trials, subject feedback was considered to be of prime importance and subject debrief questionnaires were used to gather data.

Results

The three scenarios were completed and times obtained for retrieval of all 21 survivors for each scenario. These data were first obtained using Mock-up A, and then repeated using Mock-up B on a different day. The mean retrieval times and standard deviations for each scenario were recorded. It was noted that both sets of retrieval times were faster than the preset times used in the man-movement study. Subsequently, the mean retrieval times were tested for any significant difference between Mock-up A and Mock-up B. It was observed that the improved design, Mock-up B, allowed significantly faster survivor recoveries than the previous design. Furthermore, the smaller standard deviation for Mock-up B recovery times showed that it allowed a *consistently* faster recovery time.

It was observed that with Mock-up B, the retrieving persons were able to brace their feet against the sides of the deck and so utilise power in the lower limbs when retrieving a survivor. This was not possible with Mock-up A, which appeared to be essentially an upper body exercise. The use of lower limb power on Mock-up B meant there was less need to flex the back, and meant that retrievers could get around the survivor, almost pushing them up the ramp rather than pulling entirely from behind (as for Mock-up A).

The modified Mock-up B also resulted in less need for retrievers to lean out over the water, in order to pull survivors aboard. For Mock-up B, retrievers were able to remain more within the confines of the deck, avoiding the need to lean out, with its attendant risk of losing balance and relying on a harness to prevent falling forward into the water.

The third retriever, who was very much involved in retrieving survivors during for the Mock-up A trial, was not required for the Mock-up B trial. In initially planning the trial, a third person was made available to assist with retrieval mainly to receive the survivor after they were lifted from the water and to escort them safely off the retrieval platform. For the initial trial with Mock-up A, however, they were required to hold onto the backs of the two retrievers as they leaned out to grasp and pull aboard survivors. For the Mock-up B trial, this role was not necessary, and the third person had no involvement in the actual retrieval and was free to take care of survivors as they arrived onboard.

Man movement

The issue

This trial was conducted to establish whether the proposed internal vessel layout allowed for effective casualty handling and flow from the afterdeck through to the cabin containing the seating and primary treatment areas. The priority was to assess the proposed design for the vessel in the retrieval, distribution, and securing role; included in this was the need for the crew to perform some form of basic triage to distribute the survivor's appropriately.

The apparatus

The representation of the vessel's treatment area was constructed at QinetiQ Haslar by MTMC. This static mock-up was constructed of a steel frame with plywood bulkheads and decks. The two deck mock-up had the treatment area in the lower deck and a seating area on the upper deck. Stairs at the aft end provided access to the upper deck. An afterdeck consisting of a wooden stage was constructed, and the lower deck treatment area opened out onto this deck through a doorway on the starboard side (the aft door). The afterdeck was closed in by rope fencing on its three exposed sides. This fencing was interrupted along the port side of the afterdeck to allow access to and from the mock up to simulate the area for retrieval of survivors.

The primary treatment area on the lower deck level contained 2 operating-type tables aligned parallel to the longitudinal axis of the vessel, one port and one starboard. All the emergency medical equipment was stored in the immediate vicinity of these tables, either attached to the tables, stored underneath the tables, or secured to the walls.

The secondary treatment area (the aft portion of the treatment room) contained two rows of four seats on the port side, and one row of four seats on the starboard side. Each of these rows could be collapsed to create three stretcher berths in total. In addition, there were two racks mounted on the walls, one port and one starboard, positioned above the seats, to allow a further two stretcher berths. This provided five extra stretcher berths aside from the two treatment tables in the primary treatment area.

When operational, there will be a route taken by casualties as they pass through the vessel. Therefore, the survivor is brought onboard (out of the water) at the afterdeck, they then enter the treatment room through the aft door (forward of the afterdeck and on the same level), where the extent of their injuries are assessed. Providing they are not stretcher-bound or in need of further treatment, they will exit the treatment room by ascending stairs at the aft end, which will take them onto the wheelhouse deck on the floor above, where they will be seated.

Method

The same three scenarios were used as in the previous trial. The survivor groups presented themselves at the afterdeck for recovery at the appropriate time intervals, so that they could be triaged and moved through the vessel. As previously, each scenario consisted of a total of 21 survivors for the vessel to retrieve: six with severe injuries, six with minor injuries and nine uninjured. Of the six severely injured, five had to be transported on a stretcher and the remaining one could walk but had severe chest injuries. Of the six with minor injuries, two had hand burns (and were unable to take off or put on overalls without occupying one of the crew), two had smoke inhalation (and needed to be sat downstairs near a designated oxygen supply) and two had hypothermia (and needed to be wrapped in an 'insulation' blanket). The nine uninjured were allowed to change overalls without assistance and were processed through the treatment area and up to the seats on the wheelhouse deck. They were instructed, however, to await directions at all points from the crew.

The survivors were presented to the afterdeck in an order determined by the scenario, with pauses for manoeuvring between groups of survivors. Each survivor wore a wristband with a label, which detailed his status: severely injured, minor injuries or uninjured. If injuries were present, then the nature of those injuries was detailed on the cards. The survivors presented themselves to the afterdeck as directed by the Principal Investigator and then waited until the required amount of retrieval time had passed before stepping onboard. This was 90 seconds wait for retrieval of a severely injured person and 60 or 30 seconds for retrieval of those with minor injuries or the uninjured. The clock for timing retrieval could only run when two crew were present on the port side of the afterdeck, and they were not permitted to be engaged in any other activity during this time.

Once onboard, those with severe injuries were placed on a stretcher and carried to a suitable location in the treatment room, all in real time. Walking survivors were processed through the vessel in real time also. They had to make their way into the treatment area, take off life jackets and wet overalls (simulating survival suits) and put on dry overalls (simulating 'bunny suits'), with or without the limitations of minor injuries. This change of overalls was done on the lower floor near the foot of the stairs (i.e. in the secondary treatment area). Once changed, a walking survivor would place his 'wet' overalls in a bag and make his way up the stairs to the wheelhouse deck, taking his life jacket with him, where he would find a seat.

Total time was measured for each scenario, from the start of retrieving the first casualty to the point at which the final casualty had been secured onboard. Two CCTV cameras were installed in the lower room of the mock-up and one on the afterdeck. In addition, user feedback was encouraged in informal debrief sessions.

The major issue presented by this assessment was the lack of motion, which could not be reproduced, as it would have made the whole trial unfeasible. In an attempt to compensate for this lack of motion, which it can be assumed would degrade performance to some degree, the experimenters introduced a 'call to secure' command, during which the subjects did nothing except hold on as the boat 'manoeuvred' although the clock kept running during this period.

Results

CCTV footage from the outside camera was examined for delays in flow of survivors from the afterdeck into the treatment room. For all scenarios, there was no obstruction to the flow of walking survivors into the interior. Individual task times were also recorded for key tasks e.g. for stretcher cases, the times from arrival on the afterdeck to arrival in the treatment room.

The trial demonstrated that the current design is functional yet at the same time provided several areas of insight into how the design could be improved. The key concerns were the position and size of the aft door, the respective deck levels on passing through the aft door, the presence of a storm sill in the aft door and the design of the stretcher racks. The trial showed that an optimum design for the aft door area would be key to the success of the vessel. The advantage of positioning the door on the starboard side is that it allowed the stairs to be positioned centrally, leaving the walls clear for storage of stretchers. However, as a result of this trial the position of the door was changed to a central location in the bulkhead allowing for a less obstructed passage through the vessel. The internal seating plan was amended to take account of this change and assessed at a later date. This design change also had the secondary effect of allowing more deckspace at the entrance to the vessel to cope with increased casualty flow.

Conclusion

This trial demonstrated, as far as is possible without the use of motion, that the current amended vessel layout and organisation was capable of meeting the challenges posed by a worst-case scenario, in terms of retrieving and securing survivors onboard

Surgical motion trial

The issue

This trial was conducted to establish if paramedics could safely and efficiently complete procedures required for the assessment and stabilisation of casualties, whilst exposed to extreme motion environments that are typical of the areas in which these craft are expected to operate. The most severe direct motion induced limiting performance occurs in manual tasks requiring balance and co-ordination. During rough weather working on any craft becomes more difficult and the paramedic will stop their task in hand and hold on to some suitable point to minimise the risk of injury to themselves or their patient. This means that the task takes longer to perform or in some instances the task cannot be completed at all.

Human postural stability is maintained by a complex musculo-skeletal system integrated with various control systems in the body. In fact, the musculo-skeletal system comprises an active system with over 200 degrees of freedom powered by 750 individual muscles. Clearly, an exact system identification is impracticable and so simplified postural stability models are developed as practical tools. The simplest model of postural stability is a rigid body model similar to the size and shape of the human body. Such a model has been suggested by Graham (1990) for predicting the incidence of loss of balance (an MII – motion induced interruption) as a function of lateral acceleration. This model has become widely used in the ship design community.

In general terms, the MII model predicts that a person will lose balance during a simple standing task when the tipping moment exceeds the righting moment provided by separation of the person's feet (stance). The ratio of the half stance width to the height of the person's centre of gravity is defined as the tipping coefficient. The MII uses this tipping coefficient to evaluate statistically when a tip will occur and expresses this as the number of MIIs per minute. The MII model was validated by performing postural stability experiments, Crossland *et al.* (1998), on Royal Navy volunteers, measuring the instantaneous accelerations on the body and recording the MIIs. This empirically derived tipping coefficient can be substituted for the theoretical tipping coefficient derived from purely geometric considerations.

However, as would be expected the tipping coefficients are task dependent – in the experiments undertaken by Crossland *et al.*(1998) the task routine consisted of gross motor tasks similar to those expected on a naval platform. Clearly, the paramedics are undertaking very different tasks which consist of mainly fine motor tasks, so in this instance trials were required.

Apparatus

Data from sea-keeping tests undertaken on a 1/30th scale model of the vessel were appropriately scaled (using classical hydrodynamic scaling laws) to provide motion time histories to drive the Large Motion System (LMS) at QinetiQ Bedford. This simulator is capable of 5 degrees of freedom motion (lateral and vertical displacements, roll, pitch and yaw). A proposed design for the treatment area was recreated in the motion cab. This cab included a treatment table with a range of surgical instruments and a Laerdal SimMan™ mannequin to represent a casualty. The mannequin is sufficiently realistic to simulate a variety of symptoms that can allow an objective judgement about the success or failure of a number of surgical tasks. Three cameras were placed in the cab to record the trial from a variety of angles and subjects were in continuous two way communication with the experimenters. Safety was a high priority, particularly given the displacements involved, and the subjects wore an inertial reel dampening system on their backs to prevent injury if they lost their footing.

Method

Eight surgical and assessment procedures were chosen as representing those likely to be required from the breadth of injuries and severity typically encountered by a paramedic: laryngeal mask intubation, mechanical resuscitation using a MARS

(Mechanically Assisted Resuscitation device), endo-tracheal intubation, intravenous cannulation, compilation of a trauma score, assessment and management of tension pneumothorax. These are all procedures that can be undertaken by one paramedic. Additionally, there were two procedures that required two paramedics to undertake: stabilisation of suspected cervical spine injury, stabilisation and splinting of a fractured femur. There were six subjects, all qualified offshore paramedics from AON Ltd and all but one of whom had offshore experience. All participants had a full medical prior to the trial and the trial itself complied with the Declaration of Helsinki, as adopted at the 52nd WMA General Assembly, Edinburgh, October 2000 (as with all trials in this paper).

The first day was used for familiarisation. The trial used 4 conditions of LMS motion: static and then conditions one to three being increasingly severe sea states. The presentation of these conditions was randomised, as was the order in which the tasks had to be completed by the paramedics. Due to the small sample size, the subjects completed the battery of surgical tasks twice for each exposure (the 'first attempt' and 'second attempt') , so that 12 discrete sets of timing and success/failure data were available for each motion condition. Each motion condition lasted for approximately 10 minutes. During the planning phase of the experiments, an acceptable standard was set in terms of what is a satisfactory outcome for each task. Acceptable times for completion of individual paramedic tasks have been published by Martin *et al.*(1998) and although this study does not cover all the tasks undertaken here, this published work has served as a guide. Further discussions between QinetiQ medical team and Institute of Naval Medicine allowed times to be set for all of the tasks apart from the trauma score.

On completion of each motion conditions, subjects took part in a structured debrief, which provided valuable feedback about the experience. The subjects were asked to complete a 5 point rating scale for each surgical task ranging from five: "completing the task proved no more difficult than in a stationary environment" to one: "extremely difficult: it is unlikely that the tasks will be completed". This was followed by more open ended questions to allow for more expanded feedback. These 'softer' measures were considered crucial since the subject may have successfully intubated the mannequin in less than 30 seconds, but have been of the opinion that the intubation might have been traumatic because of the motion.

Results

Mean times for all of the tasks were calculated along with standard deviations and these were tested for any statistically significant differences using ANOVA and post hoc testing. In terms of the external performance metrics, it was observed that all of the tasks but one were completed inside their respective limits. The tension pneumothorax task was affected by difficulties in use of the stethoscope due to noise, although an ANOVA analysis of this task showed no statistically significant differences in times between stationary and motion conditions, so the presence of motion did not alter performance. C-spine immobilisation and splinting of the femoral fracture were both completed in faster times than had initially been predicted. For the tasks where no external performance metric had been identified,

ANOVA analysis again demonstrated no statistically significant difference between times for stationary and motion conditions. There was no evidence of motion sickness among the subjects in any of the three conditions tested.

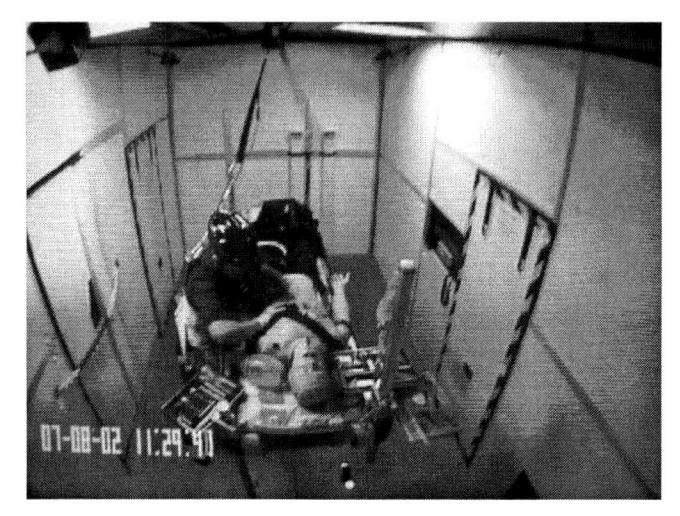

Figure 2. Medic in Large Motion Simulator

The subjects' mean ratings for the difficulty of each task were analysed using a Friedman non-parametric test for a main effect of motion. It was found that there was only a significant difference apparent in the endotracheal intubation task (between motion condition one and three, and condition two and three). However, open ended feedback confirmed that subjects did find the motion conditions more demanding than the stationary, particularly the severest condition three (7m significant wave height). Although, ultimately, the subjective data concur with the timing data that show acceptable outcomes.

The ratings results were analysed to see whether they differed between the first attempt and the second. As one would expect, the results show a general trend that the subjects began to rate the tasks as slightly easier towards the second attempt than the first, although no significance was present. Anecdotal reports indicate that it became easier to stabilise the body against ship motion as duration of exposure increased, which may in part explain these results.

Conclusion

In this trial it was shown that, when applying external performance metrics to five of the tasks, all of the tasks apart from tension pneumothorax, were completed within acceptable time periods for clinical pre-hospital practice. None of the tasks were found to take significantly longer under the conditions of motion than they did when assessed in a static environment. There is evidence from both the analysis and the subjects' feedback of learning effects in some of the tasks. It is unclear whether this is the result of subjects learning how to cope with the motion more effectively or due to learning how to complete the task more effectively. Although, given the proven experience of the subjects in performing the medical tasks, it is thought the former is most likely to be the case. Certainly the video evidence revealed that a number of bracing postures were adopted as a compensatory strategy, particularly when carrying out fine motor tasks such as IV cannulation.

In conclusion, the timing and qualitative results from this trial confirmed that the proposed vessel could be operated in a manner in which paramedics could safely and efficiently complete procedures required for the assessment and stabilisation of casualties, whilst exposed to a range of sea states.

Discussion

This paper has briefly detailed each of the three major trials used to obtain objective data about the prototype rescue vessels future casualty handling capability. The man movement study provided the timing data required for recovery and movement within the vessel. One of the assumptions made in these timing data was the time taken to recover each survivor from the water, this was then replaced by actual times from the water retrieval study (Mock-up B). The assumptions had been overestimates, so this reduced the total time taken to complete each scenario. The data gathered from the surgical trial principally established that the concept was viable in terms of fulfilling one of its primary roles of being able to deliver life-sustaining treatment to those that require it.

Timing data have been emphasised in this trial for two main reasons. Firstly, the recovery of casualties from a hostile environment such as the North Sea must be completed swiftly and secondly, the delivery of life saving procedures must be timely. However, it has also been recognised throughout the trials that the quality of the treatment provided and the usability of the system are of paramount importance. The trials provided valuable opportunities for feedback and user trialling of the sort that can only be obtained by physically undertaking the tasks in a representative environment.

Every effort was made to use a representative range of subjects (varying sizes and body strengths) for both the paramedics and the casualties throughout this series of trials. On this basis, informal observations confirm that several tasks on the vessel are more suited to those users with greater body strength. The ergonomic changes made to the vessel have ameliorated this situation, although tasks such as pulling casualties from the water, despite being proven feasible, are still physically demanding. The physical workload is acceptable on the basis that the vessel will hopefully never, or rarely, be used for its intended purpose. However, it is suggested that a selection process for the medic crew should be investigated based on the physical tasks performed in these trials.

The man movement trial and the retrieval from water did not assess the potential limitations on crew tasks imposed by sea motion. This was mainly due to practical limitations with the current equipment capability. However, the medic subjects were always aware of this issue and had previous experience of working on small boats in rough seas, experience which became particularly useful when framing opinions from the trial in the debrief sessions. That said, any trial designed to assess the effectiveness and functionality of a craft still in the design stage has limitations which can only be remedied by building the craft and testing at sea. In this sense, the trials detailed above can be considered to be part of the de-risking process.

The trials have allowed a valuable assessment of the vessels likely performance where no other data previously existed, involving the innovative use of simulation facilities to help to test a design concept for usability. Design changes made as a result of these trials have improved the functionality of the prototype which should in turn improve the eventual vessel when it goes into production.

Acknowledgements

The authors wish to thank Capt. John Gorrie - BP Exploration, and Marine and Technical Marketing Consultants - Isle of Wight. In addition, the Institute of Naval Medicine for providing input into the surgical procedures under conditions of motion trial.

References

Bridger, R.S., Oakley, E.H.N., Green, A. & Bilzon, E (2002) Sick bay design for Autonomous Rescue and recovery Craft INM Report No 2002.036. Gosport, UK: IMN.

Crossland, P. & Rich, K. (1998). *Validating a model of the effects of ship motion on postural stability*. The 8th International Conference on Environmental Ergonomics, October San Diego, USA.

Furnell, A. & Mills, M.L. (2004). *Assessment of survivor retrieval from the water onto the afterdeck of the Autonomous Rescue and recovery Craft (ARRC), using the ocean basin, Haslar, and a collapsible section of the ARRC afterdeck collar*. QINETIQ/KI/CHS/TR040848. Farnborough, UK: Qinetiq.

Graham, R. (1990). Motion Induced Interruptions as Ship Operability Criteria. *Naval Engineers Journal, 102*, 65-71.

Martin, M. , Vashisht, B., Frezza, E., Ferone, T., Lopez, B., Pahuja, M., & Spence, R. (1998). Competency-based instruction in critical invasive skills improves both resident performance and patient safety. *Surgery,124*, 313-317.

Mills, M.L. & Crossland, P. (2002). *Assessment of offshore medic performance for the Autonomous Rescue and Recovery Craft*. Report QINETIQ/KI/CHS/CR022174/1.0. Farnborough, UK: Qinetiq.

Mills, M. (2003). *Autonomous Rescue and Recovery Craft (ARRC) casualty movement study*. Report QINETIQ/KI/CHS/CR030438/1.0 Farnborough, UK: Qinetiq.

Task analysis, subjective workload and experienced frequencies of incidents in an airport control tower

Clemens M. Weikert[1,3] & Suzanne A. van Ham[2,3]
[1]*Department of Psychology, Lund University, Sweden*
[2]*Department of Psychology, Maastricht University, the Netherlands*
[3]*Swedish Centre for Aviation Research and Development, Lund University, Sweden*

Abstract

This study aims at a better understanding of the behaviour of air traffic controllers in control towers. In the tower of a major Swedish airport a task analysis has been conducted by means of observations in addition to subjective workload assessments. Based on earlier research it was hypothesized that air traffic controllers with two years or less of experience would experience higher subjective workload, than air traffic controllers with more than two years of experience. Furthermore, it was hypothesized that when traffic intensity increases, more subjective workload would be experienced. Both hypothesises were confirmed, and interesting correlations between observed variables were found. To gain insight in the rate of recurrence of unwanted incidents, a questionnaire about experienced frequencies of problems with weather with reduced visibility, the clearing of snow, system failures and the distribution of flight progress strips was administered. Major problems indicated in the responses were system failures, the distribution of flight progress strips and the clearing of snow. Suggestions for improvement and future research are presented.

Introduction

"KLM one-one-one-three, cleared for take-off, runway one-niner right". Air traffic controllers in control centres all over the world are doing the same work, they guide aircraft from destination to destination. Furthermore, they try to make efficient use of airspace. They also provide additional information to pilots, helping them to avoid bad weather conditions and give assistance in their navigational activities. In an airport control tower, air traffic controllers have a slightly different job. They assist aircraft from the ground to the air and the other way around. From the moment they leave the gate till the time they reach the boundaries of the airport airspace, they fall under the responsibility of the control tower. At the control tower of this study a team of up to nine controllers take care of all the departures and arrivals as well as ground traffic at the airport. On a busy day this can amount to around 1000 aircraft movements in 24 hours.

After September 11th 2001 there was a considerable drop in air traffic, but in spring 2004 it was almost back at the same level as before 9/11. Due to expected future increases in air traffic, there is a need for development in air traffic control. Even the most efficient systems presently in use are not able to handle the increasing traffic volume. New technological systems and higher levels of automation are needed to better utilise airspace. Automatically transponded data, more advanced radar systems and sophisticated security nets are suggested solutions (Hopkin, 1999).

Technological improvements are made mainly for two reasons. They aim at a more efficient use of airspace, and to reduce air traffic controller workload. Low workload is not always beneficial for safety, however. Weikert and Johansson (1999) found in an analysis of incidents related to air traffic control, that most incidents happen during low to moderately high traffic situations. Although this is the case for ATCC (Air Traffic Control for aircraft en route), there are indications that incidents in the TWR/TMC (tower/terminal control) happen due to overload instead of low workload. Workload in the control tower is normally more consistent, i.e. higher, than in ATCC. In the research of Weikert and Johansson (1999), another contributing factor was identified. Air traffic controllers with less than two years of experience were overrepresented among the controllers who had reported incidents. This might be explained by the time it takes for a controller to acquire the proper situational awareness. Endsley (1995) suggests a three-stage model of situational awareness. The first stage involves perceiving components of the environment. The second stage involves implementing these components into the complete picture. In the third stage the integration of the previous stages is used for planning and anticipation. Doane et al. (2004) assume that building situational awareness requires multiple cognitive resources. Working memory is one example, and it is well established that experts and novices use this type of memory differently. Experts have learned to increase their working memory capacity by chunking, thus giving them an advantage in situational awareness over novices. In the study of Doane et al. (2004), in which expert and novice pilots were tested on anticipating consequences of situational awareness, this seemed to be the case. In the study by Weikert and Johansson (1999), the controllers with more than two years of experience might have developed better working memory skills than controllers who recently finished their training. Further support comes from research by Weikert and Wempe (2000) comparing subjective mental workload of first year air traffic control students, air traffic controllers who just finished their training and experienced air traffic controllers, showing that the more experienced controllers are, the less subjective workload they experience.

When developing new air traffic control systems, the continued use of so called flight progress strips is usually questioned. Flight progress strips are paper strips in plastic badges, used to represent airplanes. Pertinent information about the aircraft's flight plan is printed on the strip. Air traffic controllers use the strips as an aid to acquire a mental picture of the traffic flow. In addition, they can use the flight progress strips as memory aids by making notes on them. When an aircraft is approaching the boundaries of responsibility of a particular air traffic controller, the flight progress strip is handed over to the next controller in charge. The active handing over of the strip is a process that supports the mental picture, but the movement itself can interfere with other activities. The system was originally developed for control towers with little space between the different controllers, but it is also used in towers where the controllers actually have to leave their position and walk to deliver the flight progress strips. Electronic strip systems have been developed (Hopkin, 1999) replacing the paper strips with electronic displays, so that the strips can be handed over by pressing a button. The strips can be generated and updated automatically. In present stripless systems, however, the controllers can not make notes on the electronic strips, and if the computer system breaks down, there is nothing left that can serve as a memory aid.

In a study by Durso et al (2004) flight progress strip interactions during operational use were examined. They investigated frequencies of markings on flight progress strips, and conducted interviews with air traffic controllers to try to understand the importance and perceived benefits of the markings they made. Thus they could establish an indication of value and functionality of different markings, which might have implications for the design of future electronic systems.

The aim of this study is to conduct a task analysis of airport tower air traffic control. A task analysis is a systematic examination of tasks, and can be useful as an evaluation method to identify problems with existing tasks (Stammers & Shepherd, 1995). In this study the task analysis was conducted in the control tower of a major Swedish airport during normal operation. A major focus of the study was to see how the time and attention of an air traffic controller is divided between different systems and tasks, and on the experienced subjective workload of the controllers. Furthermore information on frequencies of unwanted incidents was collected.

The tower in this study is modern and has only been in use for a couple of years. It has seven work positions, two tower (TWR) positions handling landing and departing aircraft, two ground (GND) positions responsible for movements on ground, a flight data assistant printing flight progress strips from the computer and distributing them to the proper control position, clearance delivery giving start up clearances to aircraft and a supervisor. In this case only the TWR and GND positions were studied. The airport has three runways in use, two parallels (01-19 R and L) and one at a right angle to these (08-26, not intersecting). For an overview of the floor plan of the control tower see Appendix.

At this control tower flight progress strips are still in use and have to be distributed over relatively long distances due to the size of the tower. Other areas of interest for this study are the weather and problems with the systems used by the controllers. Low visibility might mean that the controllers can not see the entire runway system and thus have to rely on radar and other means. This airport also has to deal with snow for a considerable period of the year. System failures are not supposed to happen, but they sometimes do, which can have major consequences. The systems in use in this particular tower are SMGCS (radar with all movements on ground), RDP (radar with an overview of aircraft in the air), AWOS (weather system), AGL (runway lighting system), and AMP (a system that deals with the expected flight plans). Aircraft are presented by dots on the radar screens with a flight information label. The information is updated on every rotation of the radar system taking between 1 (SMCGS) and 6 (RDP) seconds. It sometimes happens that labels switch between airplanes, and sometimes other things are incorrectly identified as aircraft. Controllers therefore need to check if a dot on the screen is really the aircraft with its correct label. They can verify the dot in three different ways: weather permitting, they can look outside the window and see the aircraft that is represented on the screen, they can ask for a position report from the pilot and compare to the position on the screen, or if a dot disappears on one screen and appears on the other, they are allowed to assume it is the same aircraft.

The aim of this study is to try to identify possible bottlenecks in the tower work procedures including unwanted incidents and to study the relations between subjective workload, traffic intensity and controller experience. It is hypothesized that increased traffic intensity leads to higher subjective workload and that less experienced controllers would rate their subjective workload higher than experienced controllers.

Method

Subjects

Out of a total of 55 controllers working shifts at the control tower, 24 participated in the study, i. e. the ones working during the data collection period. General experience in air traffic control varied between 1 and 33 years (mean = 11.2 years, SD = 11.1). Experience of tower air traffic control was between 0 and 33 years (mean = 10.0 years, SD = 10.6). Six of the participating controllers had two years or less of experience at the tower.

Materials and procedure

Observations were performed using observation forms. For an overview of the observed variables see Table 1. The air traffic controllers at this tower work in 3 team shifts over 24 hours. Every team has a couple of controllers at rest and at work, alternating and taking turns between different positions and having breaks. Time in position varied between 30 and 60 minutes. Observations started when a controller came back from rest and started to work, and lasted until he or she was relieved again. Mean observation time was 44.2 minutes (SD = 15.2) and observations were randomly allocated to controllers at GND and TWR positions. Two researchers conducted the observations over 5 days (from Monday the 29th of March till Friday the 2nd of April 2004), day and night, to collect data over different periods of time and different traffic intensities.

Table 1. Observation variables

Date	Walking with flight progress strips
Time beginning observation – end	Handing over strips Otherwise
Darkness: daylight / twilight / night	Rearranging strips on desk
Visibility: good / bad / none	Amount of distractions[a]
Snow: yes / no	Amount of times in distress[a]
Position: GND / TWR (arrival/departure)	[a] : specification of distraction / distress

Immediately after being relieved from a position the controllers assessed their subjective mental workload using NASA-TLX (Hart & Staveland, 1988; Swedish translation by Weikert). NASA-TLX uses 6 Visual Analogical Scales (VAS) to measure subjective mental workload, including mental load, physical load, time pressure, performance, effort and frustration level. After this assessment, the controller was additionally asked to fill in a questionnaire. The questionnaires were administered in the Swedish language, with questions about experience in air traffic control, and perceived frequencies of unwanted incidents during the last two months. The observation forms, the NASA-TLX and the questionnaires were number coded to facilitate matching the data of each single controller.

The questionnaire was available for all air traffic controllers including those not observed. Three controllers therefore filled in a questionnaire without being observed. Due to the natural work rotation in the tower some of the controllers were observed only once, and other controllers were observed a couple of times.

Results

During the week that the data were collected, the weather was sunny and clear with visibility > 50 km, which is somewhat unusual for that particular period of the year. The results of this study are thus valid for very favourable weather conditions only. The data of one controller were not included in the analysis because of refusal to fill in the questionnaire and for another controller the NASA-TLX ratings were omitted due to obvious lack of seriousness when rating.

Experience and subjective workload

For the NASA_TLX ratings a mean value over all six subscales was computed for the two groups (experienced vs. less experienced controllers). The results show that controllers with two years or less of experience, rate their subjective mental workload significantly higher than air traffic controllers with more than two years of experience (means 4.1 and 3.0 respectively; $p<.05$). When repeating the analysis using only 5 of the 6 NASA-TLX subscales, leaving out 'Performance', the outcome was the same (means 3.3 and 2.1; $p<.05$). The main reason for this was that it was found that the subscale 'Performance' did not correlate with any of the other subscales. Correlation coefficients were computed, showing significant negative correlations between 'Subjective workload' and 'Experience' both for ATC in general and TWR ($r = -.37$, $p = .019$ and $r = -.36$, $p = .022$) meaning that the more experienced controllers were in air traffic control in general and in the tower, the lower they rated their subjective workload.

Traffic intensity

Traffic intensity has been assessed using the airport's statistics of aircraft movements during the week that the data were collected. Five categories of traffic intensity were created to define a variable 'objective workload' ranging from 1 movement per 4 minutes, to 1 per 2 minutes, 1 per 1 1/3 minutes, 1 per minute and 1 per <1 minute. Objective workload thus defined, correlated significantly with Subjective workload

($r = .37$, $p = .018$). When traffic intensity increased, the subjective workload was rated higher. With higher traffic intensity the controllers have to walk more with flight progress strips (one strip = one aircraft). The observation variable 'Walking with flight progress strips' correlated significantly with 'Subjective workload' ($r = .36$, $p = .02$) but not with 'Objective workload'.

Other variables relevant to tower air traffic control

Air traffic controllers who were working in the GND positions, were looking outside significantly more often than controllers who were working in the TWR positions. Controllers in GND positions were looking outside 3,02 times (SD=1,05) per minute in average and in TWR positions 1,46 times (SD=0.65) per minute ($p<.001$). The variable 'Looking outside' correlates significantly with 'Subjective workload' ($r = .35$, $p = .025$).

Experienced frequencies of unwanted incidents

Experienced frequencies of unwanted incidents over the last two months were calculated based on the questionnaire answers. For an overview of these results see Table 2. The major part of the reported system failures were related to the ground radar (SMGCS).

Table 2. Experienced frequencies of unwanted incidents. (FPS= Flight progress strips)

Weather with reduced visibility	(times)	0	1-2	3-5	6-8	>8	-
	(%)	8,3	45,8	29,2	8,3	8,3	-
System failures	(times)	0	1	2	3	4	>4
	(%)	12,5	12,5	8,3	4,2	0	62,5
FPS distribution	(times)	0	1-4	5-8	9-12	13-15	>15
	(%)	29,2	29,2	12,5	12,5	0	16,7
Snow	(times)	0	1-3	4-6	7-10	>10	-
	(%)	16,7	45,8	20,8	4,2	12,5	-

Snow on area	Apron	Taxiway	Runway	Total
(reported times)	51	44	22	117
(%)	43,6	37,6	18,8	100

When asked to describe a particularly serious situation, 10 out of 13 controllers gave descriptions of situations involving system failures or missed approaches. The open question about the seriousness of the snow clearing problems indicated that they were mainly caused by strings of snow left on aprons and taxiways, dividing up the area and blocking aircraft. Because of this air traffic controllers sometimes divert from rules in order to get traffic in and out from the gates.

Discussion

The results confirm the hypothesis that air traffic controllers with two or less years of experience, experience higher subjective mental workload than air traffic controllers with more than two years of experience, thereby supporting results from earlier studies by Weikert and Johansson (1999) and Weikert and Wempe (2000). Furthermore, the results confirm the hypothesis that when traffic intensity increases, experienced subjective mental workload increases as well. The variables 'Objective workload' and 'Walking with flight progress strips' showed significant positive correlations with Subjective Workload. One would expect the two variables 'Objective workload' and 'Walking with flight progress strips' to correlate positively as air traffic controllers by definition have to hand over strips more often when there is more traffic. The results do not support this assumption. This might be explained by how these variables were established. 'Walking with flight progress strips' is an observation of a clear and countable behaviour in the control tower, whereas 'Objective workload' on the other hand is assessed by using airport statistics. The statistics record take-offs and landings hour by hour around the clock. To adjust this information to the observation times estimations of traffic intensity over actual observation periods were made. The number of movements per hour does, however, not reveal anything about the distribution of aircraft movements over the hour. The fact that the statistics of the airport distinguishes between arrivals and departures is beneficial for the analysis of the TWR-positions, because these controllers handle either arrivals or departures on either of two runways. GND-positions on the other hand handle all aircraft and other airport related traffic on the ground. In high traffic intensity situations two ground-positions are open instead of only one and traffic on ground is then divided between two controllers, making the estimation of 'Objective workload' more complicated. In some sense the variable 'Walking with flight progress strips' constitutes a better estimation of objective workload with its direct relation to traffic. This variable correlates positively with 'Subjective workload' but not with 'Objective workload'. A possible explanation for the non-significant correlation between 'Walking with flight progress strips' and 'Objective workload' is that only controllers handling arrivals and ground controllers have to hand over flight progress strips (to the ground controller and departure controller respectively). However, controllers handling departures do not have to hand over strips, because these aircraft contact ATCC after take off. When removing the controllers who handled departures from the analysis, the correlation increased slightly but is still not significant. Due to time constraints no analysis was made with the two levels of controller experience separated. This will be done in the near future.

The variables 'Distractions' and 'Distress' (observation form) have been omitted from the analysis, because it was found to be difficult to detect when a controller was in distress or distracted. Another variable that was omitted from the analysis was 'Darkness', because of to few observations (7 observations of 3 controllers). Weather and visibility was the same as during the daytime observations.

The results show that controllers in GND-positions are looking outside more often than controllers in TWR-positions. This can be explained by the fact that TWR-controllers have a clear stream of traffic most of the time, neatly lined up on ground or in the air. Air traffic controllers in GND-positions have the responsibility for all traffic on the ground everywhere on the airport. During observations it could be noted that controllers in GND-positions are looking outside almost 360° around, whereas controllers in TWR-positions are looking outside in a particular pattern related to their runway in use.

The questionnaire on experienced frequencies of unwanted incidents shows that controllers view system failures as a major problem. In favourable weather conditions controllers probably merely experience the system failures as inconveniencies, but in marginal weather when they have to rely on their radar and computer screens it probably adds substantially to their mental workload. The airport in this study is certified for ILS Category II (instrument landings) allowing aircraft to land with a vertical visibility of 30 m and a runway visual range of 400 m. This means that in marginal weather the controllers, sitting 85 m up in the tower, can not see the airport or only part of it. The seriousness of system failures in those situations is emphasized by the fact that the system that is most frequently reported as having malfunctions is the SMGCS (ground radar).

The distribution of flight progress strips is also quite frequently reported as a cause of problems. In a large, modern tower like the one in this study, controllers have to leave their position and walk with the strip to the position which is to receive it. Walking with strips is complicated by the fact that the controllers are connected to their positions by headset cables. During observations it happened that controllers missed a telephone call or stumbled over head-set cables when walking with strips. When handing over strips, controllers also report 'loss of focus' and 'loss of the picture of the present traffic situation', resulting in a deterioration of their situational awareness (see Endsley, 1995). There are systems available and in use where the strips are presented electronically on computer screens, which at least would solve the distribution problem. On the other hand it is not as easy to make notes on the electronic strips as on the paper strips. Informal conversations with the controllers revealed that they would like to keep the paper strips. They use them as memory aids, making notes and markings on them. In case of a total system failure they can continue operations based on the strips. As mentioned above there is recent research on the function and importance of markings on flight progress strips (Durso et al., 2004).

According to the questionnaires weather with reduced visibility does not seem to be much of a problem. As long as the systems are working correctly, and no other problems emerge, controllers indicate that they are able to do their job without looking out onto the airport. Snow seems to be more of a problem. Most of the problems with snow are reported to occur on aprons. The airport prioritises the clearing of the runways which might cause problems to accumulate on taxiways and aprons. Concerning the aprons, the problem appears to be that the snow clearing equipment leaves strings of snow, thus reducing the freedom of movement for aircraft on ground which increases the risk for congestions and delays.

The excellent weather conditions during the period of data collection notwithstanding, some problems and possible bottlenecks could be identified. The main problems appear to be related to systems failures, flight progress strip distribution and snow clearing. These problems are probably aggravated during adverse weather conditions during which a substantial increase in mental workload is expected. A continuation of this study in not so favourable weather conditions is under preparation.

Ackowledgements

The authors wish to thank the staff at the control tower for participating in the study, and the airport authorities for financial support.

References

Doane, S.M., Sohn, Y.W., & Jodlowski, M.T. (2004). Pilot ability to anticipate the consequences of flight actions as a function of expertise. *Human Factors, 46*, 92-103.

Durso, F.T., Batsakes, P.J., Crutchfield, J.M., Braden, J.B., & Manning, C.A. (2004). The use of flight progress strips while working live traffic: Frequencies, importance, and perceived benefits. *Human Factors, 46*, 32-49.

Endsley, M.R. (1995). Toward a theory of situation awareness in dynamic systems. *Human Factors, 37*, 32-64.

Hart, S.G., & Staveland, L.E. (1988). Development of NASA-TLX (Task Load Index): Results of empirical and theoretical research. In P.A. Hancock, and N. Meshkati (Eds.), *Human Mental Workload*. Amsterdam: Noord-Holland.

Hopkin, V.D. (1999). Air Traffic Control Automation. In D.J. Garland, J.A. Wise, and V.D. Hopkin (Eds.), *Handbook of Aviation Human Factors* (pp. 497-517). Mahwah, New Jersey: Lawrence Erlbaum Associates.

Stammers, R.B., & Shepherd, A. (1995). Task Analysis. In J.R. Wilson, and E.N. Corlett (Eds.), *Evaluation of human work. A practical ergonomics methodology* (pp. 144-169). London: Taylor & Francis.

Swedish Civil Aviation Administration. (http://www.lfv.se).

Weikert, C. & Johansson, C.R. (1999). Analysing incident reports for factors contributing to air traffic control related incidents. *Proceedings of the Human Factors and Ergonomics Society 43rd Annual Meeting, Houston, Texas* (pp.1075- 1079), HFES: Santa Monca, Ca, USA.

Weikert, C. & Wempe, N. (2000). Subjective workload and stress – a comparison between air traffic control students and experienced air traffic controllers. *Occupational Health Psychology: Europe 2000*, 173-175.

Appendix: Floor plan of the control tower

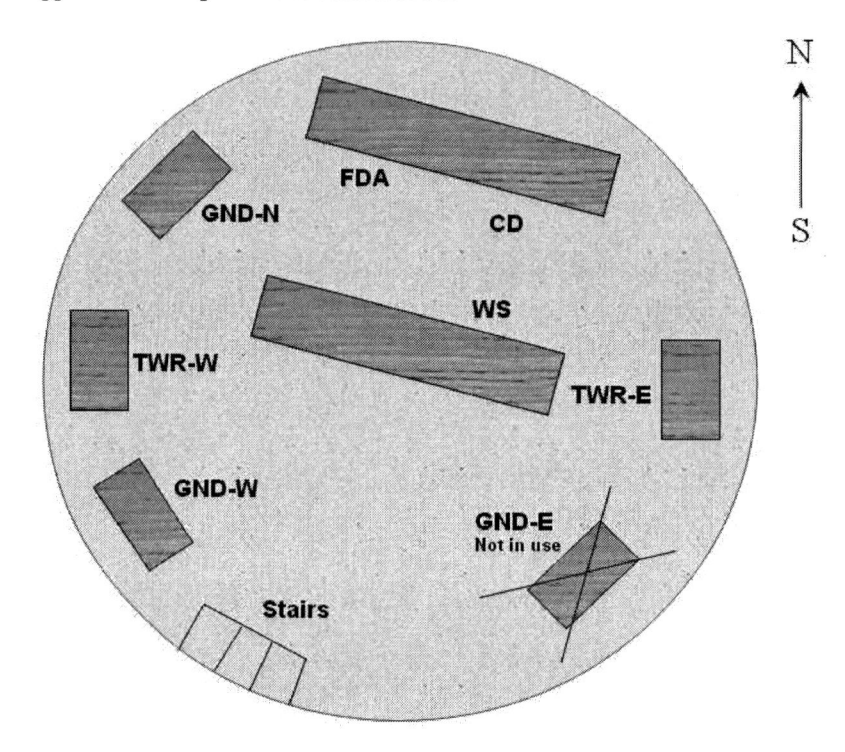

Abbreviations:

FDA	= Flight Data Assistant	GND-N	= Ground North
CD	= Clearance Delivery	GND-W	= Ground West
WS	= Supervisor	TWR-W	= Tower West
GND-E	= Ground East (not in use)	TWR-E	= Tower East

Measuring head-down time via area-of-interest analysis: operational and experimental data

Brian Hilburn
Center for Human Performance Research (CHPR)
The Hague, The Netherlands

Abstract

The so-called "Head-Down Time" (HDT) problem refers to the potential inability of an operator (e.g. a tower air traffic controller) to optimally divide attention between the primary visual field (out the tower window, for instance), and an auxiliary tool (usually in the form of a visual display screen). Recently-increased attention to the "runway incursion" problem has focused concern on the role that HDT might play in aerodrome ground safety — that is: What are the potential risks of HDT, and how are future Air Traffic Management developments likely to impact these risks? In the case of aerodrome surface movement, the question applies to three groups: tower / ground controllers; flight deck crew; and vehicle drivers.

An ongoing research project being carried out for EUROCONTROL is investigating in two phases the role and impact of HDT in aerodrome operations. An initial operational review phase identified HDT as a current and potentially important future issue, especially for tower air traffic controllers. Thus far, almost no research has been devoted to this potential problem. In the subsequent, experimental phase, a series of real-time simulations with tower controllers showed that controllers spend a majority of their time looking elsewhere than out the tower window. Not surprisingly, this was largely driven by external visibility conditions, and less so by the presence of auxiliary displays.

Introduction

The Runway Safety problem

The safety of civil aviation has improved greatly over the last half century. Improvements in terms of Communication, Navigation and Surveillance (CNS) capabilities, mechanical reliability, etc. mean that in-flight accident rates are now remarkably low. However, one persistent, and potentially growing, threat is the ground phase of operations, including landing, taxiing, and takeoff. Indeed, the most lethal aviation accident in history took place on the ground, when two loaded 747s collided on the runway at Tenerife in 1977, claiming a total of 583 lives. The chief threat to runway safety is the risk of "runway incursion," or conflict between an aircraft and either another aircraft, a vehicle, or a pedestrian.

Runway incursions appear to be on the increase[*]. EUROCONTROL reported that the number of severe (so-called categories A through C) runway incursions in Europe increased from 10 in 1999, to 44 in 2002. Similarly, the US reports a 130% increase in runway incursions from 1993 through 2000 (Jones & Rankin, 2002). Disturbingly, runway incursion risk outpaces traffic growth. A 20% growth in traffic, for instance, represents a 140% increase in runway incursion potential (Galotti, 2002). Given predicted traffic growth within Europe, it is therefore critical to address aerodrome ground safety.

The Head-Down Time (HDT) problem

Ironically, developments aimed at enhancing runway safety have in fact been identified as potential new sources of human error. The chief concern underlying the current work was that new technologies and displays might lure the operator (be it an air traffic controller, pilot, or even wheeled vehicle driver) into spending too much time "head-down."

The so-called "Head-Down Time" (HDT) problem refers to the potential inability of an operator (e.g. a tower controller) to optimally divide attention between the primary visual field (for instance, out the tower window), and an auxiliary tool (usually in the form of a visual display screen). Human factors have been linked to a vast majority (95%) of runway incursions, and failure to detect visual information has been cited as the largest single largest underlying factor (Cardosi & Yost, 2001). In aviation, effort into the HDT problem has tended to focus on the flight deck environment. Development of the "glass cockpit" over the last few decades has led to such innovations as the Flight Management System (FMS) and, more recently, graphical taxi display aids. Developments such as these have in turn fed concerns among both the R&D and operational communities that new systems might, ironically, force the crew into a "heads-down" mode, to the disregard of the primary visual input (e.g., the out-the-window view available under visual meteorological conditions). Indeed, many pilots of first-generation glass cockpit aircraft (such as the Boeing 757) reported that reprogramming the FMS in case of, for instance, runway changes, had became a demanding and distracting task during approach (especially at lower altitudes). Specific responses to the flight deck concerns have included the development of Head-Up Displays (HUDs), now being retrofitted on civil airliners. Over the years, HUDs have even made their way into passenger car production.

Human Factors of the HDT problem

There is a great deal already known about human factors of vision, including basic anatomy and optics of the eye, visual attention, and the factors influencing visual performance. A preliminary literature review drew on what is currently known in the areas of visual perception and attention, and especially on developments in various domains such as HUD development, ATC and flight deck operations, general

---

[*] This trend might in part simply reflect greater awareness of the problem

aviation, and road transport. Some of the relevant considerations included the following (c.f. Hilburn 2004):

- Complex/critical visual signals should be presented within 15 degrees of normal line of sight (Berson et al., 1981; Cardosi & Murphy, 1995);
- Angular differences large enough to require head movements (roughly 20 degrees) risk the head-down problem, and the chance of missed (or delayed responses to) critical signals (Martin-Emerson & Wickens, 1992, see Figure 1);
- Field of view is smaller vertically than horizontally, making vertical offsets potentially more critical than horizontal offsets;
- Although the risk of HDT lies in the possibility of missing critical events, absolute or percentage HDT says little about this risk;
- The aerodrome HDT problem involves not only changing point of gaze (head-up or head-down), but also readjusting (through the "near reflex") to a different viewing depth;
- Research into "innattentional blindness" (Mack & Rock, 1999) has highlighted the risk that operators can fixate without "seeing" external objects;
- That there can be too much head-up time, in cases in which more or better information is available head-down; and
- Little or no previous research has specifically addressed the head-down problem in aerodrome operations.

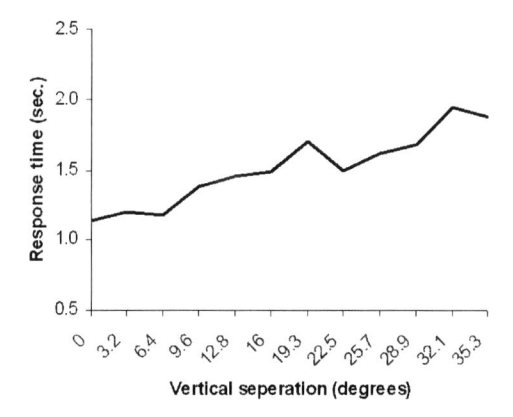

Figure 1. Response time, as a function of vertical separation between displays (after Martin-Emerson & Wickens, 1992).

Emerging aerodrome technologies

Advanced Surface Movement Ground Control System (or A-SMGCS) refers to a host of integrated technologies aimed at improving the surveillance, control, guidance, and route planning of aerodrome ground operations. Various systems are under development, and all are aimed at integrating the operation of aerodrome controllers, as well as pilots and vehicle drivers. Examples of A-SMGCS technologies include: Runway monitoring aids; In-vehicle moving map displays; Alerting systems; Vehicle transponder and tracking systems, etc. Although commercial systems are currently available or under development, the A-SMGCS concept will be evolving, both technically and procedurally, for some years to come.

The advent of A-SMGCS holds various potential human factors challenges, and might be expected to influence, for example: Attention demands and potential information overload; Greater reliance on the visual channel; and automation skills including monitoring and supervisory control. The International Civil Aviation Organisation recently identified the following aspects of A-SMGCS, which could exacerbate the head-down problem. .

- Less voice communications, and a greater mix of voice and datalink communications;
- more direct data transfer (i.e. data link) between systems;
- Seamless air and ground traffic management;
- More information sharing between controllers;
- Increased reliance on automation;
- Guidance, alerting, conformance monitoring, etc; and
- Moving map guidance

The current research sought to address the specific issue of whether aerodrome operations, and in particular future A-SMGCS scenarios, pose a substantial risk of head-down time. The research was carried out in two phases: The first consisted of a round of operational site visits, and the second a series of real-time simulations. These will now be discussed in turn.

Operational, site visits

Exploratory site visits were conducted to two major European aerodromes, each of which had fairly sophisticated SMGCS capabilities. The goals of these visits were to: Survey equipment in use; Observe vehicle, flight deck and ATC tower operations to determine the potential scope of the HDT problem; Conduct surveys and interviews on the HDT problem; and to devise and try out data collection techniques. Both visits were conducted under clear (good visibility) conditions.

Accurately tracking HDT behaviour in this context required a form of eye tracking. However, eye tracking in the various operational environments (i.e. control tower, flight deck and vehicle cabin) posed several challenges—chief among these were [1] that the system must be portable (for use in the flight deck); [2] that installation and removal time should be minimal, and [3] that the system should be non-intrusive. It was decided to use video-camera based "Area of Interest" (AOI) eye tracking (McKnight, Shinar & Hilburn, 1991; Dietz et al., 2001; Andersen & Hauland, 2000). AOI analysis provides coarser-grained scan information than do precise gaze measurement methods. The idea behind AOI is that scan pattern is related to task-specific AOIs. Example AOIs are shown for analyses of automobile driving (Figure 2a) and ATC (Figure 2b) data. AOI data are typically analysed in either the spatial domain (e.g. considering transitions between areas of the task environment (Merchant & Schnell, 2000)), or in the temporal domain (e.g. as in a timeline analysis across AOIs (Cabon, Farbos, & Mollard, 2000)).

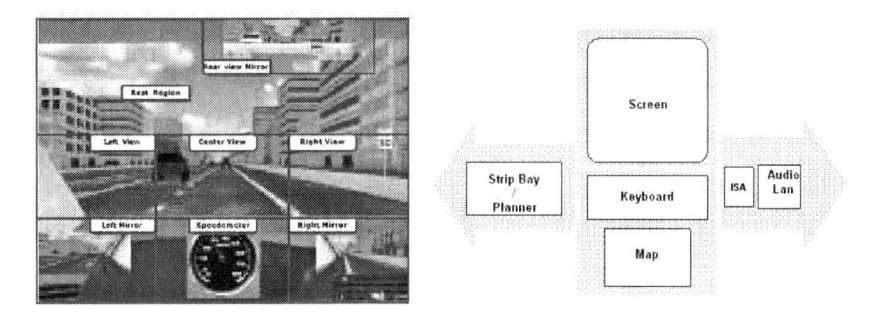

Figure 2. Visual AOIs for a simulated driving task (left, after Zheng, Tai & McConkie, 2003) and a simulated ATC task (right, after Hilburn & Nijhuis, 1999).

Method, site visits

Based on surveys and interviews of drivers, pilots and controllers, it was concluded that controllers face the greatest current and near-term threat from head-down time. Scan behaviour was therefore assessed only for controllers. A compact digital video camera was used to fixate controllers' faces, from a remote location, roughly 1.2 m in front of the controller. The camera was positioned between screens in such a way that it did not obstruct the view to either screen, nor out the tower window. A brief bit of calibration footage was next recorded, in which the controller fixated reported "hotspots" in the visual field (e.g., de-icing station, radar screen, keyboard, etc.). A recording was then made for a period of 10-30 minutes. In addition to video recording, data collection consisted of audio-recorded interviews, over-the-shoulder observations with behavioural coding and notes, physical measurements of workplace setup, etc. Video data reduction involved realtime logging (from video replay) of controller visual AOI.

Results, site visits

Survey results showed that controllers were the group most concerned with the HDT problem: some 40% of controllers, and 25% of pilots, agreed that they spend more time head-down than they used to. Vehicle drivers reported the smallest time head-down, and reported that this had not changed over time.

Controllers across the two sites looked down an average of roughly 55% of the total time. This was true for both sites. Tower controllers tended to spend slightly more total HDT than did ground controllers: 56% versus 51%, and 57% versus 52%, at the respective sites. More important than total or percentage time spent head-down, however, might be the average or maximum time spent head-down per fixation. For instance, an automobile driver who spends 50% of his time head-down might switch his glance (to look out the windscreen) every 3 seconds, or he might look out once every minute. The latter is clearly more dangerous. Based on the site visit data, controller head-down fixations averaged 13% longer than did head-up fixations (4.4s versus 3.9s), with no trend toward either a controller type (Tower versus Ground) or Site effect.

432   Hilburn

Simulator study

As part of a separate ongoing project, a series of five three-day simulations was recently conducted at the French Air Navigation Study Centre (CENA) in Athis Mons, France. The overall aim of the simulations was to evaluate the impact of A-SMGCS on tower ATC operations at Paris' two main airports, Roissy Charles de Gaulle (CDG) and Orly (ORY). In all, three CDG and two ORY simulations were carried out. The current effort piggybacked its data collection on this series of simulations.

Method, simulator study

Participants were 15 operating air traffic controllers, all licensed in both Ground and Tower operations, and currently working at either CDG or ORY. Average age was estimated at 25 years. Each three-day simulation consisted of nine 60-minute sessions. Sessions were specifically manipulated to vary within-subject the following two factors:

> *Automation level* (3)—defined as either
>   - Level 0— Baseline (current day) system;
>   - Level 1— added ground tracking, through transponder identification;
>   - Level 2—as above, plus runway incursion alerting capability.
> *Visibility (3)*—from clear to completely overcast conditions

Data recording was carried out using a total of three Digital Video (DV) recorders—one each trained on the Tower and Ground controllers, and one "scene camera" shooting over the shoulder. The three views were synchronised by flash at the beginning of each session. Data reduction and analysis involved real time video review and logging of fixation area, defined in terms of five task-relevant AOIs.

Results, simulator study

Data collection was recently completed, and data reduction/analysis is ongoing. On the basis of preliminary data analysis (using 60% of the eventual dataset), however, certain trends are emerging. For instance, controllers spent the vast majority (73.8%) of their total time head-down. Only 13.6% of their time was spent looking heads-up, out the tower window. Not surprisingly, there was a strong effect of visibility: controllers tended to fixate head-down much more under low visibility conditions. Figure 3 shows the average percentage of *head-up* time across tower and ground controllers, as a function of both visibility and automation level. Controllers spent much less time looking out the window under low visibility conditions (collapsed across visibility levels 2 and 3) than under clear conditions (5.4% and 27.9% head-up time, respectively). This is hardly surprising, and a clearly sensible strategy, if better information is available head-down (i.e. from data screens). Similarly, A-SMGCS automation (collapsed across levels 1 and 2) was associated with less total head-up time, relative to baseline (manual) conditions, at 17.8% and 11.4%, respectively. This effect was most pronounced under clear visibility conditions, though it is not clear (as shown in Figure 3) whether this is due in part to a floor effect under low visibility (i.e., in which data values were already extremely low).

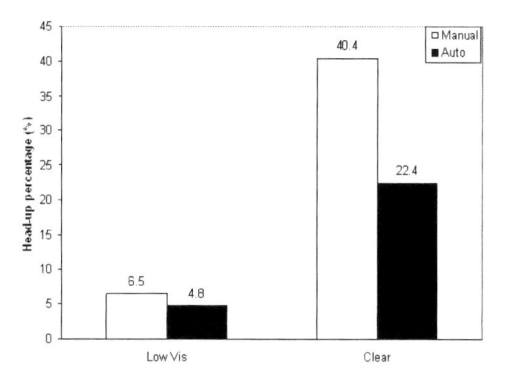

Figure 3. Percentage of head-up time, by visibility (Low vis vs Clear) and automation condition (Manual vs Auto Levels 1+2), collapsed across Tower and Ground controllers.

Another issue of interest was what influence the presence of scripted "non-nominal" (i.e. failure mode) conditions would have on HDT behaviour. Several failure episodes of seven minutes each were scripted. Each episode simulated the failure of Surface Movement Radar (SMR), with attendant loss of radar tracking data for all aircraft and ground vehicles on the aerodrome surface. Notice that during failures, better information on the location and movements of aircraft and vehicles was available out-the-window. It was therefore thought that such failure episodes might, at least under clear visibility, lead controllers to spend more time head up.

As shown Figure 4, controllers tended to go head-down during low visibility failures, but under clear weather failures did not tend to shift their attention out the window (where presumably better information would have been available). Interestingly, one effect of transient SMR failures was for controllers to increase head-down time, *after* the system was restored.

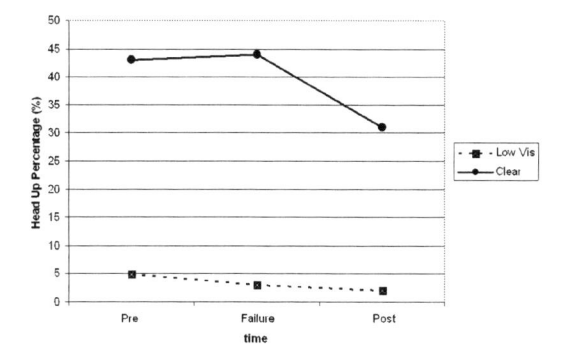

Figure 4. Effect of transient automation (SMR) display failure on percentage of controller head-up time, as a function of visibility condition (Low vis vs Clear).

There was a slight tendency for controllers to shift attention head up, when faced with automation failures under clear visibility. The effect, though, seems surprisingly small given that the failure was obvious, and controllers should have known that screen-derived information was degraded. Notice that the average head-up percentages under clear conditions (30-45%, as shown in Figure 4) are higher than generally seen in the real-time simulations. This appears to be a data artefact: First, these results were based on a small number of simulated failures; Second, pre-, failure, and post-periods were fairly small time intervals. These results, should therefore be seen as tentative. Further, the data from low visibility conditions is somewhat difficult to interpret, because of the "floor effect" under low visibility—that is, controllers under low visibility were largely already head-down at the onset.

Discussion

On the basis of site visits, it was concluded that controllers (as opposed to pilots and vehicle drivers) face the largest threat of HDT problems, both in current and likely near-term A-SMGCS scenarios. Simulation results clearly suggest that HDT is influenced by external visibility conditions. While hardly a surprising result, this does tend to validate the use of the AOI technique in the current context. On the basis of preliminary (simulation) data analysis, it also seems that automation as implemented in the simulations indeed has the potential to exacerbate the "head-down time" problem. However, several caveats seem in order. First, the absolute HDT values obtained in the simulation trials is much lower than that obtained in operational (control tower) settings. This could be evidence of a simulator effect (bear in mind that the head-down view-- real screens and flight strips—is still more realistic than the head-up simulated polygon world). Alternatively, it could be a function of the participant sample. For a variety of staffing and logistical reasons, only very young controllers were assigned to the simulation. One common conjecture among controllers—both young and old—appears to be that younger controllers tend to spend more time head-down, using the screens. Whereas the simulator data might be biased in absolute terms, it is thought that they probably do not confound the comparison between different automation conditions, or between the visibility conditions.

Obviously, some amount of HDT is desirable (if, indeed, a new display is to be used). Therefore, percentage or total HDT can tell us only part of the story of how a controller (or more generally, any operator) processes visual information. One promising avenue for answering such questions has been the use of scan path modelling, which attempts to relate scan pattern directly to task information requirements, and to evaluate scan against an optimal pattern. Modelling eye scan behaviour has often relied on the notion of information theory, which has been used since the 1950s to describe how someone should fixate different elements of a display, based on update rates, criticality of the information, and costs/benefits of missed information. Information theoretic approaches (based on pioneering work in the 1940s by Shannon and Weaver in communication theory) determine the amount of information (expressed as bits) a given scan pattern extracts, and were instrumental in establishing the traditional flight deck display layout (Fitts, Jones &

Milton, 1957). There is ongoing work in this area, and its application (while not without many practical limitations) would seem a valuable way to address such larger questions as "How much head-down time is *too* much?"

References

Andersen, H.H.K. & Hauland, G. (2000). Measuring team situation awareness of reactor operators during normal operation in the research reactor at Risø: a technical pilot study. In *Proceedings of the 3rd International Conference on Methods and Techniques in Behavioral Research*. Nijmegen, The Netherlands

Berson,B., Po-Chedley, D., Boucek,D., Hanson, D., Leffler, H., & Wasson, H. (1981). *Aircraft Alerting Systems Standardization Study, Vol 2: Aircraft Alerting System Design Guidelines*, DOT/FAA/RD-81/38/II.Washington, DC: FAA.

Cabon, P., Farbos, B., & Mollard, R. (2000). *Gaze Analysis and Psychophysiological Parameters: A Tool for the Design and the Evaluation of Man-Machine Interfaces*. Feasibility Study. Report EEC 350. Brétigny-sur-Orge, France: Eurocontrol

Cardosi, K. M & Murphy, E.D. (1995). *Human factors in the design and evaluation of air traffic control systems* (DOT/FAA/RD-95/3). Cambridge, Massachusetts: U.S. Department of Transportation, Volpe Research Center.

Cardosi, K. & Yost, A. (2001). *Controller and pilot error in airport operations: A review of previous research and analysis of safety data* (DOT/FAA/AR-00-51). Washington, DC: FAA Office of Aviation Research.

Dietz, M. et al. (2001). Tracking pilot interactions with flight management systems through eye movements. Proceedings of the *11th Annual International Aviation Psychology Symposium*. Ohio, USA: Ohio State University, Columbus.

Fitts, P.M., Jones, R.E. & Milton, J.L. (1950). Eye movements of aircraft pilots during instrument landing approaches. *Aeronautical Engineering Review, 9*, 1-5.

Galotti, V. (2002). ICAO Developments on Runway Incursions. Presented at the *ICAO/NAM/CAR/SAM Runway Safety/Incursion Conference*. Mexico City: 22-25 October.

Hilburn, B. (2004). *Head-Down Time in Aerodrome Operations: A Scope Study*. Den Haag, The Netherlands: Center for Human Performance Research.

Hilburn, B. & Nijhuis, H.B. (1999). *Eight-States Free Route Airspace Project (FRAP): Human Performance Measurement Results*. NLR Contract Report. Amsterdam: National Aerospace Laboratory NLR.

Jones, D.R., & Rankin, D.M. (2002). A system for preventing runway incursions. *Journal of ATC, July-Sep.*, 18-22.

McKnight, A.J., Shinar, D., & Hilburn, B. (1991). The visual and driving performance of monocular and binocular heavy-duty truck drivers. *Accident Analysis and Prevention, 23*, 225-237.

Mack, A. & Rock, I. (1999). Inattentional Blindness, *PSYCHE, 5(3), May*.

Martin-Emerson, R. & Wickens, C.D. 1992). The vertical visual field and implications for the Head Up Display. In *Proceedings of the 1992 Human Factors Society Meeting*. Santa Monica CA, USA: HFES.

Merchant, S. & Schnell, T. (2000). Applying Eye Tracking as Alternative Approach for Activation of Controls and Functions in Aircraft. *Proceedings of the 19th Digital Avionics Systems Conference,* Philadelphia, 7-13 October.

Zheng, X.S., Tai, Y., & McConkie, G.W. (2003). Effects of cognitive tasks on drivers' eye behavior and performance. Poster presented at the *2nd International Driving Symposium on Human Factors in Driver Assessment,* Training, and Vehicle Design, July 21-24, Park City, Utah, USA.

Abbreviations

A-SMGCS	Advanced Surface Movement Ground Control System
AOI	Area of Interest
ATC	Air Traffic Control
CDG	Charles de Gaulle aerodrome Paris
HDT	Head-down Time
HUD	Head-up Display
ICAO	International Civil Aviation Organisation
ORY	Orly aerodrome Paris
SMR	Surface Movement Radar

Aviation incident reporting in Sweden: empirically challenging the universality of fear

Kyla Steele & Sidney Dekker
The Swedish Network for Human Factors in Aviation
Linköping University, Linköping, Sweden

Abstract

To support the efforts of the Swedish Civil Aviation Authority to improve their reporting system an investigation is underway which looks at the factors which influence people's willingness to report incidents. This study consists of interviews with employees at a cross-section of airlines, as well as data gathering from the Swedish media and other sources. This paper examines the cultural issues which form the backdrop of the research. The issues of trust, fear, and blame are particularly interesting, since Sweden is a relatively non-punitive society with an extremely liberal freedom of information act. Questions about the universality or cultural dependence of fear arose because the theories on incident reporting originate mainly from English speaking countries and the structure of the reporting system in Sweden appears successful even though it defies the main tenets of those theories. Results from cultural comparisons by Geert Hofstede, and Helmreich and Merritt are interpreted as they apply to the study topic. The initiative taken by the Swedish CAA to embark on this project could be an example to other organizations to carry out broadening checks even when an existing 'safety system' appears to be successful.

Introduction

Incident reporting schemes are used in aviation as a tool for communicating near-miss experiences with other practitioners. This shared information can be used to identify and address safety concerns with the objective of avoiding accidents (see Johnson, 2000, 2003 for other potential benefits). Incident reporting systems are mandatory for the majority of commercial flight operations (including all airlines) within the current jurisdiction of the Joint Airworthiness Authority (JAA). Whereas it was (in theory) optional for European nations to join the JAA, compliance with regulations of the JAA's newly formed successor, the European Aviation Safety Authority (EASA) will be mandatory for all European Union (EU) member states.

The EASA regulation requiring individual companies to establish and run internal incident reporting systems will take effect in July, 2005 (Official Journal of the EU, 2003). The requirements provide basic guidance as to the structure and approach of the reporting system but beyond that it becomes the responsibility of the organisation to fulfil the requirements to whatever extent, and using whatever methods, they deem appropriate. The national flight safety regulatory bodies, or Civil Aviation Authorities (CAA), have no systematic means of assessing the effectiveness of such a system once it is in place – in many cases all they can do is verify whether a company has formal provisions for reporting and following up on reports, in compliance with the general guidelines.

The origins and focus of the existing knowledge base

How then, in the absence of explicit standards, should companies and CAAs proceed with creating and managing a reporting scheme? Where can they look for counsel on how to best collect and make effective use of incident data? The available literature often focuses on the solicitation of incident reports and specifically how this is influenced by fear and trust issues (Baker, 1999; Johnson, 2000, 2002b; Nicholson & Tait, 2002; Orlady & Orlady, 1999; Pidgeon, 1991; Pidgeon & O'Leary, 2000; Reason, 1997; Sullivan, 2001) and with few exceptions does not devote as much attention to issues of analysis and follow-up. These sources argue that incident reporting depends on the initiative of the individual to report occurrences, therefore any fear of repercussions for 'punishable' behaviour or actions (e.g., rule violations or errors) revealed in the reports will reduce, if not completely arrest the flow of reports, and there is historical evidence supporting this assertion (Dekker, 2003; Orlady, 1999). Thus, the literature recommends that in order to successfully facilitate incident reporting the system must be confidential, it must be administered by a neutral third-party (i.e., neither the employer nor the regulator), and there must be an official immunity policy protecting the reporter from disciplinary action or legal prosecution.

Upon reflection of the origins of the available material, one will notice a trend, somewhat typical of aviation: The published theories[*] on incident reporting originate predominately from 'Anglo' countries (e.g., the USA, the UK, Australia). The emphasis on overcoming fear in this literature may not be surprising since these nations, in particular the USA, are well known as litigious societies. It seems plausible, however, that the underlying fear of repercussions for revealing one's mistakes may be a universal human response, neither specific to national nor professional culture but this is an open question.

This same literature recommends that a blame-free, open, or 'reporting' culture is a crucial component of a strong 'safety culture' (Reason, 1997) or a characteristic of a well-functioning or generative organisation (Westrum, Lecture presentation at Linköping University, October 20, 2004). The implication here appears to be that fear is a product of the environment and thus is controllable. Since that environment is influenced by the organisational, professional, or national culture or some combination thereof (Helmreich & Merritt, 1998) then in theory it may be possible

---

[*] More correctly, the theories published in English predominately come from 'Anglo' countries. However, this situation is changing, partly due to the contributions of the recently held workshops on Investigation and Reporting of Incidents and Accidents, the published proceedings of which contain international submissions (Johnson, 2002a; Hayhurst & Holloway, 2003).

to have entire nations which are actually open and blame-free. This leads to the question of whether the theories on the influence of fear on incident reporting are generalisable to the global aviation community, or only applicable to the national cultures from whence they came?

The Swedish incident reporting system: against all odds

These question about the international applicability of organisational theory and the universality of fear arose during the literature review for an investigation of the willingness to report using the Swedish civil aviation incident reporting system (for the details of this study see Steele & Dekker, 2004). The Swedish system appears remarkable in light of the published theories because it is not confidential, it is administered by both the company and the national regulator, and there is no official immunity policy; thus it is breaking *all* of the rules! The system's very existence seems to defy logic in light of these theories, yet in recent years the Swedish CAA received over 2000 reports annually. Based on the number of reports per production flight hour, the Swedish aviation community reports at a higher rate than their American counterparts using the confidential, non-punitive Aviation Safety Reporting System (ASRS) administered by NASA, a system often used as a benchmark for success on a world scale.

The objection could be raised at this comparison that the Swedish system is mandatory, whereas ASRS is voluntary. There are at least two flaws in the logic of this argument, however. Firstly, it presumes that there is some 'respect for the law' or fear of punishment motivating Swedes to report. It is illogical to claim that fear motivates them to report while simultaneously arguing that these same people do not fear punishment if the reports reveal rule violations. They may not feel stuck between the proverbial rock and the hard place – if there is no fear then one cannot understand the issue in these terms.

Secondly, the distinction between mandatory and voluntary reporting is debatable since beyond the 'reportable incidents' explicitly listed in the JARs or EASA regulations, people are instructed to 'report situations where safety was compromised' (Official Journal for the EU, 2003). This is fundamentally negotiable, since 'safe' itself is a subjective quality, and in an inherently risky endeavour such as flying there is always some degree of risk. The old aviation saying that 'a landing is just a controlled crash' illustrates both the inherent risk and this subjective concept of safety.

Thus the concept of 'mandatory reporting' is problematic and cannot fully account for the steady reporting rate in Sweden. One explanation could be that the theories do not apply in Sweden because the national values are different. Another explanation could be the system in Sweden is not really 'working' as well as it appears to be on the surface (e.g. possibly people only feel comfortable reporting incidents of a technical nature rather than things which could incriminate oneself or a colleague, or perhaps some companies (e.g. the larger ones, or those with strong union support) may account for the majority of reports, thus giving the impression that the national system is 'working'). While there may be other explanations, this paper will focus on these two hypotheses since they could be relevant for the future work of the Swedish CAA and the European aviation community in general.

Comparing national cultures: the difference in values between Sweden and the USA

In the seminal survey of cross-cultural values done by Geert Hofstede (1980, 2001) he concludes that based on the revealed differences between national cultures organisational theories will not necessarily generalise beyond the country of origin. Reason (1997) plainly acknowledges this, essentially as a disclaimer, stating that his recommendations for 'engineering a safety culture' and a 'reporting culture' within organisations are not necessarily valid for other national cultures. Since the incident reporting theories are based on learned values like 'trust', it is relevant to understand the differences of such values between Sweden and the Anglo countries where the confidential, third-party model incident reporting system and the accompanying theories originated (i.e., the USA, Great Britain, and Australia).

Hofstede's study of IBM employees in 53 countries identifies and assesses countries based on five dimensions of culture: Power distance, uncertainty avoidance, individualism, masculinity, and long-term orientation (2001). For the purposes of this perfunctory discussion, we will use the US scores as representative of these Anglo countries, since the US, Great Britain, and Australia's scores differ by no more than 16 points (on a scale of approximately 100 points) for UAI and 5 points for the remaining four dimensions. Hofstede found statistically significant and distinct clusters for both the Anglo and Scandinavian countries.

Masculinity

Low Masculinity:

- Permissive and corrective society.
- More people see the world as a just place.
- Positive attitudes toward institutions and political establishment.
- Resolution of conflicts through problem solving, compromise, and negotiation

High Masculinity:

- Punitive society.
- More people see the world as an unjust place.
- Negative attitudes toward institutions and political establishment.
- Resolution of conflicts through denying them or fighting until the best "man" wins.

Looking closely at the values which these dimensions embody sheds some light on the cultural difference between the Swedish and Anglo cultures. Below are some of the "typical norms and values" associated with four of Hofstede's (1980) dimensions (Long Term Orientation was omitted). These are general characteristics representing the extreme ends of the spectrum and Hofstede's compiled results do not contain specific information about the applicability of each item to each country, but on a high level it should still be informative to compare country scores. These values were chosen as exemplars since they directly relate to trust and fear of blame or punishment and thus may be relevant to the willingness of people to report incidents.

The dimension of Masculinity refers to the "distribution of emotional roles between the genders" (Hofstede, 1980, p.xx). A high Masculinity ranking indicates the country experiences a high degree of gender differentiation. A low Masculinity ranking indicates that females are treated equally to males in that society. On a scale of 1 to 100 Sweden had a score of 5, the lowest of all countries surveyed, and the US scored 62 (Hofstede, 1980, p.286). This indicates that Swedish society may be more cooperative and less punitive than the United States.

Individualism

Low Individualism:

- Organizational success attributed to sharing information, openly committing oneself, and political alliances.
- Belief in collective decisions.
- Employer-employee relationship is basically moral, like a family link.
- Management is management of groups.
- Harmony: Confrontations to be avoided.

High Individualism:

- Organizational success attributed to withholding information, not openly committing, and avoiding alliances.
- Belief in individual decisions.
- Employer-employee relationship is a business deal in a "labor market".
- Management is management of individuals.
- Confrontations are normal

The dimension of Individualism is "the degree to which individuals are supposed to look after themselves" and low Individualism (or Collectivism) implies that people feel the need to "remain integrated into groups" (Hofstede, 1980, p.xx). The USA scored 91, the highest of all countries surveyed, and Sweden scored 71 (Hofstede, 1980, p.215). Both countries scored in the upper-half of the scale, however if the difference is significant then it implies that Sweden is more collectivist, believes more in information sharing, and employees may not be as fearful of their employers as in the USA.

The dimension of Uncertainty Avoidance represents "the extent to which a culture programs its members to feel either uncomfortable (higher score), or comfortable (lower score) in unstructured situations" (Hofstede, 1980, p.xix). Sweden scored below the US, with results of 29 and 46 respectively (Hofstede, 1980, p.151). The lower score of Sweden in this dimension implies that Swedes are more flexible, open, and understanding of rule violations and have more faith in the good intentions of the government and legal system than Americans.

The final dimension to be discussed is Power Distance, the "extent to which the less powerful members of [groups] accept and expect that power is distributed unequally" (Hofstede, 1980, p.xix). These scores were relatively close together and both were in the lower half: Sweden with a score of 31 and the US with 40 (Hofstede, 1980, p.87). If the difference is significant, and Sweden is certainly known as an equal society with flat organisational structures, then this has clear implications for incident reporting due to the nature of the employer-employee relationship.

Uncertainty Avoidance

Low Uncertainty Avoidance:

- Optimism about employers' motives.
- Admit dissatisfaction with employer.
- If necessary, employees may break rules.
- Openness to change and innovation.
- Most people can be trusted.
- Belief in one's own ability to influence one's life, one's superiors, and the world.
- Truth is relative.
- Few rules: if children cannot obey the rules, the rules should be changed.
- Hope of success.
- Citizens have confidence in civil service.
- Citizens may protest government decisions.
- Few and general laws and regulations.
- Citizens positive toward legal system.
- Laws usually on my side.

High Uncertainty Avoidance:

- Pessimism about employers' motives.
- Don't admit dissatisfaction with employer.
- Company rules should not be broken.
- Conservatism, law, and order.
- One can't be careful enough with other people, not even with family.
- Feeling of powerlessness toward external forces.
- Concern for Truth with a capital T.
- Many rules: if children cannot obey the rules, they are sinners who should repent.
- Fear of failure.
- Citizens lack confidence in civil service.
- Citizens' protest should be repressed.
- Many and precise laws and regulations.
- Citizens negative toward legal system.
- Laws usually against me.

Power Distance

Low Power Distance:

- Openness with information, also to non-superiors.
- Hierarchy means an inequality of roles, established for convenience.
- Superiors are people like me.
- Subordinates are people like me.
- The use of power should be legitimate and is subject to the judgment between good and evil.
- The system is to blame.
- Latent harmony between the powerful and the powerless.
- Flat organization pyramids.
- The ideal boss is a resourceful democrat; sees self as practical, orderly, and relying on support.
- Managers rely on personal experience and on subordinates.
- Consultative leadership leads to satisfaction, performance, and productivity.
- Institutionalised grievance channels in case of power abuse by superior.
- Citizens distrust press but trust police.
- Prevailing political ideologies stress and practice power sharing.

High Power Distance:

- Information constrained by hierarchy.
- Hierarchy means existential inequality.
- Superiors consider subordinates as being of a different kind.
- Subordinates consider superiors as being of a different kind.
- Power is a basic fact of society that antedates good or evil; its legitimacy is irrelevant.
- The underdog is to blame.
- Latent conflict between the powerful and the powerless.
- Tall organization pyramids.
- The ideal boss is a well-meaning autocrat or good father; sees self as benevolent decision maker.
- Managers rely on formal rules.
- Authoritative leadership and close supervision lead to satisfaction, performance, and productivity.
- No defense against power abuse by superior.
- Citizens trust press but distrust police.
- Prevailing political ideologies stress and practice power struggle

National culture: an explanation or a starting point?

In summary, many of the "norms and values" associated with Hofstede's original four cultural dimensions relate to trust, the justness and openness of a society, and the relationship between employees and management. The ones listed here all seem to support the hypothesis that Sweden and its like-minded Scandinavian neighbours may be more open, trusting, blame-free, and less punitive than the United States and the other Anglo countries. These qualities are associated with better 'safety culture' and certainly have bearing on the willingness to reveal mistakes through reporting of incidents, in order to support the larger cause of improving flight safety.

It should be kept in mind that these items were chosen subjectively by the researcher in a search for insight into trust, blame, and fear issues in the Swedish culture which could specifically explain the apparent success of the incident reporting system. The support within Hofstede's work for the hypothesis that Sweden is a more open, blame-free culture seems overwhelmingly strong and one-sided with the exception of a small number of items, most of which relate to Individualism. This is by no means a scientific finding, since this type of exercise is dependent on personal interpretations and there was no mechanism used to control for bias, however the reader is encouraged to consult Hofstede's original work and judge for him or herself.

At first glance this seems like a logical and sufficient explanation to the original question; it even makes for a happy ending. It may be tempting to accept this conclusion without deeper examination but throughout the course of the research bits of evidence accrued which both support and refute this Utopian image, giving the impression that attitudes in Swedish aviation lie closer to some middle ground.

Empirical evidence of fear, blame, and trust in Scandinavia and Sweden

The CAA's awareness of the existence of fear

The first indications to the researchers that fear might be a factor in aviation incident reporting in Sweden came from the Swedish CAA itself. They regularly survey pilots and other aviation professionals (as well as managers, but they are surveyed separately) on a whole range of aspects of the Swedish aviation system. One question asks "Why do you believe the reporting rate is not higher?" Of the available responses "Fear of repercussions from the Swedish CAA" was overwhelmingly the most popular, chosen by forty percent of respondents (more than one answer could be selected) and "Fear of my identity being known", which some feel may be interpreted as the same or related to the other 'fear' answers, had a response of twenty-four percent (Temo, 1999).

An awareness of fear was also revealed during interviews and discussions with a small number of CAA employees in which they plainly stated that company managers and individual employees are afraid of the Authority (CAA personnel, Personal communications, September 22, 2003, November 22, 2004). With respect to the CAA's power to revoke licenses one CAA staff member explained that the system is not completely non-punitive or "blame-free, but it is fair" (CAA personnel, Personal communication, November 19, 2004).

From the perspective of those outside Scandinavia[*] at least some degree of fear is not at all surprising, in fact it seems natural in light of the fact that the CAA has the power to revoke licenses and initiate legal proceedings. It may in fact be surprising for those from Anglo cultures to hear that the above survey results were not one-hundred percent, or that *all* CAA employees do not see fear as a major issue, or that in the same survey only nine percent chose "Fear of repercussions from my employer" and one quarter selected "I do not know" as a response which implies that for those respondents fear is not a major factor in the reporting decision. Although it is evident that fear exists to some degree, this type of data may hint at the quantitative difference in the extent to which it dominates the decision process.

From within the CAA the awareness and acceptance of the existence of (significant) fear is not automatic, widely accepted, nor the official "party line". This makes sense from the standpoint of the individuals in the CAA head office since they handle thousands of reports each year and they are personally not in the practice of punishing revealed transgressions. From their perspective there is no basis for any fear and it may be puzzling or surprising for them to find out that it exists. What complicates the issue is that the regional inspectors may investigate reported incidents on the CAA's behalf, so the head office cannot know the precise nature of their "public face".

Officially the CAA office staff describe their incident reporting system as open, non-punitive, and blame-free, although without a formal immunity policy in place it is non-punitive in practice only. This 'bending of the rules' by the rule-makers themselves would provide a field-day for lawyers in the USA but in Sweden there seems to be an understanding that rules are merely guidelines and common sense is needed to apply them in any situation (Sidney Dekker and CAA personnel, Personal communication, October 28, 2004).

The most convincing to the CAA that fear of reporting may be significant is the recent change in the Danish incident reporting system. After a media exposé based on misuse of the publicly available Danish incident data the reporting rate dropped to zero. The Danish parliament instituted a law legally protecting reporters and the incident data from misuse, and within the span of a few years the reporting rate now exceeds that of Sweden by a factor of nearly four. While Danish and Swedish cultures and legal systems are certainly not identical, the Scandinavian countries are more similar to each other than to other countries when viewed on a world scale. David Woods (2003) points out that organisations often do not learn important safety lessons from each other because they focus on their differences rather than their similarities. In this case it is inconsequential whether or not the Swedes feel their society is as punitive as Denmark's, the numbers are indisputable. The idea that they

---

[*] It is relevant to note that neither of the authors come from Scandinavia, and that both have lived in North America

could quadruple their current reporting rate (but is it not already 'mandatory'?) is encouraging the Swedish CAA to push for similar legal safeguards.

Fear, blame, and trust in the commercial sector

There have been occasional cases of legal action initiated against pilots in Sweden, such as the recent suit filed against a helicopter pilot for manslaughter after the death of one of his passengers (Dahl, 2004). Although these cases normally do not result in a guilty verdict they are perceived by some CAA staff as evidence that Sweden is on an irreversible course towards becoming a "more American" punitive and litigious society (CAA personnel, Personal communication, October 27, 2004).

Proposals for immunity policies which protect reporters from legal proceedings or secure the secrecy of incident data from the public will not eliminate the fear that the CAA will "show up and say (...) 'we're yanking your license'" ("Pilot#2", 2003) and during in-depth interviews with seven commercial transport pilots and three aviation company managers it became clear that this fear is something which is "very real (...) a definite reality" ("Pilot#2", 2003).

Extreme variations in attitudes towards reporting and the CAA in general were found through these encounters. Three of the pilots interviewed revealed fear of the Authority as the top factor influencing their decision to report because "You know they are looking to blame somebody. Of course they are looking for that." ("Pilot#4", 2003) and "everyone is afraid to report things. (...) you don't want to get into trouble (...) from the CAA" ("Pilot#3", 2003). Three others dismissed this reason as completely irrelevant even though one pilot explains "I always feel a little bit uncomfortable [about being blamed] but that's the way it is, you need to report it (...) we owe it to each other to tell" ("Pilot#5", 2003).

All pilots expressed a high level of trust in the non-punitive attitude of their employer, which is positive, but several also indicated that people should not report anything involving a mistake somebody made because you "wouldn't want to squeal on a friend" ("Pilot#2", 2003). A former air navigation services employee explained that the system is useful for reporting technical problems that you want fixed but it was used among controllers as an intimidation tactic, as in "do what I say or I will report you to the Authority!" (Norwegian CAA personnel, Personal communication, May 13, 2004). An inventory of the types of incidents received by the CAA reveals a predominance of technical issues (Anderson, 2004; CAA personnel, Personal communication, November 19, 2004). Whether this is due to attitudes like these above, the predominantly technical incidents listed in the regulations as mandatory to report, the way the reports are categorised after submission, or for other reasons is not clear, but it suggests that problems and hazards noticed by the flight crew which have no 'technical' outcome may be underrepresented.

Blame and 'justice' in other industries

The Swedish nuclear industry has a very relevant worst-case example of how the lack of confidentiality of reports and formal immunity policy combined with the public's right to access government documents can spell disaster for the most conscientious and safety conscious employees. An incident report was submitted after the discovery that a reactor had been operational for eight days with one of the safety back-up systems off-line. This system had not been reconnected during the preceding maintenance period due to safety concerns for the maintenance personnel, although due to a discrepancy there was signed paperwork indicating that it had been done (Israelsson, 2003). The incident report was written describing the situation since it was mandatory to do so, but according to the Swedish Nuclear Inspectorate (SKI) the intention of the reporting system is to fix the problem by removing the opportunity for repeating such errors (through procedural changes, for example) rather than punishing individuals. In this case the incident report was accessed by a member of the public who brought the matter to court. The technician who signed the paperwork was found guilty of negligently violating the laws governing Nuclear Activity.

Although it was SKI testimony which resulted in the conviction the organisation did not feel that the issue should have been brought to court. They supported the employee by paying his legal fees and the resulting fine, and he was not removed from his job (SKI personnel, Personal communications, May 25, 2004 and October 23, 2004).

The Swedish media contains many examples of nurses charged with manslaughter based on their involvement in hospital fatalities although the resulting fines or temporary license suspensions would probably be considered light sentences by North American standards.

Good but not perfect

This evidence of fear and blame found within the Swedish aviation industry and from other sources in Scandinavia contradicts the simplistic conclusion that Sweden has a perfectly blame-free, non-punitive culture. Instead the data paints a picture of a 'less-punitive' national culture (in comparison with other countries), an aviation industry with a relatively healthy 'safety-first' orientation, and a justice system tempered by common-sense.

The examples from other industries may not be considered relevant by those in aviation since pilots and other specialists may consider themselves too different from nurses or nuclear control room operators, and thus not at risk of legal consequences. The concept of distinct professional cultures with different values is worth exploring since it is clear that theoretical differences in national culture alone do not tell the whole story.

The distinctiveness of the pilot professional culture

Hofstede (1980, 1991) also analysed his IBM data according to the various occupations represented in his survey and concluded that the attitudes and values between professions or levels (such as manager or employee) can be more distinct and vary more significantly than the overall national scores. In other words, the influence of professional culture on values can be stronger than that of the national culture. A study and comparison of the values of pilots in 22 countries was done by Helmreich and Merritt (1998) based partly on Hofstede's method and using similar questionnaire items and dimensions.

An examination of the differences between Swedish pilots and the general population, and comparisons between Swedish and American pilots may offer a better theoretical explanation of the character and attitudes of Swedes towards their national aviation incident reporting system. Helmreich and Merritt's aviation study (1998) found that Individualism and Power Distance could be correlated to Hofstede's IBM results but Masculinity and Uncertainty Avoidance could not.

Individualism

The overall results scores showed a much higher level of individualism among pilots (mean of 142, standard deviation of 13 based on a linear extrapolation of the original 1 to 100 scale) than of the IBM employees in the same 22 countries (mean of 57, standard deviation of 25) (Helmreich & Merritt, 1998, p.249). The smaller range of the scores indicated to Helmreich and Merritt that for this dimension at least, pilots of different nationalities have more in common than the national groups contrasted in Hofstede's work. This is certainly the case for the Swedish and American samples, as their scores were 157 and 152 respectively. These virtually identical scores may raise a flag for the Swedish authority that the methods used by the Anglo authorities should not automatically be dismissed (Helmreich & Merritt, 1998, p.249).

Power distance

The Power Distance scores of pilots from the 22 countries were on average 23% higher than the IBM scores. Both the Swedish and American pilots scored higher than their average country scores (36 and 52 respectively) (Helmreich & Merritt, 1998, p.249), but the results did not converge in the same way as the Individualism dimension. Helmreich and Merritt suggest that this increase may be attributable to the imposed hierarchical and more autocratic command structure of a flight crew which does not allow for local adaptation.

Masculinity

In the Masculinity dimension the scores for the pilots converge very neatly and the original spread of the two scores by 57 was reduced to 6: The Swedes score increased to 23 and the American score dropped to 29 (Helmreich & Merritt, 1998, p.249). One can only speculate as to the reasons but the implications, i.e. the homogeneity of the responses, are the main concern here.

Uncertainty avoidance

The pilots' scores for the Uncertainty Avoidance dimension do not appear to fit any of the other patterns. The Swedish score dropped 20 points to 9 and the American score increased by 1 to 47 (Helmreich & Merritt, 1998, p.249).

Professional culture as a partial explanation

In three out of the four dimensions the Swedish scored increased, implying that Swedish pilots may not embody the open, trusting, blame-free values to the same extent as their non-flying countrymen. For two of the four dimensions the scores of the pilots converged with respect to their national spread, and for the other two the opposite result was seen. Thus it is difficult to make a conclusive statement about the homogeneity of pilots' attitudes and distinctness of their professional culture from these scores alone, but it may be useful to consider the norms and values which they do and do not share, and to look in greater detail at the results of Helmreich and Merritt's study to identify other 'typical pilot' characteristics.

What is not considered here is the professional culture, if such a thing exists, of aviation authorities. A look at how that differs from the general Swedish model would be relevant as well, since neither pilots nor CAAs can exist without the other. Based on the fact that many inspectors and investigators are retired pilots there may be a revealing pattern there.

Conclusion

It is neither the main focus of this study, nor within the scope of this paper to give a complete ethnographic account of the Swedish value system, however, the fact is that the Swedish incident reporting system (and those of other Scandinavian countries) has been running as an open, blame-free system without legal safeguards despite the assertions in the literature that this is not possible. This is a clear reminder of the limitations of organisational theories beyond their cultural origins since even seemingly fundamental human values are culturally influenced.

Evidence that pilots' professional culture is more similar than the general Swedish population is to the Anglo culture, and empirical examples of the presence of (some degree of) fear among Swedish aviation professionals may mean that the CAA's 'model of the world' does not exactly match the 'world they are in' (Woods, Lecture presentation at Linköping University, October 18, 2004; 2003). Thus this study served as a broadening check for the CAA and they can use this information to adapt their world model and modify their approach accordingly, ultimately resulting in an improved reporting system. This is another feather in Sweden's proverbial cap, since through the support of this research the CAA has demonstrated its openness to change and embodied the spirit of the 'learning culture' which it is attempting to impose on the organisations in its jurisdiction.

This examination of the system in Sweden has implications for future research by EASA and other EU member states to ensure that the limitations of bottled success (i.e., taking the structure of an apparently successful system and imposing it indiscriminately in another situation) are understood in the various contexts. In addition to national culture, company size could be interesting to study since there are some doubts about the usefulness of a formal reporting system in a company consisting of only a few pilots. Another area which is in need of more development is the issue of 'mandatory reporting': How successful can this approach be and what leadership or organisational factors are necessary to make it work? This leads to the difference between proactive and reactive and mandatory and voluntary reporting systems. With line operation audit type approaches being sanctioned by ICAO and at risk of becoming mandatory, it is commercially relevant to understand how these initiatives may overlap.

Research in these areas is particularly important for resource intensive safety tools such as incident reporting systems and line operation audits to prevent wasted resources, frustration, or worst of all, the potential reduction in the safety standards resulting from the inappropriate or inefficient application of such tools in the already cash-strapped aviation domain.

References

Anderson, J. (2004). *Veckostatistiken 48: Sammanställning av registrerade händelser*[Weekly Statistic Report 48: Results compiled for registered companies]. Swedish Civil Aviation Safety Inspectorate report.

Baker, S. (1999). Aviation Incident and Accident Investigation. In D.J. Garland, J.A.Wise, and V.D. Hopkin (Eds.). *Handbook of aviation human factors* (pp. 633-637). Mahwah, NJ: Lawrence Erlbaum Associates, Inc.

Dahl, L. (2004, September 3). Pilot åtalas för vållande till annans död. [Pilot charged with responsibility for another's death] *Dagens Nyheter* [Newspaper, article in archives]. Retrieved October, 2004 from the World Wide Web: http://www.dn.se/DNet/jsp/polopoly.jsp?d=1298&a=51789

Dekker, S.W.A. (2003). When does human error become a crime? *Proceedings of the 12th International Symposium on Aviation Psychology*. Dayton, OH: Wright State University.

Johnson, C.W. (2000). The limitations of aviation incident reporting. *Proceedings of the HCI Aero 2000: International Conference on Human-Computer Interfaces in Aeronautics*, (pp. 17-22). Toulouse, France. Retrieved June, 2003 from the World Wide Web: http://www.dcs.gla.ac.uk/~johnson/papers/reminders/

Johnson, C.W. (2002a) [ed.]. *Proceedings of the 2002 Workshop on the Investigation and Reporting of Incidents and Accidents.* Retrieved September 2004 from the World Wide Web: http://www.dcs.gla.ac.uk/~johnson/iria2002/IRIA_2002.pdf

Johnson, C.W. (2002b). Reasons for the failure of incident reporting in the healthcare and rail industries. *Components of System Safety: Proceedings of the 10th Safety-Critical Systems Symposium,* (pp. 31-60). Berlin, Germany. Retrieved June 2003 from the World Wide Web: http://www.dcs.gla.ac.uk/~johnson/papers/ papers/incident_problems.pdf

Johnson, C. W. (2003). *Failure in safety-critical systems: A handbook of accident and incident reporting* [book online]. Glasgow, UK: University of Glasgow Press. Retrieved October, 2003 from the World Wide Web: http://www.dcs.gla.ac.uk/~johnson/book/.

Hayhurst, K.J., & Holloway, C.M. (2003) (Eds.). *Proceedings of the Second Workshop on the Investigation and Reporting of Incidents and Accidents.* Retrieved September 2004 from the World Wide Web:

http://shemesh.larc.nasa.gov/iria03/iria2003proceedings.pdf

Helmreich, R.L., & Merritt, A.C. (1998). *Culture at work in aviation and medicine.* Aldershot, UK: Ashgate Publishing Ltd.

Hofstede, G. (1980). *Culture's consequences: International differences in work-related values.* Newbury Park, CA: Sage.

Hofstede, G.H. (2001). *Culture's consequences: comparing values, behaviors, institutions, and organizations across nations.* Thousand Oaks, CA: Sage.

Official Journal of the European Union. (2003). Directive 2003/42/EC of the European parliament and of the council of 13 June 2003 on occurrence reporting in civil aviation (English version).

Nicholson, A.N., & Tait, P.C. (2002). Confidential reporting: From aviation to clinical medicine. *Clinical Medicine.* Vol 2(3), 234-236.

Orlady, H.W., & Orlady, L.M. (1999). *Human factors in multi-crew flight operations.* Aldershot, UK: Ashgate Publishing Ltd.

Pidgeon, N.F. (1991). Safety culture and risk management in organizations. *Journal of Cross-Cultural Psychology.* Vol 22(1), 129-140.

Pidgeon, N., & O'Leary, M. (2000). Man-made disasters: Why technology and organizations (sometimes) fail. *Safety Science, 34*, 15-30.

"Pilot#2", confidential interview by K.R. Steele, November 6, 2003. Interview number P2, full transcript unpublished.

"Pilot#3", confidential interview by K.R. Steele, November 6, 2003. Interview number P3, full transcript unpublished.

"Pilot#4", confidential interview by K.R. Steele, November 11, 2003. Interview number P4, full transcript unpublished.

"Pilot#5", confidential interview by K.R. Steele, November 11, 2003. Interview number P5, full transcript unpublished.

Reason, J. (2001). *Managing the Risks of Organizational Accidents.* Aldershot, UK: Ashgate Publishing Ltd.

Sullivan C. (2001). Who cares about CAIR? *Annual Conference of the Australia and New Zealand Society of Air Safety Investigators.* Cairns, Australia. Retrieved July 2003 from the World Wide Web:
http://www.asasi.org/papers/2001/Who%20Cares%20About%20CAIR.pdf

Steele, K.R., & Dekker, S.W.A. (2004). Incident reporting from an emic perspective. *Proceedings of the 22nd Annual International System Safety Conference.* Providence, Rhode Island.

Temo, A.B. (1999). Kundmätning för Luftfartsinpektionen [Survey results for the Aviation Safety Authority].

Woods, D. (2003). Creating foresight: How resilience engineering can transform NASA's approach to risky decision making. Testimony on *The Future of NASA* for Committee on Commerce, Science and Transportation. Retrieved 7 November, 2003 from the World Wide Web:
http://csel.eng.ohio-state.edu/woods.

Effects of workload and time-on-task effects on eye fixation related brain potentials in a simulated air traffic control task

Ellen Wilschut[1,3], Piet Hoogeboom[2], Ben Mulder[1],
Eamonn Hanson[1], & Berry Wijers[1]
[1]Department of Experimental and Work Psychology
University of Groningen, The Netherlands
[2]National Aerospace Laboratory (NLR)
Amsterdam, The Netherlands
[3]Now at Institut für Arbeitsphysiologie
Dortmund, Germany

Abstract

The goal of this study was to explore the usability of eye fixation related potentials (EFRP) to measure workload and time-on-task in a simulated air traffic control (ATC) task. EFRPs were obtained by averaging the EEG locked to the point in time of maximum velocity of saccadic eye movements, derived from the EOG. During the ATC-task, participants (n=20) had to safely guide aircraft to the runway, while maintaining a minimal distance and preventing crashes. Workload was manipulated by varying traffic density and time-on-task effects were studied by comparing the first with the second hour of the task performance. Furthermore, Event Related Potentials (ERPs) on a secondary auditory oddball tasks were derived to provide an additional measure of workload using the P300 component. The P300 amplitude in the secondary oddball task was smaller with high workload and time-on-task, as expected. Both lambda response and P150 component of the EFRP in the ATC task had larger amplitudes with conditions of higher mental workload induced by traffic density. No time-on-task effects for the lambda response or the P150 were found. It is argued that EFRP was suitable for measuring mental workload effects in this study, while the results for time-on-task effects were less convincing.

Introduction

In this research the usability of eye fixation related potentials to measure workload and time-on-task effects in a simulated Air Traffic Control (ATC) task is explored. Since there are growing concerns that increased air traffic density may overwhelm air traffic controllers while compromising safety, more attention has been given to ATC workload and increase of automation in ATC (Danaher, 1985). Several psychophysiological measures have been developed which were shown to be sensitive to the cognitive requirements of complex task performance. An important source of information for measures of mental workload and time-on-task is the electroencephalogram (EEG). There are several different methods to derive information from the EEG. For example, EEG power measures and Event Related Potentials (ERPs). A less frequently studied but interesting measure is suggested by Yagi: the Eye Fixation Related Potential (EFRP; Yagi 1979).

When a person moves his eyes the electro-oculogram (EOG) shows a step-like pattern consisting of saccades and fixations. These steps can be used to determine the time points of saccade onset and offset. When an "evoked response" is obtained associated with the onset of eye fixation by averaging EEG data, this specific "evoked response" is called an EFRP. The amplitude of the EFRP components is believed to reflect changes of mental workload and time-on-task (Kazai & Yagi, 1999). The EFRP is measured by time locking to voluntary saccades, for instance during the ATC task, so no additional task or equipment is needed. This is an advantage compared to most other ERP studies that require standardized laboratory settings or a secondary task to elicit ERPs; application of such an approach when dealing with semi-realistic environments is difficult if not impossible. Additionally, in contrast to most other EEG-studies, making many eye movements is not a problem, but an advantage because of the large number of saccades per minute, resulting in a relatively good signal-to-noise ratio.

Background

ERP patterns that accompany saccadic eye movement can be classified roughly in three components: the first are the antecedent slow potentials (Kurtzberg & Vaughan, 1973) which are similar to the Bereitschaftspotential preceding voluntary movements. The second component is the spike potential, which occurs 10-40 ms prior to the saccade. The origin of the spike potential may be related to ocular muscle or ocular motor neurons discharge with or without cortical components. The third component is the lambda response or wave, which is a sharp positive component with a peak latency of about 80-100 ms from the offset of the saccade. The distribution of the lambda response on the scalp is dominant in the occipital areas and decreases towards the frontal areas (Gastaut, 1951).The lambda response was first observed in raw EEG recordings in subjects with their eyes open; the response was inhibited during eye fixation and especially in absence of contrast (Gaarder et al., 1964). Several authors (e.g., Kazai et al., 1998, 2002; Riemslag et al., 1987) have found similarities between the lambda response and the P100 component of the visual evoked potential (VEP) in the same subject. Although they found the lambda wave delayed with respect to the VEP by 50-100 ms. Current view of the mechanism underlying the lambda response is that it is related to the VEP brought about by changes in the retinal stimulation (Skrandies & Laschke, 1997).

Most research concerning the lambda response is focused on the influence of the properties of the stimulus on the lambda amplitude e.g. contrast, illumination, or the influence of saccade size and direction on the lambda amplitude and latency. A limited group of researchers, however found that the lambda amplitude might be dependent of the state of the subject: attention and arousal level (Matsuo et al, 2001), mental concentration (Yagi, 1998), (visual) fatigue (Takeda et al., 2001) and workload (Yagi, 1996, 1998).

In an experiment of Takeda (Takeda et al., 2001) another component of the EFRP is described, the P150. The P150 is the second positive peak after eye fixation. In this experiment participants were required to proof read a Japanese text on a display. The experiment with total duration of five hours consisted of two blocks in the morning and three in the afternoon. The time-on-task effects were assessed using EFRP and subjective scores of fatigue. The subjective fatigue scores were all statistically higher at the end of the block compared to the beginning. The mean amplitude of the P150 in morning blocks was significantly higher than in afternoon blocks. No differences related to time on task were found on the lambda response amplitude in this experiment.

The goal of this study is to investigate whether EFRP can be used to distinguish periods of high and low mental workload and time-on-task effects in the simplified ATC-task.

Method

Twenty-one female subjects (mean age = 22, SD = 2.0) participated in the experiment. All participants have normal or corrected-to-normal vision and hearing and are experienced computer users. They received a financial reward for their participation. All participation was fully informed, voluntary and anonymous.

Experimental procedure

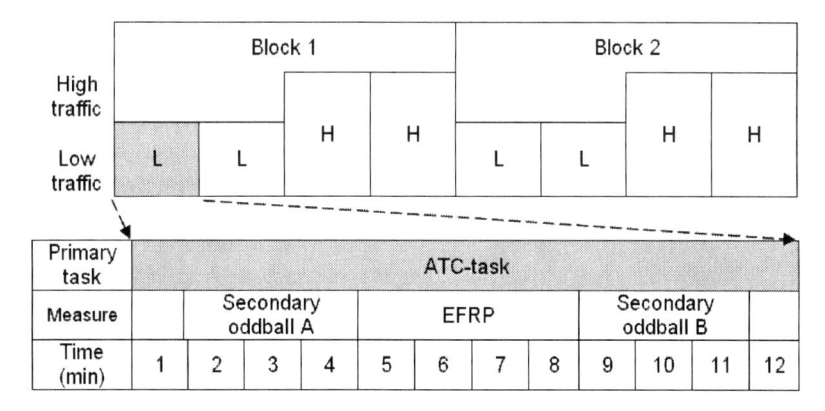

Figure 1. Figure 1. Experimental procedure: each block (60 min.) consisted of four ATC trials; task load was varied by traffic density. Each ATC trial lasted 12 min, with two periods (A &B) of 3 min. in which the secondary auditory oddball task was present. EFRP was measured when the oddball task was absent. The first and last minute were not used in the analyses.

The experiment lasted for two hours and was divided in two blocks of equal duration (see Figure 1). An ATC block consisted of four ATC trials. Each ATC trial of 12 minutes was sub-divided in three measurement periods: the first measurement period started after 1 minute with the auditory oddball as a secondary task. The second period concerned the execution of the stand-alone ATC task in which EFRPs were obtained. In the third period the secondary oddball task was performed again. The first and last minute of each 12-minute trial were not used in the analysis. There are two traffic density levels, while the ATC trial sequences were counterbalanced over subjects. The sequence in the second block was identical to the first block of the experiment to study time-on-task effects.

ATC task

The ATC task was based on an air traffic control task developed by Hoogeboom and adapted by Hoekstra (2003). The goal of the experiment was to guide incoming aircraft safely and efficient to the landing strip. The aircraft appeared from the left or the right side on the bottom of the PC-screen and have different initial headings ranging from 0 to 40 degrees when coming from the left and 320 to 360 degrees when coming from the right side of the screen. In this research it was necessary to simplify the task strongly, because it would take a long period of training inexperienced participants to perform the job of an air traffic controller at a satisfactory level. In addition, simplification was needed to obtain a well-defined experimental environment, where performance results could be related to specific aspects of the task.

Restricting the dimensions of task variation to speed and direction simplified the task enormously. All aircraft flew at the same height; this reduced the degrees of freedom considerably, and made the task easier to train in a short period of time. The second modification had to do with the minimal separation between aircraft. In reality, the minimal separation between aircraft is determined considering several properties of the aircraft for instance the size of an aircraft (especially in final approach). In this ATC task these properties of different types of aircraft were not relevant. The aircraft could be treated as one type and the minimal separation was the same for all aircraft. Third, in contrast with reality, there were no other sectors, so no communication was needed between sectors. The ATC-task simulated one final approach sector. Fourth, the communication to an aircraft was not established by voice communication. The participant had to manually guide the aircraft by selecting the aircraft by mouse-click and changing the direction and/or speed of the aircraft, as well by means of mouse-clicks. No communication delays and errors were simulated, therefore there was no need to check command implementation.

Workload was manipulated by having two levels of traffic density: low and high traffic density. The number of aircraft, which the participant had to control, was dependent on her personal level and the traffic density manipulation. The personal level was determined at the end of the training day by selecting the highest level reached at which no more than two crashes occurred.

Secondary oddball

In the auditory oddball task two types of auditory stimuli were presented, frequent (80%, 1000 Hz tones) and infrequent tones (20%, 2000 Hz tones), via speakers positioned on both sides of the participant (70 dB, 200 ms duration, 1.2 s inter-stimulus interval). The participant was instructed to react only to the infrequent stimuli by pressing the space bar on the keyboard. Performance and reaction times were registered.

EEG measures

EEG was measured on Fz, Cz, P3, P4, Pz and Oz of the 10-20 system (Jasper, 1958) with a single ear reference (right ear) and a common electrode at the top of the sternum. Horizontal EOG was recorded, in bipolar derivation, from electrodes placed lateral to both eyes. Vertical EOG electrodes were placed sub- and supra-orbital to the right eye. For EEG and EOG Ag/AgCl- electrodes were used. For EEG electrode-impedance was kept below 5 kΩ, for EOG below 12 kΩ. After electrode placement the participant was positioned in front of the computer screen at a distance of approximately 60 cm. Instructions were given not to speak and to keep movements restricted to a minimum during the task period.

The Vitaport2 system was used for the recordings. For this experiment, all data was sampled at 256 Hz and stored on a hard disk along with event markers for off-line data processing. A high-pass filter (cut-off at 0.03 Hz) and a low-pass filter (cut-off at 50 Hz) were used to filter the raw EEG data. The EOG data was recorded with a high-pass filter (cut-off at 0.03 Hz) and a low-pass filter (cut-off at 30 Hz). The online signals were monitored on a Macintosh notebook. The performance data was stored on a PC (Windows 98).

P300-component

For the processing of the data, the raw Vitaport data files were converted to HEART format (Human Factors Evaluations, data Analysis and Reduction Techniques; Hoogeboom, 2003) and the PC-based performance data of the oddball was added. Oddball performance data was synchronized with the physiological data and converted to a Brain Vision Analyzer format. In Brain Vision stimulus locked EOG and EEG segments of 200 ms pre-stimulus to 800 ms post stimulus were subjected to artifact detection. Segments with artifacts were discarded. In the next step EEG was corrected for ocular activity (Gratton & Coles, 1983). The segments were then averaged and baseline corrected. P300 was automatically detected in Brain Vision as being the highest positive peak within a time lap between 250-400 ms post stimulus on electrode Pz.

EFRP

The procedure used to determine eye fixations from the EOG channels consisted of two sequential steps. In step one, the occurrence of blinks was determined from the vertical EOG channel. In step two, the time points of saccades were determined, suppressing the detection algorithm whenever being close to a blink. The separation of blinks from saccadic eye movement was based on the observation that during a blink the measured voltage rises sharply, followed by a return to almost the same value as before the blink (Figure 2a). This behavior was different from what was observed during a saccade (Figure 2b), which was characterized by a step-like change. Besides this difference in shape, also the amplitude differs significantly;

blinks in general have larger values. Four different algorithms to determine eye fixation markers in the EEG were constructed based on the first and second derivative of the EOG. With all four algorithms an EFRP could be constructed with the same characteristics as the EFRP described by Yagi (e.g. Takeda et al. 2001). In spite of the use of different algorithms between the methods, three methods showed a similar shape and individual characteristics remained identical. The averaged EFRPs were very accurate and distinct: the spike potential, which is a component with duration of about 20 ms, showed very sharp features, indicating that the marker was a good synchronisation point.

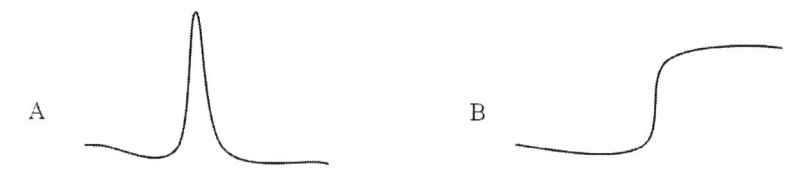

Figure 2: a characteristic blink (A) and saccade (B) in the EOG signal.

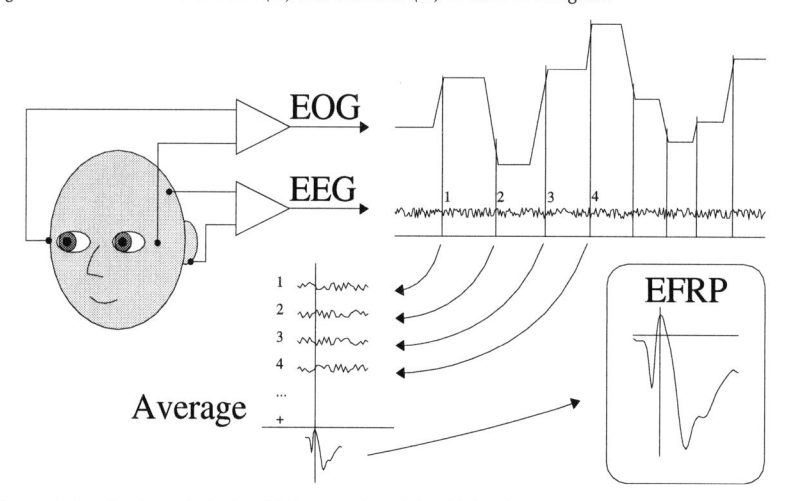

Figure 3. Derivation of the EFRP by synchronizing EEG data on time points of eye fixation, resulting in an eye fixation related potential (EFRP).

Four different labels were used: two for the horizontal EOG, making it possible to discriminate between left and right saccades and two labels for the vertical EOG to study the upward and downward saccades (Figure 3). In Brain Vision eye fixation locked segments were created of 200 ms pre-fixation to 800 ms post eye fixation. Hereafter, in the same way as ERP artifact detection for EOG and EEG was run. Segments with artifacts were discarded. In the next step EEG was corrected for ocular activity (Gratton & Coles, 1983). The artifact free and corrected segments were then averaged and baseline corrected (-200 till −100 ms). The lambda EFRP-component was automatically detected in Brain Vision as being the highest positive peak within 0 and 140 ms on electrode Oz (Gastaut, 1951). Inspection of the P150 showed, some participants had a clear P150 peak and others had a slight increase of the signal. To make analysis possible, time intervals were defined ranging from 110 until 290 ms. Each time interval consisted of about 20 ms.

Statistical analysis

The secondary oddball task was analyzed with a repeated measurement design of SPSS 11.0. When the main analysis indicated a significant interaction (p<.05) between factors, follow-up analyses were performed, adjusting error rates according to Bonferroni. The two secondary oddball periods within each ATC task (see Figure 1) were combined in the analyses of P300 amplitude and performance data. The EFRP components were analyzed in periods without secondary oddball task. The analysis of the lambda response was identical to the one applied for the P300-component. The P150 amplitude was analyzed by paired samples t-tests. Significance levels were set to $\alpha = 5\%$. Borderline significant effects were also mentioned to provide a complete view of this explorative study.

Results

Secondary oddball

Performance

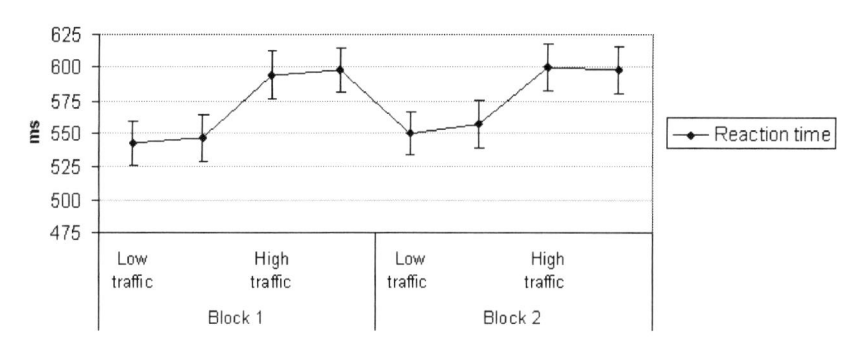

Figure 4. Average reaction time on secondary odball task

The only effect found was a main effect of traffic density ($F(3,15)=8.7$, p<.001). The reaction times of the correct responses were about 50 ms (SE =16.3) longer in high traffic density condition compared to the low traffic density trials (Figure 4 | $F(1,17)=25.3$, p<.001). Total number of infrequent high tones was sixty. On average the number of misses increased from four to nine ($F(1,17)=16.6$, p=.001) and the number of false alarms increased from 2.0 to 3.6 (Figure 5, $F(1,17)=18.4$, p=.001). No difference was found between the first and second block of the experiment, and there were no interactions.

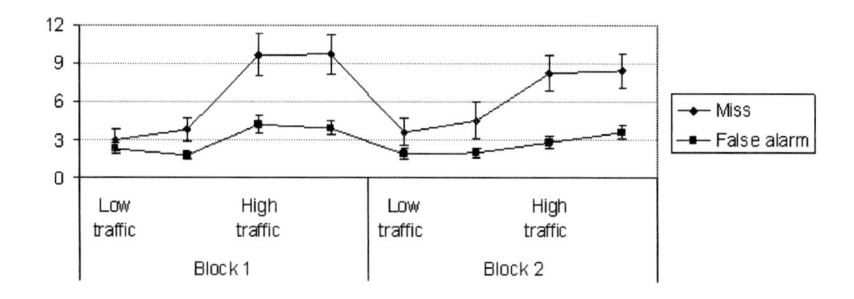

Figure 5: average number of misses and false alarms on secondary oddball task

ERP

A significant effect for traffic density was found in P300 amplitude (F (1,17) = 20.3, p< .001). During low traffic density P300 amplitude was higher (mean = 11.5, SE= 0.7) as compared to high traffic density (mean = 9.2, SE = 0.7| Figure 6). The P300 amplitude showed a significant effect for time-on-task (F (1,17) = 65.8, p< .001) with higher P300 amplitude in the first block (mean =11.6, SE= 0.6) as compared to the second block (mean = 9.1, SE = 0.7, Figure 6).

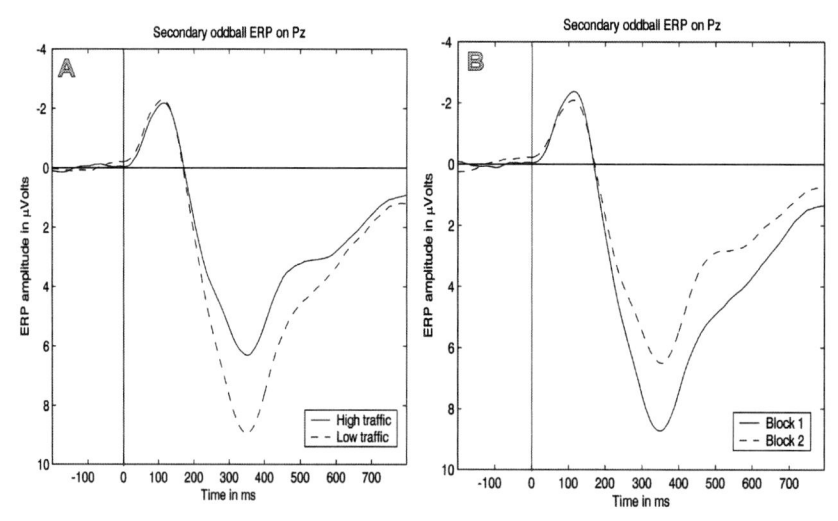

Figure 6. Grand averages secondary auditory oddball task using only infrequent tones. P300 amplitude task load effects, low versus high traffic density (A) and time-on-task effects, block 1 versus 2 (B)

EFRP

Lambda amplitude

The amplitude of the lambda response derived from horizontal saccades showed a significant main effect for traffic density (F(1,17)= 9.85, p=.006), while no main effect for time-on-task or interactions were found. In high traffic density tasks the mean lambda amplitude was higher (M = 6.9 SE = 3.7) compared to the low traffic density tasks (M = 6.1 SE =3.2). Post hoc analysis revealed that this significant difference between high and low traffic density tasks was present at all electrodes (table 1).

Table 1: Results of the lambda response amplitude for task load manipulation df (1,17).

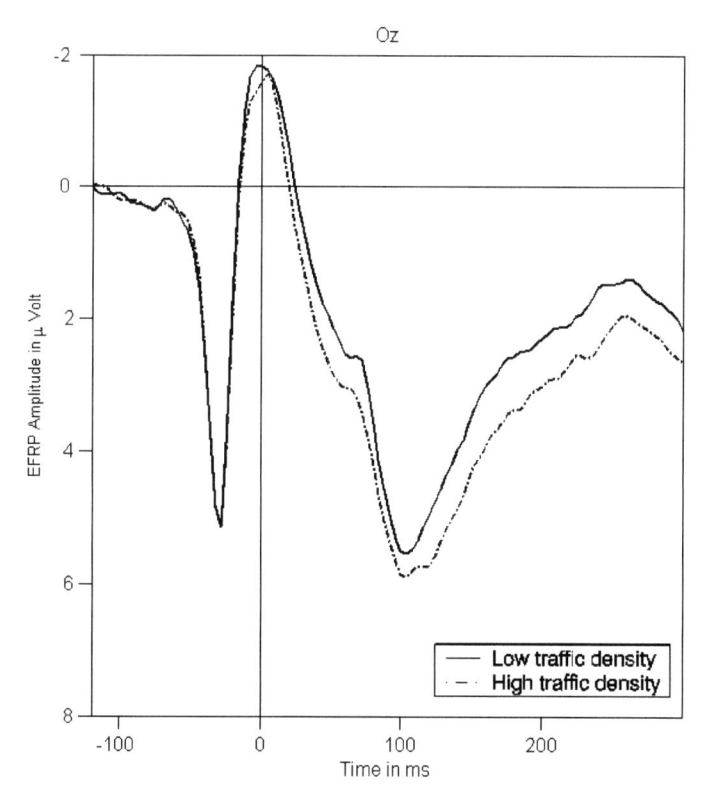

	Traffic density	Fz	Cz	P3	P4	Pz	Oz
Horizontal saccades	F= 9.85 p= .006	F=7.24 p=.015	F= 7.73 p= .013	F= 10.6 p= .005	F= 6.16 p= .024	F= 8.88 p= .008	F= 5.17 p= .036
Vertical saccades	F= 3.57 p= .076	F= 1.7 NS	F= 3.20 p= .091	F= 2.50 NS	F=.3.92 p= .064	F= 2.05 NS	F= 3.77 p= .096

Figure 7. Grand averages for high and low traffic density on Oz, lambda response visible at 100 ms (horizontal saccades).

P150-component
Visual inspection of the data showed a clearly visible positive peak after the lambda response for approximately one third of the participants, the other participants showed a slight positivity in the signal. The latency of this peak varied between 160

and 200 ms. Analysis for time-on-task showed no differences between block 1 and block 2 on Pz or Oz. However, traffic density showed a deviation of the signal at about 130 ms prolonging to 200 ms, which was clearly visible on Pz and Oz. Ten segments were evaluated of 20 ms each, differences were tested with paired t-tests (see Table 2 and Figure 7, only Pz is shown).

Results showed significant higher values during the high traffic density trials, differences were maximal between 130 and 190 ms. P150 amplitudes of horizontal saccades showed more intervals with significant results than vertical saccades did, both at Pz and Oz. Results were were more pronouced at Pz than at Oz.

Time (ms)	Horizontal saccades (Pz) Mean difference (μV)	SE	t-value (df= 83)	p-value	Vertical saccades (Pz) Mean difference (μV)	SE	t-value (df= 83)	p-value
110	0.67	0.22	3.09	.003	0.22	0.21	1.02	NS
130	0.99	0.23	4.21	<.001	0.51	0.21	2.48	.015
150	1.18	0.24	4.85	<.001	0.71	0.24	2.98	.004
170	1.28	0.23	5.63	<.001	0.68	0.22	3.17	.002
190	1.03	0.21	4.79	<.001	0.56	0.21	2.66	.009
210	0.86	0.21	4.03	<.001	0.39	0.20	1.93	.058
230	0.66	0.22	3.08	.003	0.26	0.20	1.22	NS
250	0.40	0.20	2.05	.044	0.17	0.20	0.86	NS
270	0.48	0.20	2.43	.017	0.14	0.20	0.68	NS
290	0.64	0.20	3.21	.002	0.25	0.19	1.30	NS

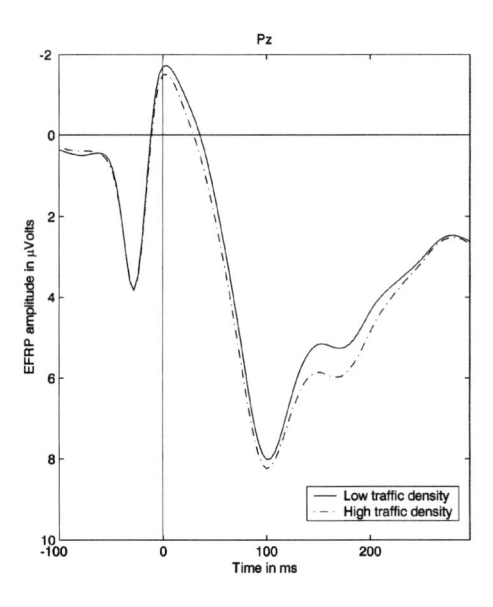

Figure 8. Grand averages for high and low traffic density at Pz (vertical saccades).

Discussion and conclusion

The results of the secondary auditory oddball task showed clear effects of the traffic density manipulation of mental workload. Participants needed more time to respond to targets of the secondary task in high traffic density trials while making more errors. The P300 amplitude was smaller when task load in the primary task was high, as was expected (Wickens et al., 1977). The results of the EFRPs based on horizontal saccades showed also changes with the different levels of task load. The lambda response and the P150 amplitudes were larger with high traffic density task, particularly on parietal sites. The lambda amplitude based on vertical saccades showed no clear effects for traffic density. Also the P150 amplitude showed more significant differences with EFRPs based on horizontal than vertical saccades.

No explanation for this difference between horizontal and vertical saccades is found in this experiment, possible factors to this difference are task characteristics, blinks or physical factors. Mental workload studied with EFRP can be related to specific task characteristics, for instance looking at the heading command might induce more mental workload than explorative scanning of the aircraft, therefore it would be interesting to divide the eye fixation into functional groups bound to task characteristics. With further investigation of the EFRP it would be recommended to have eye tracker data of eye movements next to the findings of EFRP, to provide an accurate measure of gaze direction, eye fixation and saccade duration. It would make it possible to investigate the influence of task characteristics and strategies used by the participant on the EFRP components.

Time-on-task effects are clearly visible in the oddball task; the amplitude of the P300 decreases when comparing block 1 and 2.The lambda and P150 amplitudes were not affected by time-on-task, although Takeda et al. (2001) found clear decrement of the P150 amplitude due to time-on-task.

An unresolved issue is the fact that the P300 amplitude decreases with increasing task load of the primary task, while the lambda and P150 amplitude increases. A possible factor for a larger amplitude with traffic density manipulation of workload lies in the fact the lambda response changes like the VEP with illumination (Yagi, 1998) and contrast (Yagi, 1982). During high traffic the number of aircraft on the screen is larger so the illumination from the screen will decrease and the contrast increases. This might not be a satisfactory explanation because of the different results for EFRPs elicit by vertical and horizontal saccades.

This study indicates that amplitude changes of the EFRP might be a valuable measure to investigate mental workload changes with low task intrusion, applicable in many different settings, car driving, flight simulator etc, which require saccadic eye movements. As indicated, future research should focus on which influence task characteristics and cognitive factors have on the components of the EFRP.

Acknowledgment

This study is part of the COMPANION project, which is co-funded by the Dutch Ministry of Economic Affairs (SENTER IOP MMI99002A&B).

References

Gaarder, K., Krauskopf, K., Graf, V., Kropfl, W., & Armnington, J.C. (1964). Averaged brain activity following saccadic eye movements. *Science, 146*, 1481-1483.

Gastaut, Y. (1951) Un signe electroencephalographique peu connu: les pointes occipitals servenant pendant l'ouvreture des yeux. *Review Neurology, 84*, 640-643.

Gratton, G. & Coles,M.G.H. (1983). A new method for off-line removal of ocular artifact. *Electroencephalography and clinical Neurophysiology, 55*, 468-484.

Hoogeboom, P.J. (2003). Off-line synchronization of measurements based on a common pseudorandom binary signal. *Behavior Research Models, Instruments, & Computers, 35*, 384-390.

Hoekstra, R. (2003) Een stap richting een adaptieve interface. Report NLR CR-2003-135. Amsterdam: National Aerospace Laboratory NLR.

Jasper, H. (1958). The ten-twenty elecctrode system of the International Federation. *Electroencephalography and Clinical Neurophysiology, 10*, 371-375.

Kazai, K. & Yagi, A. (1998). Location of electric current sources of lambda response estimated by the dipole tracing method. *Proceedings of the Second International Conference on Psychophysiology in Ergonomics*, Nishinomiya: Kwansei Gakuin, PIE 1998.

Kazai, K. & Yagi, A. (1999). Intergrated effect of stimulation at fixation points on EFRP. *International Journal of Psychology*, 32, 193-203.

Kazai, K., Kanamori, N., Nagai, M., & Yagi, A. (2002). Contrast dependence of EFRP and pattern-onset VEP. *Proceedings of the Second Asian Conference of Vision*. Gueongju, Korea.

Kurtzberg, D., & Vaughan, H.G. (1973). Electrocortical potentials associated with eye movements. In V. Zikmund (Ed.), *The oculomotor system and brain functions* (pp. 137-142.). London: Butterworths.

Matsuo, N., Ohkita, Y., Tomita, Y., Honda, S. & Matsunaga, K. (2001). Estimation of an unexpected- overlooking error by means of the single eye fixation related potential analysis with wavelet transform filter. *International jounal of Psychophysiology, 40*,195-200.

Riemslag, F.C.C., Van der Heijde, G.L., Van Dongen, M.M.M.M., & Ottenhof, F. (1987) On the origin of presaccadic spike potential, *Electroencephalography Clinical Neurophysiology, 70*,281-287.

Skandries, W. & Laschke, K. (1997). Topography of visually evoked brain activity during eye movements: lambda waves, saccadic suppression, and discrimination performance. *International Journal of Psychophysiology, 27*, 15-27.

Takeda, Y., Sugai, M., & Yagi, A. (2001).Eye fixation related potentials in a proof reading task. *International Journal of Psychophysiology, 40*, 181-186.

Wickens, C., Isreal, J., & Donchin, E. (1977). The event-related cortical potential as an index of task workload. In A. S. Neal & R. F. Palasek (Eds.), *Proceedings of theHuman Factors Society 21st Annual Meeting*, San Francisco, October 1977. Santa Monica (CA): Human Factors Society

Yagi, A. (1979). Saccade size and lambda complex in man. *Physiological Psychophysiology, 7*, 370-376

Yagi, A. (1982) Lambda response as an index of visual perception research. *Japanese Psychological Research, 24*, 106-110

Yagi, A. (1996). Application of eye fixation related potentials in ergonomics. In C. Ogura, Y. Koga and M. Shomokochi (Eds.), *Recent advances in event related brain potential research*. Tokyo: Elsevier.

Yagi, A. (1998) Psychophysiological Studies of Lighting Environments. *Proceedings of the first CIE Symposium on Lighting Quality*, Ottawa, Canada

Fz theta divided by Pz alpha as an index of task load during a PC-based air traffic control simulation

Matty A. Postma[1], Jan M.H. Schellekens[1], Eamonn K.S. Hanson[2], & Piet J. Hoogeboom[2]
[1]Experimental & Work Psychology, University of Groningen
[2]National Aerospace Laboratory (NLR), Amsterdam
The Netherlands

Abstract

In this investigation 21 students were trained to perform a highly simplified PC-based air traffic control (ATC) approach sequencing task simultaneously with a secondary (auditory) oddball task. Task load was manipulated by varying the number of aircraft on screen. Besides task load, time-on-task effects were investigated by monitoring the ATC task performance during two hours. The effects of the manipulations were assessed using subjective (RSME), performance and physiological (EEG spectral power, ERP P_{300}) measures. An index based on EEG spectral power in alpha and theta ranges was calculated by dividing the power of the Fz theta band by the power of the Pz alpha band. The index was tested with regard to its sensitivity for task load and time-on-task effects. The results provide support for Fz theta divided by Pz alpha as an index of task load.

Introduction

There are growing concerns that the continuous increasing level of air traffic, and hence the increasing task-levels of air traffic controllers (ATCos), may compromise safety and/or efficiency of air travel. Consequently, more attention has been devoted to measuring ATCos' workload. Two aspects of mental workload are of interest in this context. On the one hand the task load caused by the task properties on a certain moment, such as traffic density, and, on the other hand, the duration of the task. In practical settings, an index of task load should be able to distinguish the effort invested to meet the momentary demands of task load ("task-related effort") from the effort that is invested to prevent performance from degrading over time ("state-related effort"). As a general measure of task load, decreases in EEG alpha power (parietal) and/or increases in EEG theta power (frontal) have been reported in ATC simulators (Brookings et al., 1996), flight simulators (Sterman & Mann, 1995; Smith et al., 2001) and actual flight (Wilson, 2002). With increasing time-on-task a diffusely spread increase of both alpha and theta power has been reported (Schacter, 1977).

Method

Twenty-one female subjects participated in this experiment. The goal of the ATC task was to guide incoming aircraft safely and efficiently to the runway (see also Wilschut et al., 2005, this issue). Traffic density levels were alternated between high and low traffic density. Eight ATC trials of 12 minutes have been performed in succession. Within each ATC trial two dual task periods were presented with an auditory oddball task as a secondary task. This secondary task consisted of two types of auditory stimuli, frequent (80%, 1000 Hz tones) and infrequent tones (20%, 2000 Hz tones). The participant was instructed to react only to the infrequent stimuli by pressing the space bar on the keyboard. Besides task load, time-on-task effects have been investigated by monitoring the ATC task performance during two hours (block 1 and block 2). The effects of the manipulations were assessed using subjective (RSME Rating Scale Mental Effort; Zijlstra & Van Doorn, 1985), performance (hits, misses, false alarms and reaction time) and physiological (global EEG power, ERP P300) measures. An index based on EEG spectral power in alpha (8-12 Hz) and theta (4-8 Hz) ranges was calculated by dividing the Fz theta power by the Pz alpha power.

Results

The ERP P300 amplitude decreased significantly with increasing time on task ($F(1,17)=65.8$, $p<.001$; see Figure 1). Also, alpha and theta power density were highly sensitive to time-on-task. Alpha and theta both showed an increase in power diffusely spread over the scalp (see Table 1, Figures 2-3). High traffic density tasks (RSME mean= 70.1, SD= 3.5) were rated more demanding by all subjects than low traffic (RSME: mean= 55.9, SD= 3.3) ($F(1,17)=46.8$, $p<.001$). Reaction times of the secondary oddball increased with an average of 50 ms (S.E.=16.3) in the high traffic density as compared to the low traffic density tasks ($F(1,17)=25.3$, $p<.001$). Misses and false alarms increased on average from 4 to 9, and from 2 to 3.6 respectively ($F(1,17)=16.6$, $p<.001$ and $F(1,17)=18.4$, $p<.001$). ERP P_{300} amplitude of the secondary task decreased, as expected, in the high traffic density as compared to the low traffic density tasks ($F(1,17)=20.3$, $p=.001$; see Figure 4).

Table 1. Summary of within-subjects univariate effects ($F = F(1,17)$) of alpha and theta on each electrode position for time-on-task and traffic density

		Fz	Cz	P3	P4	Pz
Time-on-task	alpha	F=14.2 p=.002	F=14.8 p=.001	F=23.5 p<.001	F=15.6 p=.001	F=29.6 p<.001
	theta	F=20.5 p<.001	F=3.3 p=.09	F=5.8 p=.03	F=6.3 p=.02	F=7.7 p=.01
Traffic density	alpha	F=.63 NS	F=1.7 NS	F<1 NS	F=1.8 NS	F=4.4 p=.05
	theta	F=3.8 p=.07	F=1.6 NS	F<1 NS	F<1 NS	F<1 NS

Alpha power on Pz decreased significantly with increasing traffic density, and theta power showed a marginally significant effect on Fz (see Table 1). [LN alpha power:

low traffic density mean= 0.286, SE= 0.15; high traffic density: mean= 0.249, SE= 0.15.] Consequently, the index computed by Fz theta / Pz alpha increased significantly from low traffic density to high traffic density (F(1,17)=7.8, p=.013), but not on time-on-task (F(1,17)< 1, NS). [Index Fz theta/Pz alpha: low traffic density: mean= 0.66, SE= 0.15; high traffic density: mean= 0.74, SE= 0.15.]

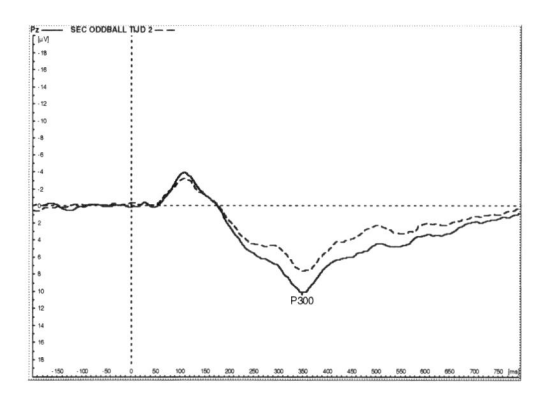

Figure 1. Time-on-task: Secondary oddball P_{300} amplitude averages of first block (——) and second block (----) of the ATC task.

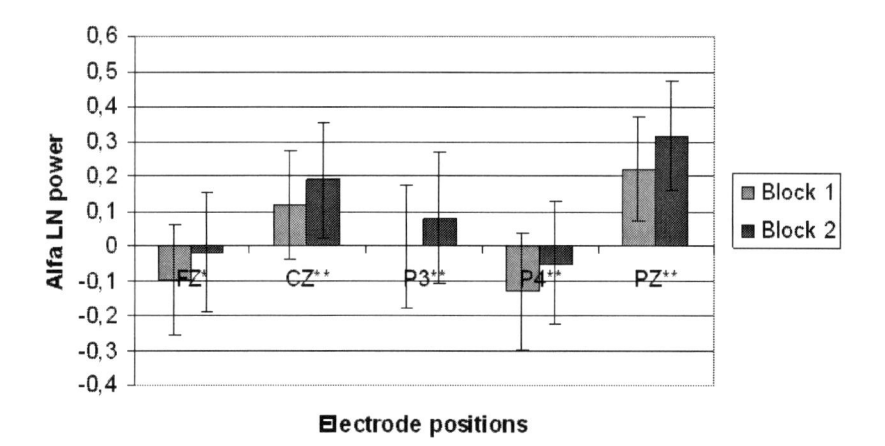

Figure 2: Alpha LN power (means and SE) on Fz, Cz, P3, P4 and Pz in first and second block (* p<.05; ** p<.001)

Conclusion

Fz theta / Pz alpha was found to be insensitive to time-on-task and sensitive to task load. As such it can be used as an index of task load. The insensitivity of the index for time-on-task makes the index highly applicable for continuous monitoring of task load in real life settings, where otherwise time-on-task would have an obscuring effect on the measurement of task load. As time-on-task effects are a relevant aspect in high demand settings (eg. ATC environment) the power in alpha and theta ranges can be monitored as indices of time-on-task effects.

Theta LN power: Block

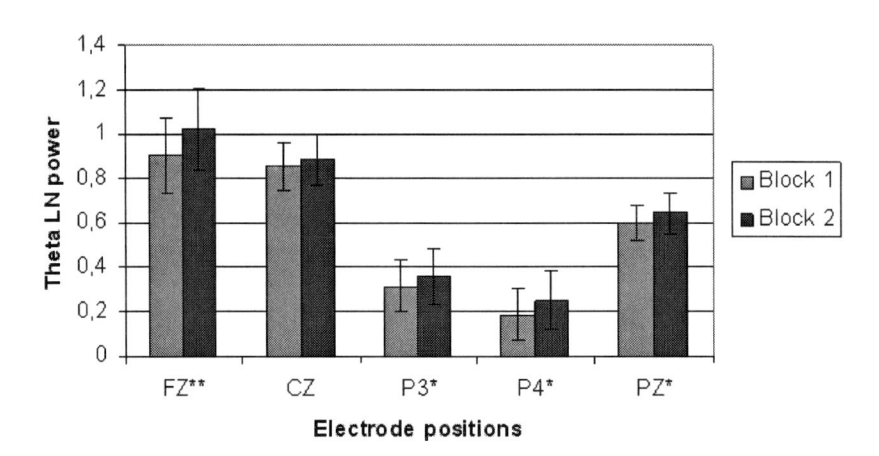

Figure 3: Theta LN power (means and SE) on Fz, Cz, P3, P4 and Pz in first and second block (* p<.05 ** p<.001)

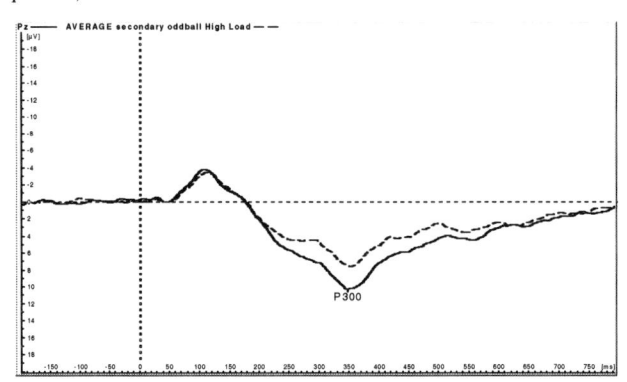

Figure 4. Secondary oddball P_{300} amplitude averages of low traffic density (——) and high traffic density (---).

Acknowledgment

This study is part of the COMPANION project, which is co-funded by the Dutch Ministry of Economic Affairs (SENTER IOP MMI99002A&B).

References

Brookings, J.B., Wilson, G.F., & Swain, C.R. (1996). Psychophysiological responses to changes in workload during simulated air traffic control. *Biological Psychology, 42*, 361-377.

Scerbo, M.W., Freeman, F.G., Mikulka, P.J., Parasuraman, R., Di Nocera, F., & Prinzel III, L. J. (2001). *The efficacy of psychophysiological measures for implementing adaptive technology*. (NASA Tech. Publication 2001-211018). Hampton, VA, USA: Langley Research Center.

Schacter, D.L. (1977). EEG theta and psychological phenomena: a review and analysis. *Biological Psychology, 5*, 47-82.

Smith, M.E., Gevins, A., Brown, H., Karnik, A., & Du, R. (2001). Monitoring task loading with multivariate EEG measures during complex forms of human-computer interaction. *Human Factors, 43*, 366-380.

Sterman, M.B. & Mann, C.A. (1995). Concepts and applications of EEG analysis in aviation performance evaluation. *Biological Psychology, 40*, 115-130.

Wilschut, E.S., Hoogeboom, P.J., Mulder, L.J.M., Hanson, E.A., & Wijers, A.A. (2005). Effects of workload and time-on-task on eye fixation related brain potentials in a simulated air traffic control task. In D. de Waard, K.A. Brookhuis, R. van Egmond, and Th. Boersema (Eds.) *Human Factors in Design, Safety, and Management* (pp. 451-463). Maastricht, the Netherlands: Shaker Publishing.

Wilson, G.F. (2002). An analysis of mental workload using multiple psychophysiological measures. *The International Journal of Aviation Psychology*, *12*, 3-18.

Zijlstra, F.R.H., & Van Doorn, L. (1985). *The construction of a scale to measure perceived effort*. Delft University of Technology, Department of Philosophy and Social Sciences.

Designing safety into future Air Traffic Control systems by learning from operational experience

Deirdre Bonini[1] & Tony Joyce[2]
[1]Integra Consult A/S, Vedbæk
Denmark
[2]Eurocontrol Experimental Centre, Bretigny sur Orge
France

Abstract

Learning from occurrence data is acknowledged as being important in a number of safety-critical domains, amongst which Air Traffic Control (ATC). The understanding derived from incidents can be used to make current operations safer, but also to improve the design of future systems, thereby creating a complete safety learning cycle that feeds back to the present and forward to the future. The SAFLearn project contributes to such a safety learning cycle in ATC, whereby lessons learned from operational experience are used to inform the design of the future systems developed at the Eurocontrol Experimental Centre (EEC). SAFLearn consists of four activities: collection, storage, analysis, and delivery. In the collection activity safety occurrence reports are first categorised according to a formal taxonomy, and then stored in a database. The analysis of safety occurrence reports by safety, human factors and operational specialists, results in the definition of lessons learned for specific EEC projects. These lessons are delivered to designers to support safety at different stages in the development of their projects, in the conceptual phase to evaluate whether current problems will be addressed, in the design phase to help prioritise design decisions, and in the testing phase to help evaluate the product. The process is first described and then illustrated through three examples.

Learning from safety occurrences

The importance of safety learning through occurrence data for safety management and safety improvement is widely acknowledged in a number of safety-critical sectors, such as civil aviation (Van Es, 2001; 2003), the chemical process and rail industry, as well as in domains such as anaesthesiology, pharmacy, and transfusion medicine (Van der Schaaf & Kanse, 2002). Through the collection, storage, analysis and exchange of safety occurrence data, safety professionals are able to build-up knowledge and gain an insight of the existing and near-term safety problems (Von Thaden & Wiegmann, 2001).

In D. de Waard, K.A. Brookhuis, R. van Egmond, and Th. Boersema (Eds.) (2005), *Human Factors in Design, Safety, and Management* (pp. 471 - 478). Maastricht, the Netherlands: Shaker Publishing.

For a number of years the ATC community has been working to improve safety through the harmonisation of the collection, storage, analysis and data exchange of safety occurrence information. Harmonised occurrence report collection and assessment schemes allow a systematic study of events and visibility of common causes, as well as the identification of appropriate corrective action and general areas where safety can be improved.

In present day ANSPs (Air Traffic Control National Service Providers), lessons are learned from "accidents that did not happen" (GAIN 2003:109), in other words situations where the safety margins were reduced below certain levels by an error, defect or design. Near-team problems are identified and operations are changed and made safer on the basis of safety occurrences.

Although Johnson (2003) argues that incident report collection can be useful to identify potential failures before an accident occurs, "many submissions do little more than remind their operators of hazards that are well understood but are difficult to avoid" (Johnson 2003:21). One of the reasons why these hazards are difficult to avoid is that operators are constrained by the system they work with, many characteristics of which cannot be easily changed. Another reason may be that the operator naturally behaves in a certain way that has not been taken into consideration when designing the system. In both cases a system that is designed differently may eliminate, or at least reduce, the impact of identified hazards. For this reason, it is useful to use lessons learned from operational experience as early as possible in the design of new concepts and systems.

The safety learning cycle, whereby lessons learned in the operational environment feed back into operations and forward into design, will not be fully developed in ATC until the lessons learned are only used to solve near-term problems. In order to fulfil the perceived need to build safety into design from a very early stage, safety information needs to be provided to the designers of future systems. Designers are considered to be all those whose decisions contribute to the development of new concepts, such as new tools and their associated working methods. Safety information is available from the occurrence databases of single ANSPs and although operational experts are part of project teams at EEC, there is no process in place that ensures that this information is relayed from the operational setting to projects in a systematic manner. It is believed that a resource that supports such a process would increase the safety of the future ATC systems developed. The SAFLearn project aims at creating such a capability, which has been conceived of as a service for projects at EEC.

The SAFLearn project: four activities

The Safety Learning (SAFLearn) project aims at creating a process that will support designers at EEC in designing safety into their projects by providing them with *lessons learned* from operational experience and safety occurrences. A lesson learned is knowledge or understanding gained by experience that has a significant impact for an organisation (ESA, 1999). The lesson learned can be 'good work practice' or a negative experience.

The SAFLearn project collects and collates operational experience from different sources, derives lessons learned of interest to EEC projects, and provides these to project teams in a format that will effectively support them in their design decisions. The project thus encompasses four activities (collection, storage, analysis and delivery), which are described in sections below. The lessons learned are illustrated through case studies that are presented to the project team and summarised in a final report. At present five projects are being supported. Reference will be made to three in this paper, as the support to the other two is at the initial stage of identification of safety issues of concern to the project. Before describing the actual activities it is useful to spend a word on the choice of case studies to learn about safety.

Case studies to learn about safety

The safety occurrence reports that are collected by SAFLearn and used to inform EEC projects of current safety issues contain sensitive information and are confidential. Thus it was necessary for the SAFLearn team to find an effective way to deliver lessons learned whilst protecting the data providers.

Expert decision makers, such as fire fighters, nurses and air traffic controllers, often transmit their knowledge and experience gained through practice using stories. The power of storytelling lies in the fact that experience can be coherently described, together with rich contextual detail, allowing a certain amount of generalisation by suggesting salient features of situations that have guided good or bad judgements (Klein, 1999).

Safety occurrence reports effectively describe an event as a story, complete with rich contextual information, main and contributory causes, lessons learned and recommendations. These reports can then be described as case studies or scenarios, containing a storyline of the development of the event, which is the result of interpretations by the actors involved and by the investigator, supported by factual evidence. Although the term 'case study' has been used traditionally in the medical and clinical psychology literature, more recently in human factors the term 'scenario' is used to refer to case studies that are a medium for participatory design, in supporting discussions between users and designers (Jarke, Bui, & Carroll, 1998). Smith et al. (1998) used a case study to identify system requirements for new ATC concepts and technology. They presented a number of stakeholders (i.e. controllers, pilots, assistants and flow managers) with a scenario that was part of an incident report, asking them to carry out a conceptual walk-through analysing all that had gone wrong and deriving requirements for a future system that would ensure the incident would not occur again. Smith and his colleagues (1998) argued that because many of the details of the design of a future system are unspecified, a scenario is the most effective way to help prototype how people and technology will coordinate in realistic operational scenarios and gain insight into issues and implications of proposed future designs. In other words, they concluded that scenarios provided an effective balance between realistic and specific detail. Furthermore, scenarios effectively capture external constraints of the context, as well as domain knowledge, which, as is typical in the workplace, is often tacit (Carroll, Rosson, Chin & Koeneman, 1998: 1167). It was thus decided to use case studies as a tool to share operational experience with researchers and support the communication and learning by project team members with different backgrounds (e.g. operational, engineering, human factors, and safety).

Activity one: collection

In compliance with ESARR (Eurocontrol Safety Regulatory Requirement) 3, all ECAC (European Civil Aviation Conference) ANSPs have to collect safety occurrences in a harmonised manner and have a learning cycle set up as part of their Safety Management System (SMS). Although all service providers are collecting, storing, analysing and using the information derived from safety occurrences, variability exists in the maturity of both their SMS and the legal framework that supports the sharing of sensitive information. The SAFLearn project is currently receiving safety occurrence reports from three service providers. By the end of 2005 a total of fifteen providers are expected to be contributing reports to the Safety Learning project.

In general, reports are being provided under three main conditions. The first condition is confidentiality. Although reports containing contextual information are provided, it is agreed that they can be distributed only after having been de-identified. Exceptions have to be agreed upon with the data provider. Reports are thus stored in a de-identified format with a reference to the original report received.

The second condition is that the reports may not be used for statistical or comparative studies. This condition is met following the understanding that ANSPs are not committed to share all their reports with the project. Furthermore, without exposure data it is not possible to weight events for their significance.

The third condition is that the reports will be used for research purposes only, and feedback on their use will be provided to the data providers. The manner in which data are provided and feedback is received is agreed on a case-by-case basis. A number of data providers have requested an annual SAFLearn activity report, others require a copy of the report given to the projects supported using their data. Many ANSPs plan to use this information to complement their own research and development studies.

Activity two: storage

The second activity of SAFLearn regards the storage of the reports. Once reports are received, the SAFLearn team analyses them according to a formal taxonomy used by incident investigators (i.e. HEIDI), enriched with terms describing future concepts. The database in which the de-identified reports are stored (SAFTool) contains the storyline of the occurrence, information on the main and contributing factors, as well as pictures to illustrate the event. Access to SAFTool is limited to the SAFLearn team. A history of the use of the report in supporting EEC projects is saved as well a reference number that refers each report back to the original provider's number. This cross-reference information is held in a separate file and is kept to allow SAFLearn to contact the data provider if additional information is required. It should be said, however, that effort on the part of both ANSPs and the EEC projects receiving the lessons learned is minimised. SAFLearn should provide a service that supports safety without requiring additional resources from the providers and users. From this perspective SAFLearn should provide a bridge between current operations and safety requirements and future concepts and solutions.

Activity three: analysis

The third activity of SAFLearn is analysis. Two kinds of analysis are carried out: pre-analysis and analysis for specific projects. As mentioned above, the SAFLearn team provide a first analysis of the safety occurrence reports that results in them being classified, de-identified and stored in the SAFTool database. This type of analysis is carried out during regular SAFLearn team sessions aimed at populating the database.

In parallel, research and development projects at the EEC that express an interest in the SAFLearn service are interviewed and asked for documentation regarding their project. It is essential that the SAFLearn team understand the aims and objectives of the client project, as well as their planning and work schedules. Projects are users, whose needs and constraints have to be captured and met. The areas that projects are interested in knowing more about with examples from operational experience are summarised under the headings of 'safety issues'. Their definition will depend on the maturity of the product of the project, as well as the understanding of safety by the project team.

The safety issues are used to select a number of relevant reports to these needs. The analysis proper then begins and the selection of de-identified reports are analysed in workshops with human factors, safety, and operational specialists. Project team members are invited and encouraged to participate in these workshops. Each workshop begins with an introduction to the aims of the session and to the focus project. The reports are read by one of the two moderations of the session and illustrated with sketches by the other one. Finally, the relevance of the report to the focus project is considered, as well as possible lessons learned. The variety in background of participants has been found to contribute to the richness of results. However, it is preferable to analyse less reports (e.g. nine in two hours) with a small group (e.g. five participants) to ensure that the appropriate atmosphere is created in the sessions. It is necessary for participants to feel free to brainstorm and keep focused on learning from the occurrence, not finding one culprit or a single reason for the development of the event. In other words, the reports should be seen as catalysts to render assumptions about new concepts, designs and working methods explicit and open to scrutiny.

The results of each workshop are summarised as lessons learned illustrated by examples, and are organised according to questions (e.g. what would the EEC project's product do in this case?), issues (e.g. this problem needs to be considered in the design of your project), and safety benefits (e.g. this example would suggest that the product of your project provides a solution to this problem).

Activity four: delivery

The final activity is that of delivery. This involves presenting the EEC project team with the results of the analyses. The project team are given examples for each safety issue that illustrate safety benefits that their project is expected to provide, as well as additional issues identified that may or may not have been addressed. A question list for the project will have also resulted from the analyses. A final report will be provided to the EEC project, summarising the lessons learned illustrated through case studies, and their feedback on the questions and issues that were raised. Feedback will also be provided to the ANSPs who provided the data. As mentioned above, the format of the feedback is agreed upon in advance with the data provider.

This final activity is the less mature of the SAFLearn activities. Two reports have been written and no evidence of the actual use of the SAFLearn results has been identified. In one of the projects supported however, the use of the results has been clearly defined. Lessons learned from examples will be used in a hazard analysis that will be conducted after the simulation, to assess the safety of the system proposed. More consideration needs to be given to finding a way to assess success and failure of SAFLearn in supporting projects learning from operational experience.

Examples of SAFLearn

The SAFLearn process is illustrated through three examples. Although desirable, it is not possible for SAFLearn to support all projects at their concept stage, thus the type of service provided is dependent first of all on the stage at which the project is, as well as on the availability of the team. In general terms, teams can be supported at their conceptual stage, during development and testing.

At the initial stages of the project SAFLearn provides general examples of issues of concern in today's systems. These examples should be used by the EEC project to brainstorm on ways in which the future can improve on today. One of the five projects that are being supported currently by SAFLearn, the Gate to Gate project, is receiving this type of service. The Gate to Gate work-package for which EEC is responsible was in its initial, conceptual, phase when SAFLearn started working with the project team. Two workshops have been held with the whole team (i.e. managers, developers, operational specialists and engineers) in which safety occurrence reports have been analysed together, illustrating a number of issues in all types of airspace. The project in fact looks at how to harmonise the overall functioning of the ATC system, from gate (of departure) to gate (of arrival). These sessions have provided a rich interaction between different team members, who have used the examples provided to evaluate proposed new functionalities of future systems. Using concrete examples has also allowed the team to explore dependencies on other projects, which were often implicit. Support to the Gate to Gate project will continue throughout the development of the technical specification of different functionalities and their testing in a simulation at the beginning of 2006. It is hoped that these workshops will allow the project team to gain a good understanding of safety issues in a variety of contexts, as well as provide SAFLearn with the experience of supporting a project throughout its development.

Currently, no project is being supported in the design or development phase. Two projects are being supported during the testing phase. The first project is concerned with the development of an Arrival Manager (AMAN), which is a tool that smoothes traffic from when it is in en-route airspace (i.e. high level) to its arrival in terminal area, before final approach to the runway. A series of examples related to the sequencing of traffic and crossing of different flows of traffic, en-route and arrivals or arrivals and departures, have been prepared and presented to the simulation team. A selection of these examples will be presented to the controllers participating in the simulation in a workshop to be held in their last week of their simulation. The examples will be used to support a brainstorming session with controllers to discuss whether similar events have occurred during the simulation or could occur, as well as mitigation means that would avoid such an occurrence taking place using the tool. In other words, both hazards and barriers will be elicited. Following the simulation both the examples used during the workshop with the controllers and those which were not used, will be provided to the project team. These will be used in discussions on safety issues that need to be addressed in future developments of the tool and during live trials simulations. These results will support a hazard analysis and help define working methods for the use of an AMAN tool.

The third example is the DOVE project, which looks at validating the operational use of data-link communications (i.e. text instead of voice communications between controllers and pilots). Although support to this project has only just started, reports are being collected and organised according to the project's safety issues. Workshops have been planned with the project team before and after the simulation, as well as with controllers during the simulation. Unlike the AMAN project, the sessions with controllers will be on a one-to-one or two-to-one basis. This should allow controllers more time to express their opinion on the relevance of the occurrence to the data-link tool as well as allow them to share their own experiences more freely.

Concluding remarks

The SAFLearn project has been conceived of as a service and described as a process comprising four activities: collection, storage, analysis, and delivery of lessons learned. The main users of this service are projects at EEC, however the vision for the SAFTool is a resource openly available to the aviation research community.

The importance of SAFLearn lies in the fact that although today operational experience is used by EEC projects through a number of channels, the provision of lessons learned from current operations is not part of a structured process. The approach used by SAFLearn to safety remains focused on expert judgment and experience, in consideration that qualitative information collected from expert decision makers remains an essential complement to more quantitative-based approaches to assuring the safety and reliability of future concepts and systems.

References

Carroll, J. M., Rosson, M. B., Chin, G., & Koeneman, J. (1998). Requirements development in scenario-based design. *IEEE Transactions on Software Engineering, 24*, 1156-1170.

European Space Agency (1999). Alerts and lessons learned: an effective way to prevent failures and problems, Workshop 29-30 September, ESTEC, Noordwijk, the Netherlands.

GAIN (Global Aviation Information Network) (2003). Major Current or Planned Government Aviation Safety Information Collection Programs, June 2003.

Jarke, M., Bui, X. T., Carroll, J. M. (1998). Scenario management: an interdisciplinary approach. *Requirements Engineering, 3*, 155-173.

Johnson, C. W. (2003). *Failure in Safety-Critical Systems: A Handbook of Accident and Incident Reporting*, London: Springer Verlag.

Klein, G. (1999). *Sources of power*. Cambridge: MIT Press.

Smith, P. J., Woods, D., McCoy, C. E., Billings, C., Sarter, N., Denning, R., & Dekker, S. (1998). Using forecasts of future incidents to evaluate future ATM system designs. *Air Traffic Control Quarterly, 6,* 71-86.

Van der Schaaf, T. & Kanse, L. (2002). Not reporting successful recoveries from self-made errors? An empirical study in the chemical process industry. In C. W. Johnson (Ed.) *Investigation and Reporting of Accidents*, GIST Technical Report G2002-2 (pp 180-183)

Van Es, G. (2001). A review of Civil Aviation Accidents Air Traffic Management-related accidents: 1980-1999. Paper presented at the 4th International Air Traffic Management R&D Seminar, New-Mexico, December 3-7th 2001.

Van Es, G. (2003). Review of Air Traffic Management-related accidents world-wide: 1980-2001. Paper presented at the FSF European Aviation Safety Seminar, March 17-19, Geneva, Switzerland.

Von Thaden & Wiegmann (2001). Improving Incident Reports using a schematic recall aid: the critical event reporting tool (CERT), Proceedings of the 45th Annual Meeting of the Human Factors and Ergonomics Society, Santa Monica CA: Human Factors and Ergonomics Society.

Acronyms

AMAN	Arrivals MANager
ANSP	Air traffic control National Service Provider
ATC	Air Traffic Control
DOVE	Data-link Operational Validation Experiments
ECAC	European Civil Aviation Conference
EEC	Eurocontrol Experimental Centre
ESARR	Eurocontrol SAfety Regulatory Requirement
HEIDI	Harmonisation of European Incident Data Initiative
SMS	Safety Management System

How do we find safety problems before they find us?

John A. Stoop
Faculty of Technology, Policy and Management
Delft University of Technology, Delft, The Netherlands

Abstract

This contribution elaborates on the safety management issues as they have been developing on a major airport in Europe. Over the past decade several subsequent safety models and concepts have been developed, each represented by different stakeholders and covering different safety issues. Starting from an event driven approach –the crash of a B747 in an apartment block- questions have been raised about the opportunities for an airport wide cooperation among stakeholders, their interfacing issues, substantive safety aspects and the required analytic potential to deal with a variety of performance indicators, including incidents. During the process, a systems approach has emerged, covering safety issues related to primary processes at the airport, combining probabilistic and deterministic safety approaches, while gaining societal support for their implementation. The developments are put in the framework of ICAO and EU developments, as well as the reintroduction of the precaution principle in the Dutch risk debate. Improving safety performance at a systems level requires timely transparency, communication between stakeholders, application of multiple performance indicators and on-line monitoring of primary processes. Safety is considered a professional as well as a societal condition for maintaining faith in the aviation sector. Such a condition requires knowledge development as well as a knowledge management strategy in order to develop, disseminate and implement new insights in the aviation community.

Introduction

Like many other countries, the Netherlands faces an era of changes. Major infrastructure projects are under construction, traffic volumes are increasing, privatization is well under way and ICT applications become widespread in the transportation industry. Systems performance has benefited from these developments, since a rapid and extensive growth has been demonstrated over the past 10 years. These developments however do have their drawbacks on safety. Despite the efforts to maintain safety at the present high levels, risk and safety issues are in the spotlights of the political debate and in the press. A small series of serious accidents has fuelled the debate on acceptability of risk. New players appear in the risk debate and decision-making arena, focusing on issues that have not taken into account before, such as victim care and rescue and emergency performance. This involvement causes considerable change in the way risk is perceived by all

In D. de Waard, K.A. Brookhuis, R. van Egmond, and Th. Boersema (Eds.) (2005), *Human Factors in Design, Safety, and Management* (pp. 479 - 491). Maastricht, the Netherlands: Shaker Publishing.

stakeholders, advocating independent accident investigation as a citizen's right and a society's duty (Van Vollenhoven, 2001). Changes in markets, context, organizational and institutional changes in the aviation sector fuel demands for a new approach beyond the level of best practices and compliance with risk standards.

First, a transfer of responsibilities in risk management has taken place from governmental agencies to privatized companies such as Amsterdam Airport Schiphol. The management of the airport is confronted with the fact that their safety efforts must be made visible and taken into account quantitatively not only to demonstrate their compliance, but at the same time should accommodate the desired growth in traffic movements and planned increase in passenger and freight volumes.

Second, incorporating scientific knowledge and operational expertise in design and operational practices should facilitate an improved performance of the system since best practice approaches have reached a safety level plateau (Hillestadt, 1993). Historically, implementation of knowledge has been achieved by retrospective analysis of system failures and drafting recommendations, mainly based on accident and incident analysis and specific scientific research into critical aspects of the systems failure. However, availability of such explanatory variables may provide a systems analysis with insight into the systems functioning, but does not necessarily provide the systems management with control variables to enhance the systems safety performance and change variables to cope with system deficiencies.

The issue can be stated as: how to deal with system failure and where to intervene in the system in order to achieve safety enhancements. Transparency of the primary process and explicit knowledge management strategies become necessary. The Schiphol airport case will highlight this issue of knowledge management, risk control and system change.

The precaution principle

By nature, design and technological development are dealing with uncertainties. Based on proven deficiencies in the systems design and operation a timely adaptation of the system is required to cope with operational demands. In case of uncertainty, risk management is merely 'precaution' because the hazards themselves hardly can be identified qualitatively, let alone quantitatively (Petroski, 1992). Also during long-term change in system operations, deviation may occur from designed specifications and intended use. Such deviant operations outside the design envelope may induce failure.

In particular in the transportation sector where a demand for continuity and synchronization is critical, the exclusion of a failing component, function or loss of performance is essential. Rather than taking the risk of failure, the precautionary exclusion of a failure is essential under the motto: better safe than sorry. To prevent reoccurrence of an unexplained failure, designers and operators have to act before understanding. After taking precautionary measures, a further investigation into the failure is necessary to understand the failure mechanism.

Dealing with failure

In considering the two origins of failure -uncertainty about the systems behaviour and operating loads deviant from design specifications- two different concepts to deal with failure are available (McIntyre, 2000).

1. *understanding failure*

   Understanding failure requires several steps, starting with establishing the precise sequence of events –expressed in specific accident scenario descriptions- identification of specific failure mechanisms -such as Controlled Flight Into Terrain, runway incursions, midair collisions or loss of control in final approach- and the development of scientific evidence in the explanation of the failure mechanisms. Uncertainty regarding such failure distinguishes two types of uncertainty: a qualitative uncertainty regarding the scenario description and a quantitative uncertainty regarding the statistical frequency and severity of the consequences (RIVM, 2003). Understanding failure provides a basis for intervention in the characteristics and conditions, eliminating deficiencies in the systems design and operation. Accident investigation is a critical instrument in this concept, dealing with a fact-finding phase before analysis, drawing up of recommendations and implementation of systemic changes may take place.

2. *prediction of failure*

   The probability of failure may be managed by reducing the uncertainty on the likelihood of failure. Such reduction can be established by developing knowledge on the technical endurance, critical threshold values, acceptable load patterns, tests and benchmarking methods and establishing the operating envelope of the system. Specific knowledge is required on loads, aging, prediction on failure frequencies, residual performance, and reliability of systems and their components. The emphasis is on prevention, prediction of system behaviour based on generic failure and analogies with historical data on comparable systems. Reliability and maintenance are critical instruments in this concept. An understanding of failure by single accident investigations is replaced by an approach, which relies on generic failure typologies, systems modelling and availability of large sets of reliable data. After initially modelling technical systems, developments have emphasized the modelling of non-technical system components such as the human factor, management, organization and culture.

Expert knowledge

The need to achieve a satisfactory explanation of failure and accidents based on scientific knowledge and notions has created new scientific areas of interest (Carper, 2001). Such a focus however fades on a relatively short notice, once a public interest is gone. On the long term, such a focus dissipates into the mainstream of the interest of a scientific discipline or disappears with the superannuating of its advocates if there is no structural need for knowledge within the sector itself. During a first phase of technological development learning from system deficiencies is necessary to eliminate teething troubles from these new developments. During the maturity phase, of technology and systems a permanent reflection is required on standards and regulations in order to prevent the occurrence of unnoticed deviations from design specifications by adaptations to changes in the operational requirements or environment. Essentially, the need for independent investigations has put requirements on investigations with respect to quality and credibility assurance (Smart, 2004). These qualification needs have stimulated training and certification of aviation accident investigators. It has led to the establishment of professional organizations, the International Society of Air Safety Investigators ISASI and its maritime equivalent Maritime Accident Investigation International Forum MAIIF, creating a critical mass and continuity in skills and experience in aviation and maritime accident investigations.

ICAO Annex 13
During the close of the Second World War, the USA, Canada and the UK took the initiative to establish an international organization for harmonizing international civil aviation and setting standards and procedures for developing the sector on a global scale. At the foundation of ICAO, the International Civil Aviation Organization, a series of Annexes was drafted, including accident and incident investigation, the ICAO Annex 13. The precaution principle and a timely feedback of findings are pivotal. From the beginning on there is a clear distinction between blame and causation for the benefit of taking rapid and necessary measures. A strict separation is maintained between Technical Investigations and Judicial Inquiries. Annex 13 settles the agreements and cooperation between states which are involved in an aviation accident, being the State of Occurrence, Operations, Registry and Manufacturing and the provides harmonization of investigation procedures. This harmonization receives ample consent and gains the status of pseudo-legislation for the ICAO member states (ICAO, 2001).

Separation from blame
Before the Second World War already, the aviation industry concluded that learning from deficiencies had to take place on an international, sectoral level rather than on a basis of a single operator, manufacturer or state. In order to keep public faith in the aviation industry, a common process of learning without allocating blame was deemed necessary (Cairns, 1960). In order to provide a timely feedback to all stakeholders in the sector, accident investigations had to be separated from judicial procedures, which by nature, are very time consuming and focus on individual responsibilities and liability.

This approach is until today based on the principle of precaution:

> *in order to protect the aviators from threats of serious or irreversible damage, a timely response is required, while a lack of full scientific certainty shall not be used as a reason for postponing cost-effective measures to prevent further degradation.*

Consequently, two types of investigations emerged, which could be conducted parallel to each other: an independent technical investigation into the causes of an accident and a judicial inquiry into responsibilities and liability.

Technical investigations into the failure of designing and operating aircraft have seen an impressive development. Based on a limited number of 'showcases' design principles were developed, such as fail-safe, safe life, damage tolerance, crash worthiness, situation awareness or graceful degradation. Several cases such as the De Havilland Comet, Tenerife, Mount Erebus, TWA-800, Valuejet and Swissair have become typical for specific deficiencies in the aviation system. They have lead to many practical changes as well as new expertise on specific academic areas varying from as metal fatigue to human failure, crew resource management or life-cycle maintenance. Within this precaution context, developing knowledge is a prerequisite for understanding failure and enhancing system safety performance (Stoop, 2003).

Generalizing and combining the concept, a process approach

Focusing on system deficiencies requires structuring of the attention. A focus towards the primary processes requires a subdivision of these processes in several steps. Such steps mark the transition between flight phases, the actors involved, their supporting equipment and procedures, transfer of responsibilities and information needs (Van Mierlo, 2000). Geographically, these flight phases for inbound aircraft can be distinguished in four areas: approach, landing, taxi lanes and aprons. A hands-over procedure for Air Traffic Control distinguishes Area Control Centers (CTA), Approach Control Facility (TMA) and Tower Control (CTR). The sequence may be repeated for departing aircraft, distinguishing eight flight phases.

Each flight phase requires specific information regarding heading, altitude, speed, flight conditions, surrounding traffic and specific instructions. During approach, the aircraft is guided towards the airport. Communication aims at the positioning, navigation and communication with surrounding aircraft. Communication is transferred from Area Control Center to Approach Control Facility (ACF), allowing the aircraft to enter the first phase. The boundary between phase one and two is the responsibility of ACF and Aerodrome Control Tower (ACT) for a further guidance during the landing phase. In the landing phase the aircraft touches down, reduces speed and leaves the runway. Hands-over of the aircraft to Ground Control occurs during taxiing after which a gate is allocated. In the third phase, the aircraft is the responsibility of Ground Control, which coordinates all aircraft movements in the taxiing area. The fourth phase covers the handling of the aircraft during docking, in which landside logistics, maintenance and dispatch activities are performed. The departure phases cover the same 4 phases of approach, landing, taxiing and docking in a reversed manner, starting with the pushback from the gate. During each of the phases, specific hazards can be allocated to activities and tasks, taking into account characteristics of equipment, people and procedures. A risk assessment can be made, based on a variety of risk performance indicators, including the potential of damage and injury. However, other performance parameters can be added in the process, referring to other systems aspects and functions such as punctuality, quality, reliability and costs. An overall assessment of the safety performance can be incorporated in the system performance. In addition, the processes and phases can be discriminated with respect to the decision making level at which safety could or should play a role. A distinction is possible between operational, tactical and strategic levels in the decision making, each requiring a dedicated assessment of safety issues. Such an encompassing system does however not yet exist (Van Mierlo, 2000). It is one of the challenges of the 'causal' model for the airport risk-modelling project.

All these primary processes and their inherent events can be depicted in a flight handling model of Schiphol airport, linking causal factors to the primary process of flight handling (Stoop 1993, based on Hillestadt 1993, see Figure 1).

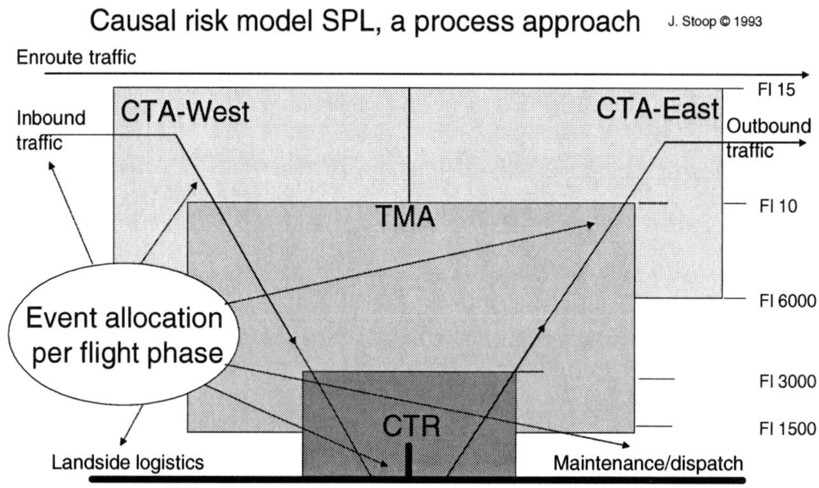

Figure 1. Primary processes and events during flight handling

The Schiphol Airport case study

Despite the strong reduction of the number of aviation accidents per million aircraft movements over the period 1965 until 1985, the aviation community is dissatisfied with the stagnation of the accident rate. Over the past 10 years, the decrease in accident rate has stabilized at an almost constant level. In 1998, leading aviation organizations and countries provided a strong impulse for the improvement of aviation safety. The Gore Commission chaired by the Vice President of the USA and the Federal Aviation Administration (FAA) launched the Safer Skies-A Focus Agenda program. This program aims at a five-fold reduction of the aviation accident rate within 10 years. In Europe, the Joint Aviation Authorities (JAA) launched the JAA Safety Strategy Initiative (JSSI) as a basis for an American-European co-operation. The motive for this massive initiative is in the reduction of major aviation accidents, irrespective of the worldwide growth in aviation traffic volumes. In addition, an ICAO (International Civil Aviation Organization) initiative focuses on the Safety Oversight Program and the European Civil Aviation Conference (ECAC) stimulates teams, assigned to SAFA (Safety Assessment of Foreign Aircraft). In addition to focusing on specific accident types envisaged in this program, such as controlled flight into terrain, loss of control runway incursions or weather-related accidents, the Dutch Government has given priority to the development of policy instruments.

This policy orientation has emerged from a crash of an El-Al Boeing 747 in an apartment block in the Bijlmermeer near Amsterdam in October 1992 (Hillestadt, 1993). As a part of the revision of the Dutch aviation safety policy, an extra set of measures was initiated in the Nota Veiligheidsbeleid Burgerluchtvaart (civil aviation safety document, MoT, 1999). One of these measures defined a new approach by designing a new measuring system for the external safety of the airport. This system comprises of three principal elements: measuring, calculation modelling and data collection. The objective is the availability of parameters, indicators and causal models which are not only appropriate for measuring safety, but also to supply instruments for measuring the effects of a safety policy, preferable in a prognostic manner. A causal model should supply insight into cause-effect relations within the network of functions in the aviation system, qualitatively as well as quantitatively. International co-operation is considered to be crucial. A first start has been made with a Dutch–UK investigation into the causes of full freighter accidents, since their accident rate is four times higher than passenger carrying aircraft over the 1980 to 1996 period is. Simultaneously, an Integrated Safety Management System has been implemented at Schiphol Airport, focussing on the co-operation between all stakeholders at the airport regarding safety. The intended 'causal model' should establish a more direct causal relation between measures which enhance the internal safety, the effects on safety in general and residents in the area in particular (MoT, 1999). Finally, a set of measures in case of emergencies should intend to reduce the consequences of an air disaster as much as possible. The new 'causal' approach therefore addresses safety more integral: internal safety, external safety, rescue, and emergency are all involved in the approach.

A cause for improved risk modelling?

Deficiencies in the risk debate can be observed with respect to apparent discrepancies in risk perception among stakeholders and a lack of consensus on the methods and risk mitigation strategies. Such a consensus is required to realize policy goals and stakeholders ambitions or to complete an infrastructure project without endless design modifications and construction delays.

Risk perception

Some observations can be made concerning the risk and safety deficiencies as they appear in the newspapers. First, disasters seem to come as a complete surprise to everybody (Ale, 2003). Immediately, the public and governmental responses express their disbelief that such an event could have been happening and request an in-depth investigation.

However, many of the precursing factors, which may lead to such an 'unforeseeable' event, are present and known by the experts in the sector from previous, similar occurrences. The probability of such an event is only zero if the activity is eliminated from the site. Expert judgements on the frequency of such events as 'negligible' prove to be disputable. Second, the extend of events is beyond imagination. The failure mode and sequence is said to be unforeseen. The consequent effect and impact is therefore also unforeseen and proves to extend far beyond the limits of acceptable impact. Accident scenarios are incorrectly excluded from regular risk assessment procedures (Hendrickx, 1991). Third, several defenses have failed. Once the activity was accepted in the area, regulations and enforcement proved to be not fail-safe. Quantities were exceeded, inadequate maintenance was tolerated, enforcement lacked and fire-fighters and rescue services were not informed about the nature and extend of the possible event. Questions rose after major disaster focus on size of the event and the perceived consequences, questioning credibility of the risk policies and decision making.

Deficient risk modelling?

Current practices in quantitative risk analysis for major infrastructure projects and other hazardous undertakings are based on generic applicable risk models. Since no specific systems descriptions are available, because they conflict with the generic nature of the model, no specific scenarios will be incorporated in the risk assessment. A further reduction of scenarios may take place if a scenario is ranked with a very low probability and therefore neglected, irrespective of the possible consequences. Risk models and scenarios have their origins in the hazardous material sector and have thereby an inherent characterization of the hazards involved. Other hazards, originating from the characteristics of another sector, such as transportation processes, are less likely to be incorporated. Hazards origination from the transportation of large amounts of passengers and goods have different characteristics, dealing with rescue and emergency activities. At the same time, major infrastructure projects generally have a unique character by their size and nature. This may well legitimize a specific modelling and risk assessment.

If a major accident occurs, the consequences may not be in the absolute numbers of fatalities alone, but in the overall numbers of fatalities, injuries and individuals exposed to the hazards; the population at risk. At present the assessment of a major event is measured by the 'Group Risk' definition, expressed in an fN diagram (frequency versus Numbers). This diagram relates the number of instantaneous fatalities to the probability of the event and may compensate for the 'risk aversion' within a society by adding an exponential factor to the slope of the curve. A 'population at risk' approach however covers a larger number by establishing the overall population which is exposed to hazards and may express the risk aversion more adequately.

Runway incursions, a test case for a combined approach?

In the debate about a more integral safety enhancement strategy in the Netherlands, an attempt became necessary to fulfil the requirements of all major players in the aviation safety arena. The Ministry of Transportation initiated a project to develop a more 'causal' model to compensate for the apparent deficiencies in the conventional probabilistic risk assessment methodology. The case of runway incursions has been selected by the author to illustrate the potential of such a new approach. The risk of runway incursions and the causal chain of events leading to their occurrence are more closely taken into account.

Runway incursions, a risk problem?

A fundamental question regarding runway incursions is whether they pose an actual risk to the safety of aviation. For our purpose a runway incursions is defined as an event in which an airplane in its early departure or final approach, finds the runway blocked by other airplanes or objects that may jeopardize its safety with catastrophic consequences. Are runway incursions really so frequent that they should be prioritized as a threat to the functioning of the aviation system? Based on accident data, only very few runway collisions with disastrous consequences have occurred. The most well known accidents occurred at Tenerife in 1977 (AAIB, 1979) and Taiwan in 2000 (see Table 1).

Table1. Runway collisions

Tenerife	killed	injured	unhurt	total
KLM	248	–	–	248
Panam	326	55	15	396
Taiwan	83	45	51	179

On march 27th, 1977 a collision occurred at the airport of Los Rodeos at Tenerife (Spain) between two B747 aircraft of KLM and Panam, during take-off of the KLM Boeing, while the Panam Boeing was still taxiing on the runway in thick fog in the opposite direction. Nobody of the KLM 234 passengers and 14 crew members survived, while in the Panam Boeing 9 out of the 16 crew members and 317 out of the 380 passengers were killed, 7 crew members and 48 passengers were injured and 15 passengers remained unhurt. The general summary of the investigation established that: 'the accident was not due to a single cause' (AAIB, 1979). The misunderstanding arose from generally used procedures, terminologies and habit-patterns. The unfortunate coincidence of the misunderstanding with a number of other factors nevertheless resulted in a fatal accident. In the operation of the KLM crew, nor in those of the tower controller or the Panam crew, actions can be indicated which should be considered as serious errors. However, in varying degrees, a non-optimal functioning can be recognized with all parties'. The cause of the accident discriminates between human factors of both crews and tower controller, radio communication using ambiguous terminology and coincidence of a number of circumstances which directly influenced the course of the events and ultimately resulted in the collision.

In Taiwan, on October 31, 2000, at approximately 23.18 local time, a Singapore Airlines Flight SQ006 Boeing 747-400 airplane entered the incorrect runway at Chaing-Kai-Shek Airport. Heavy rain and strong wind from the typhoon "Xiang Sane" prevailed at the time of the accident. The airplane was destroyed by its collision with the runway construction equipment and by post impact fire. Prior to the accident, a NOTAM (Notice to Airmen) was issued indicated that portion of the runway 05R was closed for construction. There were a total of 179 people on board with 159 passengers, 3 flight crewmembers and 17 cabin crews. At the time, 83 people died and 45 people were injured. The final accident investigation report still has to be issued.

On December 10th 1998, a runway incursion occurred at Amsterdam Airport Schiphol between a Delta Airlines flight 039 Boeing 767 and a towed KLM Boeing 747. The incident did not result in a serious accident. Nevertheless, it was reported to and investigated by the Dutch Transportation Safety Board (DTSB, 2001). The findings and recommendations in the final report, issued in January 2001, bear some striking resemblance wit the two other occurrences. At the time of the serious incident low visibility and a low cloud base made visual control from the tower impossible. Low visibility procedures were in force. DAL 39 had been cleared for take-off from runway 24. Almost at the same time, a KLM Boeing 747 being towed and accompanied by a yellow van was cleared to cross runway 24. During the take-off roll the pilots of the DAL 39 observed the towed Boeing 747 crossing the runway. The take-off was aborted and the aircraft brought to a standstill before reaching the position of the tow. Contributing factors to the incident were: low visibility weather conditions, inadequate information during radio communication, misinterpretation of position and movement of the tow, take-off clearance without positive confirmation and insufficient teamwork and supervision in the tower.

The similarities between the cases raise some questions:

- are these events rare and unique deviations from operational standards under poor conditions
- or are these events indicating system deficiencies which hardly result in accidents with very serious consequences

Apparently, the use of accident data alone does not provide enough information. An extension towards incidents may be necessary to identify the actual frequency of runway incursions, irrespective of the outcomes.

Runway incursions, frequent incidents?

The frequency of runway incursion incidents and accidents in the USA alone has raised concern with the National Transportation Safety Board (NTSB), expressed by the testimony of its chairman Carmody before a Committee of the House of Representatives on March 28, 2001 (NTSB, 2001). According to FAA, the number of air travellers will increase from 604 million in 2000 to over 926 million by 2012. In addition, the FAA projects that aircraft operations at air route traffic control centres will increase from 46 million to about 61 million and that the number of passengers on foreign flag carriers travelling to and from the USA is expected to increase from 140 million to 267 million in the same period (see Table 2).

FAA data show that there were 429 runway incursions in the USA last year, more than twice the 200 incursions occurring in 1994 and a significant increase from the 322 in 1999. The runway incursion rate per 100.000 operations was .63 in 2000, up from .47 in 1999. It should be taken into consideration that small business aircraft accounted for a major contribution to these numbers. Since 1993, the NTSB has issued almost 100 safety recommendations to the FAA regarding runway incursion issues. In 1991, the FAA stated that the cornerstone of its runway incursion efforts was the development and implementation of the Airport Movement Area Safety System, or AMASS. The system works on audible and visual alert to controllers when an aircraft or vehicle is occupying a runway and the arriving aircraft is close to the threshold or a departing aircraft is detected on the runway by the system. However, the alert parameters were not based on human performance studies but empirically determined on a prototype of the AMASS system.

Table 2. growth in air traffic

Air traffic	in 2000	in 2012
Air travellers in millions	604	926
Aircraft operations in millions	46	61
Foreign flag passengers to USA	140	267

It has been nearly 10 years after the NTSB issued its recommendations on developing and implementing an operational system to alert controllers to pending runway incursions at all terminal facilities that are scheduled to receive airport surface detection equipment. So far, none of the systems has been commissioned for full operational use at any airport in the USA. Criteria for installation of airport ground surveillance systems and commitment to a specific date for completion of the acquisition and delivery of the systems are still lacking (NTSB, 2001).

In conclusion, runway incursions can be considered a low risk event with even very low risks for large commercial aircraft if their frequency, aggravating circumstances and their disastrous consequences are taken into account. If the frequency is taken into account irrespective of the outcomes and circumstances, the risk is much higher, especially when the risk is related to the growth in traffic volume and traffic density. System deficiencies at the level of developing and implementing vital support systems become apparent and reoccurrence of causal chains seem to appear with respect to procedures, human factors, communication and equipment (Stoop, 1997).

Conclusions and discussion

Quantitative risk modelling has laid the basis for enhanced levels of risk decision-making in The Netherlands. Due to new developments, new areas emerge, calling for a more 'causal' approach. Experiences, based on some major projects and major accidents in The Netherlands, have indicated three areas for debate to develop the insights in dealing with risk:

- major infrastructure projects may benefit from a combined effort in probabilistic and deterministic risk modelling to facilitate rescue and emergency aspects
- scenario definition should incorporate system or project specific scenarios to legitimize the specific nature of such systems or projects
- risk assessment modelling should not only take into account compliance with risk standards, but should be linked with risk management strategies as well.

At present, external risk models are based on probabilities and consequences of a limited number of accident scenarios, derived from a worldwide representative sample of accidents. Airport safety management systems however are based on airport specific sets of multiple performance indicators, including accidents and incidents. The paper has described a conceptual framework for a 'causal' model for the integral safety of a major airport. The framework takes into account a multi-actor setting, multiple performance indicator data and primary phases of the flight process at an airport, including surface movements, ground handling and communication. The use of accident an incident data in the model is indicated. An analysis of some major accidents is made to allocate risk-contributing factors to specific deficiencies in the system.

References

AAIB, (1979). Verdict of the Aircraft Accident Inquiry Board. Regarding the accident on March 27, 1977 at Los Rodeos Airport, Tenerife (Spain) involving the aircraft PH-BUF of Royal Dutch Airlines N.V. (KLM) and N736PA of Pan American World Airways Inc. (PANAM), *Aircraft Accident Inquiry Board, The Hague,* July 1979.

Ale, B. (2003). *That will not happen to us.* Inaugural lecture Delft University of Technology. Faculty of Technology, Policy and Management. (In Dutch: *Ons overkomt dat niet.* Inaugurele rede Technische Universiteit Delft, Faculteit Techniek, Bestuur en Management). 17 september 2003

Cairns, C. (1960). Report of the Committee on Civil Aircraft Accident Investigation and License Control. Ministry of Transportation. United Kingdom 1960

Carper, K. (2001). *Forensic Engineering.* London: CRC Press.

DTSB, (2001). Final report 98-85/S-14, N 193 DN, Boeing 767, 10 December 1998 Amsterdam Airport Schiphol. The Hague, the Netherlands: *Dutch Transportation Safety Board.*

Hendrickx, L. (1991). *How versus how often.* The role of scenario information and frequency information in risk judgement and risky decision making. Doctoral Thesis Rijksuniversiteit Groningen, november 1991

Hillestadt R., Solomon, K., Chow, B., Kahan, J., Hoffman, B., Brady, S., Stoop, J., Hodges, J., Kloosterhuis, H., Stiles, J., Frinking, E., and Carrillo, M. (1993). *Airport Growth and Safety.* A study of the External Risks of Schiphol Airport and Possible Safety-Enhancement Measures. EAC-RAND Santa Monica

ICAO (2001). Annex 13 to the Convention on International Civil Aviation, Aircraft Accident and Incident Investigation, Ninth Edition, July 2001

McIntyre, J. (2000). *Patterns in Safety Thinking.* Aldershot, UK: Ashgate

MoT (1999). Safety Policy Civil Aviation. (In Dutch: Veiligheidsbeleid Burgerluchtvaart). Tweede Kamer, vergaderjaar 1998-1999, 24 804, nr 21. (*Letter from the Minister of Transportation (MoT) to the Dutch Parliament*)

NTSB, (2001). *Testimony of Carol Carmody.* Acting Chairman, National Transportation Safety Board before the Committee on Appropriations, Subcommittee of Transportation and Related Agencies, House of Representatives, Regarding Aviation Safety. March 28, 2001.

Petroski, H. (1992). *To engineer is human.* The Role of Failure in Successful Design. Vancouver, WA, USA: Vintage Books.
RIVM, (2003). *Sober Dealing with Risk.* (In Dutch: *Nuchter Omgaam met Risico's*). Report 251701047/2003. Bilthoven, the Netherlands: RIVM.
Smart, K. (2004). Credible investigation of air accidents. *Special Issue of the Journal of Hazardous Materials.* Papers from the JRC/ESReDA Seminar on Safety Investigation of Accidents, Petten, the Netherlands, 12-13 May, 2003. Vol 111 (2004), pag 111-114.
Stoop, J.A. (1993). Causal risk model Schiphol, a process approach. In: Hillestadt R., Solomon, K., Chow, B., Kahan, J., Hoffman, B., Brady, S., Stoop, J., Hodges, J., Kloosterhuis, H., Stiles, J., Frinking, E., and Carrillo, M. (1993). 1993 *Airport Growth and Safety.* A study of the External Risks of Schiphol Airport and Possible Safety-Enhancement Measures. EAC-RAND Santa Monica
Stoop, J.A. (1997*).* Accident scenarios as a tool for safety enhancement. In Hale, Wilpert, and Freitag. *After the Event. From accident to organisational learning.* (pp 79-94). Oxford: Pergamon Press.
Stoop, J.A. (2003). Divergence and convergence, trends in accident investigation. *Second Workshop on the Investigation and Reporting of Incidents and Accidents.* IRIA 2003. NASA/CP-2003-212642
Van Mierlo, J. (2000). *Developing Safety Performance Indicators in order to improve the integral safety at Schiphol Airport.* Graduation thesis Faculty of Systems Engineering, Policy Analysis and Management, Delft University of Technology. (In Dutch: *De ontwikkeling van Safety Performance Indicatoren ter verbetering van de integrale veiligheid op de luchthaven Schiphol.* Afstudeerverslag Technische Bestuurskunde Technische Universiteit Delft).
Van Vollenhoven, P. (2001). Independent Accident Investigation: Every Citizen's Right, Society's Duty. *Third Annual Lecture European Transport Safety Council.* Brussels, 23th January 2001

Developing a safety culture in a research and development environment: Air Traffic Management domain

Rachael Gordon & Barry Kirwan
EUROCONTROL Experimental Centre
Bretigny-sur-Orge, France

Abstract

Measuring safety culture has been undertaken in many industries (e.g. oil, nuclear, aviation) over the past twenty years, as a proactive method of collecting safety information about the current level of safety in the organisation. However, there has been little work undertaken to develop the safety culture of the designers of these technological systems, to ensure that their designs are endeavouring to reach the highest levels of safety. A tool was developed to measure the current level of safety culture of designers in an air traffic navigation R&D organisation which contains 21 sub-sections under the following four main headings: i) Management demonstration of safety; ii) Planning and organising for safety; iii) Communication, trust & responsibility for safety and iv) Measuring, auditing and reviewing. The findings indicated that the main areas for improvement are: i) the safety management system; ii) team integration and; iii) responsibility for safety. Based on the survey findings some changes were undertaken in an attempt to improve the safety culture at the centre and a repeat survey is planned for April, 2005 to assess any improvements. This paper will describe the survey method and findings, the safety improvement plan, preliminary findings from the follow-up survey and lessons learnt during the change process.

Introduction

Safety culture in ATM operations

Air Traffic Management (ATM) is currently seen by other industries as a 'High Reliability Organisation' (HRO), although it is not fully understood why ATM is so safe. Safety, at the levels seen in ATM, is something of an 'emergent property', built on the professionalism within the industry, and decades of trial and error in evolving best practices and procedures. It is obviously desirable that ATM retains this hard-won HRO status. The most likely way it could lose this characteristic is via fundamental change, i.e. changes at the core of ATM (since change is one of the main generalised causes of accidents). A recent study found that around 50% of incidents and accidents have a root cause in design (EUROCONTROL, 2004), thus indicating that it is important at the development stage of new systems that safety is considered to ensure that such changes will not result in losing the emergent property of safety.

There is an obvious need for good safety culture in operations, where controllers are directly involved in ensuring that aircraft are separated from each other by 5 nm (nautical miles) horizontally and 1000 ft vertically. The need for having 'informed', 'just', 'reporting', 'flexible', 'learning' (Reason, 1997) and 'wary' cultures (Hudson, 2001`) in an operational centre is apparent, however, this requirement is not always obvious to those who are not at the operational 'sharp end', such as the developers of new technologies, systems and procedures.

The need for a good safety culture in R & D

Many high reliability industries around the world are showing an interest in the concept of 'safety culture', as a way of reducing the potential for large-scale disasters. Organisations have certain characteristics which can be called its 'culture'. These are generally invisible to those within the company, and yet quite transparent to those from a different culture. Safety culture is a sub-set of organisational culture which has been described as: *'who and what we are, what we find important, and how we go about doing things around here'* (Hudson, 2000). Researchers have shown that organisations with good safety cultures tend to have fewer accidents. However there is little work on safety culture in research organizations, and what exactly it might mean at a working level in such organizations.

In practical terms, if safety is considered and is believed to be important when new systems are being developed, this will affect how the system is developed. Inevitably there are many decisions and 'trade-offs' that must be made during early development and design stages, however these may occur too early in the development stage to come under formal safety assessment scrutiny. Thus, if there is not a particularly positive and active safety culture amongst the developers, safety may not be considered during such decisions and trade-offs. Safety can simply be seen to be someone else's problem, or something that can be fixed later on.

It is important therefore that those working at this design and development end of the operational spectrum understand safety, and know how their decisions can affect and curtail real safety later on. This is not an easy process, especially as it can be seen to constrain designer 'freedom' and creativity. However, at a practical level, the following types of safety questions have been identified through discussions with developers at the Eurocontrol Experimental Centre (EEC) as desirable in a research/design organization for a safety critical industry:

- For the system I am working on, what types of incidents and accidents have occurred?
- How could a new design avoid such problems?
- Could my new system add new problems, alone or in concert with other developments?
- Could my system solve problems in other areas?

safety culture in the air traffic management domain 495

- Will the operator (in this case a controller) be able to deal with events such as system failures, whether revealed or unrevealed, and other anticipatable events?
- Will the system be robust and supportive enough that the future operator can deal with future events that I cannot at this point anticipate? etc.

Such thinking, and resources to support such thinking (e.g. knowledge of incidents and failure paths etc.), should help generate more safe systems.

Since safety is an essential aspect of research and design in ATM it is therefore important that the EEC organisation strives for higher levels of safety awareness, more positive attitudes and commitment to safety. For this reason, it was thought to be appropriate to measure and improve safety culture in the EEC.

Measuring safety culture

Safety culture (or safety climate) in the workplace has traditionally been measured using questionnaire surveys, and has focused on the perceptions and attitudes of the workers. More recently, organisations' safety culture have been measured in terms of their level of maturity, which can be described on a line from 'emergent' to 'continually improving', from worst to best. The Safety Culture Maturity Model[1] (Fleming et al., 1999) was originally designed for the offshore oil industry, but the structure has been used in other high reliability organisations, including ATM and to measure the level of maturity of safety in design (Sharp et al., 2002) in the offshore oil industry. The Safety Culture Maturity Model[1] contains 5 iterative stages of maturity (see Figure 1), where organisations can progress sequentially by building on their strengths and removing the weaknesses.

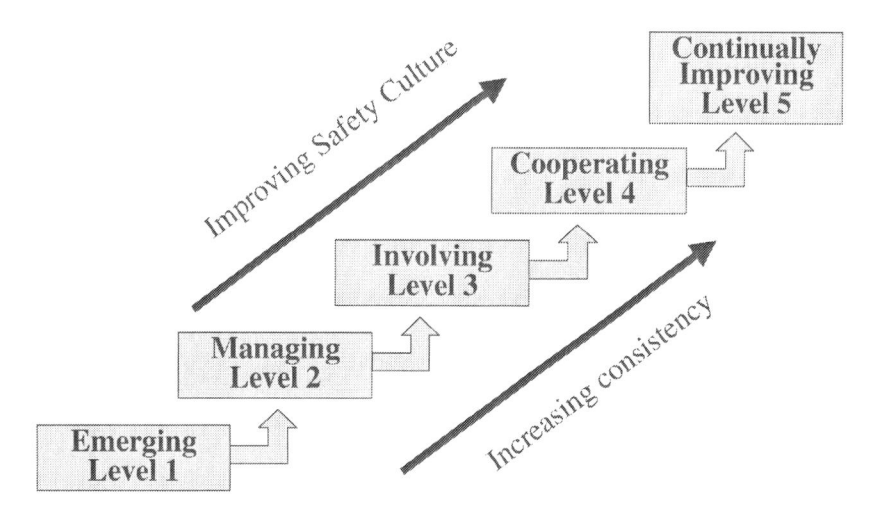

Figure 1. The Safety Culture Maturity Model®[2] (from Fleming et al., 1999)

---

[2] Safety Culture Maturity is a Registered Trademark of the Keil Centre Ltd, 2003.

In the early stages of a safety culture (Levels 1 and 2), top management believes accidents to be caused by stupidity, inattention and, even, wilfulness on the part of their employees (in an operational environment). In a design organisation, management do not believe that their organisation can influence the safety of future operations (such as ATM). Many messages may flow from management to the 'shop floor', but the majority still reflect the organisation's primary production goals, often with 'and be safe' tacked on at the end.

At the "Involving" stage (Level 3), the foundations are laid for acquiring beliefs that safety is worthwhile in its own right. By constructing deliberate procedures, an organisation can force itself into taking safety seriously. At this stage the values are not yet fully internalised, the methods are still new and individual beliefs generally lag behind corporate intentions. However, a safety culture can only arise when the necessary technical steps and procedures are already in place and in operation. Level 4 means that the organisation really gets to grips with safety issues with commensurate resources, and at Level 5 the organisation is largely controlling and managing safety effectively but without complacency, and is continually improving its efforts.

Safety culture in other research and development organisations

In a paper by Vecchio-Sadus & Griffiths (2004), a mineral processing and metal production research and development organization (similar in many ways to the EEC) was used as a case study to show how occupational health and safety marketing strategies can be used to influence behaviour and promote management commitment and employee empowerment to enhance safety culture. In order for health & safety promotions to have an impact on people's behaviour, Vecchio-Sadus & Griffiths (2004) think it is important to use marketing strategies. In fact, they report improvements in:

- safety culture (accountability and commitment by management; an increase in the number of employees taking ownership of their work environment and knowledge improvement);
- improvements in risk management (increase in the number of risk assessments; better job safety procedures; improvements to the workplace, plant & equipment)
- improvement in overall performance (substantial decrease in lost time injury rate; decrease in compensation claims; improvement in the investigation and documentation of incidents; winning research contracts because of the safety systems and culture)

The aims of carrying out safety culture maturity surveys in organization in general are to:

- assist in informing senior management about cultural or behavioural issues and developing effective safety improvement plans
- for developing management's thinking about the type of organisation they are managing and where they want to go

- encourage managers to develop the organisation's safety maturity
- provide a practical framework for developing improvement plans and selecting appropriate interventions (as the level of safety maturity influences the appropriateness and effectiveness of different safety improvement techniques).

The objectives of this project were to

- develop a measure of safety culture that is relevant to ATM R&D organisations,
- measure the safety culture at the EEC and
- develop some plans for improving the safety culture.

Method

Development of a safety culture measure for R & D

Six organisations have developed and tested safety culture using maturity levels (ISO, 1990 ; Fleming et al., 1999 ; Sharp et al., 2002 ; Nickelby et al., 2002 ; NATS, 2002 ; Hudson, 2001), mainly in the offshore oil industry. The EEC safety culture survey is based on these measures and adapted to ATM and R&D. The questionnaire has 5 levels of maturity where organisations can progress sequentially by building on their strengths and remove their weaknesses. The following 5 maturity levels labelled by the Keil Centre have been used for the purposes of the EEC safety culture survey (SCS):

Level 1 – Emerging - safety defined as technical & procedural solutions and compliance with regulations; safety not seen as key business risk; accidents seen as unavoidable
Level 2 – Managing - safety seen as a business risk; safety solely defined in terms of adherence to rules & procedures; accidents seen as preventable
Level 3 – Involving - accident rates relatively low (reached plateau); management think frontline employees are critical to improvements; safety performance is actively monitored
Level 4 – Proactive - managers/staff recognise that a wide range of factors cause accidents; organisation puts effort into proactive measures to prevent accidents
Level 5 – Continually Improving - sustained period of no recordable or high potential incident, but no feeling of complacency; constantly striving of finding better ways of improving hazard control.

The SCS, developed at EUROCONTROL Experimental Centre, has 21 elements which are contained within 4 main elements:

- Management Demonstration
- Planning and Organising for Safety
- Communication, Trust and Responsibility
- Measuring, Auditing and Reviewing

Table 1 provides some examples of the 21 Elements, giving a flavour of what is contained in the "Emerging" and "Continually Improving" levels. These elements

Table 1. Examples of Statements from the Safety Culture Survey Tool

	Element	Lower Levels of Element (1-2)	Higher Levels of Element (4-5)
Management Demonstration	1 Management Commitment to Safety*	Safety is considered an employee responsibility. Lip service is paid by senior management to the importance of safety commitment	Senior management demonstrate commitment to safety. Management & staff frequently discuss safety. Good safety behaviour is recognised
	2 Safety Performance Goals*	Safety goals are only assigned as an ad hoc response to an incident and tend to be based on previous experience only	Goals are set with reference to external benchmarks and internal history. Improvement targets are set
	3 Impact	There is no mechanism for results of safety activity to influence management or design decisions	Business processes ensure that safety has authority to enforce design changes or stop a project
	4 Investment & Resource Allocation*	Little or inappropriate provision is made for resources or facilities to conduct safety activity	Organisation makes strategic investments in developing organisation wide safety processes
	5 Policy & Strategy on Safety	No organisation wide safety policy or strategy. Safety activities occur in an unsystematic, unplanned way	Documentation and accessible organisation-wide policy of safety. Monitoring of policy and strategy forms an integral part of the organisation's business processes
	6 Safety versus Productivity*	Safety assessments are not undertaken because they often interfere with getting the work done	Staff are encouraged to take account of safety in design, which is fully resourced and supported and it is encouraged over and above getting the project completed on time
Planning & Organising	7 Safety Planning	Safety is not pre-planned ad occurs in an ad hoc, unsystematic manner	Strategic planning for safety is automatically initiated as part of core business processes
	8 Training & Competence	Staff are assigned to safety activities based on their availability, rather than on having training or relevant experience	A comprehensive safety training programme exists. Competence standards are used. The effectiveness of training is measured
	9 Knowledge of ATM Risks*	Employees are unaware of the impact that their work has on future ATM safety	All staff are aware of new and recurring ATM risks and fully understand how their work impacts safety
	10 Risk Assessment & Management	Risk assessment is reliant on individual experience from specialists or experienced managers	Risk information is routinely used in planning. There is wide employee involvement in risk assessment
Communication, Trust & Responsibility	11 Communication	There is no feedback to staff regarding ATM safety issues	Staff regularly bring up project safety concerns and feel confident to raise them with management
	12 Integrated Teams*	Safety effort is external to project teams	Safety personnel have a core role in project teams and have status at relevant meetings
	13 Involvement of Employees*	There is limited employee attendance in safety activities/meetings	Employees are heavily involved in contributing to the design, implementation and measurement of safety related changes
	14 Relationship w/ External Regulator	The objectives of the EEC and the safety regulator are diametrically opposed	Regular audits are undertaken by the regulator and viewed as constructive monitoring of safety
	15 Involvement of Stakeholders	No formal provision is made for gaining access to stakeholders	Appropriate and representative stakeholders are engaged at the right time on projects
	16 Trust & Confidence*	Trust and confidence of employees is assumed by management	Employees are confident that complete pictures of safety performance and progress against targets are communicated
	17 Responsibility for Safety	Safety specialists undertake safety activity in isolation from staff	Everyone in the organisation believes and accepts that safety is their responsibility
Measuring, Auditing & Reviewing	18 Organisational Learning	No process to assess or feedback safety learning; Information is shared on a "need to know" basis	A safety information system promotes sharing of safety issues and learning through the effective presentation of information
	19 Safety Management System / Audit	The concept of a SMS is not recognised; Isolated policies and procedures exist	A comprehensive SMS exists and covers all aspects of safety and is designed to be practical and achievable for all employees
	20 Achievement of Safety Targets	Criteria for determining whether safety targets have been achieved are applied ad hoc and tend to be inappropriate	Business processes actively identify the achievement of safety targets as criteria for the success of a project and they form important milestones
	21 Test of safety in design	Test and evaluation of safety issues tends to occur ad hoc in response to specific incidents and may not be appropriate	Information critical to H&S management is fed back from the safety tests and evaluations across the organisation

were developed based on the research of five research groups (Fleming et al., 1999; Sharp et al., 2002; Nickelby et al., 2002; NATS, 2002; Hudson, 2001), who have published the key elements which they believe to describe the safety culture of the organisations they were measuring. Three of the researchers were measuring safety culture maturity in operational environments (Fleming, 1999; NATS, 2002; Hudson, 2001); Sharp et al. (2002) was measuring the safety maturity in design; and Nickelby et al. (2002) were measuring human factors maturity in design. Surprisingly, there was quite a lot of difference in the key elements examined by the different researchers. However, there were a small number of common elements for each of the maturity models, such as Training and Organisational Learning. The first draft of the questionnaire was shown to four 'key' people in the organisation to determine the relevance of the items to the ATM research and design organisation. Some comments were made regarding the relevance of some of the statements in the questionnaire, such as the use of incident and accident reports, as they are not so relevant in an R&D centre compared to an operational centre.

Measuring safety culture

A cross-section of 40 staff within the EEC were targeted for the main survey. The participants were from different projects, at different levels in the organisation and the sample included contractors. The participants were selected from the personnel list to ensure that a cross sectional sample was chosen. The selected participants were chosen from different research areas, had different types of expertise and different levels of responsibility. The participants were initially contacted by e-mail and provided with information about the study. Participants were then asked to join a group of about 10 others (for a designated meeting) to complete the questionnaire individually. At this meeting, participants were provided with the background and purpose of the study and instructions on how to complete the questionnaire. Participants were encouraged to ask questions before and during the session when necessary. The survey was carried out in March, 2003 and a total of 36 participants responded. For each of the 21 elements, respondents indicated which level of maturity they believed the EEC was currently at, based on the statements provided in the questionnaire. The scores were entered into an Excel spreadsheet for analysis. For each element, the respondents scores were summed and averaged across the whole sample.

Results

The overall results indicated that the elements with the highest average scores all came from the 'Management demonstration' category, including: "Management commitment to safety"; "Safety performance goals" "Investment and resource allocation" and "Policy and strategy on safety". The elements with the lowest average scores included: "Safety management system"; "Responsibility for safety", "Integrated teams" and "Risk assessment and management". The average standard deviation between respondents was +/-0.87, indicating a fairly wide variety of responses.

Five 'best' areas identified by the safety culture survey

The following five key elements were chosen as the most positive out of the 21 key elements.

- Management commitment to safety
- Safety performance goals
- Policy & strategy on safety
- Trust & confidence
- Test & evaluation of safety in design

These key elements were thought to be between levels 2 and 3 by the respondents, indicating that although they were not the worst performing elements, they are by no means indicating that the EEC has a good safety culture. To have a good safety culture, the EEC would need to reach above level 3.

Five main problem areas identified by the safety culture survey

The following paragraphs describe the definitions of the levels for each of the above 5 key elements and the next level should be the next aim for the EEC.

1. Safety management system/ Auditing safety

The element thought to be weakest in the EEC was *Safety management system /Auditing safety* – from the fourth section: Measuring, auditing and reviewing. The standard deviation (+/-0.82) indicates that there was only a little variation between respondents.

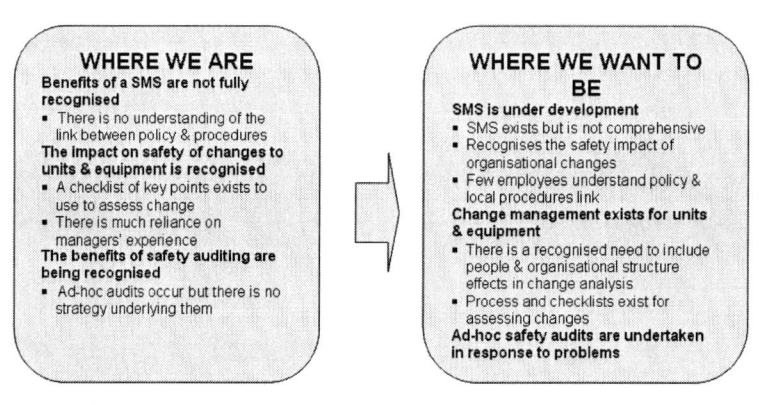

WHERE WE ARE
Benefits of a SMS are not fully recognised
- There is no understanding of the link between policy & procedures

The impact on safety of changes to units & equipment is recognised
- A checklist of key points exists to use to assess change
- There is much reliance on managers' experience

The benefits of safety auditing are being recognised
- Ad-hoc audits occur but there is no strategy underlying them

→

WHERE WE WANT TO BE
SMS is under development
- SMS exists but is not comprehensive
- Recognises the safety impact of organisational changes
- Few employees understand policy & local procedures link

Change management exists for units & equipment
- There is a recognised need to include people & organisational structure effects in change analysis
- Process and checklists exist for assessing changes

Ad-hoc safety audits are undertaken in response to problems

2. Integrated Teams

The element thought to be 2nd weakest in the EEC was *Integrated teams*, which is under the main third section of Communication. The standard deviation (+/-0.86) indicates that there was no variation between respondents.

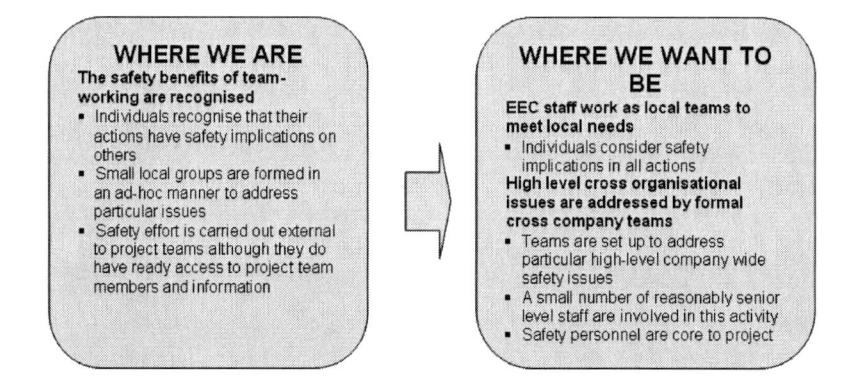

WHERE WE ARE
The safety benefits of team-working are recognised
- Individuals recognise that their actions have safety implications on others
- Small local groups are formed in an ad-hoc manner to address particular issues
- Safety effort is carried out external to project teams although they do have ready access to project team members and information

→

WHERE WE WANT TO BE
EEC staff work as local teams to meet local needs
- Individuals consider safety implications in all actions

High level cross organisational issues are addressed by formal cross company teams
- Teams are set up to address particular high-level company wide safety issues
- A small number of reasonably senior level staff are involved in this activity
- Safety personnel are core to project

3. Responsibility for safety

The element thought to be 3rd weakest in the EEC was *Responsibility for safety* – from the Communication section. The standard deviation (+/-0.83) indicates that there was only a little variation between respondents.

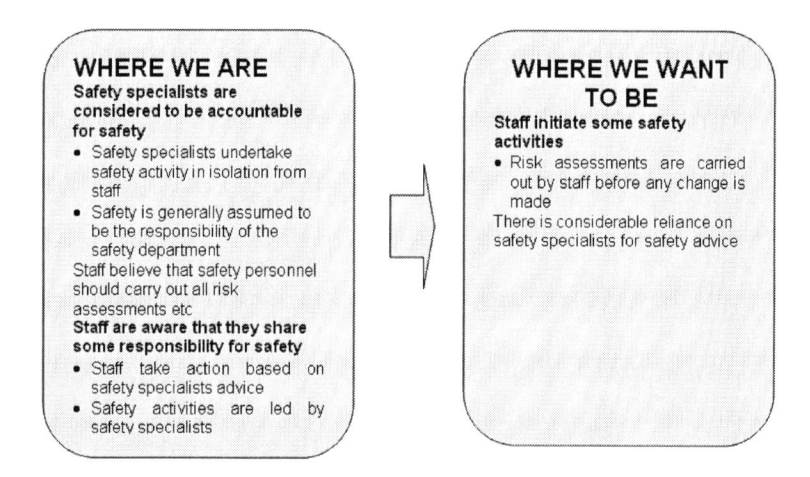

WHERE WE ARE
Safety specialists are considered to be accountable for safety
- Safety specialists undertake safety activity in isolation from staff
- Safety is generally assumed to be the responsibility of the safety department

Staff believe that safety personnel should carry out all risk assessments etc

Staff are aware that they share some responsibility for safety
- Staff take action based on safety specialists advice
- Safety activities are led by safety specialists

WHERE WE WANT TO BE
Staff initiate some safety activities
- Risk assessments are carried out by staff before any change is made

There is considerable reliance on safety specialists for safety advice

4. Risk assessment and management

The element thought to be 4th weakest in the EEC was *Risk assessment and risk management* – from the section 'Planning and organising for safety'. The standard deviation (+/-0.81) indicates that there was only a little variation between respondents.

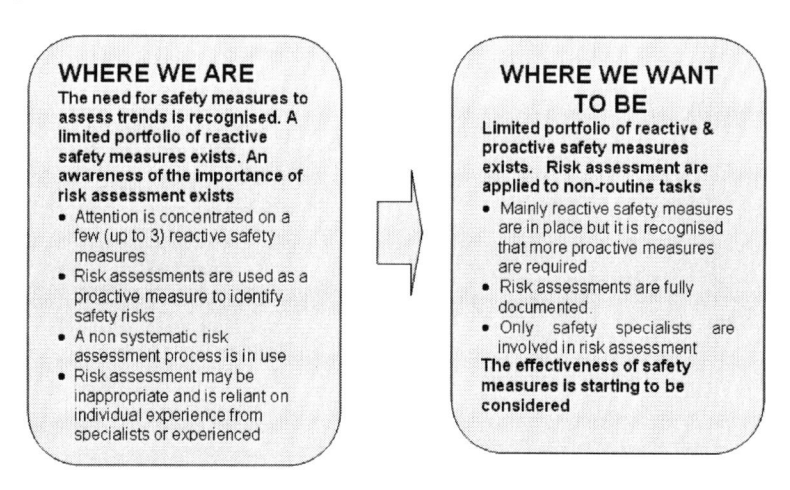

WHERE WE ARE
The need for safety measures to assess trends is recognised. A limited portfolio of reactive safety measures exists. An awareness of the importance of risk assessment exists
- Attention is concentrated on a few (up to 3) reactive safety measures
- Risk assessments are used as a proactive measure to identify safety risks
- A non systematic risk assessment process is in use
- Risk assessment may be inappropriate and is reliant on individual experience from specialists or experienced

WHERE WE WANT TO BE
Limited portfolio of reactive & proactive safety measures exists. Risk assessment are applied to non-routine tasks
- Mainly reactive safety measures are in place but it is recognised that more proactive measures are required
- Risk assessments are fully documented
- Only safety specialists are involved in risk assessment

The effectiveness of safety measures is starting to be considered

5. Training and Competence

The element thought to be 5th weakest in the EEC was *'Training and competence'* – from the section 'Planning and organising for safety'. The standard deviation (+/- 0.64) indicates that there was only a little variation between respondents.

WHERE WE ARE
A limited safety training programme exists
- Safety training is provided as needs arise on an ad-hoc basis on specific projects or activities.
- There is a reliance on transfer of skill / knowledge from one worker to a trainee.
- Advice tends to be based on past activities and experience
Staff may be assigned to work on safety activities based on their availability, rather than having training or relevant experience, though they will usually be supervised by someone who is qualified

WHERE WE WANT TO BE
An employee safety training programme exists
- Mainly focused on classroom training.
- Competence standards are not being developed.
- Training is often provided in response to problems.
- Front-line staff receive training as required.
- Staff assigned to carry out safety activities will have experience in areas related to safety

Conclusions

Implementation Plan and Accomplishments

The overall results from the first safety culture measurement of the EEC showed that the EEC has a reasonable degree of safety culture, but that there is room for improvement. The five main problem issues identified by the SCS were used to develop an implementation plan, which was adopted by "SAGE" (Safety Awareness Group in the EEC). The goal of SAGE was to (i) determine the priorities for improvement and (ii) develop a suitable action-plan based in part on the issues thought to require the most improvement. The implementation plan includes the following items and accomplishments so far are depicted in italics.

1. Safety Management System. A SMS needs to be developed more fully. The concept and benefits of a SMS needs to be fully recognised by EEC; a link between the SMS policy and the local tasks needs to be understood by employees. Initially, information (in the form of discussions/workshops) needs to be provided for project managers in order for them to become more familiar with what the SMS is and how this will impact their projects. It was decided both at a very high level (Eurocontrol Agency Board), and within the SAGE group, that a SMS for the EEC should be developed consistent with the SMS being developed by Eurocontrol Headquarters (Safety Management Unit - SMU). The main differences between the two SMS's will be policies for simulations; key risk areas and roles and responsibilities. The key part of the SMS issue was thought to be the link between the SMS policy and the local task needs to be understood by the staff and contractors. *This is currently being undertaken and is on schedule.*

2. Auditing Safety. A strategy for auditing safety within projects needs to be developed based on the structure of the SMS. The safety audit will be a proactive means to assess what safety issues have (and have not) been addressed in the project. Initially, safety audits could be used in response to safety problems, although the aim will be to carry them out on a more regular basis. *This has been delayed until 2006, as it was thought other issues needed to be dealt with beforehand.*

3. Cross-company Teams. Cross-company (cross-discipline) teams should be set up to address particular high-level company wide safety issues. A number of cross-discipline/company safety groups have recently been put in place in the EEC, internally they include SAGE; externally they include the Safety R&D review group; and Eurocontrol (HQ) Review. However, it was thought that further additional internal groups should be formed in order to spread safety awareness through the EEC. *This has been started initially with Safety Assessment User Group meetings.*

4. Risk Assessment & Management. More emphasis should be placed on the quality and objectiveness of risk assessments, as well as a more systematic risk assessment procedure. This is currently being undertaken in SAND (Safety Assessment for New Designs). *The SAND work has increased in size and scope, and more people are now being trained in safety assessment skills in the EEC.*

5. Training & Involvement in Safety Activities. Team members should be encouraged to use and act on advice given by the safety specialists. In addition, team members should be involved in safety activities (such as risk assessments) with the support of safety personnel and should be trained to carry out these safety activities. Some senior staff should be involved in the safety activities. The first part of this issue is being covered within the SAND project. Safety training is also being undertaken for more general safety assessment methods. Further safety awareness training sessions are planned and have been occurring for the general staff Weekly Information Corners in the form of video presentations on ATC incidents. *A training package for SAND has been produced and is currently being delivered internally.*

6. A further task to be undertaken to improve the EEC safety culture questionnaire, based on comments from respondents. The maturity of the EEC safety culture will be measured again April 2005, to see if any change has occurred. *This is currently being discussed, and further refinements to the tool have been initiated from comments by participants.*

Final Comments

The survey gains a 'grass roots' perspective on safety where the questionnaire method seems to be still the most appropriate way to gain an honest assessment of safety culture, although perhaps interviews might enhance the accuracy (by clarifying questions etc.) and reduce the variation of responses. The survey also takes the issues of safety culture 'out to the troops', involving them and not only the 'chain of command'. The SCS was undertaken as a questionnaire survey, because of the limited time to carry out interviews with 40 people, and because it was thought that if the participants were interviewed by a member of EEC staff, the survey results may be influenced. With some additional time spent on group discussions with SCS participants after the survey, it may be possible to examine the important issues in more detail and come up with more focused and relevant issues to the EEC.

Acknowledgements

The authors would like to express their sincere gratitude to all participants in the study.

Caveat

The opinions expressed in this paper are those of the authors only and do not necessarily reflect those of the EEC or parent or related organisations.

References

BSI (2000) ISO 9001 Quality Management Systems – Guidelines for Performance Improvement, London.

EUROCONTROL (2004) Review of Root Causes of Accidents due to Design. EEC Note 2004-14, Bretigny-sur-Orge, France: EUROCONTROL Experimental Centre (available at: http://www.eurocontrol.int/eec/public/standard_page/2004_note_14.html)

Fleming, M. (1999) Safety Culture Maturity Model. UK HSE Offshore Technology Report OTO 2000/049, Norwich: HSE Books.

Hudson, P. (2001) Aviation Safety Culture. Paper presented at Safeskies 2003 conference, Canberra, Australia.

National Air Traffic Services (2002) The Safety Culture Assessment Framework for UK National Air Traffic Services.

Nickelby (2002) Framework for assessing human factor capability. UK HSE Offshore Technology Report OTO 2002/016. Norwich: HSE Books

Reason, J. (1997). Managing the Risks of Organisational Accidents. Hants, England: Ashgate Publishing Ltd.

Sharp, J.V., Strutt, J.E., Busby, J., & Terry, E. (2002) Measurement of organisational maturity in designing safe offshore installations. Proceedings of 21st International Conference on Offshore Mechanics and Arctic Engineering, June 23-28, Oslo, Norway.

Vecchio-Sadus, A.M. & Griffiths, S. (2004) Marketing strategies for enhancing safety culture. *Safety Science, 42*, 601-619.

Moving closer to Human Factors integration in the design of rail systems: a UK regulatory perspective

Deborah Lucas
HM Railway Inspectorate, HSE
Manchester, UK

Abstract

Major rail accidents in the UK, including the Southall and Ladbroke Grove train crashes, have shown the key role of human factors in the prevention and mitigation of catastrophic risk on the railways. The human factors community appreciates how user centred design approaches can eliminate opportunities for human error and improve the mitigation of potential human failures. Whilst the concept of Human Factors Integration (HFI) is well established in the aviation sector, it has only recently been introduced in the UK rail sector and is still gaining acceptance amongst engineers. This paper outlined how regulatory intervention on a major rail construction project in the late 1990s was a driving force for new human factors standards on HFI. Case studies from the rail sector showing both initial successes and perceived barriers to the use of such approaches are given.

Introduction

Her Majesty's Railway Inspectorate (HMRI) is the health and safety regulator for UK railways. It currently employs over 200 staff predominantly operational and specialist inspectors at offices throughout the UK. HMRI inspectors investigate incidents and complaints on the railways. As the rail safety regulator HMRI inspectors enforce relevant UK health and safety legislation. This includes the powers to inspect and monitor safety compliance of members of the rail industry, to issue improvement and prohibition notices and to initiate prosecutions for breaches of health and safety legislation. The rail safety regulator is also responsible for the approval of all rail safety cases. Under existing legislation HMRI also grants approval for new works, plant and equipment. This latter function is changing with alterations to European and UK legislation.

During 2003 HMRI recruited two human factors rail specialists. This reflected an increased awareness of the need to promote better management of human factors issues in the rail industry as well as the recognition of the need for more accessible specialist advice within HMRI. The need for improved management of human factors issues in the rail industry was clear from a number of serious rail accidents and their investigations. For two of the more serious train crashes in the UK in recent years the public inquiries took evidence from psychologists as to the possible causes of 'signals passed at danger' (SPADs). (This is where a train driver goes through a signal set to red or stop i.e. 'danger'.)

Human Factors in UK rail public inquiries

The inquiry into the train crash at Southall on 19th September 1997 where 7 people were killed and 150 injured considered possible human factors causes that might have explained why the passenger train collided with a freight train. The evidence given to Professor John Uff's inquiry (Uff, 2000) included information on causes of inattention, visual search patterns, reliance on safety devices, and the impact of a second person in the cab. The inquiry's recommendations included one for reliable research into human behaviour studies relating to (train) driver performance.

There was a further public inquiry by Lord Cullen (Cullen, 2001) into a train crash at Ladbroke Grove on October 5th 1999, where two passenger trains collided head on resulting in the death of 31 people and 244 injured. This inquiry took evidence not only on the training of drivers, the responses of signallers, and safety culture issues but it also addressed design issues affecting human performance. In particular, evidence was given on the difficulties that train drivers had in sighting signals on this part of the network. Signal SN109, one of a number of signals on an overhead gantry, had been passed at red on eight occasions since August 1993. Lord Cullen stressed the need for the rail industry to *'redouble its efforts to provide a system of direct management and training that is secure against ordinary human error whilst endeavouring to reduce the incidence of such human error to an absolute minimum'* (Cullen, 2001, page 168).

A third public inquiry into train protection and warning systems and the prevention of SPADs (Uff & Cullen, 2001) was unusual in that it considered broader questions of safety on the railways rather than a specific train accident. The inquiry requested the preparation of a document on 'human factors principles' by all those who had presented expert psychological evidence to the Southall and Ladbroke Grove inquiries. The joint report on human factors is published in annex 7 of the inquiry report in full. The joint report lists a set of consensus principles that are relevant to reducing the number of signals passed at danger. The principles include matters relating to: design issues, signal sighting, warning devices, training and route knowledge, alertness and fatigue, developing human factors capability and awareness, incident investigation, confidential reporting, and driver management. Two of the consensus principles that relate to incorporating human factors in design are as follows:

'Human factors aspects of train driving cover not only the characteristics of the driver (e.g. route knowledge, alertness, etc.) and the equipment (e.g. signalling, train controls and instruments, etc.), but also the interface between the driver and the equipment. Full evaluation of this human-machine combination is vital for the reduction of opportunities for human failures. Human errors do not cause problems unless the characteristics of the equipment allow them to. This analysis should be done by incorporating knowledge about human information processing, human reliability, and good ergonomic principles into the design and evaluation process. It

is a serious weakness if well established principles of ergonomics are ignored.' (Uff & Cullen, 2001, annex 07, page 2)

'The design process for new equipment e.g. cabs, interfaces of equipment in control rooms, should consider human factors issues explicitly. Building the needs of the users of systems into design prevents human errors from arising. This requires early incorporation of human factors thinking and involving of future users of the equipment in the design process.' (Uff & Cullen, 2001, annex 07, page 2)

Human Factors in design

Why was there a need to state explicitly what engineers and human factors specialists in some other industrial sectors were already doing i.e. incorporating human factors in design? There are at least three reasons.

Firstly, the rail industry had not looked to human factors good practice in sectors such as defence and aviation. There was no equivalent of the US defence industry's 'Manprint' initiative or the UK's military 'Human Factors Integration (HFI) Standard' (Defence Standard 00-25). There were isolated historical instances of incorporating human factors into the design of some driver cabs or signalling systems. However, typically, relevant ergonomic expertise was sought late in the process as a result of the initiative of an innovative engineer (often when human-machine interface issues had already emerged) rather than being formally mandated by a process or standard.

Secondly, human factors were predominantly seen as a research rather than an applications area for the UK rail sector. Research is clearly needed but then it should lead to the application of good practice for human factors in the design of systems, in risk assessment, in incident investigation, in health and safety management systems and in the management of change.

A third reason was that there was no consistent request from the safety regulator for evidence of human factors integration when approving new equipment or systems. High level principles advising of the need to consider design issues in relation to human failures existed (e.g. HSE, 1999; Noyes, 2001) but without industry standards or regulatory intervention these principles were not widely used in the UK rail sector. Textbooks existed providing models for HFI (e.g. Redmill & Rajan, 1997) but without experienced HF rail professionals such models were not regularly applied.

Regulatory intervention on human factors in design

HMRI made a significant intervention on human factors in design in the case of the extension to the Jubilee Line underground service during 1999. There were inspections during the building and installation of station operations rooms and equipment, together with identification of emerging problems at a temporary signalling centre. This highlighted problems with how human factors principles and good practice had been incorporated into the system design (Timothy & Lucas,

2001). The company concerned (London Underground Limited) responded positively by putting into place a team of human factors specialists to collate, remedy and close out issues. The issues were resolved but at a cost. To avoid future problems the company mandated a company standard on the integration of human factors into systems development with the business objective of ensuring that systems are developed which will deliver required operator health, safety and performance characteristics. This standard (LUL, 2000) is based on the UK defence standard on HFI and is supported by a manual of good practice (LUL, 2002).

The military standard includes the need for a HFI Plan, the incorporation of human factors needs into systems requirements, building human factors into the physical design of a system, and considering the issue in system assessment and evaluation. Six domains are covered: manpower (number of staff required for the system), personnel (aptitudes, experience and other human characteristics), training, human factors engineering, system safety, and health hazards.

Whilst there is no overall rail industry standard on HFI this LUL standard represents current good practice and has been welcomed by HMRI. The signs so far are that the standard is leading to an increased awareness amongst suppliers and procurers in the rail sector of the need to address human factors issues in the design process. A recent initiative from Network Rail has been for them to prepare their own company standard on ergonomics in design. The principles of identifying human factors issues during design and installation and having a system for closing out such issues before commissioning have been evident in recent signalling centre replacement work on the West Coast route. Generally HMRI human factors specialists have seen much more evidence of the planning and resourcing for HFI in projects with approaches for information and discussion being made to us earlier in the design process. This is a very positive development. Railway Safety has recently published an application note on human error within its on-line Engineering Safety Management series (Railway Safety, 2003).

Requests from HMRI for evidence of the consideration of human factors in the design of new driving cabs also appears to be giving rise to an increased awareness of the need to provide such information. In particular, the use of training simulators to run assurance tests for the use of new equipment in cabs has been a new innovation. Whilst awareness is increasing there is still the need for industry to adopt a standard that would mandate including human factors in design of new equipment. The human factors community also needs to do more to raise the level of knowledge amongst system suppliers of the problems that design can cause and how to address these. Ergonomists and psychologists need to work within design teams to ensure that they propose specific solutions to identified problems rather than giving generic advice which is often hard for a designer to use. This means that the human factors design specialists will need a good knowledge of the rail domain and will be able to apply methods and standards for HFI. Knowledge of risk assessment techniques and the use of human reliability assessment methods are also desirable if the human factors specialist is to be integrated into the design or safety case team. Whilst the number of applied psychologists and ergonomists with these skills is growing there is still a shortage of suitably experienced human factors professionals with safety critical system design and assessment experience in the UK rail sector.

Indicators of successful Human Factors Integration

HMRI has observed a number of projects with poor HFI and now some with more positive treatment of human factors. Some key elements have emerged and these are used as informal indicators by HMRI to establish if human factors has been adequately considered within the project. The elements are in three groups: resources and project commitment, the design process, and assurance and testing. Each of these is outlined below.

1. *Resources for human factors integration and project commitment for HFI*
   HMRI looks for evidence of high-level project commitment to HFI, together with a planned process of considering human factors issues throughout the entire design process. It is positive if there are some relevant performance measures given in the specification. The effort given to human factors should be proportionate to the size of the project, the novelty of any new functions, and the consequences of human failures. We look at whether competent ergonomists, human factors professionals or HCI specialists are used and what their role and relationship is to the project team. A key critical success indicator is whether there is suitable project authority for human factors specialists.
2. *Design process including user involvement.*
   HMRI would expect to see a description of the users, their main tasks and goals together with an appreciation of the context of use including relevant normal, degraded and emergency scenarios. For systems with new functions a realistic operational concept would explicitly state any new user goals and tasks. If similar systems already exist then capturing the lessons learnt is important. HMRI look for the participation of users in design. We would expect to see a list of relevant ergonomic standards used in the design including standards such as ISO 11064 'Ergonomic design of control centres', BS EN ISO 13407 'Human-centred design processes for interactive systems' and ISO 9241 'Ergonomic requirements for office work with display terminals (VDTs). A key area is alarm presentation and handling (see EEMUA, 1999). Serious issues with alarm handling are often indicative of wider human factors integration problems within the project. Another critical success indicator is the process for managing project human factors issues including the arrangements for logging, prioritising, and closing these out in a timely manner. Closing out human factors issues often involves a discussion and agreement between the designers or suppliers and the future operators of the system to ensure appropriate risk controls are in place. These risk controls would include operational arrangements for training, staff numbers, supervision, and changes to documentation. The human factors specialists within the project often play a key role in facilitating such discussions
3. *Assurance and testing.*
   HMRI look for a statement of the risk controls that are necessary including personnel characteristics, staffing numbers, required training and procedural controls. A positive element of some new systems has been evidence of

assurance testing where a range of scenarios have been played out to identify whether the system is robust from a human reliability perspective. Ideally the scenarios include high levels of workload, combinations of operational problems, and equipment or system failure. For some safety systems independent assessment of the human factors design aspects are presented to the regulator. This may include formal human reliability assessment to demonstrate that controls have reduced the risk to as low as reasonably practicable (ALARP).

Conclusion

Over the last 4 years the UK rail sector has moved a long way towards effective consideration of human factors in design. The emergence of formal company standards of HFI is a key driver for this move and the emphasis on human factors by the safety regulator is helping to push this forward. Whilst some barriers to such integration still exist there is a general increase in awareness amongst many rail companies of the need to consider ergonomic and psychological issues in design. It is important that the current momentum continues to be encouraged to reduce the potential occurrence of human failures with serious consequences in the rail sector.

References

British Standards Institute (1999). *Human centred design process for interactive systems*, BSI EN ISO 13407. London: British Standards Institute.
Cullen, D. (2001). The Ladbroke Grove Rail Inquiry Part 1 Report. Norwich, UK: HSE Books.
Defence Standard 00-25 (1988) *Human factors for designers of equipment* (issue 2, amendt 1). Glasgow, UK: MoD Defence Standards.
EEMUA (1999) *Alarm systems, a guide to design, management and procurement.* Engineering Equipment and Materials Users Association Publication No 191.
HSE (1999) *Reducing error and influencing behaviour.* (HS(G)48) Norwich, UK:HSE Books.
International Organisation for Standards (1999) *Ergonomic Design of Control Centres.* ISO 11064.
International Organisation for Standards (dates depend on part) *Ergonomic requirements for office work with display terminals (VDTs)* ISO 9241.
London Underground Ltd, Chief Engineer's Directorate, (2000). LUL Standard E1035 A1 Human Factors Integration in System Development.
London Underground Ltd, Chief Engineer's Directorate, (2002) LUL Manual of Good Practice M1035 A1 Good Practice in Human Factors Integration.
Noyes, J. (2001) *Designing for Humans*. Hove, UK :Psychology Press.
Railway Safety (2003) Application Note 3 Human Error: Causes, Consequences and Mitigations, Issue 1.0 Engineering Safety Management Yellow Book 3. www.yellow-book-rail.org.uk Accessed 31 March 2005.
Redmill, F., & Rajan, J. (1997) *Human factors in safety-critical systems*. Oxford: Butterworth-Heinemann.

Timothy, D. & Lucas, D. (2001). Procuring new railway control systems: A regulatory perspective. *People in Control conference* (PIC2001). Manchester, UK June 2001.

Uff, J. (2000). *The Southall Rail Accident Inquiry Report*. Norwich, UK: HSE Books.

Uff, J. & Cullen, D.(2001). *The Joint Inquiry into Train Protection Systems*. Norwich, UK: HSE Books.

The views expressed are those of the author.

Problems of limited context in redesign of complex situations in infrastructures

Ellen Jagtman & Erik Wiersma
Safety Science Group
Delft University of Technology
Delft, The Netherlands

Abstract

Society changes continuously which puts pressure on transport infrastructure to accommodate the consequences of these changes. Design of infrastructure does not always consider risks that are a result of the changed situation. Three cases are discussed that are caused by a too narrow focus of designers. The first case shows how the focus on the initial problem in seeking an alternative layout, without paying due attention to the context of the transport system, contributed to an accident. The second case discusses the temporary layouts users met during reconstruction works. The temporary nature caused designers to accept sub-optimal solutions. The difficulties for users in understanding constantly changing circumstances were also not taken into account. The final case discusses how the implementations of new technologies might influence redesign of the physical infrastructure. Although methods exist that cover these problems implicitly, there are no simple solutions to tackle them. Risks included in the design in these cases had to be dealt with by the road users. Awareness of these issues is the most important factor to ensure that they are addressed sufficiently for new and temporary situations. The traffic HAZOP introduced in this paper makes such discussions explicit.

Introduction

Redesign of road infrastructure is a common phenomenon. Redesign may have various underlying reasons, such as increase in use, required maintenance or new conditions for certain traffic situations. Demands for transport are still growing which may cause bottlenecks in transport networks. Moreover, demands may change as a result of new development areas in the neighbourhood of current infrastructure. Such areas could involve large living areas[*], which cause mainly additional car traffic and public transport. Other developments are business or science parks, which not only attract additional personal transport but also change the demand for goods transportation. Apart from changes in transport demands, changing transport policies

---

[*] In the Netherlands so-called VINEX locations are developed in which within a period of 5-10 years hundreds of new houses and apartments are built

In D. de Waard, K.A. Brookhuis, R. van Egmond, and Th. Boersema (Eds.) (2005), *Human Factors in Design, Safety, and Management* (pp. 513 - 526). Maastricht, the Netherlands: Shaker Publishing.

and related policy goals may require redesign of traffic systems. One of the first of such changes resulted in the introduction of motorway environments at which only motorised traffic is allowed. This however has been the result of coping with increased motorisation rather than safety (Koornstra et al., 1992). Other examples of changes to the infrastructure which relate to safety goals are the introduction of roundabouts, 30 km/h areas and 'woonerfs'. Since the 1990 northern European countries have introduced more integrated policy plans in which the traffic system including infrastructure, vehicles and regulation have to be tuned to the users (DETR, 2000; Koornstra et al., 1992; Ministerie van Verkeer en Waterstaat, 2000; SNRA, undated). Apart from changing demands and policies, the necessity for road maintenance can initiate redesign of the local situation. Maintenance has a recursion character in the life cycle of infrastructures. The necessity to pave or asphalt a road again creates an opportunity to consider changes of the current system and to give priority to parts of the traffic system that require regular maintenance. As various reasons may initiate redesign of the infrastructure, a variety of problems may come along. In this paper the authors distinguish redesign problems caused by the new layout after redesign, problems of temporary layouts during construction works and problems of changes of use of existing infrastructure with new technologies. Each category will be briefly explained.

Redesign of infrastructure often focuses on the initial reason for considering changes at a specific location, without paying enough attention to human factors. Redesign is not so much risk and problem driven, but instead is mainly solution driven. For instance, if the redesign is meant to extend the capacity, the alternatives from which designers select their solution are driven by optimising traffic flow. Often these alternatives do not solve other relevant problems present in the old situation, and may introduce additional problems in the renewed situation. Examples of such problems are: difficulties to gain a clear view; complexity caused by conflicting manoeuvres; mixing of different vulnerable and motorised road users. Designers should take the context of both the old and the alternative situations into account to prevent that the new situation is a sub-optimal solution.

The reconstruction phase itself introduces additional (temporary) problems, and therefore requires special attention. A characteristic feature of the reconstruction phase is the use of various temporary layouts. For instance, reconstruction in order to extend a motorway from two to three lanes requires that traffic during the road construction activities is temporarily shifted to the lanes in the opposite direction. As a result traffic from both directions has to share one carriageway for a period of time. After one lane is finished traffic is shifted to the other carriageway. During reconstruction different layouts succeed one another. Because these layouts are only used temporarily, the measures used are inferior to the measures applied in the final situation. This increases risk levels of such temporarily used layouts. Control of these increased risk levels is often put onto the individual road users, for instance by imposing reduced speed limits and constraints to overtaking. The temporary nature of construction in large-scale projects sometimes becomes semi-permanent; situations may last for months, or even a number of years. This undermines the acceptability of these sub-optimal layouts.

redesign of complex situations in infrastructures 515

A third phenomenon in redesign relates to innovation, which provides new opportunities for design alternatives. New technologies can be introduced in different elements of the traffic system. It can be implemented into the infrastructure, the vehicles or both. Technological opportunities increase rapidly. As a result improvements of, adjustments to and expectations about new developments often succeed each other quite fast, certainly faster than changes are made to the traditional infrastructure. The new technologies used in today's traffic may however require considerations of redesign of the traffic system including its infrastructure. The changes made to the suspension in truck cabins result, for example, in truck drivers getting fewer (or hardly any) stimuli when driving too fast through a curve. Either adjustments to the infrastructure or additional technologies (e.g., stability programs, ESP) can assist drivers in overcoming this problem. The limited context of in this case the new technology again provides a sub-optimal solution. If technology requires redesign of the infrastructure the difference in life cycle of the traditional infrastructure and the new components in the traffic system should be kept in mind.

This paper illustrates the problems of redesign described above and discusses a possible approach to better cope with these problems. Three cases will be discussed to elaborate on the redesign problems addressed in this introduction. After discussion the cases, some inadequacies of current practise in redesign of traffic systems are addressed and an approach to try to overcome these is discussed.

Redesign issues: three cases of too narrow focus in solution seeking

In the introduction three categories of redesign problems have been addressed. This section provides examples to illustrate each of these. The first case describes the successive redesign choices made that contributed to a train-bus collision. The second case discusses long term construction works and a number of the sub-optimal solutions implemented during reconstruction works. The final case discusses the implications of using adaptive cruise control in daily traffic.

Redesign as contributing factor to train-bus collision

On October 25th 1995 an express commuter train heading for Chicago collided with a school bus at the village of Fox River Grove, Illinois. The collision occurred while the school bus was waiting on Algonquin Road to enter US route 14 (see Figure 1). The collision caused seven fatalities amongst the bus passengers and 24 serious to minor injured passengers. None of the train passengers were injured. The accident was analysed by the NTSB (National Transport Safety Board, 1996).

The NTSB report addresses a number of changes to the local traffic situation both to physically layout and to the control system of the junction. These adjustments to the traffic situation had influence on the possibility of this accident to occur. The first important modification made to this situation involved redesign of the US route 14 in 1989. The US route 14 was widened and traffic lights were installed to control road traffic on the intersection of Algonquin Road and the US route 14. As a result of the reconstruction the space between the railway tracks and parallel road (US 14) had been halved. Large vehicles such as trucks and busses no longer fitted between the stop line for the traffic lights and the railway tracks (see 'separation area' in Figure 1). Detection of the traffic control system near the road junction however demanded vehicles being first in line to drive up to the stop line across the tracks. In 1994 the control system of the lights were slightly adjusted. Later in 1995 the train warning system was changed. Instead of 30 seconds prior to arriving at the railway crossing the control system alarmed 25 seconds before arriving. This modification had not been passed on to the controllers of the traffic lights controlling the road junction. The traffic control system was therefore kept the same. The changes caused the bus on the day of the accident to have too little time to leave the tracks before the train arrived.

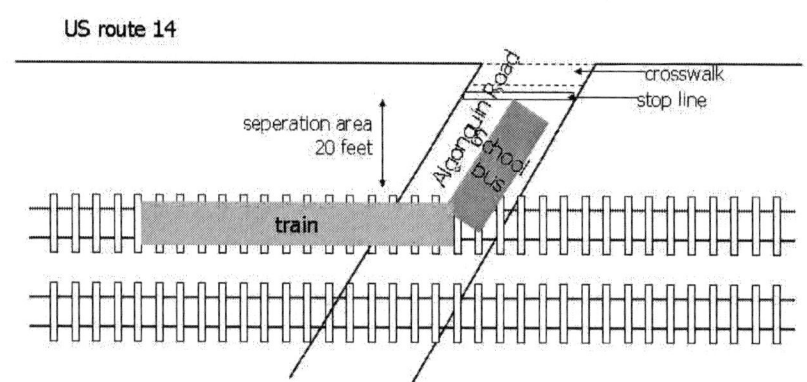

Figure 1. Situation diagram of train – school bus collision

Figure 2. Modifications after accident showing additional warnings for Algonquin Road (left) and warning signs for pedestrians (right).
Figures/pictures available in colour at http://extras.hfes-europe.org

After the accident some modifications were made to the railway crossing and the traffic light control system (see Figure 2). The location of the traffic lights for the crossing of the two roads was kept similar, namely the vehicles have to cross the railway junction to activate the detection system. In order to prevent vehicles waiting in line on the railway tracks, additional traffic warning signs were placed. This included signs to increase awareness of specific risks in this situation and signs to inform on a penalty for not paying attention to the rules. For the pedestrian crossing parallel to the US route 14 the green-phase in case a train is approaching was shortened. This is noticed to pedestrians with a sign as well (see Figure 2). The latter modification should give vehicles waiting for a green light on Algonquin Road more time to clear the railway tracks in case a train is approaching. The modifications made after the accident however still allow traffic situations similar to the one that caused the accident in October 1995. If a truck or a bus is the first vehicle approaching the traffic lights, the vehicle has to drive up to the stop line to activate the detection system for the traffic lights, even though these vehicles will remain partly on the railway track.

Problems during temporary situations

Redesign has caused temporary sub-optimal situations that often confuse road users. A junction between a road and a railway in the area close to where the authors work was explored to discuss the problems during reconstruction works. The situation is located nearby The Hague, The Netherlands, where a motorway and railway tracks are situated parallel (east-west). At the location drivers leaving the motorway in southern direction have to cross the railway after which a junction with traffic lights is located (see Figure 3). This location shows similarities with the Fox River case. Traffic lights at this location have been installed to control traffic from a new side road, which is used by construction vehicles for houses that are built in the neighbourhood. Reconstruction has started in 2002, and has to result in a situation in which the road-railway crossing will be replaced by a crossover around 2006. Apart from installing a crossover, the railway is to be extended with two additional tracks and the exit and access road to the motorway will be moved about 300 meters.

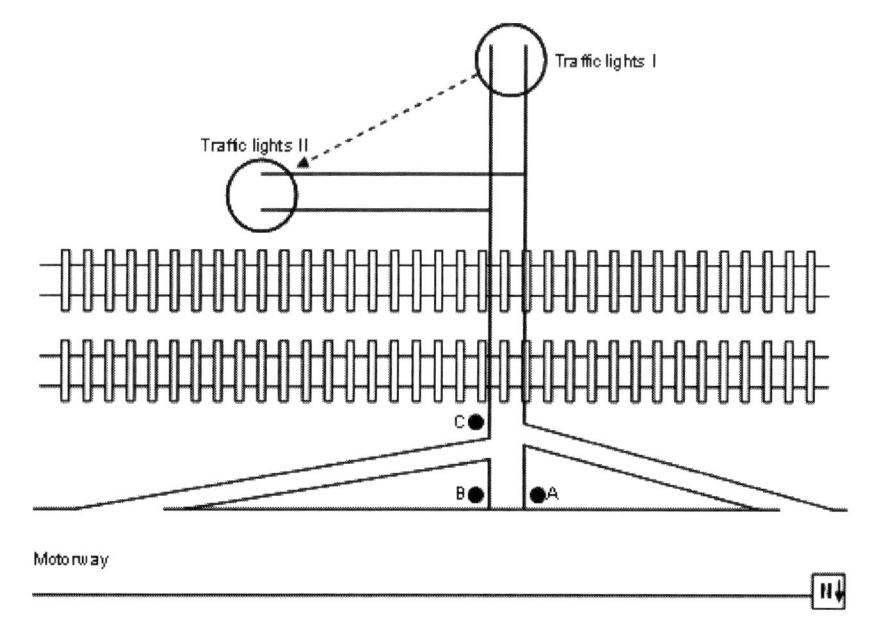

Figure 3: Layout(s) of road-railway crossing during reconstruction

In the period between 2002 and 2006 several changes have been made to the local situation in order to be able to create the final situation. During the reconstruction works, road users have been confronted with at least four different temporary layouts. In three different stages the railway crossing has been in use during the road works, namely: at the start of reconstruction (November 2002), after shifting of the road and the traffic lights (February 2003) and after modification of the railway crossing (April 2004) (see Figure 4). During the fourth stage in December 2004 the railway level crossing has been closed for all road traffic.

Figure 4. Temporary situations (a) November 2002; (b) February 2003, (c) April 2004. Photos available in colour at http://extras.hfes-europe.org

Picture a shows the situations after the start of the reconstruction in 2002. This picture was taken from the camera position A marked in Figure 3. The traffic lights were located at location I (see Figure 3). At this moment attention was especially paid to the extension of the railway. Meanwhile demand for traffic on this road was growing caused by houses being built close to the railway crossing during this phase. As a result the railway crossing was often blocked by traffic waiting for the traffic lights to turn green. Different from the situation in Fox River there was no communication between the temporary installed traffic lights installation and the railway gates.

A specific problem at the location was the communication and decision making between different stakeholders. This resulted in poor tuning of each other's redesign intentions. Road users at this location were confronted with various changing situations. Shortly after the start of construction works at the railway crossing, the local road authorities decided that the main road had to be closed for all motorised traffic. Therefore a new road was opened parallel to the railway (see traffic lights II in Figure 3). As a result of the modifications drivers were confronted with a sharp turn to left immediately after crossing the railway junction (see picture b in Figure 4). This picture was taken from camera position B marked in Figure 3. Again often vehicles were standing still op the railway tracks. However leaving the tracks when a train was approaching became even harder. Secondly, because of the sharp turn, a lot of drivers did not pay any attention to cyclists who used the same road at this location.

The third situation was a result of negotiations. At the start of the project in 2002 it had been planned that first the railway track was to be closed, than the crossover for vehicle would be realised and finally a tunnel would be constructed for cyclists. This choice implied that cyclists for about a year would not be able to cross the railway at this location and have to make a detour of about a kilometre. During construction this decision was recalled in which the problems with the sharp bend were also taken into account. This resulted in the third situation (picture c in Figure 4). This picture was taken from same camera position as picture a, position A (see Figure 3). The road was temporary relocated in order to make space for construction of the tunnel for cyclists. From this point on vehicles passed the railway tracks diagonally and as a result the curve is less sharp. The problem of blocking the track however remained.

During the reconstruction works twice an accident resulting in a collision between a bus and a train occurred. In both cases a device in the bus switched into a mode for letting passengers on and off the bus. In one of the cases a prior collision with an element (used to mark the temporary route) possibly caused this system to switch on (Raad voor de Tranportveiligheid, 2004). Apart from these two known accidents often temporary unclear situation have been recorded. Figure 5 shows some pictures of an obstruction of one driving direction near the crossing to do night time works. The first and second picture was taken from camera position C (see Figure 3) with the back towards the railway crossing. The third overview picture was taken again from camera position A. The sign in the first picture (see Figure 5) clearly stated that drivers had to enter the motorway to the right. However, the sign a bit further along the road, closer to the construction (Figure 5 second picture) had an unclear arrow to the left. Following this arrow would either mean entering the exit of the motorway or following the road that was blocked. The result is seen in the left picture in which some users chose the motorway, but others made a u-turn in front of the construction site or tried to use the blocked road (as the driver marked in Figure 5 who tried to use the cycle path). This temporary solution showed users can be confused resulting in a lot of unwanted traffic situations.

In winter 2004/5 the fourth stage was realised. At that moment the railway crossing was closed and the tunnel for cyclists was opened (although cyclists have to walk down a stairway with their bicycle). From this moment on vehicles had to make a detour. The crossover will not be ready before approximately the beginning of 2006. After various temporary situations in which risks near the railway tracks increased, motorised traffic in this fourth stage have to drive through the residential area to reach another access to the motorway. The fact that the railway closed before the crossover is ready has nothing to do with physical limitations but was caused by the different stakeholder targets. The railway organisation closed the railway crossing to meet their targets in removing crossing on lines that are used intensively. On the other hand, the redesign of the road traffic system was prioritised for a year after the railway organisation intended to close the junction. As a result users are forced to find an alternative.

Figure 6. Unclear signs near construction site including vehicles 'not obeying' rules (21 October 2004). Photos available in colour at http://extras.hfes-europe.org

Problems with use of new technologies on existing infrastructure

The traffic system does not only involve fixed physical infrastructure. Changes to vehicles using the infrastructure may require modifications of the physical infrastructure. More and more information and communication technologies (ICT) are designed to be implemented into our traffic system. Such systems may install

ICT in the infrastructure (e.g., dynamic information panels), in vehicles (monitoring functions – fuel, tires; assisting drivers – distance, speed or lane keeping support) or combined in infrastructure and vehicles (informing or assisting drivers).

In the early 1990s, research programs in Europe and the US directed towards development of automated highway systems with components in both the infrastructure and in vehicles (Vahidi & Eskandarian, 2003). Due to financial and other practical implications focus in the short term shifted from the automated highway concepts to in-vehicle supporting systems. However, for innovations in the long term the concepts including intelligence in the physical infrastructure are still under development. The implications of the use of new technologies have been tested in various experiments. The results of experiments with similar longitudinal assistance technologies seem to contradict to one another (e.g., Hoetink, 2003; Jagtman et al., 2001; Vahidi & Eskandarian, 2003). Although the benefits and shortcomings are not well understood yet, many of these functions have already become available to the market (Vahidi & Eskandarian, 2003). In particular a distance keeping device known as Adaptive Cruise Control (ACC) has been introduced first to top models and has already become available on middle class vehicles. ACC combines conventional cruise control with a collision avoidance system. During the development of this system, such as is customary with all new devices, at a certain point during design, a market introduction for the product was considered. Desirability of market introduction was not only determined based on the safety of the product, but also involved other criteria (Marsh, 2003). As a result the system was considered well enough developed for market introduction by manufactures, even though there were often still imperfections in the system. At the same time, manufactures were working on redesign of the first generation ACC devices to enlarge the functionalities of the devices (e.g., Venhovens et al., 2000).

The ACC system in some perspectives might be considered to be not matured and therefore puts demands on the users. Drivers have to cope with many inadequacies of the first generation of the ACC device. These problems can be categorised in three areas:

- known but accepted shortcomings such as not detecting dirty vehicles in front and the ACC vehicle speeding in curve since ACC no longer detected the vehicle in front;
- experiences that differed from pre-introduction studies, such as a tendency to follow the vehicle in front instead of overtaking it (the latter was found in several driver simulator studies);
- unknown and unexpected behaviour of the system and the driver. A driver for example experienced the ACC vehicle to suddenly accelerate while a vehicle with a higher speed merged in the gap between the ACC vehicle and the vehicle it was following.

In Jagtman and Wiersma (2003) these three areas are illustrated in more detail. A test ride in an ACC vehicle in normal traffic for only a few hours showed that still many safety issues are still unsolved. ACC provides the user with an additional supervisor task while driving. The driver is made responsible to observe if the device can be safely used under the traffic circumstances in which he or she is travelling. Problems such as the first area, known but accepted shortcomings, are the result of a narrow design focus. The first generation ACC systems are all designed for straight clear roads to be used in low dense traffic conditions, which often are quite unrealistic traffic conditions. On the other hand the driver can switch the system on anywhere under any condition if the vehicle is moving with a speed of above approximately 40 km/h. The operability range of the device gives the user the opportunity to switch it on in conditions that are not covered within the design scope at all. Consequently, this technology can be operated for instance on curved roads located in urban areas. The introduction of in-vehicle technology with such imperfections has to be controlled by other system components. Hence, the design of the physical infrastructure can be used to influence the driver in either using or not using support systems. During the test rides for example the ACC was switched on driving at 50 km/h in an urban area where the road was both wide and paved with asphalt. The device was not switched on where the road had narrowed parts, humps or was paved using bricks.

Although system designers are improving the first generation ACC systems, the current systems will stay available, since the life cycle of a vehicle is likely to be larger than of the new devices. These innovations require consideration of alternative designs of the physical infrastructure. The infrastructure that is redesigned or renewed today will remain for probably at least a decade. In the mean time certainly other new more or less matured new technologies will be introduced into the market. Since modifications to the physical infrastructure are not made on yearly basis and are very expensive, additional requirements of new technologies have to be considered.

How to deal with these problems?

There are no straightforward, easy to use methods available to completely solve the problems described in this paper. Probably the most important aspect in finding solutions to the problems described here lies in involving different actors in the design process. Different actors have a different perspective on the problem and incorporating these different perspectives may help identifying a wider range of problems that need to be addressed in (re)design. Both designers of new situations and policymakers play a role in this process, but often work at different fields, without communicating about particular situations. Bringing together these two fields can already be an important step in bringing a multi-actor perspective into design.

Jagtman (2004) has developed a method that brings together designers and other stakeholders, such as policymakers to systematically discuss designs. The method that is based on the HAZOP methodology uses a structured brainstorm for identifying all sorts of safety problems that may occur in traffic systems. The method is used for identifying hazards and operability problems that prevent efficient operation. It is based upon the assumption that most problems (deviations) are missed because the system is complex rather than because of lack of knowledge on the part of the design team (Lawley, 1974). Risk and safety related aspects are special phenomena, since these, different from for example optimising capacity, often deals with effects resulting from processes that differ from the intended traffic processes. They current approaches have a lack of attention to the issues relating to unintended effects caused by designs choices or use. Identification of all kinds of unintended effects in the current approaches is either fixed by the use of checklists or only is only done implicitly. The HAZOP has been introduced to make safety problems that may occur explicit. The HAZOP uses combinations of parameters and guidewords to initiate the discussion on a potential deviation. Each of the combinations is discussed during the brainstorm focussing on whether the deviation can occur in the traffic system and what may cause the deviations. The brainstorm results in a large variety of potential safety problems. The final step in the traffic HAZOP involves discussion on the expectations of road users based on the deviations that have been discussion during the brainstorm. The discussion is focused on expectations rather than 'conventional' parameters such as speed, headway or conflicts, since expectations play an important role in decisions of road users while participating in traffic.

The use of HAZOP has opportunities for each of the three cases. In this paragraph a number of examples are addressed which result from student assignments using the HAZOP approach and from the case studies performed for the development of the method (see Jagtman, 2004). The layouts choices made in the Fox River Grove case have been driven by the possibilities first to extend the road capacity and secondly to decrease the time the railway gates are closed. A HAZOP study for extending the US route 14 will focus on the implications of the new layout. Various combinations of parameters (location, distance, attention and violation) and guide words (wrong, different, no) will identify deviations that describe problems of trucks or busses. For example: "if a bus is the first in line, the bus is blocking part of the railway tracks when waiting for a green traffic light signal" and "trucks have to cross the railway tracks to activate the detection systems of the traffic lights, in order not to block the tracks the truck has pass the stop line and wait partly on the pedestrian crossing". The modifications made to train warning system could be discussed using parameters as attention, (travel) time, flow rate. Examples of discussions in a HAZOP are: "driver of vehicles that passed the railway tracks will pay attention to the traffic light instead of possible closing of the railway gates", "inappropriate time available for traffic on Algonquin Road to leave the railway tracks". In case of the second case, the temporary locations, the HAZOP study can focus on various temporary layouts. Before construction work starts a number of these temporary layouts can be discussed. The HAZOP for example can focus on the timing of for example building of the cyclist tunnel, the use of a curve directly after a railway crossing and entirely closing of the railway junction. Examples of deviations are: "there is no warning system available when cyclists continuing to cross the tracks, although the junction is officially closed", "when approaching the location attention is mainly focused at changes (in comparison to the previous temporary situation) instead of focusing on the traffic present at that moment" and "new bottlenecks in the traffic flow caused re-routing of the original road, especially in sharp curve or near narrowed passage". For the third case the operation of ACC in other circumstances than the operation for which it is intended will be part of a HAZOP discussion. This can involve identification of location related deviations, such as "users will try the ACC system on roads where a speed above 40 km/h is allowed", "ACC used in urban environments has problems detecting other "obstacles" than moving vehicles (pedestrians, cyclists and animals)" and "drivers cruising with ACC may forget to take over if vehicles in front are stopping for a red traffic light". Moreover, the discussion can include the system and traffic circumstance, such as "Effect of the deceleration force of the ACC device depends on the characteristics of the pavement and may vary depending on the weather conditions", "gaps between ACC vehicles based on the current defined safe distance will be larger than the average gaps in Dutch traffic" and "larger gaps may result in large number of vehicles getting in front of the ACC vehicle after a lane change".

Discussion

Redesign of road infrastructure is a common activity in modern society. Changed demands on infrastructure and maintenance are a constant phenomenon in modern traffic. These changed infrastructures often have to be fit in an already existing situation. This increases the complexity of the redesign, since not only the risks of the new traffic situation have to be managed, but also the risks of activities that interact with the road traffic. A second complexity issue is caused by the introduction of new technologies that can set additional requirements to the road infrastructure. This paper shows how accidents can happen if the context of a situation is not taken into account enough when redesigning infrastructure.

The case studies show how a too narrow design focus leads to inappropriate traffic situations or accidents. The train-bus collision in Fox River Grove shows how a number of modifications create a situation in which larger vehicles are partly blocking the railway tracks while waiting for a traffic light to turn green. The sub-optimal solutions during construction works show similarities of blocking tracks. Temporary infrastructural layouts add even more constraints on possible solutions. Besides the fact that traffic activities have to go on as much as possible during reconstruction, there also are fewer budgets available for temporary measures. On top of that the users of the temporary infrastructure get less time to get used to situations, because the situations regularly change, with increased risks as a result. The third example explains how a too narrow focus of introduced new technologies influences the demands on the user and the road infrastructure. In all three cases designers eventually give the task of dealing with the problems they created to drivers, by putting constraints on their behaviour in the form of warnings and instructions. Instead of creating inherently safe situations, the result of the designers' efforts is more dangerous traffic.

The solution this paper proposes is to involve all actors in the process. This has two main advantages. First the different actors bring different perspectives to the problem that may lead to improved designs, which take into account a broader range of aspects of a traffic situation. Second, the involvement of both designers and policymakers into the design process may help raise awareness of all parties of the difficulties of dealing with the complexity of modern infrastructure design, even when the topic is a situation which will not last for more than a couple of days.

References

DETR. (2000). *Tomorrow's roads: Safer for Everyone - The Government's road safety strategy and casualty reduction targets for 2010*. London, UK: Department of the Environment, Transport and the Regions.

Hoetink, A.E. (2003) Advanced cruise control in the Netherlands: a critical review. Paper presented at the Solutions for today and tomorrow - *10th World conference on Intelligent Transport Systems*, 16-20 November 2003, Madrid.

Jagtman, H.M. (2004). *Road Safety by Design: A decision support tool for identifying ex-ante evaluation issues of road safety measures*. Delft: Eburon.

Jagtman, H.M., Marchau, V.A.W.J., & Heijer, T. (2001). Current knowledge on safety impacts of Collision Avoidance Systems (CAS). Paper presented at the *Critical Infrastructures - 5th International Conference on Technology, Policy and Innovation*, Utrecht.

Jagtman, H.M., & Wiersma, E. (2003). Driving with adaptive cruise control in the real world. Paper presented at the Improving safety by linking research with safety policy and management, *Proceedings of the 16th ICTCT workshop*, Soesterberg.

Koornstra, M.J., Mathijssen, M., Mulder, J.A.G., Roszbach, R., & Wegman, F.C.M. (1992). *Naar een duurzaam veilig wegverkeer: nationale verkeersveiligheids-verkenning voor de jaren 1990/2010 (Towards a 'sustainable safe' road traffic: the National exploration of traffic safety for the years 1990/2010)*. Leidschendam: SWOV.

Lawley, H.G. (1974). Operability Studies and Hazard Analysis. *Chemical Engineering Progress, 70* (4), 45-56.

Marsh, D. (2003). Radar reflects safer highways. *EDN Europe 2003(March)*, 21-28.

Ministerie van Verkeer en Waterstaat. (2000). *Van A naar Beter - Nationaal Verkeers- en Vervoersplan 2001-2020*. Den Haag, The Netherlands: Ministerie van Verkeer en Waterstaat.

National Transport Safety Board. (1996). *Highway/Railway Accident Report - Collision of Northeast Illinois Regional Commuter Railroad Corporation (METRA) train and Transportation Joint Agreement School District 47/155 School Bus at Railroad/Highway Grade Crossing in Fox River Grove, Illinois on October 25, 1995* (No. PB96-916202): NTSB.

Raad voor de Tranportveiligheid (2004). *Twee bus/trein-botsingen op overweg bij Nootdorp - verkort onderzoek (Two bus/train-collisions on railway junction near Nootdorp – reduced investigation)*. Den Haag, The Netherlands: Raad voor de Transportveiligheid.

SNRA. (undated). *Vision Zero - from concept to action* (No. VV88223). Stockholm: Swedish National Road Administration.

Vahidi, A., & Eskandarian, A. (2003). Research advances in intelligent collision avoidance and adaptive cruise control. *IEEE transactions on intelligent transportation systems, 4*, 143-153.

Venhovens, P., Naab, K., & Adiprasito, B. (2000). Stop and Go Cruise Control. *International Journal of Automotive Technology, 1,* 61-69.

Acknowledgement to reviewers

The editors owe debt to the following colleagues who helped to review the manuscripts for this book:

Min An, The University of Birmingham, Birmingham, UK
Yvonne Barnard, EURISCO International, Toulouse, France
Birgitte M. Blatter, TNO Work and Employment, Hoofddorp, the Netherlands
Willem Bles, TNO Defence, Security and Safety, Soesterberg, the Netherlands
Paul Boase, Transport Canada, Ottawa, Canada
Anne Bolling, VTI, Linköping, Sweden
Marie-Christine Chambrin, Université de Lille, Lille, France
Hein Daanen, TNO Defence, Security and Safety, Soesterberg, the Netherlands
Thomas Degré, INRETS, Arcueil, France
Tania Dukic, Arbetslivsinstitutet, Göteborg, Sweden
Michael Falkenstein, Institute for Occupational Physiology, Dortmund, Germany
Jean Michel Hoc, Centre National de la Recherche Scientifique, Nantes, France
Jettie Hoonhout, Philips Research, Eindhoven, the Netherlands
Wiel Janssen, TNO Defence, Security and Safety, Soesterberg, the Netherlands
Christian Knoll, Institut für Arbeitswissenschaft und Technologiemanagement, Universität Stuttgart, Germany
Peter Ladkin, University of Bielefeld, Bielefeld, Germany
Daniel McGehee, Human Factors and Vehicle Safety Research Division, The University of Iowa, Iowa City, USA
Kathryn Mearns, University of Aberdeen, Aberdeen, UK
Jan Moraal, Den Haag, the Netherlands
Ben Mulder, University of Groningen, Groningen, the Netherlands
Björn Norlin, Swedish Defence Research Agency, Linköping, Sweden
Jan Noyes, University of Bristol, Bristol, UK
Jan Joris Roessingh, National Aerospace Laboratory NLR, Amsterdam, the Netherlands
Bert Ruitenberg, Schiphol Tower Approach, Schiphol, the Netherlands
Mark Scerbo, Old Dominion University, Norfolk, USA
Angelia Sebok, Micro Analysis & Design, Boulder, Colorado, USA
Jan Skriver, Scandpower Risk Management AB, Uppsala, Sweden
Frank Steyvers, University of Groningen, Groningen, the Netherlands
Jean Vanderdonckt, Université Catholique de Louvain, Louvain-la-Neuve, Belgium
Andras Vereczkei, Medical School University of Pécs, Pécs, Hungary
Nicholas Ward, University of Minnesota, Minneapolis, USA

In D. de Waard, K.A. Brookhuis, R. van Egmond, and Th. Boersema (Eds.) (2005), *Human Factors in Design, Safety, and Management* (pp. 527). Maastricht, the Netherlands: Shaker Publishing.